# Process Chemistry
# in the
# Pharmaceutical Industry

# Process Chemistry in the Pharmaceutical Industry

edited by

**Kumar G. Gadamasetti**

*The Bristol-Myers Squibb Company*
*New Brunswick, New Jersey*

*and*

*Amgen, Inc.*
*Thousand Oaks, California*

Library of Congress Cataloging-in-Publication Data

Process chemistry in the pharmaceutical industry / edited by Kumar G. Gadamasetti.
      p.       cm.
   Includes bibliographical references and index.
   ISBN 0-8247-1981-6 (alk. paper)
   1. Pharmaceutical chemistry.  2. Chemical processes.  I. Gadamasetti, Kumar G.
   RS403.P765   1999
   615'.19--dc21

99-26854
CIP

This book is printed on acid-free paper.

**Headquarters**
Marcel Dekker
270 Madison Avenue, New York, NY 10016
tel: 212-696-9000; fax: 212-685-4540

**Eastern Hemisphere Distribution**
Marcel Dekker AG
Hutgasse 4, Postfach 812, CH-4001 Basel, Switzerland
tel: 41-61-261-8482; fax: 41-61-261-8896

**World Wide Web**
http://www.dekker.com

The publisher offers discounts on this book when ordered in bulk quantities. For more information, write to Special Sales/Professional Marketing at the headquarters address above.

Current printing (last digit):
10 9 8 7 6 5 4

10 06014723

**PRINTED IN THE UNITED STATES OF AMERICA**

To my brother Ram

# FOREWORD 1

Chemists play major roles in the pharmaceutical industry. Some design and synthesize potential medicinal compounds and are often referred to as being in the "discovery" groups. Other chemists work in "development" groups, devising practical ways to make the compounds once their biological properties show that they have real medicinal potential. This work can lead to the synthetic schemes needed to make large amounts of a potential drug for animal and human testing, and then to the practical approach for manufacture of a compound that is approved for human or veterinary use.

In consulting with various pharmaceutical companies for almost 40 years, I have been impressed by the difference in chemical approach needed in these two areas. For drug discovery, a procedure is needed to produce the desired target compound rapidly, essentially without regard to cost or manufacturing practicality. A synthesis will be preferred if it proceeds through an intermediate that can branch to generate an entire family of potential drugs, in order to permit their evaluation. Normally, no time would be spent optimizing a reaction or inventing new reactions to achieve the goal. That time would have been wasted if the target compound turned out to be biologically uninteresting.

In development research, the target compound is known to be of significant biological interest, a potential product. Thus time must be spent to optimize the synthetic scheme, so it can be used to make large quantities of the target. Even early in the development stage the chemists are already considering whether the synthetic scheme—normally different from the one first used during the discovery phase—might be used for manufacturing if the candidate compound did survive further testing. The cost of reagents, and the cost of handling waste products, are kept very much in mind.

In both activities (drug discovery and development research), a firm grounding in synthetic organic chemistry is essential. In discovery research this should be supplemented by some knowledge of pharmacology and increasingly by some familiarity with computer design of molecules. Discovery chemists work at the chemistry/biology interface.

Chemists doing process research in the development groups must also have a good grounding in physical organic chemistry, as well as synthetic organic chemistry. They will be called on to change reactions so as to improve them, even to invent new reactions based on fundamental reasoning. They are working at the forefront of chemistry.

In the best companies, discovery and development are both very strong. The discoverers might have the excitement of actually inventing a drug; however, some beautiful chemistry can lead to a medicinal failure. Many thousands of compounds are made for every one that succeeds as a useful medicine, so there is a good chance that someone could spend an entire career in discovery without coming up with a commercial product. The developers work on projects with a much higher likelihood that their efforts will contribute to an actual medicinal product. Also, they can afford to be truly inventive in their chemistry.

Both activities are enjoyable and challenging, and both are necessary parts of the great contribution that chemistry makes to health. This contribution needs to be better appreciated. I am glad to see that Dr. Gadamasetti has organized and contributed to this book that illustrates so well the world of process chemistry in the pharmaceutical industry. It is important reading for all those who want to understand how the miracle of modern medicinal chemistry is performed, especially for those considering a career in the pharmaceutical industry.

**Ronald Breslow**
University Professor and Professor of Chemistry
Columbia University
New York, NY 10027, USA

# FOREWORD 2

When historians record the key scientific achievements of the twentieth century, they surely will note the role of pharmaceuticals in extending life expectancy and improving our quality of life. They undoubtedly will highlight the pharmaceutical industry's brilliant discoveries and the creative chemical and biological insights that led to them. An interesting question is whether or not historians will recognize the critical role that process chemistry has played in bringing these discoveries to the patient. This part of the "new drug" story often is less visible than the discovery part. Yet present in every drug discovery achievement is an exciting story of chemical process development. The successful launch of a new therapeutic agent requires the timely development of an economically feasible chemical process. The problems that must be solved to achieve this end are intellectually challenging and usually require creative chemistry. In the era of "green chemistry" and sustainability, the importance and value of world-class process chemistry can only rise in the pharmaceutical industry. *Process Chemistry in the Pharmaceutical Industry* tells a series of compelling stories that should help everyone to appreciate the importance of process chemistry in drug discovery and development.

**Paul Anderson**
DuPont Pharmaceuticals
P. O. Box 80500
Wilmington, DE 19880-0500, USA

# THE CHALLENGES OF PROCESS R&D

The translation of a laboratory method for making an organic chemical on a milligram scale to a production process on a kilogram to tonne scale while maintaining high quality, reproducibility, and minimum cost, is an extremely challenging task. Most organic chemists working in this field of process R&D find it intellectually stimulating, as well as very satisfying, when the process is eventually successfully operated on a large scale. As we move toward the end of the century, however, the challenges of process R&D are even greater, particularly in the pharmaceutical industry, for the following reasons.

**The molecules are becoming more complex.**

New drugs seem to have higher molecular weights, a wider variety of heterocyclic functionality, more chiral centers, and more complex structural features than drugs developed in the 1970s and 1980s. These drugs can only be made selectively by using the armory of new methodologies initially devised in academic laboratories but adapted for industrial processes on a larger scale.

**The specifications and analytical chemistry requirements for new chemical entities are becoming increasingly narrow.**

New drugs of the 1990s may be extremely pure (>99.9%) or their impurities have to be characterized when present in levels of 0.1% and below. These specifications place stricter requirements on intermediate, starting material, reagent and solvent quality such that analytical chemistry is a more important part of process R&D than in the past.

**The time scale for developing new routes and new processes is much more compressed.**

Since drug companies have compressed the time to market from a typical 8–10 years in the 1980s down to 4–6 years in the 1990s, the pressure on process R&D is now intense. Process research is often the rate-limiting step in moving a drug forward into trials—the need to generate kilograms of complex drug substance within months of the project being started means that organic process chemists must be even more innovative. The pressure to retain the chemistry devised by discovery chemists and simply scale it up is always present but, in my view, should be resisted. Time spent on developing a robust synthetic route suitable for large scale operation will pay for itself in the longer term, even if the effort encounters initial delays.

**The legislative requirements of processes impose additional tasks on process R&D.**

The Food and Drug Administration and other regulatory bodies have imposed (quite correctly, in my view) the concept of validation of processes on the pharmaceutical industry. The process R&D chemist is faced with extra tasks which require a more detailed understanding of each chemical reaction in the sequence, but this has the result that this understanding leads to better, more reproducible processes.  However, additional requirements regarding safety in scale-up and environmental issues all impact on the process research.

These challenges are being met by process chemists all over the world.  In this book, case studies from process R&D in the pharmaceutical industry will show how these challenges have been addressed.  Each company will approach the challenges in a slightly different way, and each new molecule brings its own problems, which require innovative and creative solutions.

The reader of this book will, I am sure, be stimulated by the chapters describing successful strategies for making these complex molecules efficiently, and in many cases on a tonne scale. Dr. Gadamasetti is to be congratulated on persuading so many industrial chemists to find the time to write up their work and that of their colleagues (after all, process R&D is very much a team effort) for the benefit of readers, both from industry and academia. This book will provide valuable teaching material for universities and will help students to understand the fascination of organic process R&D, hopefully persuading the best chemists to enter this most rewarding profession.

**Trevor Laird**
Scientific Update, UK
(Editor, *Organic Process Research & Development*)

# PREFACE

On our return flight from the conference on Chiral Discrimination (1993) in Montreal, Canada, Conrad Kowalski (SmithKline Beecham Pharmaceuticals) and I were engaged in sharing some general views about process development in the pharmaceutical industry. We reached a consensus that chemical development, unlike the rest of the sciences, is not taught in academic institutions. Hence, the fresh graduates looking to pursue careers would less likely consider process research and development. This prompted me to organize a volume on process chemistry in the pharmaceutical industry. The conception came to reality due to the encouragement and support of my mentor, Chris Cimarusti, from Bristol-Myers Squibb Co.

The chapters in this volume have stemmed from in-depth discussions with several experts in the pharmaceutical industry and in academia. The principles and the practices related to process research and development are spread throughout the chapters in the volume. Each chapter has been contributed by experts in the specialized field. No attempt has been made to discuss the principles and the definitions of process research and development in a separate chapter. A conscious effort was made to choose the topics of the chapters and the division of the chapters so as to bring to the reader a complete picture of the chemical process development in the global pharmaceutical industry. Hopefully, the reader will have an opportunity to understand and apply the basic principles involved in the development of the active pharmaceutical ingredient of potential drugs to improve the quality of human life.

**Kumar G. Gadamasetti**
Amgen, Thousand Oaks, CA

# ACKNOWLEDGMENTS

I wish to acknowledge Chris Cimarusti, Jerry Moniot, and Neal Anderson from Bristol-Myers Squibb Co. for their encouragement and continuous support throughout this project. I would also like to thank Ronald Breslow from Columbia University, Martin Kuehne from the University of Vermont, Tim Osslund from Amgen, Burt Chritensen (formerly from Merck Research Laboratories), Seemon Pines (formerly from Merck Research Laboratories), Paul Anderson (DuPont Pharmaceuticals) and Trevor Laird (Scientific Update) for their valuable ideas and suggestions. My special thanks to all my friends in the pharmaceutical industry for their input and support throughout the evolution, progress and completion of this volume. The main credit goes to all the contributors for their dedication and effort. I am indebted to all the reviewers for their time, constructive criticism and input. Finally, I would like to thank my wife, Vidya, who gave tremendous moral support during the preparation of this volume.

# CONTENTS

## Enzymatic Intervention and Phase Transfer Catalysis

## Asymmetric Synthesis and Enantioselectivity

## Drug Substance Final Form and Process Safety

**Design of Experiments and Automation**

# CONTRIBUTORS

**Anderson, Benjamin A.**
Lilly Research Laboratories, A Division of Eli Lilly and Company, Chemical Process Research and Development Division, Indianapolis, IN 46285
e-mail: anderson_benjamin_a@lilly.com

**Anderson, Paul S.**
DuPont Merck Pharmaceutical Co., P. O. Box 80500, Wilmington, DE 19880-0500
e-mail: Paul.s.anderson@dupontmerck.com

**Authelin, Jean-René**
Process Chemistry, Rhône-Poulenc Rorer, 94403 Vitry Sur Seine Cedex, France

**Ayers, Timothy A.**
Hoechst Marion Roussel, Inc., 2110 E. Galbraith Road, Cincinnati, OH 45215

**Blaser, H.U.**
Scientific Services, Novartis AG, CH-4002 Basel, Switzerland
e-mail: hans-ulrich.blaser@sn.novartis.com

**Bray, Brian L.**
Chemical Development Department, Glaxo Wellcome Inc., Research Triangle Park, North Carolina 27709

**Breslow, Ronald C. D.**
Department of Chemistry, Columbia University, 3000 Broadway, New York, NY 10027
e-mail: rb33@columbia.edu

**Carlson, John**
Chemical Development, Novartis Pharmaceuticals Division, Summit, NJ 07901-1398

**Cimarusti, Christopher M.**
Process Research and Development, The Bristol-Myers Squibb Pharmaceutical Research Institute, P.O. Box 191, New Brunswick, New Jersey 08903-0191

**Confalone, Pat N.**
Chemical Process Research & Development, The DuPont Merck Pharmaceutical Company, Chambers Works, Deepwater, NJ 08023

**Dewitt, Sheila**
Orchid Biocomputer, 201 Washington Road, Princeton, NJ 08543-2197
e-mail: sdewitt@orchidbio.com

**Farina, Vittorio**
Department of Chemical Development, Boehringer Ingelheim Pharmaceuticals, 900 Ridgebury Road, Ridgefield, CT 06877
e-mail: vfarina@bi-pharm.com

**Federsel, Hans-Jürgen**
Chemical Process Development Laboratory, Astra Production Chemicals AB, 151 85 Södertälje, Sweden
e-mail: Hans-Jurgen.federsel@chemicals.se.astra.com

**Gadamasetti, Kumar G.***
Process Research and Development, The Bristol-Myers Squibb Pharmaceutical Research Institute, P.O. Box 191, New Brunswick, New Jersey 08903-0191
* Current address: Small Molecules Process Chemistry, Amgen Inc., 1840 DeHavilland Drive, Thousand Oaks, CA 91320-1789
e-mail: kumarg@amgen.com

**Gamboni, R.**
Chemical and Analytical Development Pharma, Novartis AG, CH-4002 Basel, Switzerland

**Giannousis, Peter**
Chemical Development, Novartis Pharmaceuticals Division, Summit, NJ 07901-1398
e-mail: peter.giannousis@pharma.novartis.com

**Grozinger, Karl**
Department of Chemical Development, Boehringer Ingelheim Pharmaceuticals, 900 Ridgebury Road, Ridgefield, CT 06877

**Halpern, Marc E.**
PTC Technology, Suite 627, 1040 N Kings Hwy, Cherry Hill, New Jersey 08034
e-mail: halpern@phasetransfer.com

**Hansen, Marvin M.**
Lilly Research Laboratories, A Division of Eli Lilly and Company, Chemical Process Research and Development Division, Indianapolis, IN 46285
e-mail: mmh@lilly.com

**Hoffmann, Wilfried**
Parke-Davis Pharmaceutical Research, A Division of WarnerLambert Company, Chemical Development, Freiburg, D-79090 Freiburg, Germany

**Jacobsen, Eric N.**
Department of Chemistry and Chemical Biology, Harvard University, Cambridge, MA 02138

**Jakupovic, Edib**
Chemical Process Development Laboratory, Astra Production Chemicals AB, 151 85 Södertälje, Sweden

**Laird, Trevor**
Scientific Update, Wyvern Cottage, High Street, Mayfield, East Sussex TN20 6AE, UK
e-mail: scientificUpdate@dial.pipex.com

**Lantos, Ivan**
Synthetic Chemistry Department, SmithKline Beecham Pharmaceutical R&D, King of Prussia, Pennsylvania 19406-0939

**Leimer, Marius**
Chemical Development, Novartis Pharmaceuticals Division, Basel, Switzerland, CH-4001

**Martinelli, Michael J.**
Chemical Process R&D, Lilly Research Laboratories, A Division of Eli Lilly and Company, Lilly Corporate Center, Indianapolis, IN 46285-4813
e-mail: martinelli_michael_j@lilly.com

**Matsuoka, Richard**
Synthetic Chemistry Department, SmithKline Beecham Pharmaceutical R&D, King of Prussia, Pennsylvania 19406-0939

**McFarlane, Ian M.**
Dagenham Research Centre, Rhône-Poulenc Rorer Central Research, Rainham Road South, Dagenham, Essex, Essex RM10 7XS, UK

**Mueller, Richard H.**
Process Research and Development, The Bristol-Myers Squibb Pharmaceutical Research Institute, P.O. Box 4000, Princeton, New Jersey 08543-4000
e-mail: mueller_r@bms.com

**Müller-Bötticher, Hermann**
Geschäftseinheit Feinchemikalien, Boehringer Ingelheim KG, D-55216 Ingelheim, Germany

**Newton, Christopher G.**
Dagenham Research Centre, Rhône-Poulenc Rorer Central Research, Rainham Road South, Dagenham, Essex, Essex RM10 7XS, UK

**Okabe, Masami**
Process Research, Chemical Synthesis Department, Hoffmann-La Roche Inc., Nutley, New Jersey 07110-1199

**Owen, Martin R.**
Chemical Development, Glaxo Wellcome, Gunnels Wood Road, Stevenage, Herfordshire, SG1 2NY, England
e-mail: mro1220@ggr.co.uk

**Partridge, John J.**
Chemical Development Department, Glaxo Wellcome Inc., Research Triangle Park, North Carolina 27709

**Pilipauskas, Daniel R.**
Chemical Sciences, Searle, 4901 Searle Parkway, Skokie, Illinois 60077

**Pitchen, Philippe**
Cawthorn Centre, Rhône-Poulenc Rorer Central Research, 500 Arcola Road, Collegeville, PA 19426

**Pugin, B.**
Scientific Services, Novartis AG, CH-4002, Basel, Switzerland

**RajanBabu, T. V.**
Department of Chemistry, The Ohio State University, Columbus, OH 43210
e-mail: rajanbabu.1@OSU.edu

**Rihs, G.**
Scientific Services, Novartis AG, CH-4002 Basel, Switzerland

**Roth, Gregory P.**
Department of Inflammatory Diseases and Medicinal Chemistry, Boehringer Ingelheim Pharmaceuticals, 900 Ridgebury Road, Ridgefield, CT 06877
e-mail: groth@bi-pharm.com

**Schaub, B.**
Chemical and Analytical Development Pharma, Novartis AG, CH-4002 Basel, Switzerland

**Schmidt, E.**
Chemical and Analytical Development Pharma, Novartis AG, CH-4002 Basel, Switzerland

**Schmitz, B.**
Chemical and Analytical Development Pharma, Novartis AG, CH-4002 Basel, Switzerland

**Sedelmeier, G.**
Chemical and Analytical Development Pharma, Novartis AG, CH-4002 Basel, Switzerland

**Senanayake, Chris H.**
Chemical Research & Development, Sepracor Inc., 111 Locke Drive, Marlborough, MA 01752

**Spindler, F.**
Scientific Services, Novartis AG, CH-4002 Basel, Switzerland

**Sudhakar, Anantha R.**
Chemical Process Research and Development, Schering-Plough Research Institute, Union, NJ 07083

**Thiruvengadam, T. K.**
Chemical Process Research and Development, Schering-Plough Research Institute, Union, NJ 07083

**Thompson, Michael D.**
Process Chemistry, Rhône-Poulenc Rorer, Collegeville, Pennsylvania 19426

**Varie, David L.**
Chemical Process R&D, Lilly Research Laboratories, A Division of Eli Lilly and Company, Lilly Corporate Center, Indianapolis, IN 46285-4813
e-mail: varie@lilly.com

**Vicenzi, Jeffrey T.**
Lilly Research Laboratories, A Division of Eli Lilly and Company, Chemical Process Research and Development Division, Indianapolis, IN 46285
e-mail: vecenzi_jeffery_t@lilly.com

**Walker, Derek**
Chemical Development, Schering-Plough Research Institute, Union, NJ 07083

**Waltermire, Robert E.**
Chemical Process Research & Development, The DuPont Merck Pharmaceutical Company, Chambers Works, Deepwater, NJ 08023
robert.e.waltermire@usa.dupont.com

**Wetter, Hj.**
Chemical and Analytical Development Pharma, Novartis AG, CH-4002 Basel, Switzerland

**Wu, Guangzhong**
Chemical Process Research and Development, Schering-Plough Research Institute, Union, NJ 07083

**Zmijewski, Milton J.**
Lilly Research Laboratories, A Division of Eli Lilly and Company, Natural Products Discovery Development Research, Indianapolis, IN 46285
e-mail: zmijew@lilly.com

# Process Chemistry
## in the
# Pharmaceutical Industry

# OVERVIEW

# Process Chemistry in the Pharmaceutical Industry: An Overview

**Kumar G. Gadamasetti**[1]
*Process Research and Development, The Bristol-Myers Squibb Pharmaceutical Research Institute, P.O. Box 191, New Brunswick, New Jersey 08903-0191 USA*

Key Words: process research, process development, green chemistry, chemometrics, chirality

## Introduction

The term "process" is, in general, misinterpreted as scale-up work by the overwhelming majority of the chemists entering the synthetic community. Scientists engaged in the various aspects of pharmaceutical process research and development have a highly refined appreciation for the challenges of large scale synthetic chemistry. Process chemistry reaches its full potential in the development work that is carried out in the pharmaceutical industry in pursuit of safe, efficient, economical, and environmentally friendly syntheses of complex molecules for use in treating human disease.

The mission of process chemistry in the pharmaceutical industry is to provide documented, controlled synthetic processes for the manufacture of supplies to support the development programs and future commercial requirements for an active pharmaceutical ingredient (API) or the drug. The mission represents a tremendous challenge to the synthetic skills of the process scientists as the requirements for drug supply progress from milligrams to metric ton quantities.

The concept in organizing this volume is to share with the reader some of the approaches to synthetic development, some solutions to real challenges faced by the pharmaceutical process research and development groups in terms of process chemistry. In most cases, the focus of process development efforts include any and all kinds of syntheses, separations, and establishment of appropriate physical forms, introduction and maintenance of chirality (chiral pool, resolution or enzymatic intervention), effective process control, hazard minimization, and cost containment. Hopefully, the reader will gain an insight into the logic that a variety of process chemistry scientists/laboratories employ to meet these synthetic challenges. Equally, it is hoped that the volume may serve as a tool to render guidance to those interested in pursuing a career in process chemistry related fields. In the aggregate, the volume is intended to help the reader grasp the nature of the synthetic chemistry challenges and understand that scale-up of a reaction is but one facet of the effort in process development.

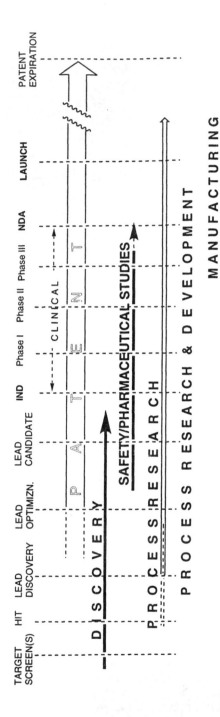

FIG. 1

## Big picture

In most cases of diseases, including cancer, the majority of patients prefer chemotherapy over surgery or radiotherapy. The in-depth understanding of the existing diseases and discovery of new diseases is an ongoing process in the field of medical sciences. The rapid rate of discoveries and the fundamental understanding of the diseases generate the need for new and improved pharmaceutical drugs to treat and enhance the quality of human lives. In the pharmaceutical industry, the urgent need to invent and launch the drugs in the market in a timely fashion imparts an added challenge to the process development scientists.

Innovation in the manufacturing process is more and more critical to new products and sustained financial returns. According to an article published in Harvard Business Review[2] that reported on 23 major development projects at 11 U.S. and European pharmaceutical companies, several companies either delayed a product launch or inhibited the commercial success of the product once on the market due to process development problems. Noted in the same article was the impact of process development on cost improvement and the importance of process research on process development lead times. Early and effective interaction of process research personnel with medicinal chemists and early innovations in process development is believed to shorten the IND and the NDA timelines, respectively. These overall reduced timelines would allow an early launch of the drug into the market (Fig. 1).

As noted by Chris Cimarusti in the second chapter, the decade of the '90's has been a turbulent one for the pharmaceutical industry and as we approach the new millennium, development chemists can expect to see more drug candidates and less time to progress these drug candidates. An excellent discussion of the process research and development strategy for accelerated projects has been provided by Cimarusti in that chapter.

## Green chemistry

At the turn of the century, one of the most challenging issues for the chemical communities is the continued environmental concerns in developing chemistry and chemical processes. Despite the fact that environmental consideration in the past entailed more engineering than chemistry, by looking at the end-of-process treatments to eliminate pollutants, there has been a growing movement to address environmental concerns by calling on chemistry. This research has been termed "green chemistry". Joseph Breen[3] (Chair, Committee on Environmental Improvements) comments, "chemists working today to prevent pollution tomorrow are doing green chemistry". As Professor Breslow pointed out,[4] the biblical injunction, "hurt not the earth, neither the sea, nor the trees," is guiding a lot of chemical effort. Professor Barry Trost's "atom economy," concept[5] evaluates processes by how many of all the atoms of the reagents end up in the desired products. The reader is encouraged to identify and appreciate the practical examples pertaining to green chemistry addressed by several authors in the case studies presented in this volume.

Of the several citations in the recent years related to green chemistry, the following two examples may give some idea about the approach to environmentally benign chemistry. Not only the processes are high yielding but the biproduct is water. Noyori and coworkers[6] have developed an efficient, environmentally friendly method for oxidizing primary and secondary alcohols (Scheme 1). The Japanese scientists use aqueous hydrogen peroxide as oxidant, a lipophilic quaternary ammonium hydrogen sulfate as a phase-transfer agent and a tungsten catalyst. Marko and coworkers[7] at the Catholic University of Louvain in Belgium have developed a catalytic

Scheme 1. Noyori's oxidation of alcohols

examples:

640 mmol                    93% (isolated)

925 mmol                    87% (isolated)

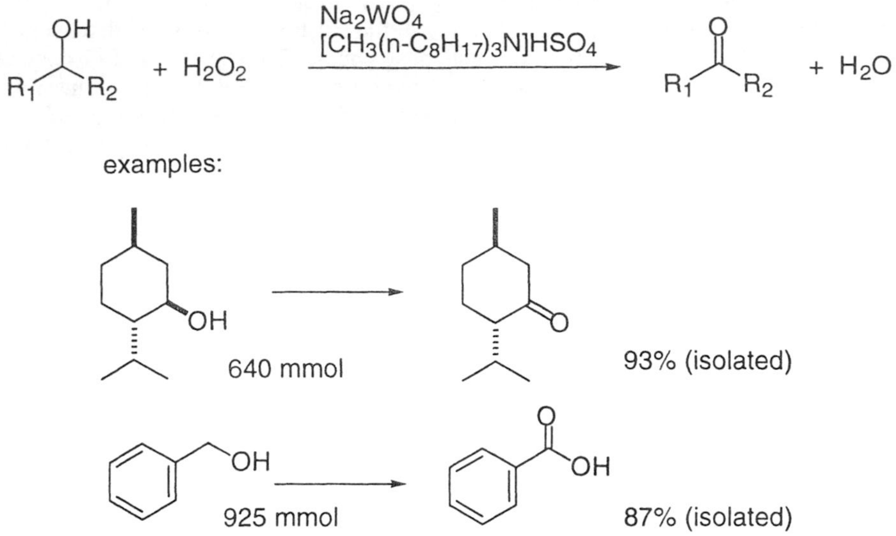

Scheme 2. Marko's oxidation of alcohols

examples:

88% (isolated)

92% (isolated)

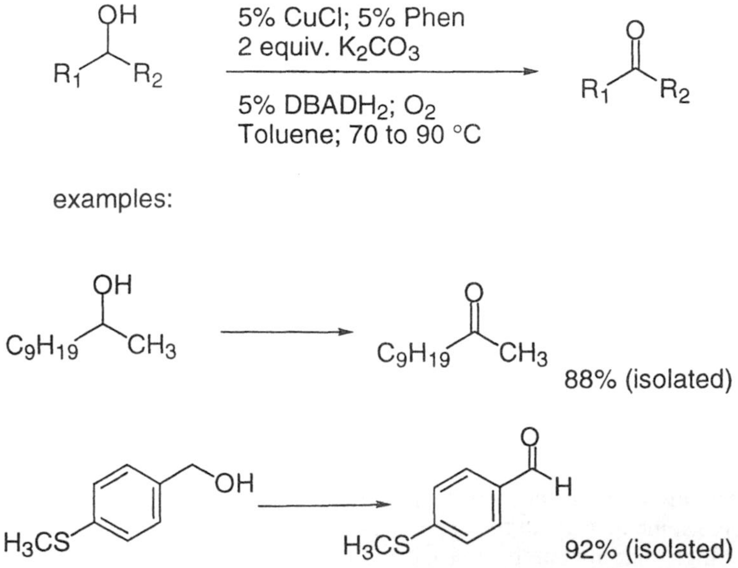

system that also selectively oxidizes alcohols to aldehydes and ketones under mild basic conditions (Scheme 2). The oxidizing system consists of copper(I) chloride complexed to phenanthroline and an azo compound or a hydrazine. The complex is adsorbed on a potassium carbonate support. Oxygen or air is used in the system as the oxidizer. It is believed that in the immediate future, "Green chemistry" will be more prevelant in all sectors of chemical industries.

## Prescription Pharmaceutical Drugs

The top 100 prescription drugs by worldwide sales in 1997, according to the list published in Med Ad News,[8] generated collectively $85.13 billion in sales compared to $73.44 billion in 1996. The top therapeutic categories included (in descending order by sales) gastrointestinal, cardiovascular, anti-infective, cholesterol lowering, psychotherapeutic, respiratory, and anti-cancer. The active pharmaceutical ingredient (API) listed in top drugs included chiral, racemic and achiral molecules.

### The Top 30 Prescription Drugs by Worldwide Sales

The list shown below ranks the top 30 prescription drugs by worldwide sales in 1997. The information includes the chemical structure, chemical name, brand name, marketer and the indications for the drug. The number on the left indicates the rank by worldwide sales as of 1997. The structures of the prescription drugs are reproduced from the Merck Index[9] to facilitate the reader grasp the diversity of functionality, substitution pattern and the nature of the molecule.

A = Brand name
B = Chemical name
C = Marketer
D = Worldwide sales 1997 (US $ in millions)
E = Indications for the drug

1

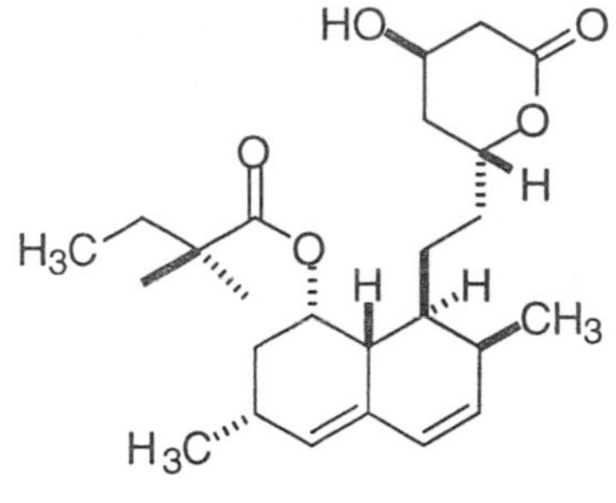

(optically active)

A. Zocor
B. Simvastatin
C. Merck & Co.
D. $3,575.0
E. Cholesterol reduction

2

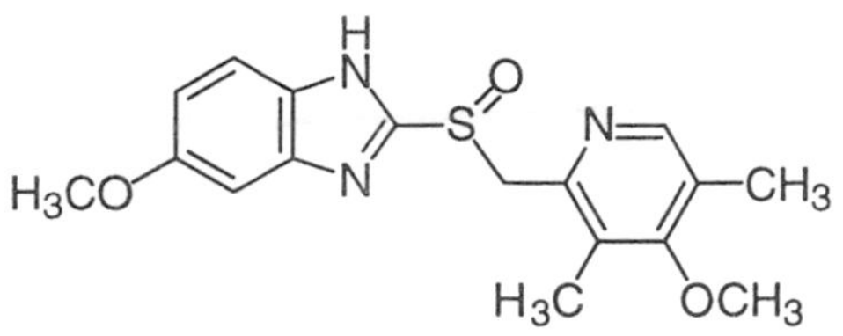

A. Losec/Prilosec
B. Omeprazole
C. Astra/Astra Merck
D. $2,815.8
E. Ulcers

3

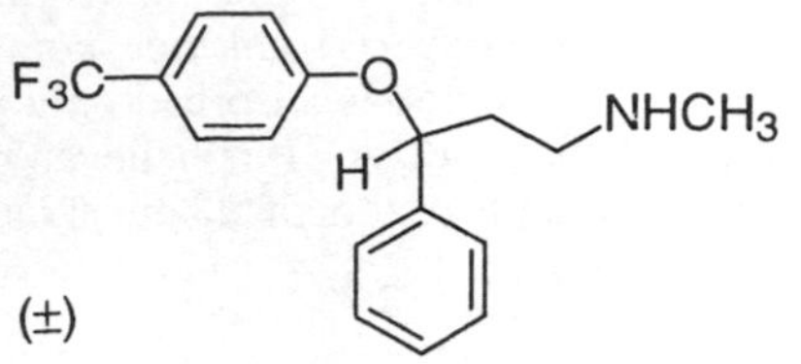

(±)

A. Prozac
B. Fluoxetine
C. Eli Lilly & Co.
D. $2,559.0
E. Depression

4

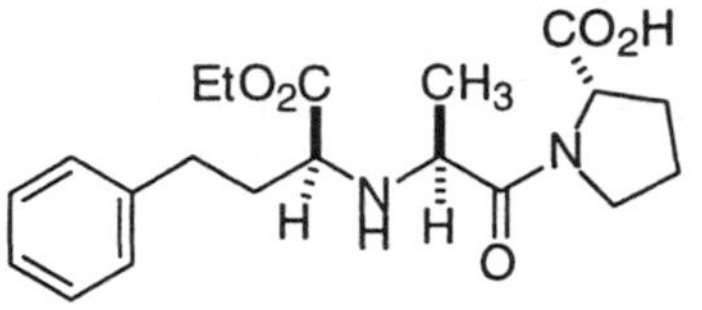

(optically active)

A. Renitec/Vasotec
B. Enalapril
C. Merck & Co.
D. $2,510.0
E. Hypertension, heart failure, left ventricular dysfunction

5

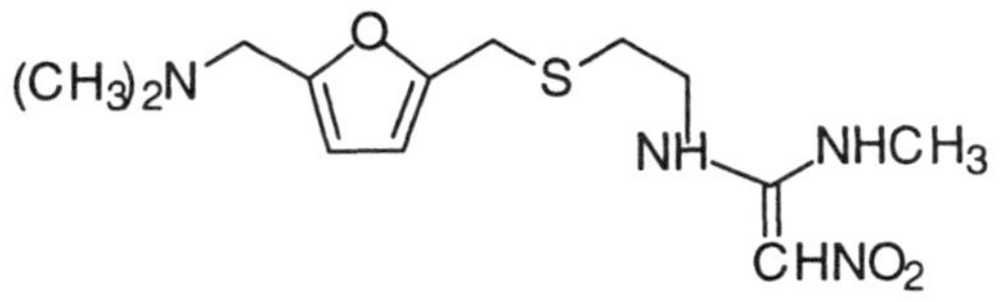

A. Zantac
B. Ranitidine
C. Glaxo Wellcome
D. $2,255.0
E. Ulcers

6

STRUCTURE
[same as #2]

A. Prilosec
B. Omeprazole
C. Astra Merck Inc.
D. $2,240.0
E. Ulcers

7

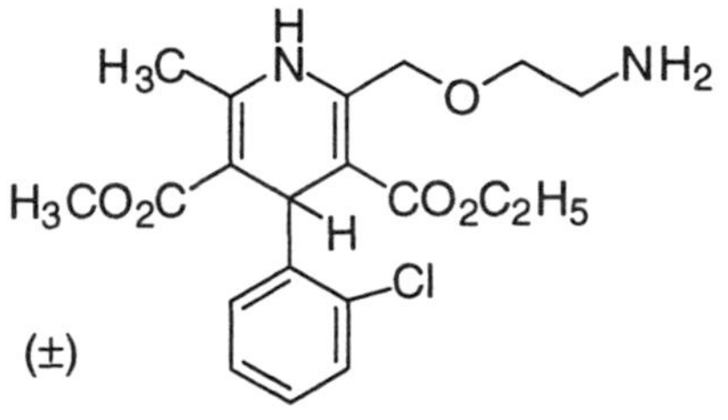

(±)

A. Norvasc
B. Amlodipine
C. Pfizer
D. $2,217.0
E. Hypertension, angina

8

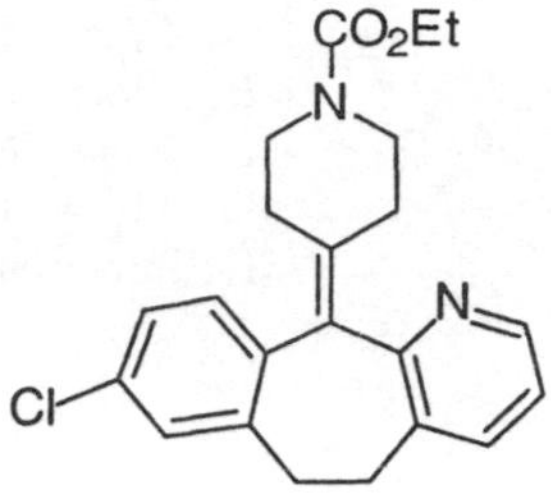

A. Claritin
B. Laratadine
C. Schering-Plough Corp.
D. $1,726.0
E. Allergies

9

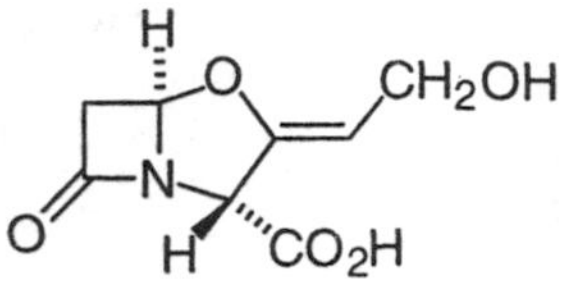

(optically active)

A. Augmentin
B. Clavulanic acid
C. SmithKline Beecham
D. $1,517.0
E. Infections

10

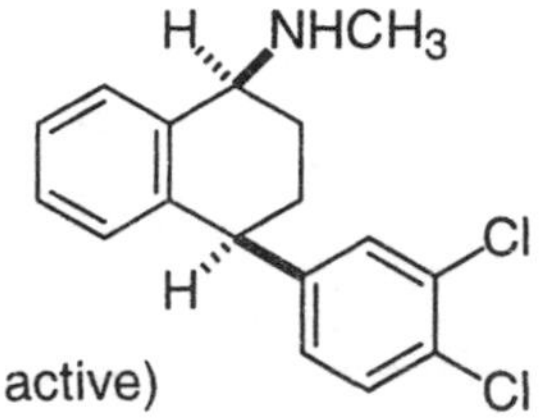

(optically active)

A. Sertraline
B. Zoloft
C. Pfizer
D. $1,507.0
E. Depression

11

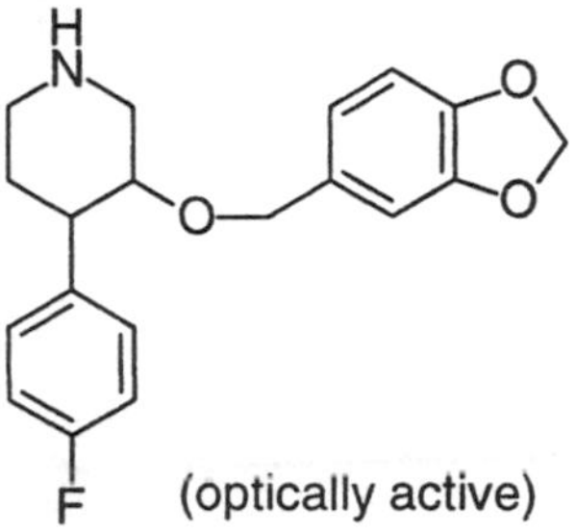

(optically active)

A. Paroxetine
B. Paxil
C. SmithKline Beecham
D. $1,474.0
E. Depression

12

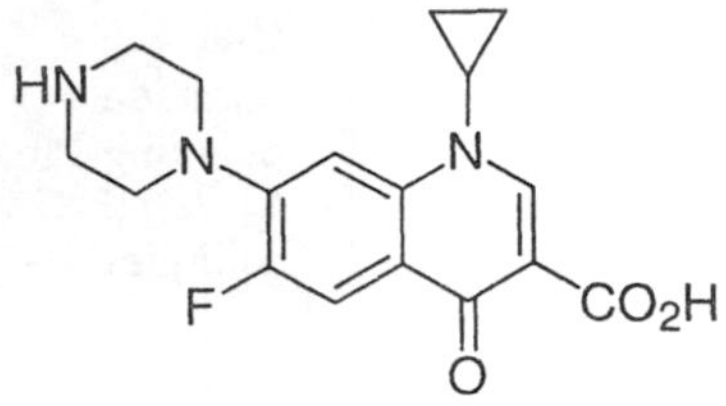

A. Ciprofloxacin
B. Cipro
C. Bayer Corpn.
D. $1,441.1
E. Infections

13

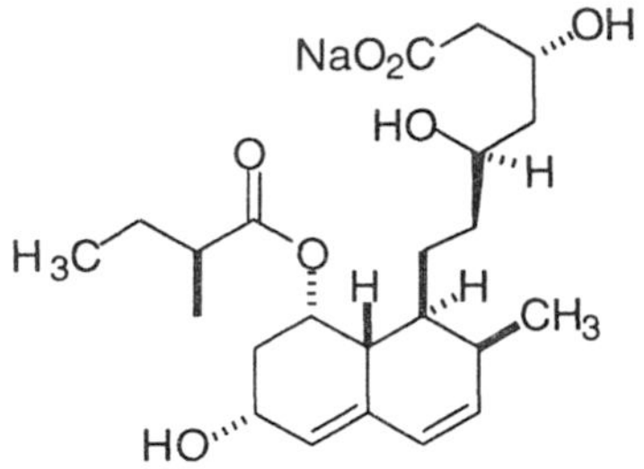

(optically active)

A. Pravastatin
B. Pravachol
C. Bristol-Myers Squibb Co.
D. $1,437.0
E. Cholesterol reduction

14

STRUCTURE
[same as #13]

A. Pravastatin
B. Mevalotin
C. Sankyo Co. Ltd.
D. $1,406.7
E. Cholesterol reduction

15

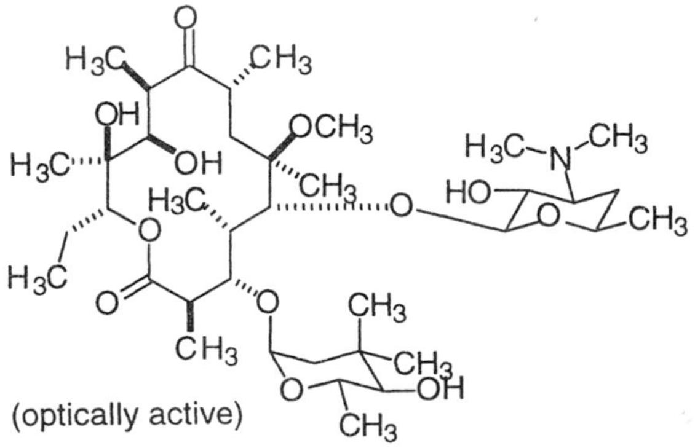

(optically active)

A. Clarithromycin
B. Biaxin
C. Abbott Laboratories
D. $1,300.0
E. Infections

16

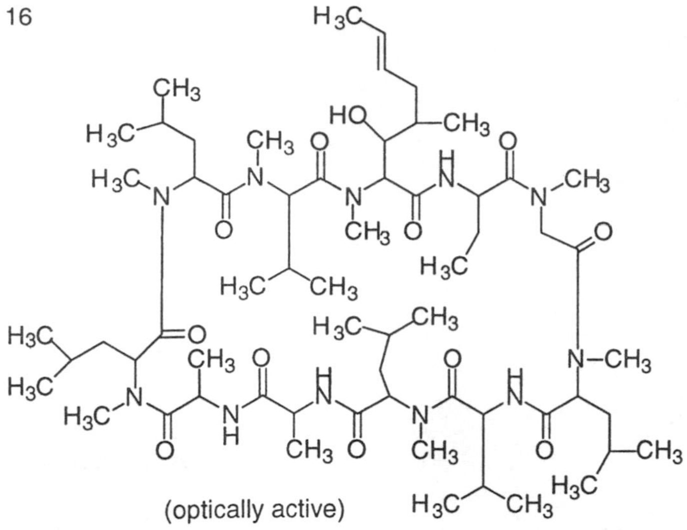

(optically active)

A. Cyclosporin
B. Sandimmun/Neoral
C. Novartis
D. $1,254.0
E. Oral rejection prevention

17

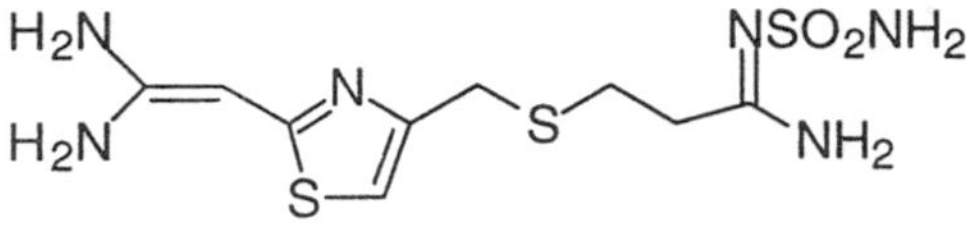

A. Famotidine
B. Pepcid
C. Merck & Co.
D. $1,180.0
E. Ulcers

18    PROTEIN

A. Erythropoietin (Epoetin alfa)
B. Epogen
C. Johnson & Johnson
D. $1,169.0
E. Red blood cell enhancement

19    PROTEIN

A. Erythropoietin (Epoetin alfa)
B. Epogen
C. Amgen
D. $1,160.7
E. Red blood cell enhancement

20

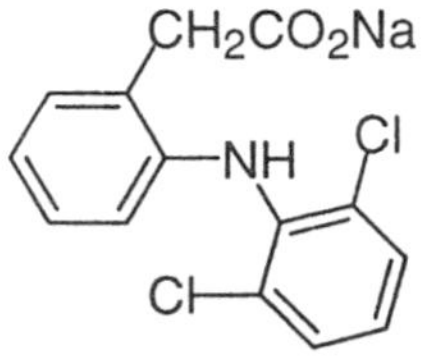

A. Diclofenac Sodium
B. Voltaren-XR
C. Novartis
D. $1,105.8
E. Arthritis

21

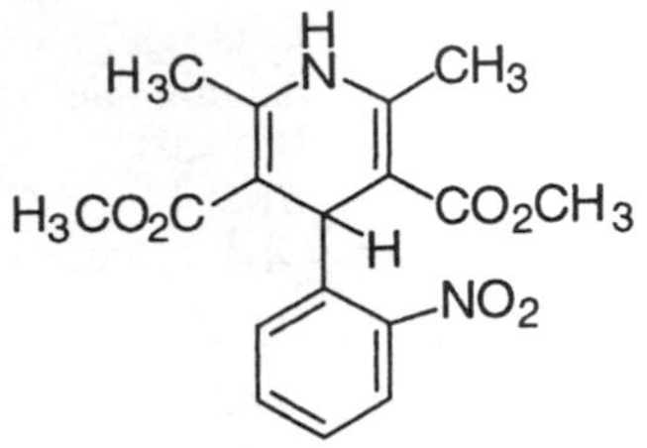

A. Nifedipine
B. Adalat
C. Bayer Corpn.
D. 1,101.0
E. Hypertension, angina

22

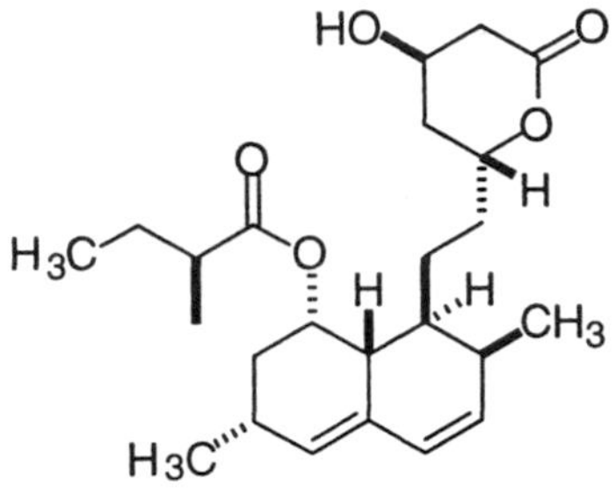

(optically active)

A. Lovastatin
B. Mevacor
C. Merck & Co.
D. $1,100.0
E. Cholesterol reduction

23

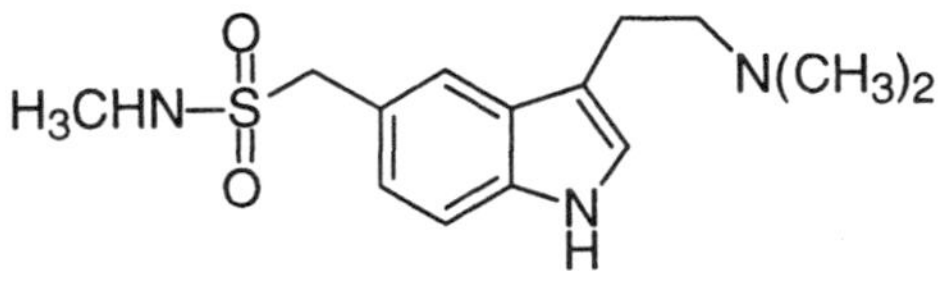

A. Sumatriptan
B. Imitrex
C. Glaxo Wellcome
D. $1,085.7
E. Migrain

24          PROTEIN

A. Filgrastim
B. Neupogen
C. Amgen
D. $1,055.7
E. Restoration of white
   blood cells

25

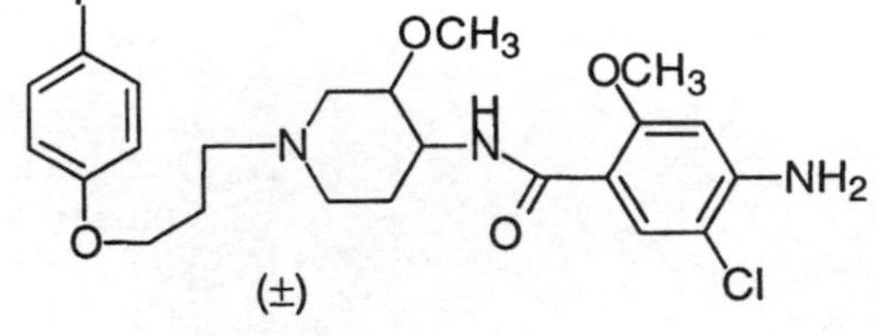

(±)

A. Cisapride
B. Prepuulsid
C. Johnson & Johnson
D. $1,045.0
E. Gastro-oesophageal reflux
   disease

26

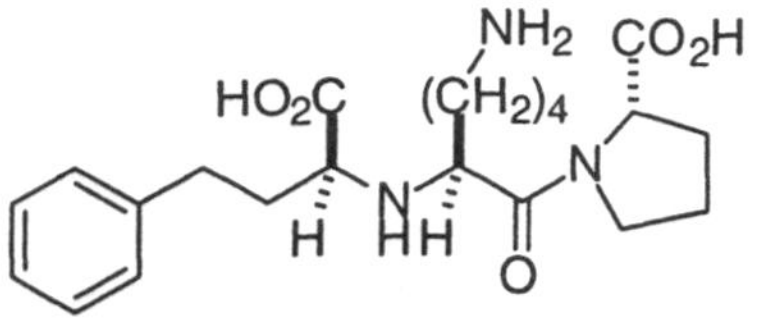

(optically active)

A. Lisinopril
B. Zestril/Zestroretic
C. Zeneca
D. $1,035.0
E. Hypertension, congestive heart failure

27

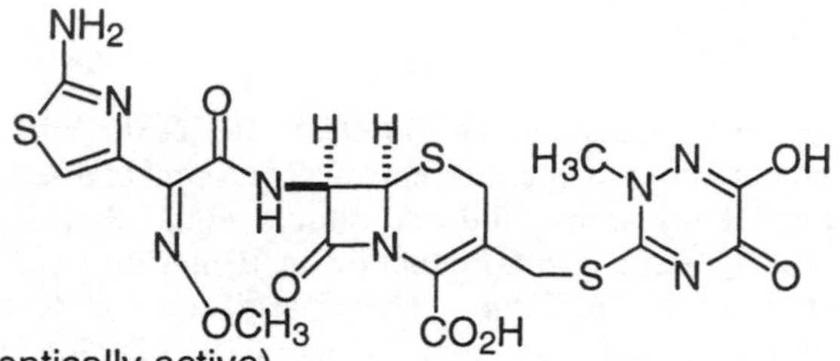

(optically active)

A. Ceftriaxone
B. Rocephin
C. Roche
D. $1,011.4
E. Infections

28

5-oxoPro-His-Trp-Ser-Tyr-Leu-Leu-Arg-Pro-NHC$_2$H$_5$

(optically active)

A. Leuprolide
B. Lupron
C. TAP Pharmaceuticals
D. $990.0
E. Prostate cancer and endometriosis

29

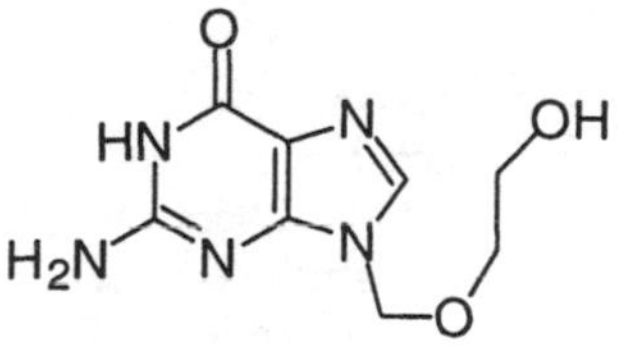

A. Acyclovir
B. Zovirax
C. Glaxo Wellcome
D. $951.2
E. Herpes

30

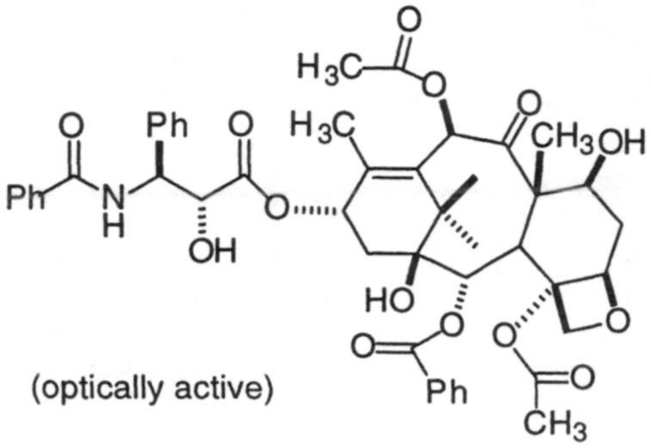

(optically active)

A. Paclitaxel
B. Taxol
C. Bristol-Myers Squibb
D. $941.0
E. Ovarian and breast cancer

## Process Chemistry

### Background

Acetylsalicylic acid, most commonly known as aspirin, was developed from identifying the glycoside salicin in the bark of willow trees (*Salix sp.*). The natural product extracts were known for the antipyretic action. Aspirin, a synthetic derivative of salicin that is available without prescription, is consumed at an estimated rate of 35 tons daily in the United States alone. Ever since the early synthesis[10] of aspirin in 1853, the synthetic community realized the demanding need for optimizing the chemical processes for manufacturing. Thus, process chemistry has been and will continue to be instrumental in producing cost-effective and adequate quantities of complicated active pharmaceutical ingredients.

The late Max Tishler[11] (1906-1989), president at Merck Sharp & Dohme Research Laboratories, was one of the early practitioners of the concept of developmental research. He emphasized that the quality of industrial research could and should equal that of academic research. He developed techniques for the large scale production of Penicillin G, as well as a manufacturing process for the production of Cortisone. The Pharmaceutical Research and Manufacturers of America honored him with the Discoverers Award in 1989 for his outstanding contributions.

A quick glance at the chemical literature has revealed the lack of adequate exposure and the desired number of citations related to drug process research and development. However, in recent years there has been an increased interest directed towards publishing[12-15] and organizing the scientific meetings under the banner of process chemistry. Several papers pertaining to process chemistry have been and are being published in reputed synthetic chemistry journals. The first International Conference on Organic Process Research and Development was organized by Scientific Update[16] in November 1997. The journal: "Organic Process Research and Development"[17] was first published in January 1997.

### Need for and the Role of Process Chemistry

In a pharmaceutical company various departments play specific roles in furthering the drug development programs. While the medicinal or discovery chemists identify the new drug candidates to treat or prevent a particular medical indication, the process chemists are responsible initially to supply the API in kilogram quantities for various studies needed to file the Investigational New Drug (IND) Application to Food and Drug Administration (FDA).[18] The IND studies include: animal toxicology, pharmaceutical and formulation studies, drug substance and drug product stability, metabolism and pharmacokinetics. Upon IND approval, process scientists must reliably produce multiple kilograms of API for clinical trials. The paradigm shift (1990s) of increasing number of compounds entering the development has rendered a tremendous impact on chemists responsible for preparing supplies of these new drug candidates. Details of this topic are outlined in Chapter 2.

The focus of process chemistry differs from routine organic chemistry. It emphasizes on optimization and defines the controls to make the sequence of reactions amenable to scale up. A good process should reliably yield high purity product made by a patent unencumbered process. In short, the overall thrust of the scientists engaged in the process chemistry is to develop the shortest, least expensive, safest and most environmentally friendly process in kilogram quantities.

## Case Studies

Chapters 3 through 14 are case studies which provide an excellent detailed accounts of the processes to synthesize the drug substance from the exploratory level to manufacturing stage. The complicated synthesis of the oxabicyclic [2.2.1] system (Ifetroban sodium) involving a chiral imide route discloses the challenge of elevating the overall yield from 3% (23 steps) to 28% in 12 steps by the scientists at the process laboratories in Bristol-Myers Squibb Company (chapter 3). Carbon-carbon bond formation via classical Grignard chemistry and the successful scale up processes are presented in chapters 3, 7 and 13. An efficient 6-step enantioselective synthesis based on a novel nucleophilic addition-intramolecular cyclization reaction is elaborated in chapter 4. The difficult, yet highly processable synthesis of Vitamin D involving photolysis (chapter 5), by the Hoffmann-La Roche Process group and the enantioselective route to azetidinones comprising an aldol reaction and a novel fluoride catalyzed cyclization (Chapter 13; Schering-Plough Research Institute) are noteworthy.

A highly optimized, large scale synthesis of a purine bronchodilator involving catalytic transfer hydrogenation (chapter 6) by the Astra group in Sweden and the large scale enantioselective hydrogenation reaction in the preparation of *(R)*-Levoprotiline (chapter 11) by the Novartis scientists from Basel, Switzerland describe cost-effective methods. The evolution of the ultimate manufacturing process utilizing unique alkylation strategy with chiral auxiliary (Ontazolast; Boehringer Ingelheim) and an excellent account of the route selection to scale up the $5HT_{1a}$ receptor antagonist by the Lilly Research Laboratories are described in chapter 7 and chapter 9, respectively. A practical approach to scale up the Diels-Alder reaction (Chapter 11), the Vilsmeier Haack reaction (chapter 4), the Friedel Crafts reaction (chapter 4) and the Friedel Crafts acylation (chapter 9) are presented in addition to the Grignard and the aldol reactions to construct the key intermediates.

Separation of the desired isomers from the mixture of racemates involving the classical resolutions are described in several examples throughout the case studies. The work by the Schering-Plough Research group to synthesize Dilevalol (chapter 8) led to a simple and high yielding process. The large scale synthesis using radical carbamoylation reaction by the process group at Novartis (Summit, New Jersey and Switzerland) is described in chapter 10. An outstanding account of the process R&D accomplishment of practical and large scale synthesis of the HIV protease inhibitor (cyclic urea) has been described in chapter 12 by the DuPont Merck Pharmaceutical Company.[19] Synthesis of the common lactone moiety of HMG-CoA reductase inhibitors utilizing asymmetric synthesis by the Rhone-Poulenc Rorer group in UK and USA is presented as a case study in chapter 14.

In summary, the chemistry described in the case studies provide a wealth of practical approaches to large scale syntheses.

## Special Topics

The topic of 'chiral drugs' has gained a considerable attention in recent years in the pharmaceutical industry. The first wave of the chiral drugs is at the verge of losing the seventeen- to twenty-year term of patent protection. It is estimated that two-thirds of the drugs in development are chiral compounds with $73 billion global market potential.[20] Process scientists in the chemical and pharmaceutical industries are challenged with the task to develop cost-effective processes to build the chiral building blocks and the molecules for global needs. An outstanding review on the 'Manufacturing processes and the applications of optically active compounds' was documented in two volumes entitled, "Chirality in Industry".[15]

About half of this volume is dedicated to the special topics. The introduction and the maintenance of chirality (chiral pool, resolution or enzymatic intervention) and asymmetric syntheses are presented in three different chapters. Most of the larger pharmaceutical companies have a group of scientists responsible to build the synthons or the chiral building blocks mediated by microbial or enzymatic technology. An interdisciplinary approach between chemistry, biocatalysis and engineering to manufacture the benzodiazepine drug candidate (LY 300164) is presented in chapter 15 by the scientists at the Lilly Research laboratories. A detailed account of the practical catalysts for asymmetric synthesis has been outlined by Professor Rajanbabu (Ohio State University) and Ayers (Hoechst Marion Roussel Inc.). An updated account of the asymmetric epoxidation using chiral (salen)Mn(III) complexes is summarized by Professor Jacobsen (Harvard University) and Senanayake (Sepracor Inc.) in chapter 19. An elegant selection of work on enantioselectivity is written by the group from Glaxo-Wellcome Inc. The other special topics chapters include: phase transfer catalysis (PTC Technology) drug substance final form selection (Rhone-Poulenc Rorer), process safety (Parke-Davis Pharmaceuticals Research, Germany), design of experiments (Searle) and automation (Glaxo-Wellcome and Orchid Biocomputer).

## Future Trends

In the decade of the '90's, the pharmaceutical industry is undergoing significant changes in consolidating resources, restructuring, increasing overall efficiency and accelerating the drug development processes. As a result, one would expect an increased number of development candidates and INDs which would demand cost effective, environmentally friendly, faster chemical processes to be developed by the process scientists. **Acceleration** of the development timelines will be the driving force for the process personnel in the new millennium. **Integration** of the activities at various phases (process research, development, engineering and operations, etc.) will be one of the major considerations for the aggressive future. Continuing emphasis on more effective **experimental design** and laboratory **automation** for method development, process optimizations, and screening the reactions (as illustrated in chapters 22 and 23), adapt to new and improved **skillsets** (chemometrics, personnel and scientific), **consolidation** of resources and the skillsets and **outsourcing** (wherever necessary) are, perhaps, a growing trend to move us into the 21st century.

## References

1. Current address: Small Molecules Process Chemistry, Amgen Inc., 1840 DeHavilland Drive, Thousand Oaks, CA 91320-1789.
2. Pisano, G. P.; Wheelwright, S. C. The New Logic of High-Tech R&D, *Harvard Business Rev.* **Sept.-Oct. 1995**, 93.
3. Breen, J. J. *Chem. & Eng. News.* **Dec. 22, 1997**, 47.
4. Breslow, R. "The Greening of Chemistry", *Chem. & Eng. News*, **Aug. 26, 1996**, 72
5. (a) Trost, B. M. *Science*, **1991**, *254*, 1471. (b) Trost, B. M. *Angew. Chem. Int. Ed. Engl.* **1995**, *34*, 259.
6. Sato, K.; Aoki, M.; Takagi, J.; Noyori, R. *J. Am. Chem. Soc.* **1997**, *119*, 12386.
7. Marko, I. E.; Giles, P. R.; Tsukazaki, M.; Brown, S. M.; Urch, C. J. *Science*, **1996**, *274*, 2044.
8. *Med Ad News*, **May 1998**, *17(5)*, 14.
9. *The Merck Index*, Eleventh Edition, Ed.: Susan Budavari, **1989**, ISBN No. 911910-28-X, Published by Merck & Co., Inc., Rahway, N. J., U.S.A.

10. Prepn.: Gerhardt, C. *Ann.* **1853**, *87*, 149.  Manufacturing from salicylic acid and acetic anhydride: Faith, Keyes and Clark's *Industrial Chemicals*, Ed.: Lowenheim, F. A.; Moran, M. K. Wiley-Interscience, New York, 4th ed., **1975**, pp. 117-120.
11. Part of the information on Late Max Tishler was provided by Seemon H. Pines, Ex-Senior Vice President, Merck Research Laboratories.  Tishler joined Merck Laboratories in 1937 and was the president of the company in 1957.
12. *Drug Process Research and Development*; Lewis, N.; Mitchell, M. *Chem. & Industry*, **June 3, 1991**, 374.
13. *Process Development: Fine Chemicals from grams to kilogram*; Lee, S. A.; Robinson, G. E. *Oxford Chemistry Primers Series #30*, **1995**, Oxford Univ. Press, Walton St. Oxford OX26DP, UK.
14. *Principles of Process Res. and Chem. Dev. in the Pharmaceutical Industry*; Repic, O. John Wiley & Sons Inc., **1998**.
15. (a) *Chirality in Industry: The Commercial Manufacture and Applications of Optically Active Compounds*, John Wiley & Sons, **1992**, *Vol. I*, Ed.: Collins, A. N.; Sheldrake, G. N. and Crosby, J.
    (b) *Chirality in Industry: The Development in the Manufacture and Applications of Optically Active Compounds*, John Wiley & Sons, **1997**, *Vol II*, Ed.: Collins, A. N.; Sheldrake, G. N. and Crosby, J.
16. 1st International Conference on Organic Process Research & Development; Conference organized by Scientific Update (Trevor Laird, UK); Nov. 1997, San Francisco, USA.
17. *Organic Process Research & Development* : An International Journal Published jointly by the American Chemical Society and The Royal Society of Chemistry, Ed.: Trevor Laird, Scientific Update, Wyvern Cottage, High Street, Mayfield, East Sussex TN20 6AE, UK.  First publication came out in **Jan. 1997**.
18. In the United States of America, the pharmaceutical company submits the results with an IND application to the U.S. FDA.  Globally there are similar central government authorities and processes for drug approval and marketing.  However, many countries outside the U.S. require negotiation of price and reimbursement with the government prior to marketing.
19. Currently known as DuPont Pharmaceuticals.
20. Stinson, S. C. *Chem. & Eng. News.* **Oct. 20, 1997**, 38.

# STRATEGY

# Process Research and Development Strategy for Accelerated Projects

**Christopher M. Cimarusti**
*Process Research and Development, The Bristol-Myers Squibb Pharmaceutical Research Institute, P.O. Box 191, New Brunswick, New Jersey 08903-0191 USA*

Key Words:  process research, process development, aztreonam, fosinopril, captopril

The decade of the '90's has been a turbulent one for the pharmaceutical industry;  the only constant has proven to be change.  Chemistry as practiced within the pharmaceutical industry has changed at least as much as business strategy or any of the other key disciplines that contribute to new drug development.  Discovery chemists have incorporated combinatorial chemistry, automated synthesis and virtual screening into their search for new drug candidates.  These changes have been undertaken to decrease the time required for the discovery phase of drug development, to increase the number of new drug candidates and to decrease the resources (including amount of compound used) applied during the discovery phase.

Thus, development chemists can expect to see more drug candidates and, as we approach the new millennium, less time to progress these drug candidates will be allotted by management.  These circumstances have forced a review of the underlying strategies that development chemists use to guide their work.  The paradigm described in this article has evolved during the last two decades at Bristol-Myers Squibb.  It had its origins in the development of the monobactam antibiotic aztreonam and matured during the development of the ACE inhibitor fosinopril.  This ten-year period (1979-1988) saw the involved staff grow from a handful to more than 200 individuals.  Following the merger with Bristol-Myers in 1989, significant time and effort were expended in reexamining strategy and organization.  The redeployment of chemistry capacity and capability between "R&D" and "Technical Operations" in 1990 enabled each group to proactively support its mission.  Further valuable insight was gained in the development and manufacturing introduction of the semi-synthetic paclitaxel process.  Finally, the current emphasis on time-based competition and portfolio management has led to additional optimization of strategy and tactics and, most importantly, the realization that continual improvement is the one appropriate constant in times of change.

## Time-based Competition

One key influence on drug-development strategy is time-based competition. Pharmaceutical companies have announced very aggressive targets: Glaxo is striving to remove 4-5 years from their process while Novo-Nordisk is seeking to halve its development time. Several years ago, the time of development averaged 15 years and the cost of development was estimated to be 350 million dollars. The $2660 per hour spent on development can be viewed as a powerful incentive to reduce time.

A second incentive is the diminishing period of exclusivity that results from a development time of 15 years, a patent time of 20 years from filing and the demonstrated ability of generic competitors to rapidly enter and transform the market place. Weeks after Bristol-Myers Squibb's Capoten® patent expired, generic captopril was selling for $0.05 per tablet (the wholesale price of a Capoten tablet was $0.56). Within a few more weeks, generic competitors were giving away generic captopril as part of bundled deals. This type of rapid market transformation is likely to be the rule in the future. Though faster development will not change the results, it will provide more revenue before patent expiration. A year's faster development is expected to provide a year's revenue in the marketplace created by the innovator, not by the generic competition.

## Portfolio Management

### Early Development

One of the major changes in the conduct of development projects has been the concept of exploratory or early development. Many companies have adopted the strategy of evaluating compounds quickly and efficiently in "early" development to answer a basic question: "Is this likely to be a drug?" Since only one in ten compounds entering development is approved, this strategy seeks to eliminate the other nine at early project times and with a minimum of resources. Expending 90% of resources on the 90% of projects that fail is the extreme that this strategy seeks to avoid. For process chemists, this strategy means producing sufficient amounts of many drug candidates to answer the key question with a minimum of time and effort. As we shall see later, this can be done with very limited resource input.

### NCE Approvals/Year

A second facet of portfolio management that dramatically influences the work pallette of process chemists is the increasing number of compounds entering development. Several pharmaceutical companies have shared their targets for new product approvals per year. Hoechst Marion Roussel has targeted two per year while Glaxo wishes to introduce "three significant new medicines per annum". Using the benchmark of one successful project for each ten entering development, these goals require the entry of 20-30 compounds per year into the development pipeline. The impact of this throughput on chemists responsible for preparing supplies of these new drug candidates will be tremendous.

## Aztreonam

### Background

In 1978, when Richard Sykes and his colleagues identified a bacterium that produced a novel β-lactam, there was no process research function within the Squibb Institute for Medical Research.

There was a Chemical Process Development department within the Institute that had recently completed the introduction of the captopril process[1] into manufacturing. Without a strong process research contribution, the original process (Scheme 1) for captopril **4** involved adding a racemic side chain **1** to proline **2** and selectively crystallizing the desired diastereomer **3**. This proved to be a practical process but did not qualify as an efficient one. Fortunately, the correct enantiomer of the side chain became available from external suppliers and this simple sequence became a very efficient process that eventually saw considerable optimization in the manufacturing plant.

Scheme 1. Original Captopril Process

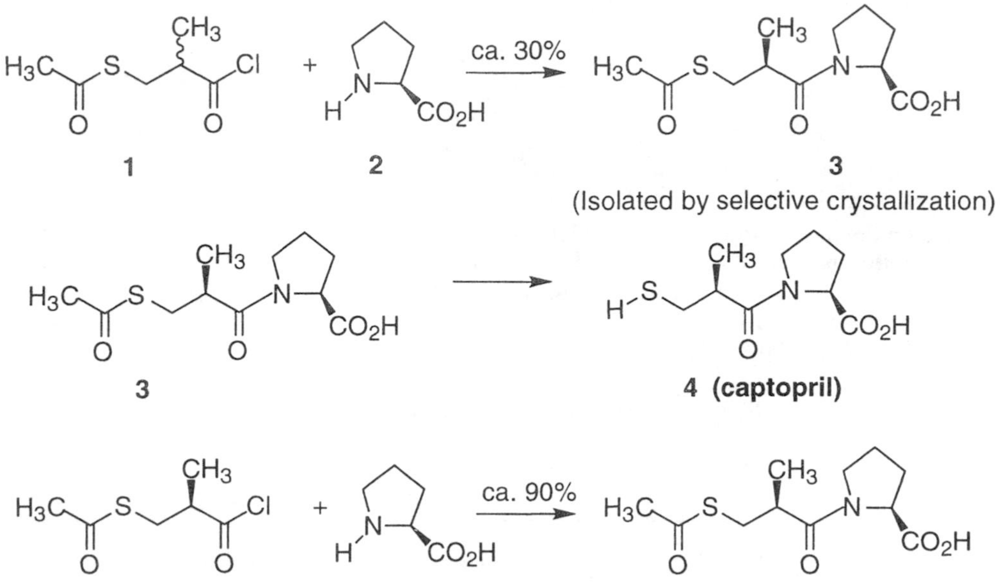

(Isolated by selective crystallization)

**3**

**4 (captopril)**

When the structure of the first monobactam (SQ 26,180, **5**) was elucidated,[2] it was apparent that process development alone would not be sufficient. A process research effort to devise synthetic access to monobactams of diverse structure would be necessary to support the discovery program and (it was hoped) the resulting development project. A portion of our synthetic antimicrobial group was therefore diverted to process research.

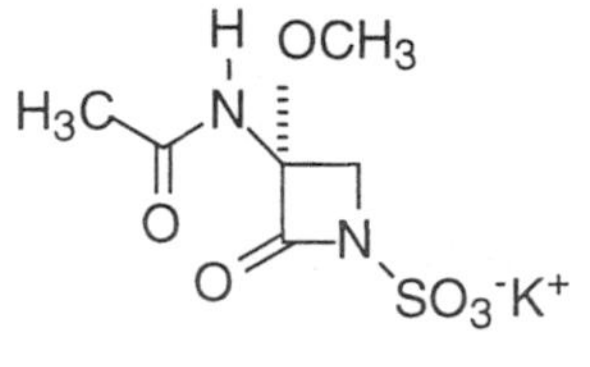

**5 (SQ 26,180)**

The initial goal was access to the nucleus of the monobactams since it was expected that identification of the best "side chain" would take significant time. The first processes (Scheme 2) were based on conversion of the bicyclic beta lactam 6-aminopenicillanic acid (6-APA) to monocyclic azetidinones that were sulfonated to provide monobactams.[3] These processes, though cumbersome, enabled detailed structure-activity work to begin. At best, these processes can be characterized as "expedient".

Scheme 2.  Expedient Synthesis of Monobactams

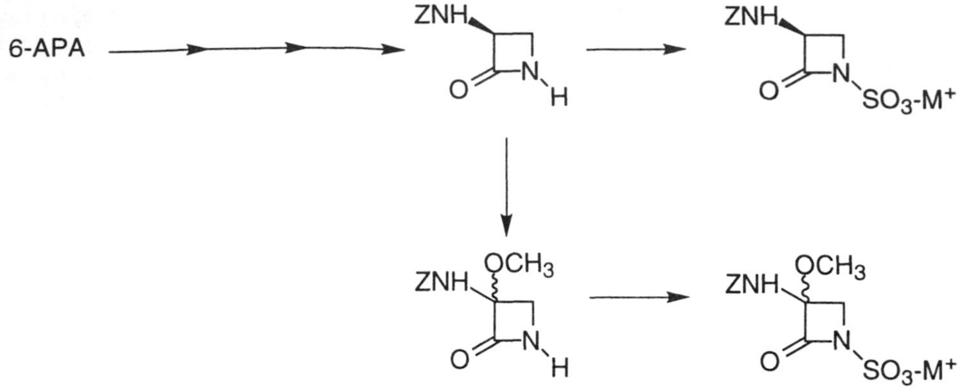

Based on the pioneering report by Miller,[4] we were able to devise practical access (Scheme 3) to the 4α-methylmonobactam nucleus.[5] This involved the conversion of threonine to the 4α-methyl azetidinone by a modification of Miller's hydroxamate methodology.  The crystalline nature of intermediates **6-8** contributed significantly to the rapid scale-up of this process in our New Brunswick pilot plant and timely support for the IND filing.

Scheme 3.  Practical Synthesis of Monobactams

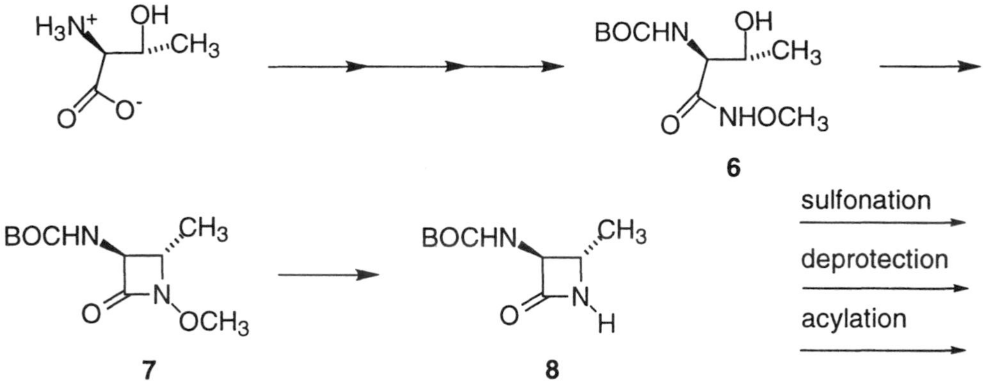

While this scale-up was under way, process research chemists provided an efficient process (Scheme 4) to monobactams.[6]  Conversion of threonine to the amide mesylate **9**, followed by sulfonation and *in situ* cyclization provided simple, direct and cost-effective access to the desired 4α-methyl monobactam moiety.  This route was ready to support the NDA development phase of aztreonam.

Scheme 4. Efficient Access to Monobactams

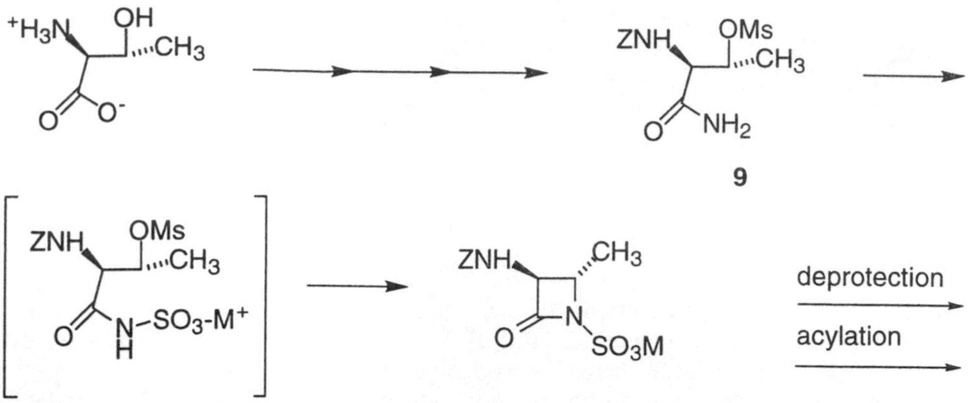

## Side-chain Construction

As the structure-activity relationships unfolded, it became apparent that α-oximino-aminothiazolylacetyl side chains such as **A** would be part of the eventual drug candidate. These side chains had reached prominence with the discovery of cefotaxime **10** and there was a wealth of compound and process priority in the patent literature to confound route selection for development of a similarly substituted monobactam such as aztreonam **11**.[7]

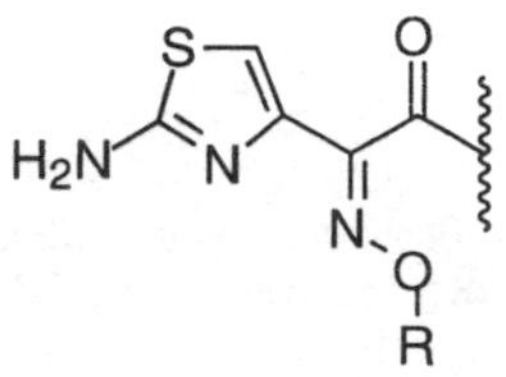

**A**

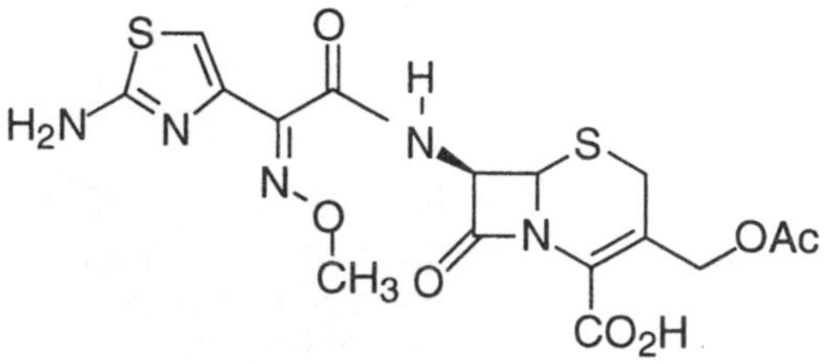

**10** (cefotaxime)

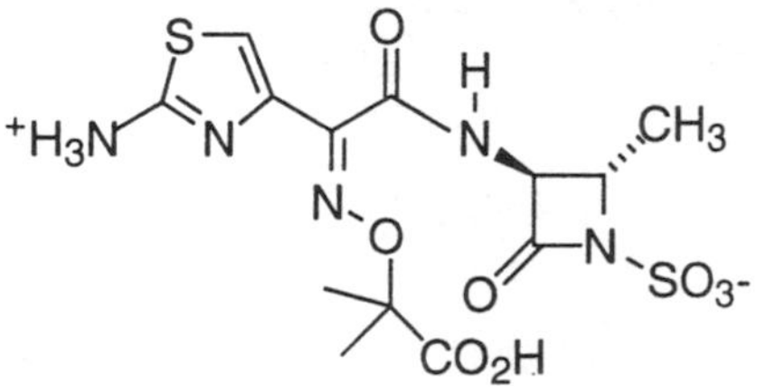

**11**

The work described above on nucleus process research had provided access to 4α-methyl-3-aminomonobactamic acid **12** and structure-activity relationships soon provided the conviction that the first drug candidate would be based on this 4α-methyl nucleus. Thus, the first synthesis of the eventual drug candidate aztreonam **11** was accomplished by coupling **12** with the mono-protected diacid **13**. Though suitable for small quantities, this procedure (Scheme 5) had several drawbacks.

Scheme 5. Expedient Synthesis of Aztreonam

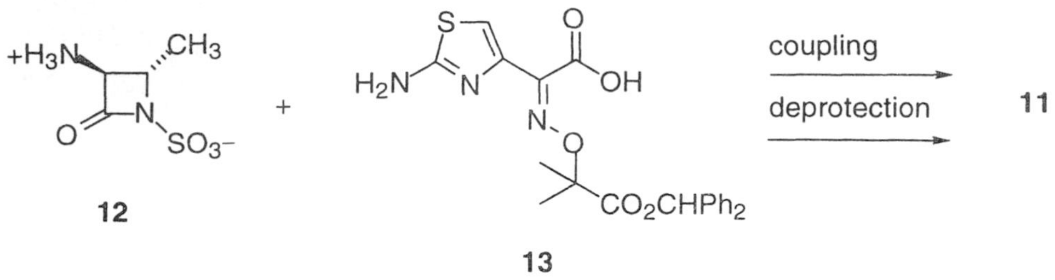

The deprotection required trifluoroacetic acid that was expensive and difficult to access in large quantities (in the cephalosporins, mineral acid could be used; the reactivity of the monobactam nucleus did not allow this). The work up of this reaction involved the use of ether to precipitate the crude aztreonam as a filterable solid (ether is not a welcome solvent in the pilot-plant environment). In addition, and most importantly, there were several patent applications that claimed **13** and processes utilizing it. In order to assure a rapid and commercially unencumbered development of aztreonam, we decided to avoid the use of **13**.

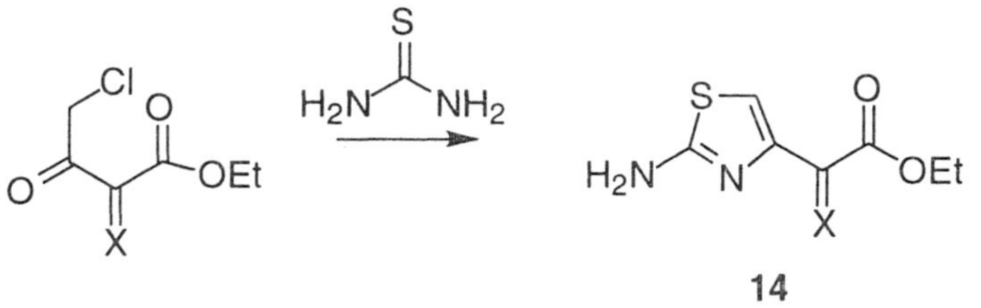

The preparation of these side chains from derivatives of ethyl chloroacetoacetate and thiourea was well known and, in contrast to cases where X = O or NOR, the unsubstituted parent compound (**14**, X = H$_2$) was commercially available and patent-unencumbered. Therefore, we designed and developed a process (Scheme 6) that proved to be practical at laboratory, pilot-plant and plant scale based on coupling of **11** to **15**, followed by oxidation to the α-ketoamide, oximation, and deprotection.[8] Though the Z/E ratio in the oximation step was only 85/15 after deprotection, the resulting Z-isomer (aztreonam) could easily be separated from the E-isomer by simple crystallization.

Scheme 6. Practical Synthesis of Aztreonam

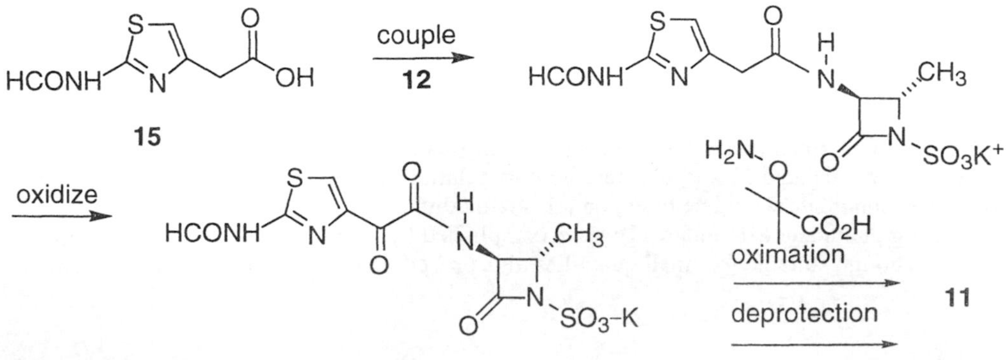

This process supported Phase-III development, NDA/MAA filing and approval, and launch of aztreonam. It was not, however, as convergent as we would have liked. A major step towards a more convergent synthesis was taken by using the α-ketoacid **16** (Scheme 7). Appropriate commercial arrangements were reached that allowed us access to this substance. Coupling of **16** with **12**, followed by oximation/deprotection, provided a more convergent, "efficient" process to aztreonam. Since the deprotection of the formamide moiety could be accomplished with mineral acid, it was actually a superior procedure to the more convergent process utilizing benzyhydryl ester **13** (one less step but requiring trifluoroacetic acid). The one remaining vexing issue was the 85/15 ratio of Z/E isomers formed in the oxidation step.

Scheme 7. Efficient Process to Aztreonam

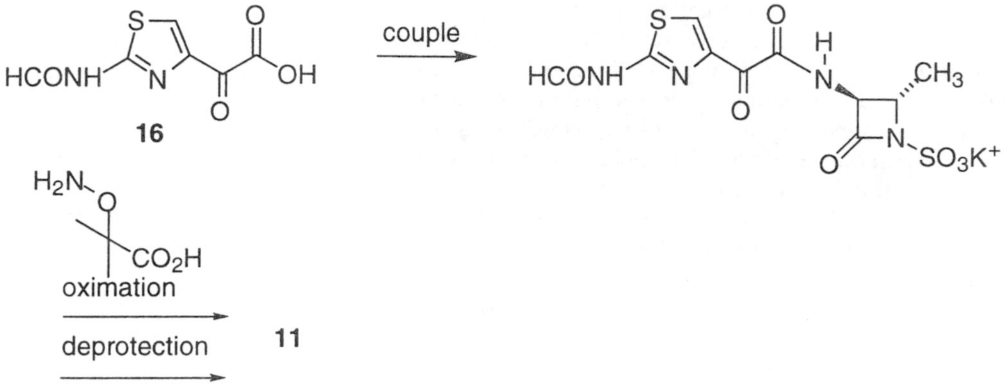

Further detailed examination of this reaction provided a significant advance: addition of copper (II) acetate modified the ratio to ca. 95/5 and provided a significant yield enhancement.[9] Thus, several years after introduction into manufacturing, the second of two process changes resulted in an optimized process (Scheme 8) for aztreonam.

Scheme 8. Optimal Process to Aztreonam

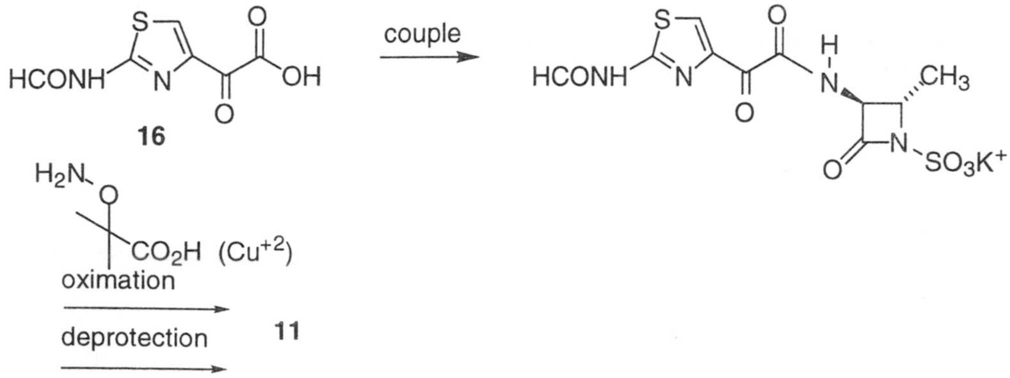

### Lessons from Aztreonam

Based on this involvement, a process research group was started within the organic chemistry department of the Squibb Institute for Medical Research. In order to enable the group to prepare material at significant scale, our scale-up laboratory was affiliated with the group. The close collaboration with discovery chemists that proved useful in this case was made a cornerstone of group strategy. The collaboration with the process development department that extended over several years and resulted in moving from expedient to efficient processes was similarly incorporated. Finally, the interaction with the manufacturing group, while successful, was recognized as an area for improvement.

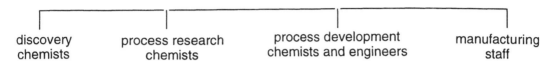

discovery chemists     process research chemists     process development chemists and engineers     manufacturing staff

On a different level, the desire to accelerate the clinical evaluation of this new class of β-lactams had forced us to use a step-wise approach to chemical support of the project. The expedient synthesis used during the discovery phase was judged incapable of supporting rapid development. The practical synthesis that supported full development and launch was judged incapable of supporting cost-effective manufacture for the long term. Finally, the efficient synthesis based on use of an α-keto acid allowed us to progress to an optimal procedure with only minor adaptation on the manufacturing floor. Conceptually, therefore, (for future projects of significant complexity) we were prepared to support different stages with different processes (Table 1) and to have a smooth transition between them.

Table 1. Process Characteristics

| Stage | Process Characteristic |
| --- | --- |
| Discovery | Expedient |
| Development | Practical |
| Launch | Efficient |
| Marketing | Optimal |

### Fosinopril

During the late phase of the discovery program to find a second-generation ACE inhibitor, 4(trans)-cyclohexylproline **17** emerged as the likely "nucleus". At about the same time, a series of phenylalkylphosphinic acids were identified as outstanding moieties to ensure active-site zinc binding. The combination of these moieties provided the diacid **18** that had pharmacokinetics in animals consistent with once-a-day dosing in man. Since diacid **18** was poorly absorbed in animals, a program to devise an appropriate prodrug strategy was begun. While this was under way, process research chemists began to examine improved access to **17**.

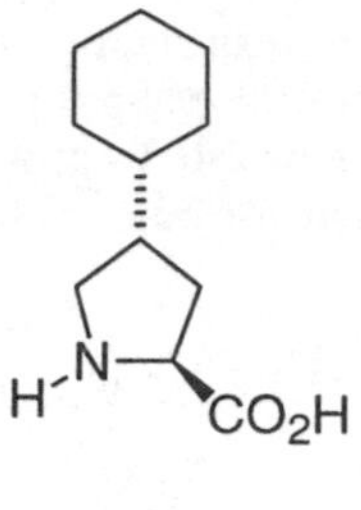

**17** 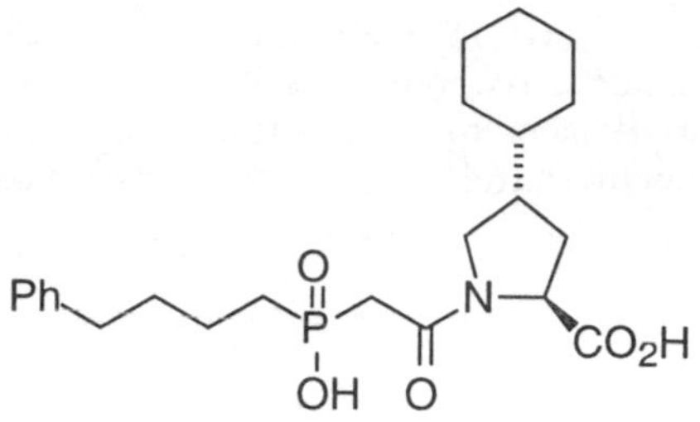                    **18**

## 4(trans)-Cyclohexylproline

A previous discovery program to improve captopril by modifications to the proline moiety had utilized Z-ketoproline as an intermediate (Scheme 9). Reaction with the phenyl grignard reagent proceeded to give a poor yield of carbinol **19** due to competing enolization. Elimination followed by lithium/NH$_3$ reduction provided a 1:1 mixture of cis and trans-4-phenylproline that were easily separated by virtue of the insolubility of **20**. Hydrogenation of the materials provided initial access to trans **17** and cis 4-cyclohexylproline. This process could be improved by substitution of triphenylcerium (almost _no_ enolization) for the Grignard reagent and elimination of alcohol from the lithium ammonia reduction (6:94 cis/trans ratio). However, difficulties in reproducibily drying cerium trichloride precluded practical use of this process.

Scheme 9. Expedient Synthesis of 4(trans)-Cyclohexylproline

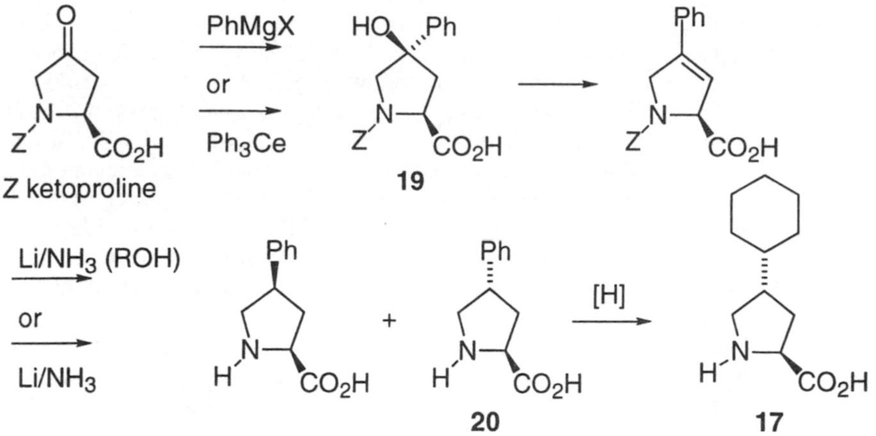

Another line of investigation led to a practical process (Scheme 10) that enabled us to file an IND for fosinopril and demonstrate that it was likely to be an effective, once-a-day antihypertensive. Reaction of the phenyl cuprate reagent with tosylate **21** gave the trans-4-phenylproline derivative **22** via intermediate formation of an aziridinium ion (one inversion) followed by attack of the phenyl moiety (second inversion).[10] This unexpected result provided access to 4(trans)-phenylproline from the naturally occurring 4(trans)-hydroxyproline. This sequence could be scaled up in the kilolaboratory with one modification that provided a second, unexpected finding. In the laboratory, the cuprate reaction was accomplished in ketyl-distilled diethyl ether. Upon scale-up, tetrahydrofuran (THF) was added to reduce the volume of diethyl ether to practical levels; this resulted in the production of a variable amount (depending on the ratio of diethyl ether to THF) of the enantiomeric cis compound **24**.

In the presence of THF, the lithium anion of **22** is deprotonated to give a dianion; reprotonation on the sterically more accessible face (away from the phenyl group) provided **24**. Though practical for kilolaboratory use, this process was not amenable to pilot-plant use (diethyl ether as solvent; dimethylsulfide as the recrystallization solvent for cupric bromide (the cuprate precursor)).

Scheme 10. First Practical Synthesis of 4(trans)-Cyclohexylproline

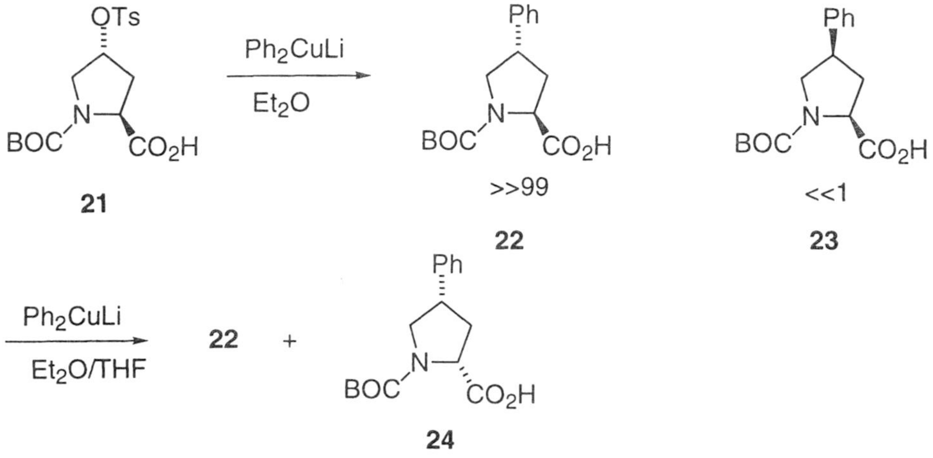

A second laboratory synthesis of 4(trans)-cyclohexylproline was quickly developed in the laboratory,[11] based on alkylation of the pyroglutamate derivative **25** (Scheme 11). Generation of the lithium anion with LDA followed by alkylation with cyclohexenyl bromide gave an excellent ratio (2:98) of cis to trans product. LAH reduction followed by hydrogenolysis, provided prolinol **28** that could be directly oxidized (Jones reagent) to 4(trans)-cyclohexylproline. This process could be modified for pilot-plant use. The lithium anion was generated with lithium amide (generated *in situ*) and alkylated in a mixture of ammonia and toluene. This alleviated the need to prepare LDA with butyl lithium on large scale. The original LAH reduction in ether was replaced by an LAH reduction in toluene with a small amount of THF added to provide a much safer reaction solvent. The direct Jones oxidation of **28**, which required extensive manipulation to remove metal ions, was replaced by a three-step sequence (benzoylation, oxidation, deprotection).

This process allowed us to prepare several hundred kgs of **17** and, subsequently, several hundred kgs of fosinopril. In our judgment, however, this process was not suitable for introduction into manufacturing. It was long (nine steps) and featured the use of LAH (a reagent we have not yet utilized in a manufacturing plant).

Scheme 11. Second Practical Synthesis of 4(trans)-Cyclohexylproline

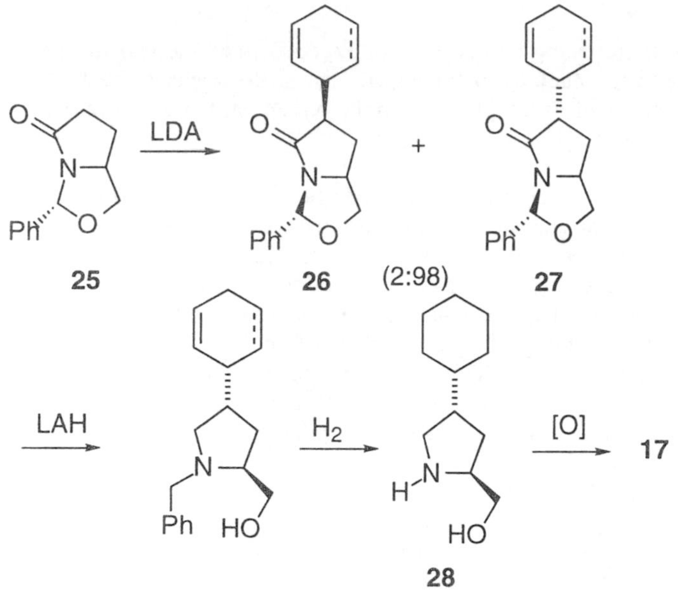

Another line of investigation (Scheme 12) led to an efficient synthesis that was suitable for introduction into manufacturing. This involved, as the key step, a Friedel-Crafts alkylation of benzene by the cis-mesylate **29**.[12]   After an initial demonstration of feasibility, a detailed investigation revealed that N-benzoyl, mesylate, and free acid were the most appropriate ensemble for the nitrogen protection, leaving group and acid moieties, respectively.  This sequence of six steps proved to be practical in that all intermediates were crystalline, yields were uniformly high, and (with appropriate engineering controls) the handling of benzene as solvent as well as reagent proved not to be problematic.  After evaluation of all available options, this process was chosen for introduction into manufacturing.  Its startup and subsequent use at plant scale were successful and it remains in use today.

Scheme 12. Efficient Synthesis of 4(trans)-Cyclohexylproline

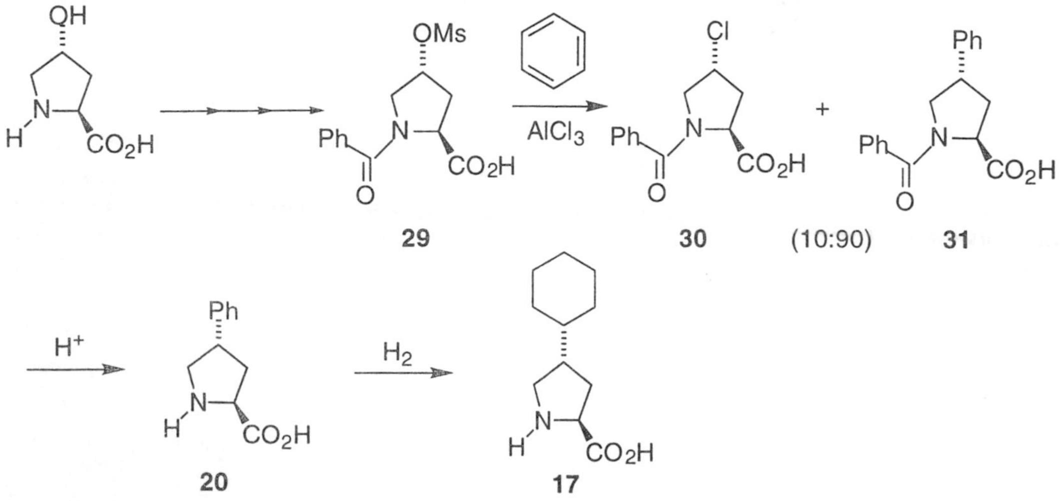

**Lessons from Fosinopril**

The progression of aztreonam chemistry from expedient to practical to efficient processes was repeated for access to the nucleus of fosinopril. Since these cases, we have been committed to supporting various stages of development with the process at hand, rather than delay development while a better process is sought.

As fosinopril process development unfolded, the complexity of both side chain and nucleus construction led us to reconsider our process introduction. We decided that proactive support for process introduction could best be achieved by a development group and pilot plant at a small manufacturing plant reserved for new process introductions and launch support. An appropriate plant was available within our manufacturing network, and resources were made available to start a development group (consisting of equal numbers of analysts, chemists and chemical engineers) at the site and outfit a pilot plant. This group proved invaluable in serving as a bridge between US-based development groups and the European plant staff and enabled the smooth transition into manufacturing of the fosinopril (and subsequent) processes.

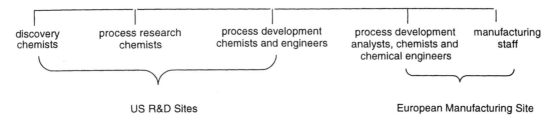

**Post-merger Integration**

The merger of Bristol-Myers and Squibb in 1989 occurred soon after the fosinopril manufacturing startup. It brought together two companies that valued development, but had organized it in different ways. Within Bristol-Myers, chemical development was split between R&D and manufacturing; unfortunately, neither group had the capacity to address the project portfolio of the R&D division. Within Squibb, Chemical Development was part of Technical Operations; however, as I have just described, it had been provided the resources to proactively support the discovery portfolio at the same time it addressed process introduction and optimization of processes in manufacturing. After a detailed analysis of options, process development capacity and capability was allocated to both the Pharmaceutical Research Institute (R&D) and Technical Operations (Manufacturing). The mission of these two development groups was then aligned with the mission of the organizations they served. The R&D development group's mission included support of discovery programs and rapid initiation, progression, filing and approval of development projects. The Technical Operations' development group's mission included support and optimization of manufacturing, preparation of bulk drug during Phase-III for clinical supplies and registrational stability, and process introduction into manufacturing. During the critical Phase-III "full development" stage of projects, therefore, these groups collaborate in supporting projects whose probability of success is much better than 50%.

## Paclitaxel Development and Process Introduction

The first test of this organizational paradigm was provided by the paclitaxel project-and it worked beautifully! Institute development groups looked at a wide range of technologies to assure access to paclitaxel that was not based on its isolation and purification from the bark of the Pacific yew. The decision was made early in the project that this was a viable short-term strategy (it was expedient and supported launch), but that rapid progression to a more practical and efficient process was mandatory. Institute process research chemists evaluated many options and selected a semisynthetic strategy based on using an activated β-lactam and a more readily available taxane "nucleus" (10-DAB) as key synthons. In collaboration with their process development colleagues, they chose an ensemble of protecting groups and produced kg-scale batches in an R&D pilot plant. As this activity was in progress, Technical Operations development groups introduced enzymatic access to the chiral β-lactam into a plant (where they had a strong presence) and participated in the design and construction of a new plant to accommodate the process at a European site (where they also had a strong presence). Finally both Institute and Technical Operations groups participated in the successful manufacturing start-up of the complete process.

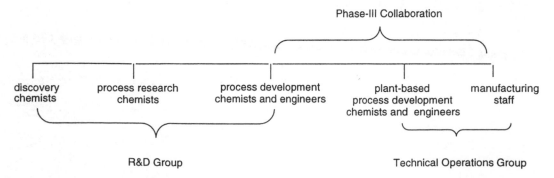

### Lessons from Paclitaxel

Paclitaxel development provided a test of the organizational paradigm that we had selected during post-merger integration. It proved ideally suited to the rapid identification (R&D group), and development and manufacturing introduction (joint effort of R&D and Technical Operations groups) of a complex process under great time pressure and regulatory scrutiny. Based on this success (and others, more recent ones) we are confident that, in our company culture, this paradigm of collaboration and capability aligned with organizational mission is an optimal one.

## Acknowledgments

The author wishes to acknowledge the following Bristol-Myers Squibb staff that he has had the privilege of collaborating with during the past 20 years. The approach described in this chapter was developed, in large measure, by his learnings from them: M. Borrero, D. Bouzard, K. Brady, R. Carroll, T. Denzel, P. Diassi, L. DiFazio, J. Douglas, W. Ferguson, D. Floyd, R. Hecht, J. Heikes, A. Kilbane, W. Koster, D. Kronenthal, G. Michel, J. Moniot, R. Mueller, U. Nager, R. Partyka, R. Schwartz, P. Singh, P. Sprague, L. Steele, L. Szarka, J. Thottathil, N. Tejera, C. Tollinche, W. Winter.

## References

1. Ondetti, M; Cushman, D. U.S. Patent **1977** US4046889.

2. Parker, W. L.; Koster, W. H.; Cimarusti, C. M.; Floyd, D. M.; Liu, W. C.; Rathnum, M. L. *J. Antibiot. 182, 35,* 189-195.

3. Cimarusti, C. M.; Applegate, H. E.; Change, H. W.; Floyd, D. M.; Koster, W. H.; Slusarchyk, W. A.; Young, M. G.. *J. Org. Chem.* **1982**, *47*, 176-178.

4. Miller, M. J.; Mattingly, P. G.; Morrison, M. A.; and Kerwin, T. F. Jr. (1980). *J. Amer. Chem. Soc.* **102**, 7026.

5. Floyd, D. M.; Fritz, A. F.; Pluscec, J.; Weaver, E. R.; Cimarusti, C. M. *J. Org. Chem* **1982**, *47*, 5160-5167.

6. Floyd, D. M.; Fritz, A. F.; Cimarusti, C. M. *J. Org. Chem* **1982**, *47*, 176.

7. Aztreonam, A Synthetic Monobactam; Sykes, R. B.; Philips, I.; Ed; Supplement E to Volume 8; *J. Antimicrobial Chemotherapy* **1981**.

8. Unpublished results, Bristol-Myers Squibb Pharmaceutical Research Institute.

9. Sedergran, T. C.; Anderson, C. F.; U.S. Patent **1987** US4675398.

10. Thottathil, J. K.; Moniot, J. L.; *Tetrahedron Letters* **1986**, *27,* 151.

11. Thottathil, J. K.; Moniot, J. L.; Mueller, R. H.; Wong; M. K. Y; Kissick, T. P.; *J. Org. Chem.,* 1986, 51, 3140-3143.

12. Kronenthal, D. R.; Mueller, R. H.; Kuester, P. L.; Kissick, T. P.; Johnson, E. *Tetrahderon Letters,* **1990**, *31,* 1241.

# CASE STUDIES

# A Practical Synthesis of Ifetroban Sodium[1]

**Richard H. Mueller**
*Process Research and Development, The Bristol-Myers Squibb Pharmaceutical Research Institute, P.O. Box 4000, Princeton, New Jersey 08543-4000 USA*

Key Words: ifetroban, BMS-180291, synthesis, Grignard reaction, chiral imide

Ifetroban sodium (BMS-180291, **1**) is a long-acting, orally bioavailable, highly selective thromboxane A2 receptor antagonist with potent antithrombotic and anti-ischemic properties.[2] Cardioprotective in myocardial ischemia and reperfusion injury in animal models,[3] ifetroban sodium currently is in human clinical trials. It is the last in a series of lead compounds and drug candidates which emerged from a medicinal chemistry program spanning more than a decade. It is a single enantiomer and possesses four stereogenic centers. The original synthetic sequence for **1** comprised some 23 steps and proceeded in an overall yield (via the longest linear sequence) of less than 3%. Subsequent process research and development efforts resulted in a more efficient 12-step synthesis and an overall yield of 28%. These accomplishments will be described in the context of some of the paradigms, exigencies, and dichotomies encountered in any process research and development endeavor.

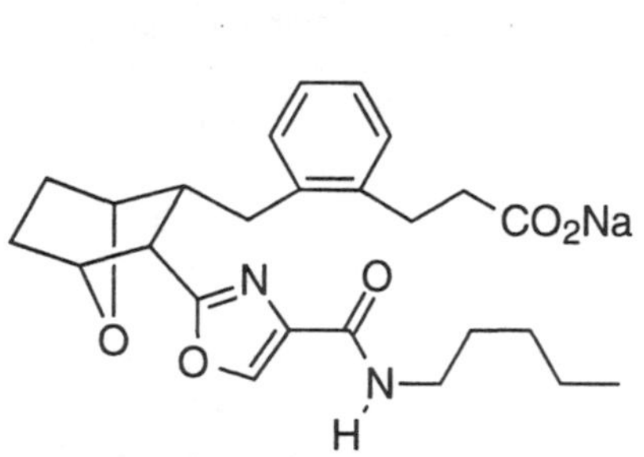

**1** BMS-180291
ifetroban sodium

The first drug candidate in the thromboxane A2 receptor antagonist program was SQ-28668 (**2**), followed by SQ-30741 (**3**), lead compounds SQ-33742 (**4**) and SQ-33961 (**5**), and finally BMS-180291 (**1**). Common to this series of compounds, all of which are single enantiomers, is the oxabicycloheptane nucleus with two sidechains appended in an *exo* orientation so as to provide four contiguous stereogenic centers.

2 SQ-28668

3 SQ-30741

4 SQ-33742

5 SQ-33961

For those readers unfamiliar with how things operate in the pharmaceutical industry, a brief description is in order. The goal of the medicinal, or drug discovery, chemist is to identify a new drug candidate to treat or prevent some particular medical indication. Once a new drug candidate has been selected, the next important step is the filing of an Investigational New Drug Application (IND) with the Food and Drug Administration (FDA). On approval of the IND, the compound can be administered to humans for the first time in Phase-I clinical studies. Studies which are reported in an IND include animal toxicology studies, pharmaceutical and formulation studies, drug substance and drug product stability studies, metabolic studies, and pharmacokinetic studies. This work, of course, requires relatively large amounts of the drug substance, substantially more than was prepared during the course of the medicinal chemistry program. Generally, it becomes the responsibility of a process chemistry group to produce this material, within, of course the shortest feasible timeframe. Pharmaceutical process chemistry groups are organized in different ways. The Bristol-Myers Squibb paradigm allocates this function among Process Research, Process Development, and Manufacturing. Process Research is responsible for providing the supplies of bulk drug substance required for an IND filing, including the initial clinical batch (or batches). Typically, the initial supplies are prepared by some combination of a modified medicinal chemistry synthesis and new, alternative synthetic processes more suitable for scale up. At Bristol-Myers Squibb, this responsibility is shared by process research chemists and Kilo Lab chemists, the Kilo Lab being responsible for the scale up of processes provided by the process research chemists. Kilo Lab scale up work typically is performed in up to 22-L glassware. Process Development, on the other hand, is responsible for further scale up in the Pilot Plant and typically takes over the preparation of bulk drug substance beyond the initial Phase-I clinical supplies.

The original synthesis of SQ-28668 started with the Diels-Alder reaction between furan and maleic anhydride to provide anhydride 6. Catalytic hydrogenation produced 7, and reduction with sodium borohydride gave 8. DIBAL reduction resulted in conversion to racemic 9, which was resolved by crystallization of the *d*-menthol acetal 10 followed by hydrolysis to give 11 as a single enantiomer. A one-carbon Wittig chain extension and hydrolysis provided the six-ring hemiacetal 12, and a second Wittig reaction produced 13, suitable for further elaboration not only to SQ-28668 but also SQ-30741 and SQ-33742. Note that reaction of 11 with an aryl Grignard reagent provided access to later compounds in the series, SQ-33961 and BMS-180291, via 21.

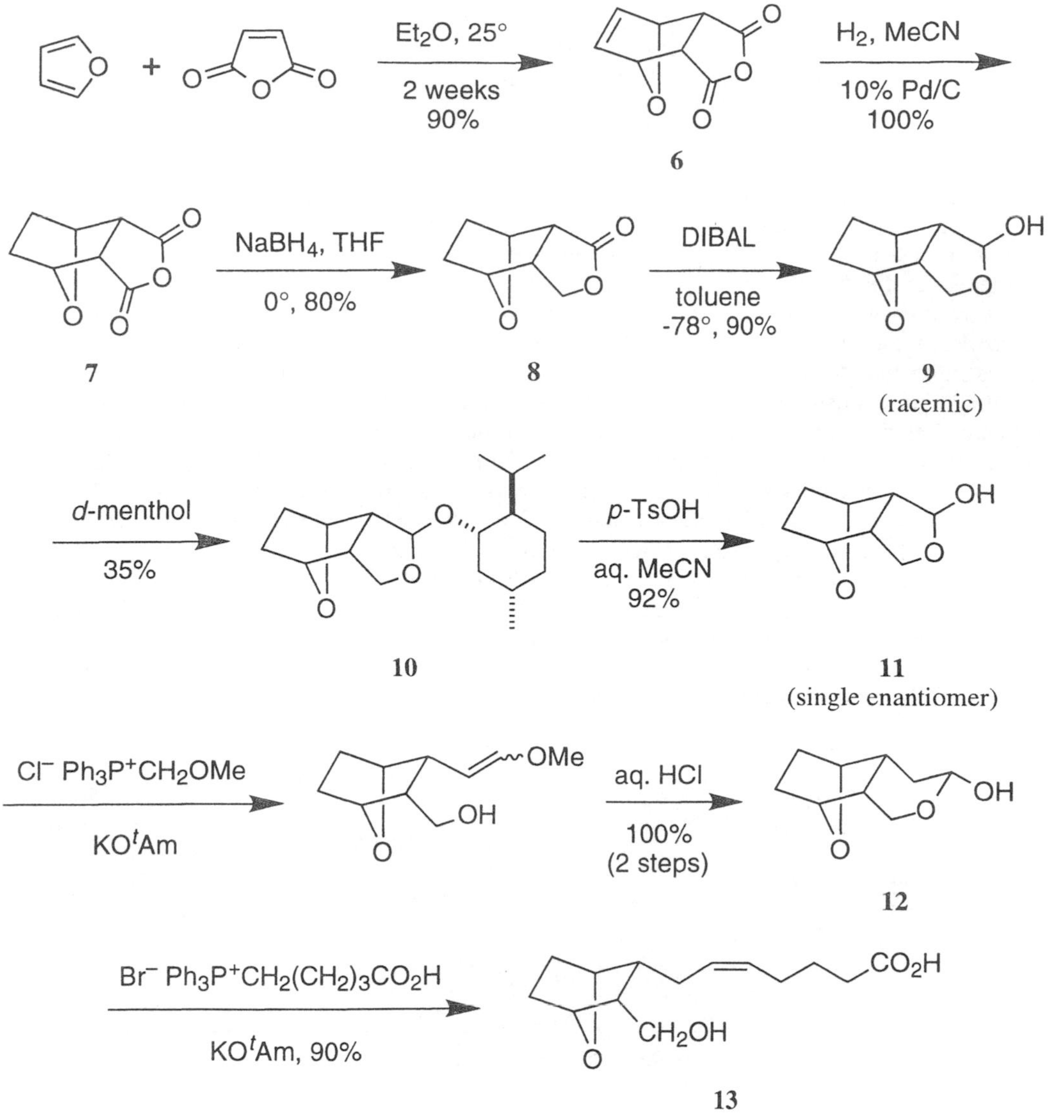

When the decision was made to progress the first drug candidate SQ-28668 to IND project status, it became the responsibility of the Process Research group to provide the required supplies of bulk drug substance. Numerous batches of **12** already had been prepared by the Kilo Lab in support of the drug discovery program, so Kilo Lab personnel were well-experienced with the synthesis of **12**, which, although lengthy and labor intensive, was well in hand on scales up to 100 g. The decision was made to use the existing synthesis of **12** and to concentrate process research efforts on its conversion to SQ-28668, rather than to devise an entirely new, more efficient synthesis. Decisions of this type typically are driven by expediency--the sooner bulk is available, the sooner the IND can be filed and, if the compound is ultimately successful, the sooner it can reach the marketplace. Also, considerable resources would have to be invested in the identification of a new synthesis but, if the compound fails in the clinic, these efforts would have been for naught. Generally, more compounds fail to reach the market than succeed, and so the usual *modus operandi*, particularly for long and involved projects, is to proceed in the Kilo Lab with a "quick-fix" version of the original synthesis. Process research efforts to identify and demonstrate a new and improved synthesis may be initiated simultaneously or at some later date, depending on a number of factors, including the likelihood of success of the project, the priority of the project, and resource availability. Thus, early on in a project, "Make stuff!" wins out over "Learn to make it better!"

The above sequence was scaled up in the Kilo Lab to produce 2 kg of **11** and ultimately 1.2 kg of SQ-28668. Also, in anticipation of success in the clinic, work was initiated to implement the existing process in the Pilot Plant, but first a number of concerns needed to be addressed. The first was the use of diethyl ether in the first step, an extreme explosion hazard risk. EtOAc and toluene both proved to be satisfactory solvents for the reaction. The second concern was the use of anhydrides **6** and **7**, particularly as isolated, dry solid intermediates. Earlier experience had demonstrated that **6** and **7** were potent irritants, causing inflammation and severe swelling on accidental skin exposure (*cf.* cantharidin, **16**). The inhalation risk associated with handling these compounds as dry (and dusty) solids, especially on large scale, was especially intimidating. An alternative sequence was developed which bypassed anhydride **7** and minimized the opportunity for exposure to **6**. This was accomplished by reversal of the hydrogenation and borohydride reduction steps in the original synthesis and isolation of **6** as a filtered wet cake, which was carried directly into the next step. Thus, **6** was reduced to **14** with NaBH$_4$ followed by double bond hydrogenation to give **8**. The workup of the DIBALH reduction was modified to increase the workability. Acetal formation was accomplished using an acidic ion exchange resin; this allowed removal of the acid catalyst by filtration instead of requiring an aqueous, extractive workup. Hydrolysis of the acetal to afford **11** was accomplished again using an acid ion-exchange resin under carefully controlled conditions to minimize formation of the dimeric species **15**.

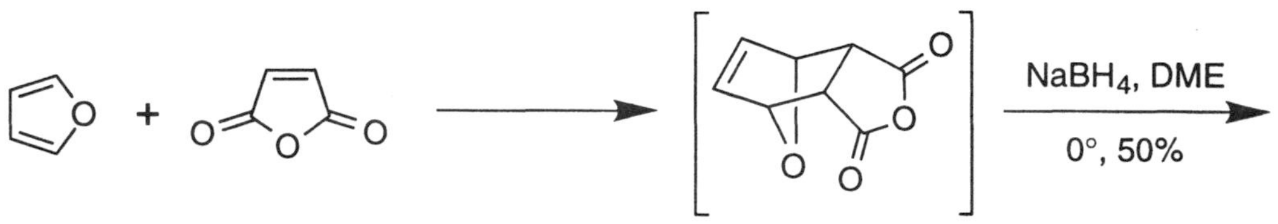

6

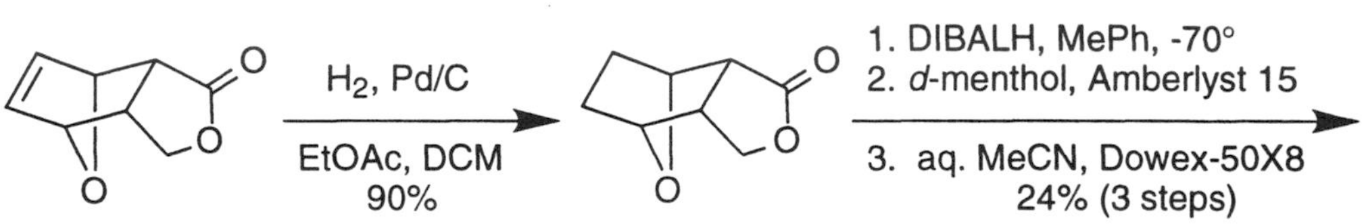

14                              8

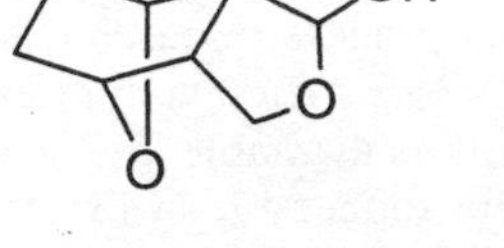

11
(single enantiomer)                    15 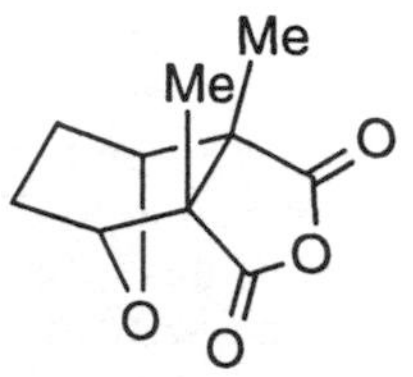                              16

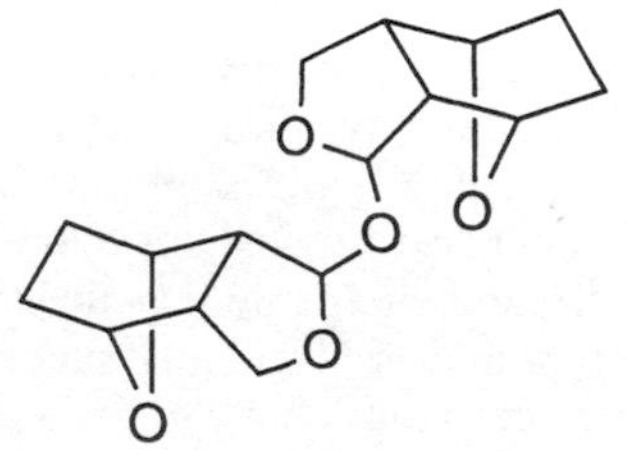

Unfortunately, SQ-28668 failed in the clinic due to limited potency and duration of action. Subsequently, SQ-30741 was advanced as the next clinical candidate. The Kilo Lab was utilized to prepare initial supplies of the compound using **11** on hand from the initial Pilot Plant campaign, and the Pilot Plant was mobilized to prepare additional supplies. In initial Phase-I studies, however, SQ-30741 exhibited extensive first-pass metabolism in the clinic and further work was halted, but not before a large quantity of **11** had been prepared in the Pilot Plant. Some 17 kg of **11** remained unutilized and was put into cold storage. Eventually, BMS-180291 came to the fore as the next clinical candidate, and the remaining **11** proved very useful in jump-starting the initial Kilo Lab campaign to prepare material.

As mentioned earlier, the original synthesis of BMS-180291 (**1**) proceeded via reaction of the five-ring hemiacetal **11** with the Grignard reagent derived from **20**. Bromo compound **20** was prepared starting from *o*-bromobenzaldehdye (**17**) by sequential Wittig reaction, catalytic hydrogenation, DIBALH reduction, and protection as the dimethylthexylsilyl ether. The use of LAH in the reduction of **18** resulted in significant cleavage of the aromatic halide. Generation of the Grignard reagent followed by addition to **11**, pretreated with one equivalent of a sacrificial Grignard reagent to conserve the more precious **20**, provided **21**. Hydrogenolysis of the benzylic alcohol in Grignard product **21** provided **22**. The primary alcohol of **22** was converted to the acetate and the silyl ether directly oxidized with Jones reagent. Subsequent treatment with methanol and acid esterified the carboxylic acid with the simultaneous liberation of the primary alcohol to give **23**, thus setting up for a second Jones oxidation to provide key intermediate **24**.

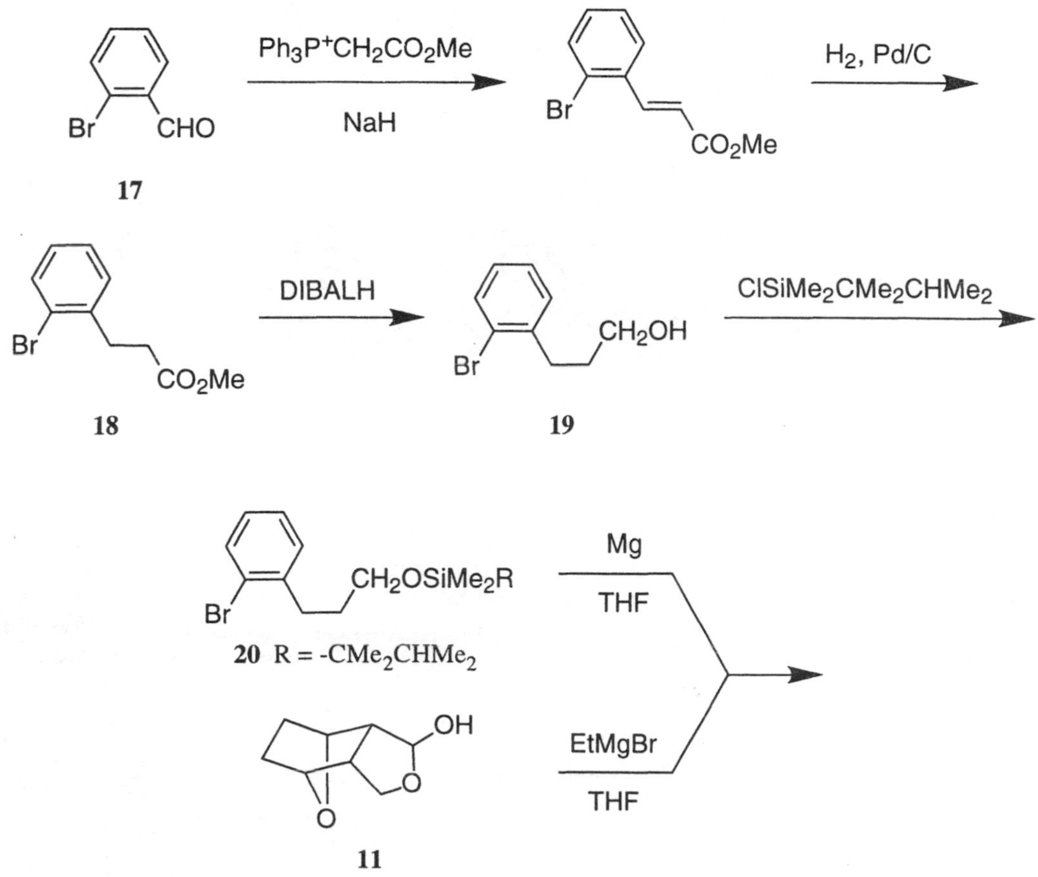

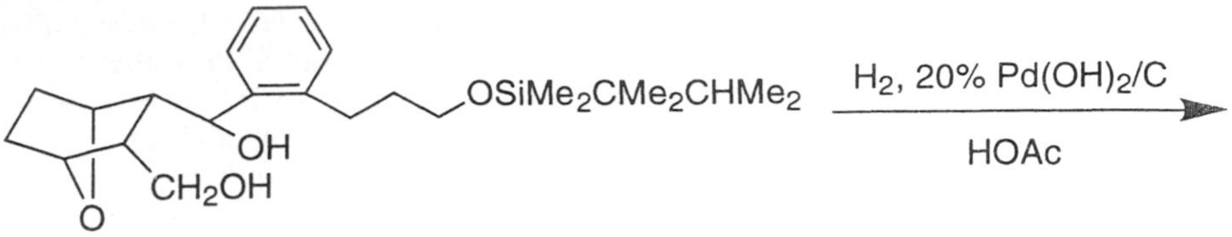

**21**

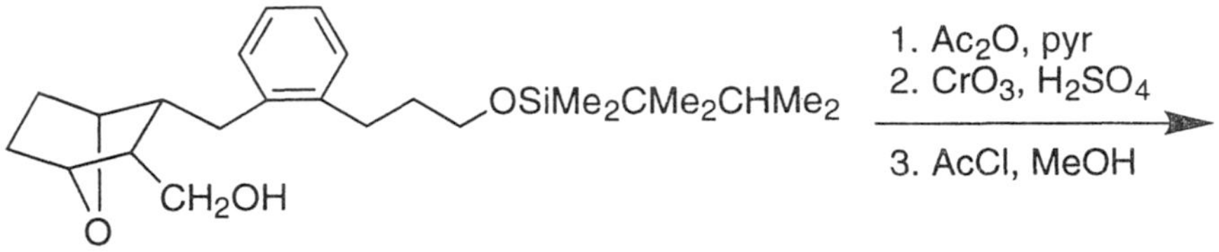

**22**

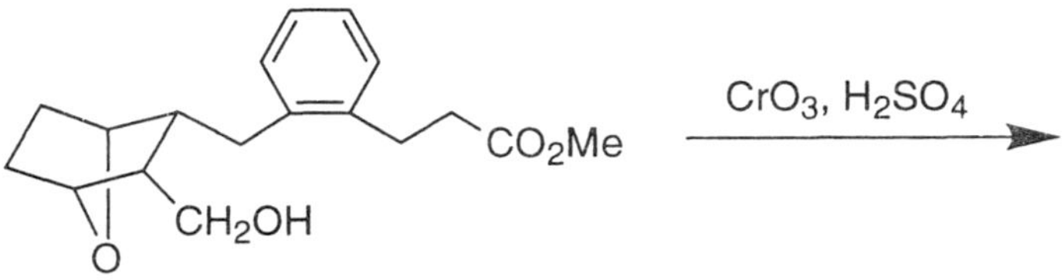

**23**

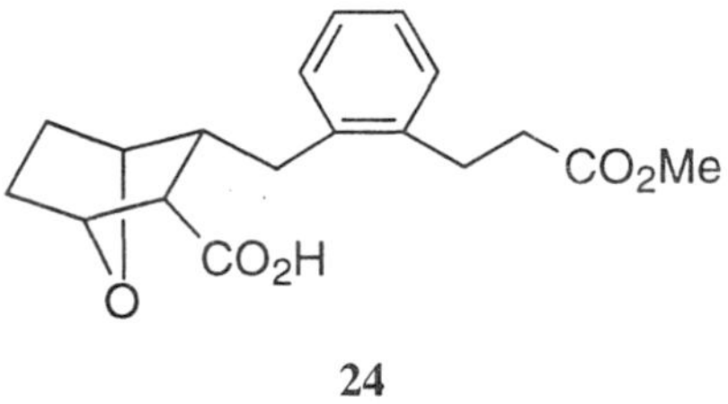

**24**

The last stage of the synthesis involved introduction of the oxazole side chain. This was accomplished by coupling of **24** with amide **25** (derived from L-serine as shown) to give **26**. Cyclization to oxazoline **27** was performed by mesylation and treatment with base. Conversion to the oxazole was performed with cupric bromide and DBU, a significant improvement over the classic literature methodology for this transformation using nickel peroxide. Finally, hydrolysis of the ester was accomplished using NaOH in aqueous THF; acidification produced **1** as the free acid.

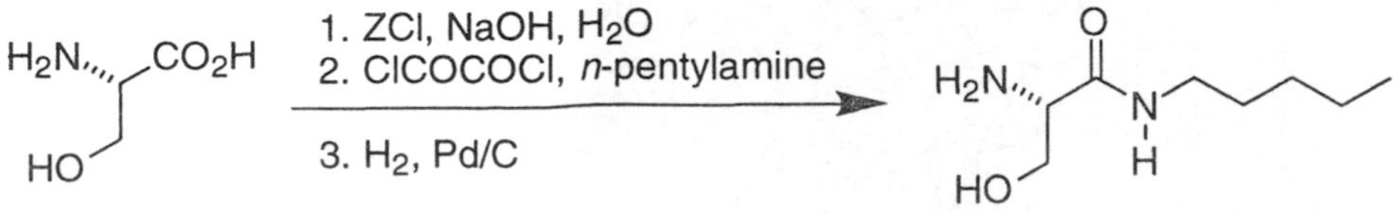

L-serine                                                                        **25**

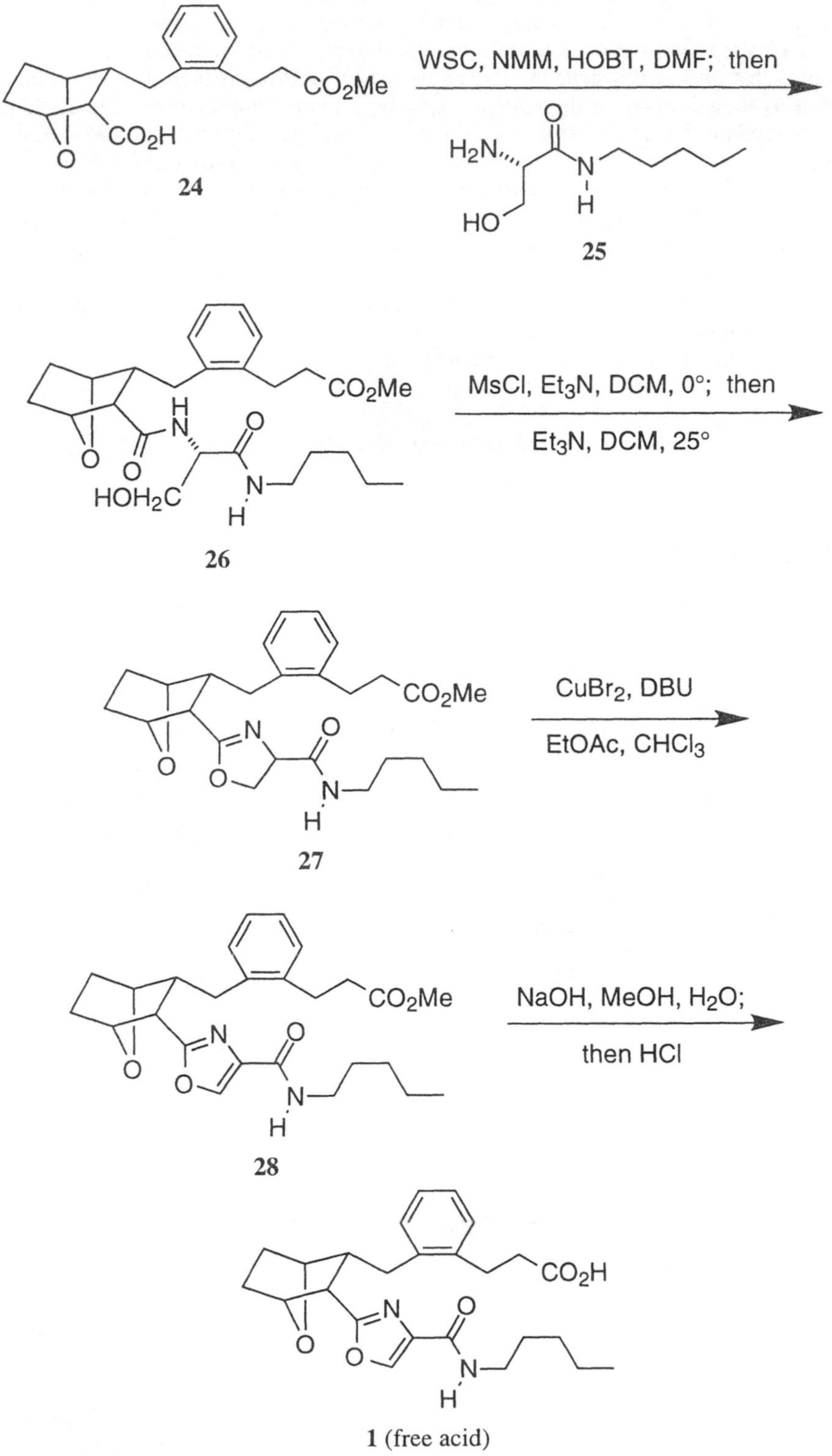
CO2Me
CO2H
24
WSC, NMM, HOBT, DMF; then
H2N
HO
25
CO2Me
HOH2C
26
MsCl, Et3N, DCM, 0°; then
Et3N, DCM, 25°
CO2Me
27
CuBr2, DBU
EtOAc, CHCl3
CO2Me
28
NaOH, MeOH, H2O;
then HCl
CO2H
1 (free acid)

A condensed schematic of the overall original synthesis of BMS-180291 is provided in Scheme 1. Immediately apparent is its length, 23 steps. Although the synthesis is convergent, the longest linear sequence is 16 steps, and the overall yield via the longest linear sequence is less than 3%. Of particular note is that ten[4] of the steps involve an oxidation state adjustment (three oxidations, seven reductions). Although many of these steps might be considered necessary or unavoidable, four of them are redundant in that they involve reduction of a carboxylic acid oxidation state to a primary alcohol and then later, after some intervening chemistry, reoxidation back to the carboxylic acid oxidation state (see **18 → 19** followed later by **22 → 23** and **7 → 8** followed later by **23 → 24**). Another inefficiency in the original synthesis was that a resolution process was involved in the procurement of a single enantiomer of **11**. Resolutions are inherently inefficient, in that the maximum theoretical yield, in the absence of an efficient recycling process, is 50%. The most problematic step in the sequence was the CuBr$_2$-DBU oxidation of oxazoline **27** to oxazole **28**; even with this improved procedure, reaction was slow and often required a second and sometimes even a third charge of reagents to effect completion.

**Scheme 1: Original Synthesis of BMS-180291**

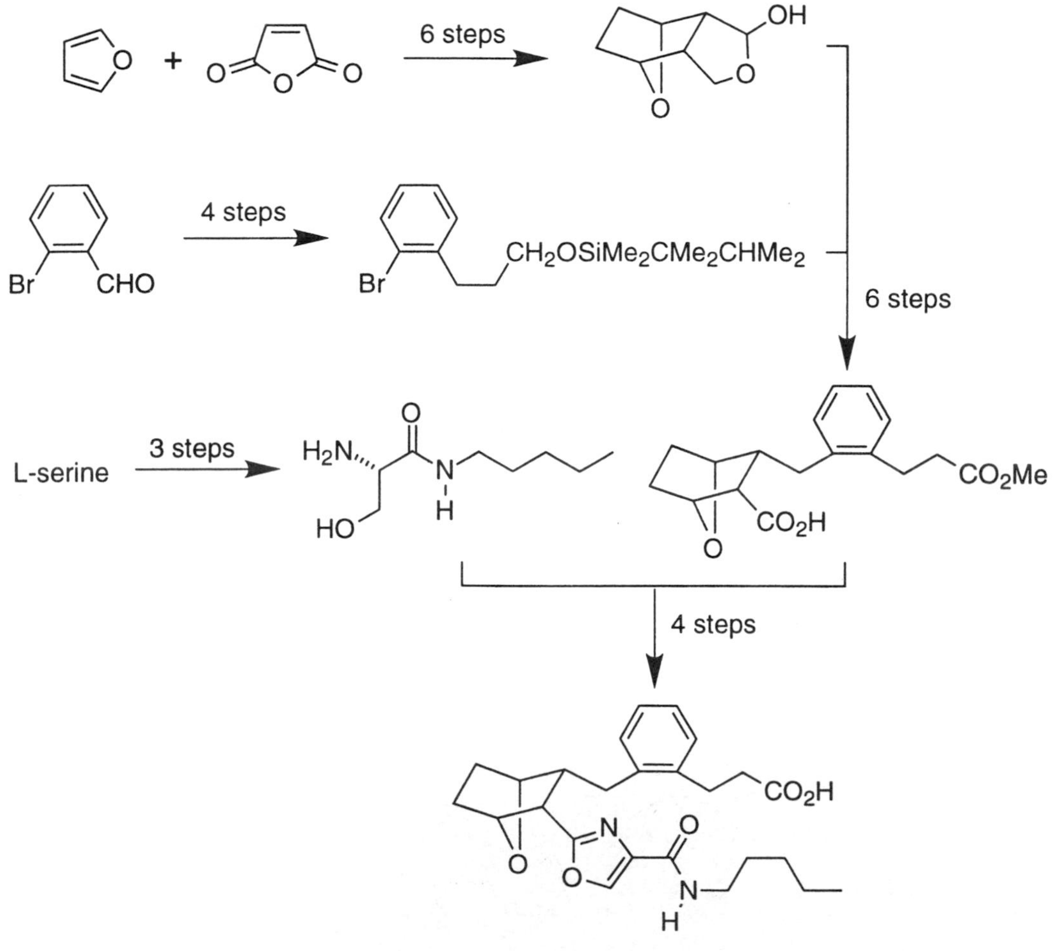

Upon elevation of BMS-180291 to drug candidate status, it was readily apparent that the existing synthesis would have to be implemented in order to supply the material needs within any sort of reasonable timeframe, this decision driven in part by the availability of **11** remaining from the aborted SQ-30741 Pilot Plant campaign. Process research efforts to implement a "quick fix" to the subsequent steps of this synthesis were initiated, and within a relatively short time some minor improvements were introduced. The dimethylthexylsilyl protecting group in the Grignard substrate **20** was replaced with the less costly and more readily available dimethyl-*t*-butylsilyl group with no detriment to the yield or reagent stability. The order of the steps in conversion of the resulting Grignard product **29** to **23** was changed. Acetylation of both hydroxy groups of **29** was performed to give **30**, which was then oxidized (Jones reagent) to give **31**. Hydrogenolysis in acidic MeOH then accomplished three things: hydrogenolysis of the benzylic acetate, esterification of the carboxylic acid, and hydrolysis of the primary acetate, to give **23**, thus saving one step over the original synthesis.

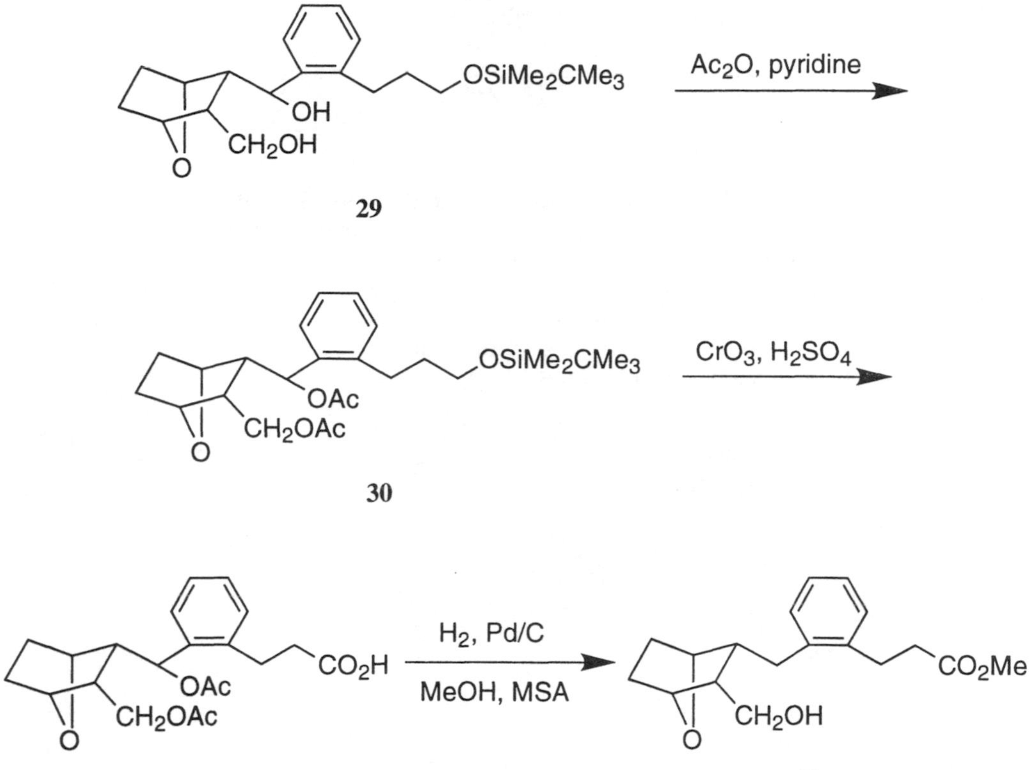

Another change was to perform the coupling of **24** with **25** via the acid chloride instead of with water-soluble carbodiimide (WSC). The most significant improvement resulted from a reinvestigation of the most problematic step in the sequence, oxidation of oxazoline **27** to oxazole **28**. Several alternative oxidation conditions were investigated. NBS-AIBN[5] bromination was complicated by the presence of four benzylic hydrogen atoms and did not prove useful. Oxidation attempts using $CuBr_2$, TEMPO, and DBU gave the oxazole in only 10% yield. As mentioned previously, the $CuBr_2$-DBU oxidation of **27** to **28** in $EtOAc$-$CHCl_3$ was slow and often required a second and sometimes even a third charge of reagents to effect completion, which often required 4-5 days and proceeded in variable yields (50-80%). Also, the continued use of $CHCl_3$ would become problematic on scale up. The use of other bases and solvents in this transformation was examined further. The observations that triethylamine (TEA) and diisopropylethylamine (DIPEA) gave only low yields of the desired product and that aged (4 days) solutions of CuBr$_2$-DBU were ineffective suggested the possibility that the amine could be reducing the copper, either by a single electron or an ionic hydride transfer process. Thus the use of several amines [*t*-BuNH$_2$, DABCO, and hexamethylenetetramine (HMTA, **32**)] expected to be poor hydride donors was examined. When used alone with CuBr$_2$, these amines proved relatively inefficient, presumably because the resulting copper complexes were only poorly soluble. However, in combination with DBU the solubility was increased, and the reaction proceeded well; HMTA gave the best results. Thus reaction of oxazoline **27** with 4 mole equivalents each of CuBr$_2$, DBU, and HMTA in DCM at room temperature for 5 hours (instead of several days) consistently produced **28** in >80% isolated yield; additional charges of reagent no longer proved necessary.[6] Interestingly, substitution of Cu(OTf)$_2$ for CuBr$_2$ in this oxidation also proved effective, indicating that the mechanism does not occur via bromination and subsequent elimination.

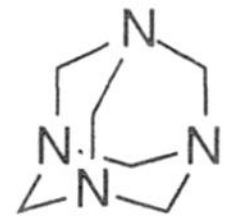

32

An alternative oxazole synthesis also was demonstrated.[7] Diamide **26** was converted to its mesylate and then to the corresponding chloride **33**. Elimination provided olefin **34**, which was treated with bromine and then TEA to afford bromoolefin **35** as a 1:9 mixture of E and Z-isomers. Attempted cyclization using TEA (toluene, 70-75°) resulted in no reaction, while NaOMe in MeOH formed the *trans*-oxazole acid **36** exclusively. However, cesium carbonate in dioxane produced the desired oxazole **28** in 63% yield. This sequence offered no particular advantages over the CuBr$_2$-DBU oxidation methodology.

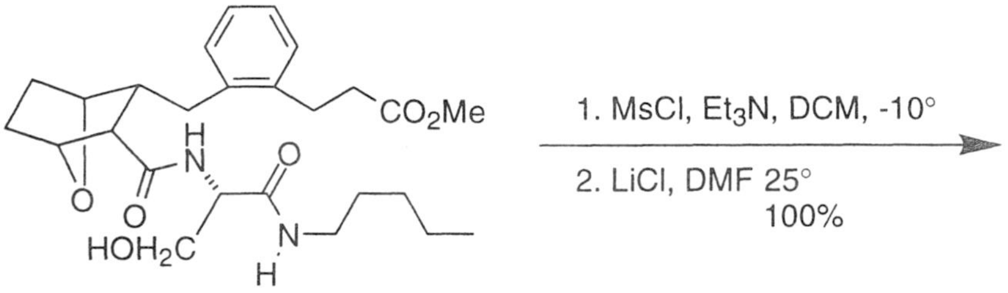

1. MsCl, Et$_3$N, DCM, -10°

2. LiCl, DMF 25°
100%

26

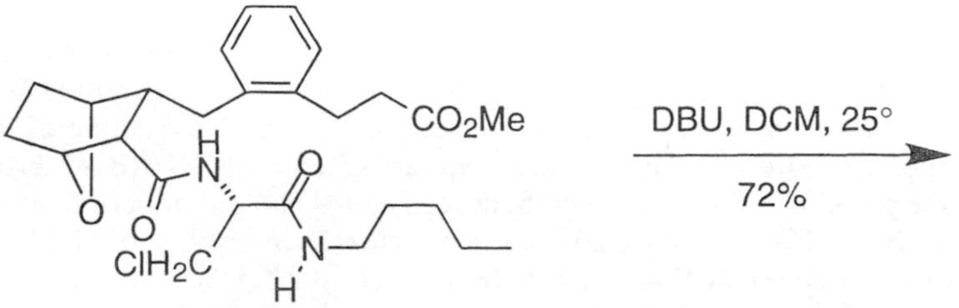

DBU, DCM, 25°

72%

33

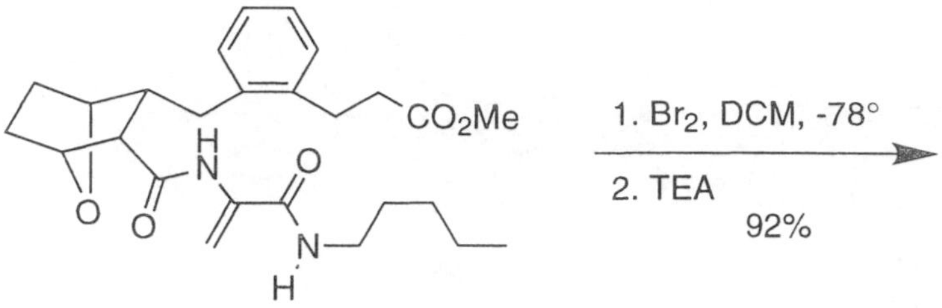

34

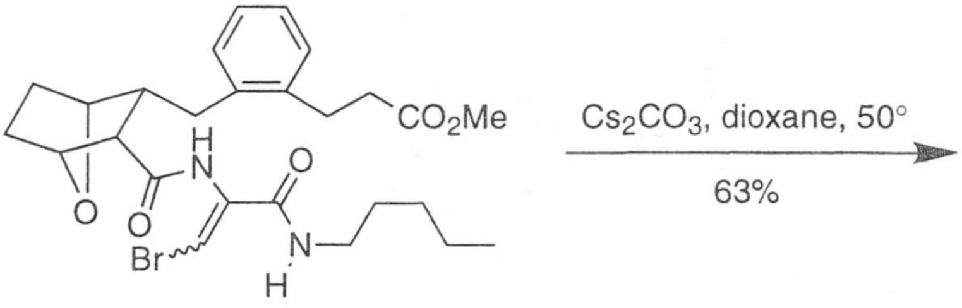

1. Br$_2$, DCM, -78°

2. TEA
92%

35

Cs$_2$CO$_3$, dioxane, 50°

63%

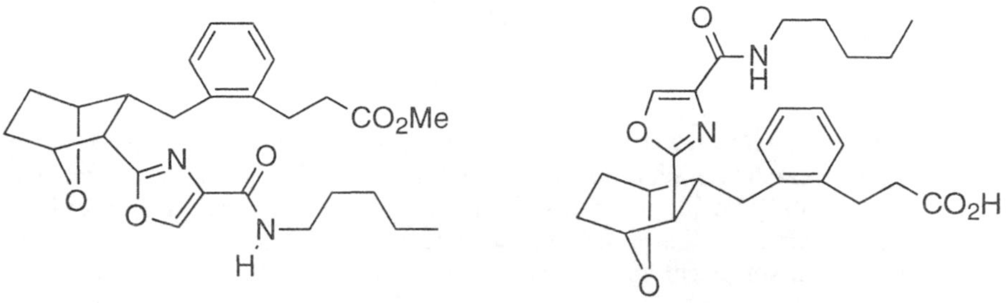

28

36

The modified oxidation methodology was utilized to prepare a total of ~20 kg of BMS-180291 by a combination of Kilo Lab and Pilot Plant efforts. Concurrently with this work, process research activities to identify a better route were underway. The goal was two-fold: to achieve a synthesis involving chiral induction and to obviate the necessity for redundant oxidations.

One avenue which was explored involved the use of alternate Diels-Alder substrates. Furan is notoriously selective in terms of the partners with which it will react; typically, two carbonyl activating groups are required. Reaction of **37** (as a model) with furan provided cycloadduct **38** in 99% yield. Catalytic hydrogenation produced the *cis-endo* derivative **39**. Selective epimerization of the ketone sidechain with DBU in DCM gave **40** which was reduced with NaBH$_4$ to afford **41** as a 7:3 mixture of isomers. Treatment with DBU in hot toluene resulted in epimerization of the ester group to the exo face, lactonization driving the otherwise unfavorable equilibrium to the *cis-exo* lactone **42** (as a 7:3 mixture of isomers at the benzylic carbon atom). However, this sequence would require the synthesis of a more elaborate derivative of **37** to allow for introduction of the carboxylic acid chain.

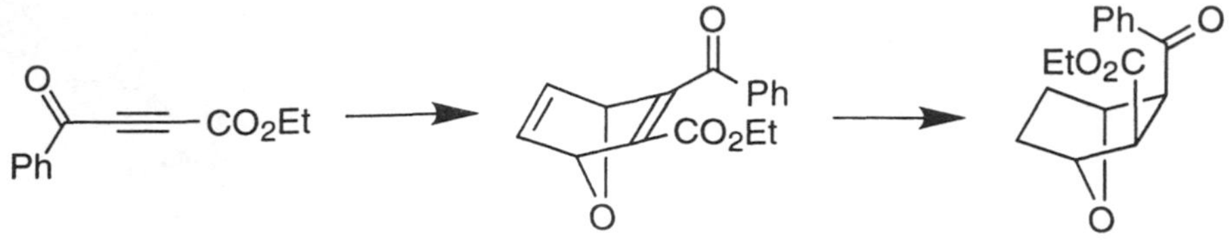

37                                    38                                    39

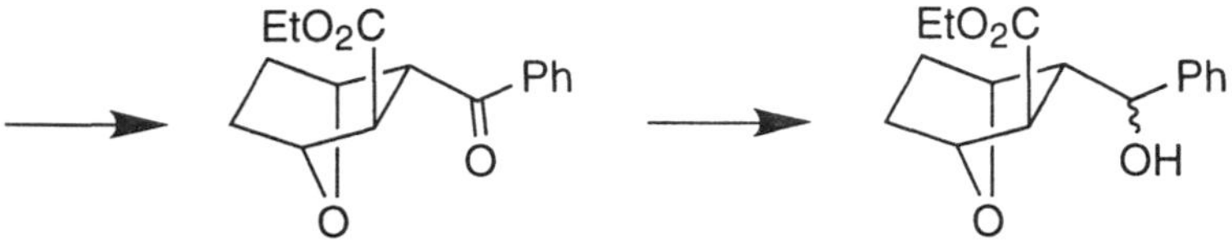

40                                              41

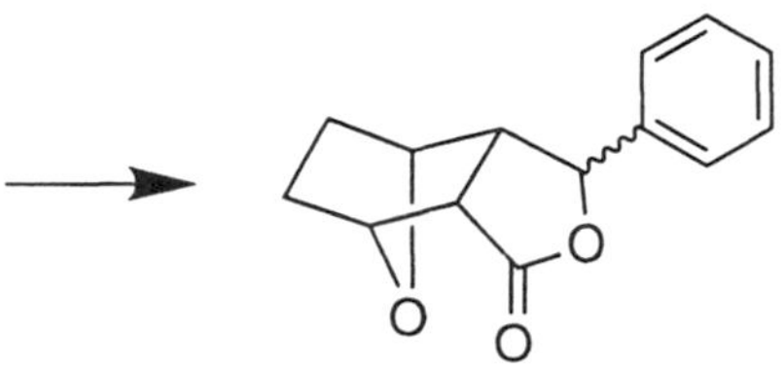

42

A more successful approach to the problem involved the desymmetrization of *meso*-anhydride **7** by a chiral Grignard reagent.[8]  Such desymmetrizations were known using chiral heteroatom nucleophiles[9] and chiral hydride reagents;[10] however, no examples using chiral carbon nucleophiles with anhydrides were found.  Of concern was the potential high cost or multistep preparation of the precursor chiral auxiliaries associated with the chiral carbanions;  instead, a readily available and inexpensive auxiliary was desired.   Thus, the ephedrines and pseudoephedrines were attractive as potential chiral auxiliaries, especially since  chiral *ortho*-metalated aminals,[11] acetals,[11,12] and oxazolidnes[13] have been added to aldehydes with moderate to high diastereoselection.  Condensation of *ortho*-bromobenzaldehyde with (-)-ephedrine produced oxazolidine **43**, isolated as a single diastereomer after crystallization.[14] Transmetallation afforded the corresponding lithium reagent **44**, and further reaction with $MgBr_2$ provided the Grignard reagent **45**.  Addition of anhydride **7** to the Grignard reagent produced a mixture of ketones **46** and **47**. *In situ* reduction was effected by addition of MeOH followed by $NaBH_4$, and finally treatment with acid gave a mixture of desired lactone **48** and its enantiomer **49** in  a ratio of 17:83 (ee 66%), respectively.

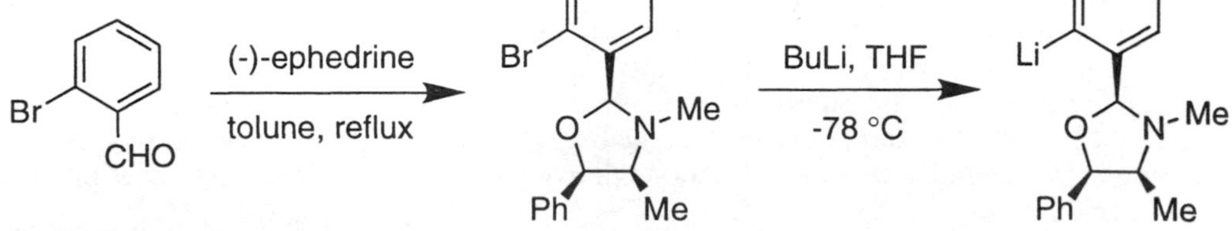

43                                              44

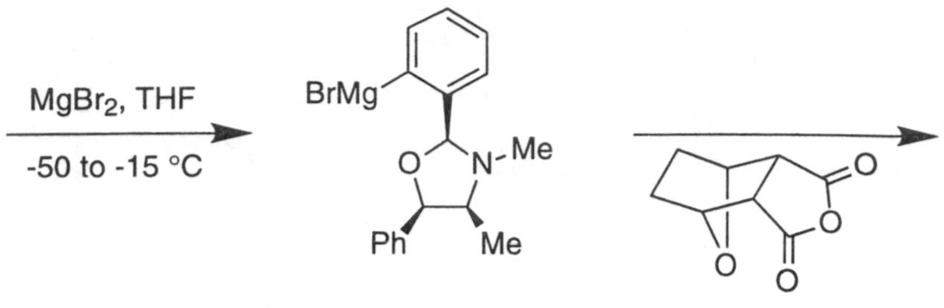

**45**            **7**

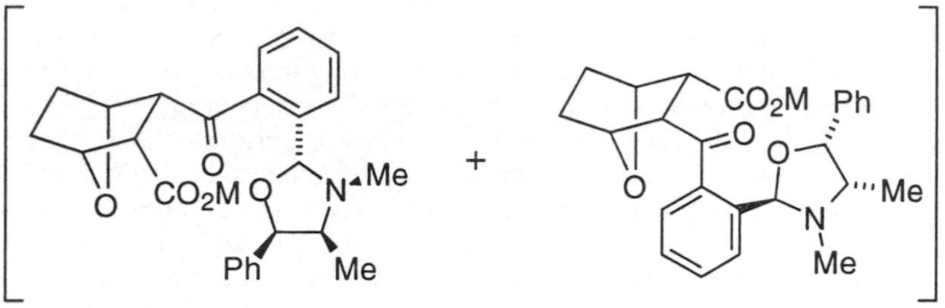

**46**        **47**

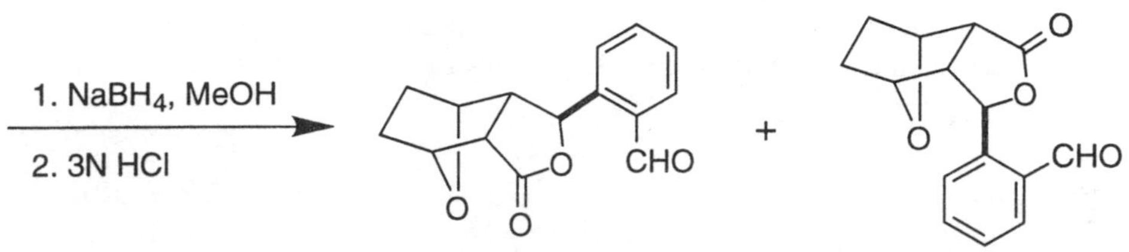

**48**        **49**

Much better results were obtained with the Grignard reagent derived from (-)-pseudoephedrine. Reaction of *o*-bromobenzaldehyde as above with (-)-pseudoephedrine gave **50** as an oil in quantitative yield as a 38:1 mixture of diastereomers, used as is in the next step, generation of the Grignard reagent **51**. Subsequent transformations as described above produced the desired lactone **48** in higher yield (64%) and with greater ee (98.3%); crystallization provided material with ee 99.2% (>200:1).

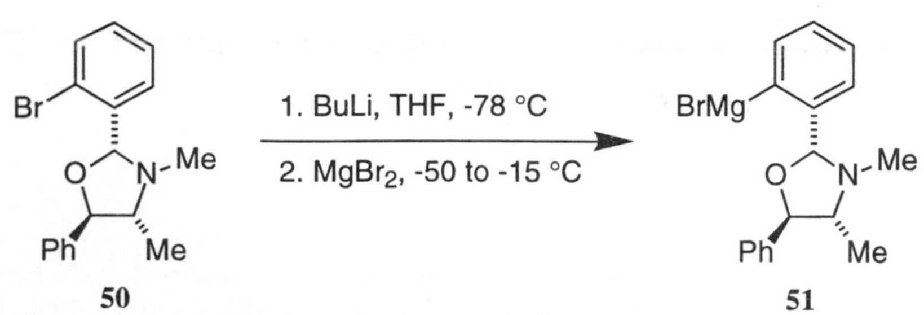

**50**        **51**

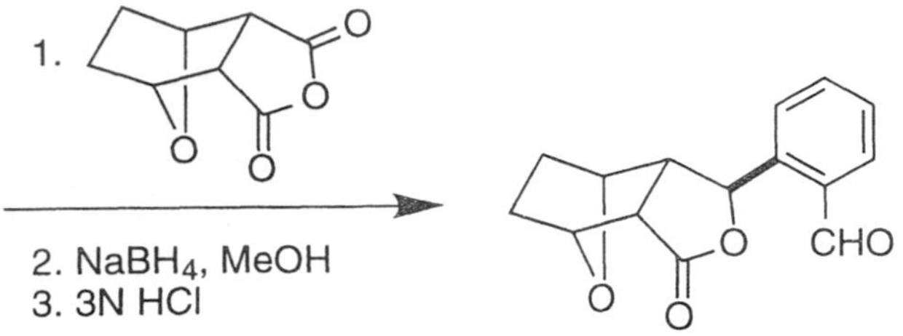

1.

2. NaBH₄, MeOH
3. 3N HCl

48

Several additional points of interest emerged from these studies. The first is that lithium reagent **44** produced **49** as the major product (ee 26%), opposite to that obtained with the corresponding Grignard reagent **45**. Another item is the obtention of **48** as essentially (~99.7:0.3) a single isomer at the carbon atom bearing the aromatic substituent, indicating that the borohydride reduction had occurred with very high diastereoselectivity. This selectivity is a consequence of the presence of magnesium ion during the reduction; since the reduction was performed *in situ*, magnesium ion was present from the Grignard reaction. In the absence of magnesium ion during the reduction step, due either to an intervening water workup or *in situ* reduction of a lithium reagent reaction, **48** typically was obtained as a 60:40 mixture of epimers at the benzylic carbon atom. Although this stereocenter is subsequently removed later in the synthesis, the importance of obtaining a single isomer is not lost on a process chemist. The ease and efficiency of isolation by crystallization of a single isomer is much greater than had a mixture of isomers been obtained.

The utility of **48** in the synthesis of BMS-180291 was demonstrated by conversion to **24**, an intermediate in the original synthesis. Horner-Emmons methodology, as modified by Masamune and Rousch,[15] provided the unsaturated ester **52**, which was converted to **24** under hydrogenation conditions.

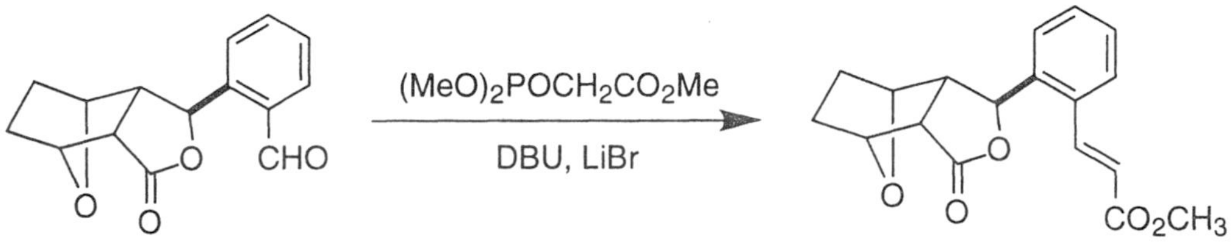

(MeO)₂POCH₂CO₂Me

DBU, LiBr

48                                                                    52

H₂, Pd/C

solvent?

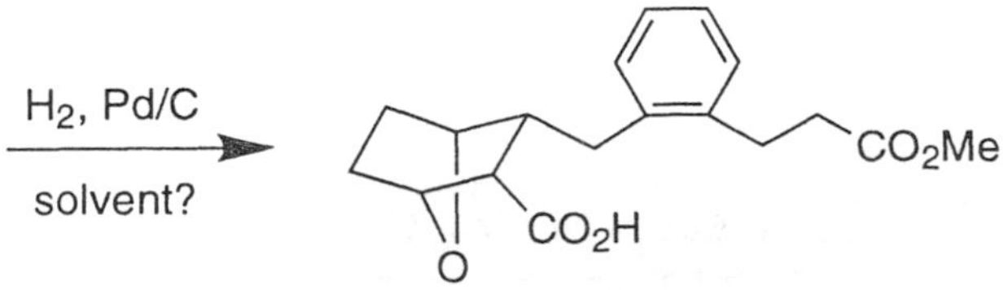

24

Although the *meso*-anhydride desymmetrization sequence to **24** was considerably more efficient than the original route (7 steps overall to **24** instead of 16 steps), the use of isolated anhydride **7** raised safety concerns, and the yield in the Grignard to lactone conversion was lower than was considered practical. However, another approach was under investigation in which the chirality of the reacting partners was reversed: reaction of an achiral Grignard reagent with a chiral imide. Ultimately, this approach proved very fruitful. Reactions of achiral imides[16] and chiral imides derived from tartaric acid[17] and malic acid[18] with Grignard reagents have been reported in the literature, but no examples were found in which the chirality is present in the amine moiety. Treatment of the chiral imides **53** or **54** prepared from S-valinol or S-phenylglycinol, respectively, first with one equivalent of EtMgBr followed by one equivalent of Grignard reagent **56**[19] produced a complex mixture, presumably due to equilibration between the keto and hemiamidal isomers of the product. In order to simplify analysis, the crude reaction mixture was reduced *in situ* by addition of EtOH and NaBH₄ followed by sequential acidification, solvent exchange to toluene, and heating to form lactone aldehyde **48** in ~65% yields with ee 72% and, as above, with high diastereoselectivities at the benzylic carbon atom. More importantly, the less structurally complex imide **55**, derived from the more readily available (and considerably less expensive) S-methylbenzylamine, produced the same enantiomer of **48** with greater selectivity (94:6 in the reaction mixture, >99:1 after crystallization directly from the reaction mixture) and in higher yield (89%). Note that the phenyl group of **54** has stereochemistry opposite to that of **55**, but **54** and **55** both produce the same enantiomer of **48**.

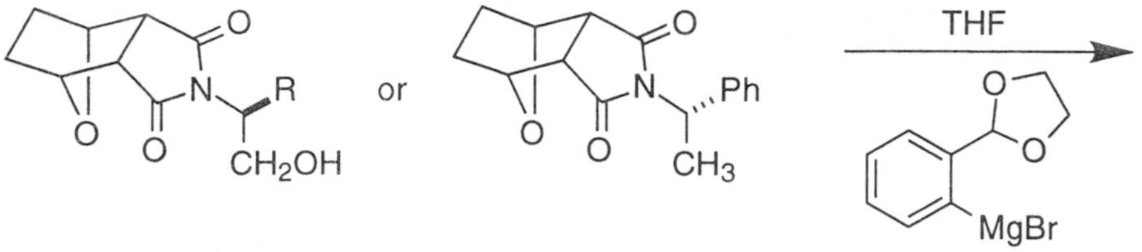

53  R = iPr
54  R = Ph

55

56

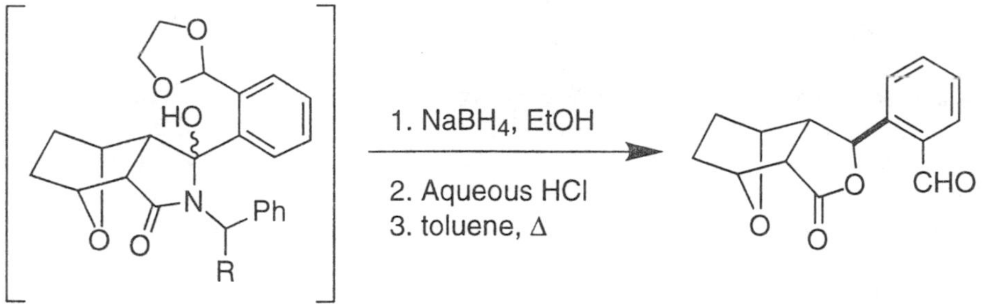

1. NaBH₄, EtOH

2. Aqueous HCl
3. toluene, Δ

48

Even more efficient would have been the use of a chiral imide derived from serine, since the oxazole side chain is constructed by formation of a serine amide (see **26**) later in the synthesis. However, reaction of **57** or **58** with Grignard **56** followed by conversion as above to lactone **48** (for analytical purposes) produced the lactone in 53 and 35% yields with ee 71 and 96%, respectively. Although the ee of the product from **58** was satisfactory, attempts to increase the yield proved fruitless.

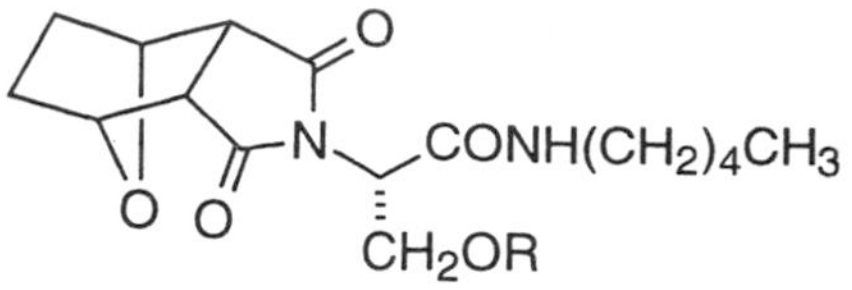

**57** R = H
**58** R = TBDMS

With improved methodology in hand for preparation of the desired single enantiomer of key intermediate **48**, attention was next focused on the development of a more practical and efficient synthesis of imide **55**. One approach proceeded via reaction of maleic anhydride with S-methylbenzylamine to give maleimide **59**[20] in 64% distilled yield. Diels-Alder reaction with furan in the presence of 0.3 equivalents $AlCl_3$ provided the desired *exo* isomer **60** in 88% crystallized yield; the presence of the $AlCl_3$ was necessary in order to effect equilibration of the kinetic ~3:1 *endo:exo* isomer mixture to the thermodynamically favored *exo* isomer. Hydrogenation then gave imide **55** in 95% crystallized yield.

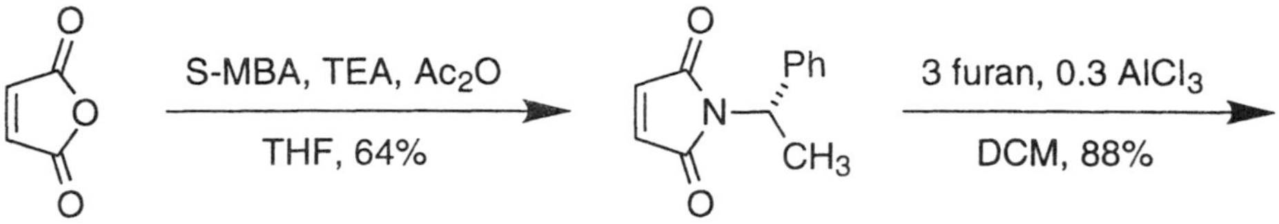

**59**

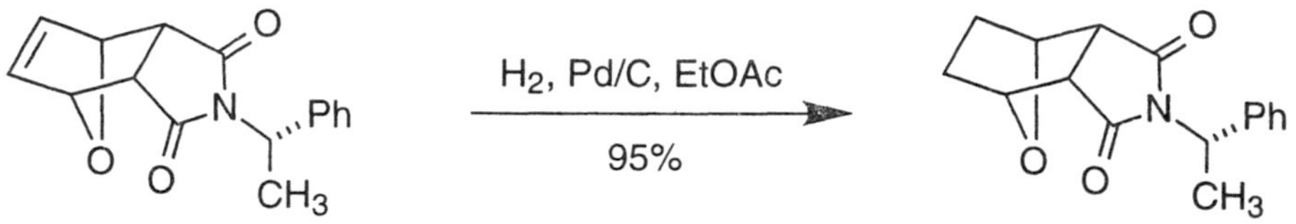

**60**                                                                 **55**

A second approach involved Diels-Alder reaction followed by treatment of the resulting anhydride **6** (*endo-exo* equilibration occurs without catalyst at room temperature) *in situ* with S-methyl-benzylamine and TEA in *n*-BuOAc to generate **61** (interestingly, as a 1:1 mixture of diastereomers). Formation of the salt stabilized **61** toward retro-Diels-Alder reaction. *In situ* hydrogenation in *n*-BuOAc followed by filtration of the catalyst and cyclization in refluxing *n*-BuOAc gave **55** in 85% crystallized yield overall. Use of the same solvent for all the reactions provided obvious processing advantages.

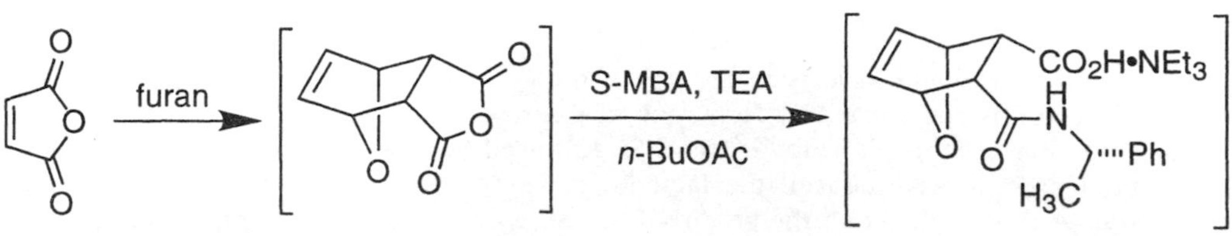

**6**                                                                 **61**

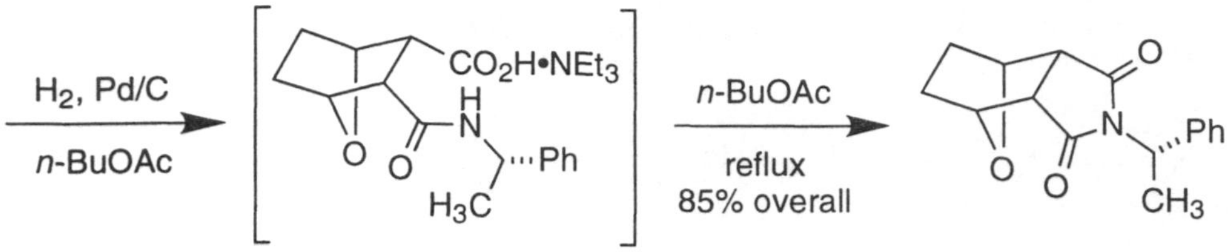

**55**

A third approach involved Diels-Alder reaction between maleic anhydride and furan in EtOAc or toluene, decantation of the supernatant from the crystalline **6**, addition of acetonitrile, and hydrogenation to give **7**. Reaction with S-methylbenzylamine then gave **55** in 80-95% crystallized yield. These latter two sequences, although consisting of several operations, are considered as a single step each, since they can be performed in one pot and involve isolation only of imide **55** and not of any of the intermediates.

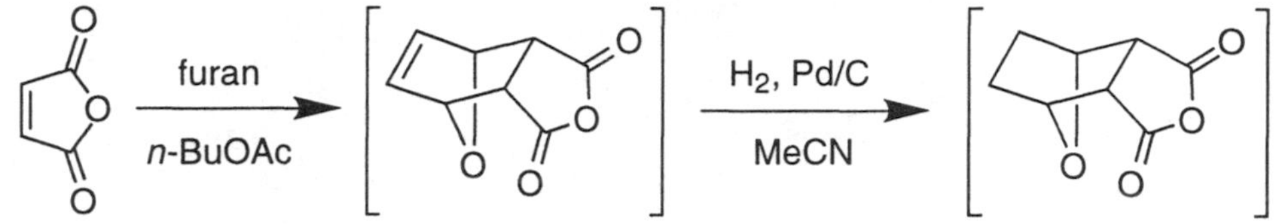

**6**                                                                **7**

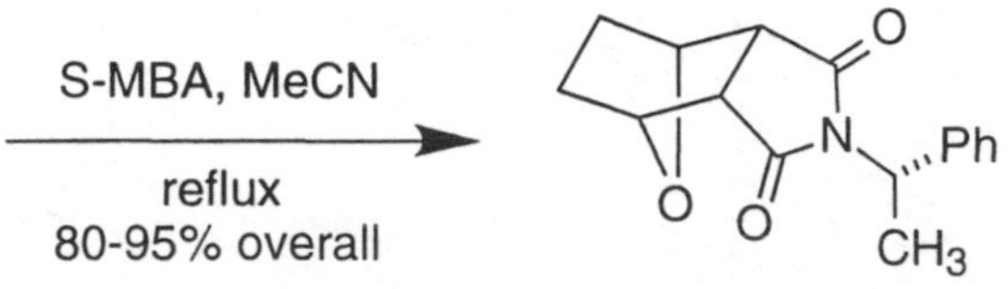

**55**

A condensed schematic of the overall alternative synthesis of BMS-180291 is provided in Scheme 2 below. Worthy of note is that of the ten oxidation-reduction reactions present in the original synthesis, only three remain.[4] Thus, the synthesis of key acid ester intermediate **24** has been reduced dramatically from 16 steps (see Scheme 1) to 4 steps, and the yield has been increased ten-fold from 5% to 52%; the overall synthesis of ifetroban sodium has been cut virtually in half (from 23 steps to 12), and the yield (via the longest linear sequence) improved from 3% to 28%. Finally, as any process research and development chemist can attest, there are, and always will be, opportunities for further improvements.

**Scheme 2: Alternative Synthesis of BMS-180291**

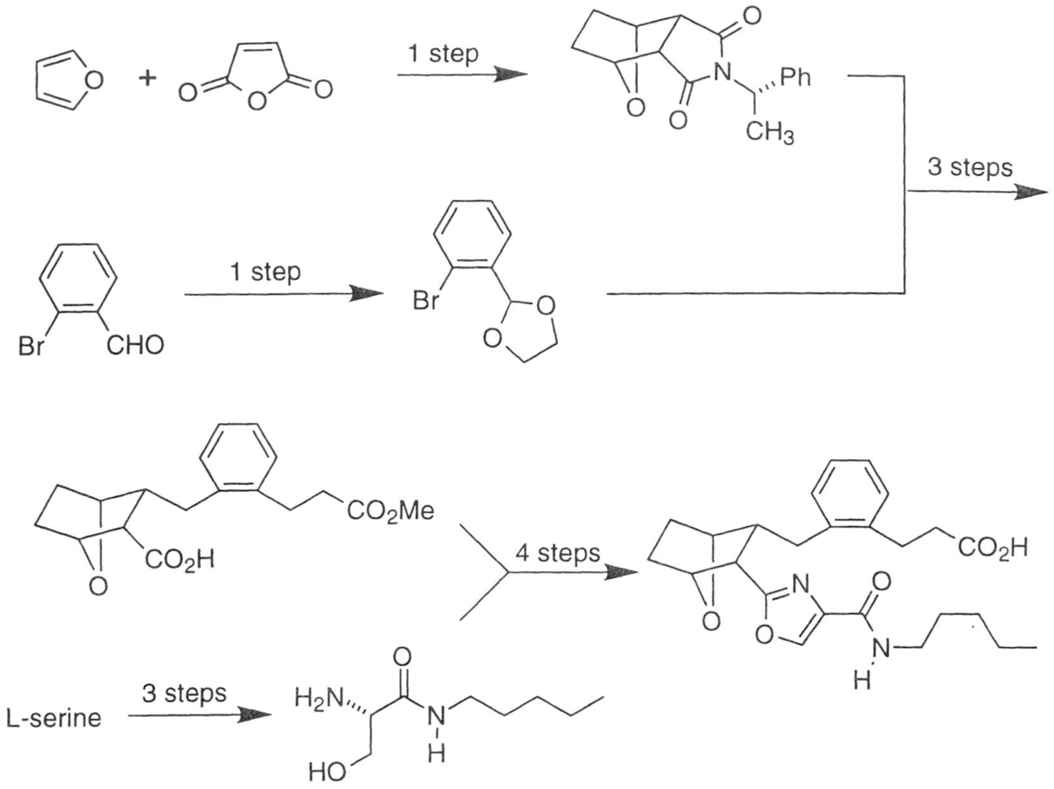

### Acknowledgments

Contributors to the work described in this chapter, which the author has been privileged to present, include members from the Departments of Chemical Process Research: B. Boyhan, T. Denzel, P. Dunn, R. Fox, K. Gadamasetti, J. Godfrey, J. Heikes, M. Humora, J. Janotti, T. Kissick, D. Kronenthal, D. Kucera, J. Moetz, J. North, P. Pansegrau, C. Papaioannou, Y. Pendri, M. Poss, A. Pullockaran, S. Real, J. Reidy, D. Rosen, P. Schierling, M. Schwinden, J. Singh, W. Szymanski, J. Thottathil, E. Vawter, T. Vu, S. Wang, M. Wong, H. Wu, and J. Wurdinger; Chemical Process Development: N. Anderson, T. Ary, R. Bakale, J. Berg, P. Bernot, R. Bogaert, J. Bush, C.-K. Chen, D. Chen, K. Colapret, A. Dadiz, P. deMena, R. Deshpande, J. Grosso, J. Harris, C. Jaeger, P. Jass, D. Kacsur, S. Kiang, D. Kientzler, V. Kortan, B. Lotz, D. Lust, J. Metzger, M. Miller, J. Moniot, H. Murphy, K. Natalie, B. Phan, G. Powers, V. Rosso, A. Rusowicz, T. Sedergran, M. Shipp, M. Shiver, A. Singh, C. Spagnuolo, S. Srivastava, D. Thurston, J. Venit, L. Vivona, R. Weaver, C. Wei, and W. Winter; and Discovery Chemistry: J. Barrish, J. Das, J. Gougoutas, W.-C. Han, J. Reid, and S. Spergel.

## References

1. Portions of this work have been described previously: Mueller, R. H.; Wang, S.; Pansegrau, P. D.; Janotti, J. Q.; Poss, M. A.; Thottathil, J. T.; Singh, J.; Humora, M. J.; Kissick, T. P.; Boyhan, B. *Org. Process Res. Dev.* **1997**, *1*, 14.
2. Misra, R. N.; Brown, B.B.; Sher, P. M.; Patel, M. M.; Hall, S. E.; Han, W.-C.; Barrish, J.; Floyd, D. M.; Sprague, P. W.; Morrison, R. A., Ridgewell, R. E.; White, R. E.; DiDonato, G. C.; Harris, D. N.; Hedberg, A.; Schumacher, W. A.; Webb, M. L.; Ogletree, M. L. *Bioorg. Med. Chem. Lett.* **1992**, *2*, 73.
3. Gomoll. A.W.; Schumacher, W.A.; Ogletree, M.L. *Pharmacology* **1995**, *50*, 92. Gomoll. A.W.; Grover, G. J..; Ogletree, M.L. *J. Cardiovascular Pharmacology* **1994**, *24*, 960.
4. The Grignard reaction is not included in this count.
5. Kashima, C.; Arao, H. *Synthesis* **1989**, 873; Meyers, A. I.; F. Tavares, F. *Tetrahedron Lett.* **1994**, *35*, 2481; Tavares, F.; Meyers, A. I. *Tetrahedron Lett.* **1994**, *35*, 6803
6. Barrish, J. C.; Singh, J.; Spergel, S. H.; Han, W.-C.; Kissick, T. P.; Kronenthal, D. R.; Mueller, R. H. *J. Org. Chem.* **1993**, *58*, 4494-4496.
7. Das, J.; Reid, J. A.; Kronenthal, D. R.; Singh, J.; Pansegrau, P. D.; Mueller, R. H. *Tetrahedron Letters* **1992**, *33*, 7835.
8. Real, S. D.; Kronenthal, D. R.; Wu, H. Y. *Tetrahedron Lett.* **1993**, *34*, 8063.
9. a) Suda, Y; Yago, S.; Shiro, M.; Taguchi, T. *Chem. Lett.* **1992**, 389 and references therein. b) Ward, R. S. *Chem. Soc. Rev.* **1990**, *19*,1 and references therein. c) Thottathil, J. U.S. Patent **1987**, US 4743697.
10. Matsuki, K.; Inoue, H.; Takeda, M. Tetrahedron Lett. 1993,34,1167 and references therein.
11. Commercon, M.; Mangeney, P.; Tejero, T.; Alexakis, A. *Tetrahedron: Asymmetry* **1990**, *1*, 287.
12.. Kaino, M.; Ishihara, K.; Yamamoto, H. *Bull. Chem. Soc. Jpn.* **1989**, *62*,3736.
13. Takahashi, H.; Tsubuki, T.; Higashiyama, K. *Synthesis* **1992**, 681.
14. a) Soliman, S. A.; Abdine, H.; El-Nenaey, S. *Aust. J. Chem.* **1975**, *28*, 49. b) Agami, C.; Rizk, T.; *Tetrahedron* **1985**, *41*, 537 and references therein.
15. Blanchette, M. A.; Choy, W.; Davis, J. T.; Essenfeld, A. P.; Masamune, S.; Roush, W. R.; Sakai, T. *Tetrahedron Letters* **1984**, *25*, 2183.
16. Hitchings, G. J.; Helliwell, M.; Vernon, J.M. *J. Chem. Soc., Perkin Trans. 1,* **1990**, 83. Nishio, T.; Yamamoto, H. *J. Heterocyclic Chem.*, **1995**, *32*, 883.
17. Yoda, H.; Katagiri, T.; Takabe, K. *Tetrahedron Lett.* **1991**, *32*, 6771. Yoda, H.; Kitayama, H.; Yamada, W.; Katagiri, T.; Takabe, K. *Tetrahedron Asymmetry* **1993**, *4*, 1451.
18. Ohta, T.; Shiokawa, S.; Sakamoto, R.; Nozoe, S. *Tetrahedron Letters* **1990**, *31*, 7329.
19. Tomcufcik, A. S.; Wright, W. B. Jr.; Meyer, W. E. U.S. Patent 4,892,885, Jan. 9, 1990.
20. Braish, T. F.; Fox, D. E. *Synlett.* **1992**, 979.

# Synthesis of 5-Lipoxygenase Inhibitors

**Ivan Lantos and Richard Matsuoka**
*Synthetic Chemistry Department, SmithKline Beecham Pharmaceutical R&D, King of Prussia, Pennsylvania 19406-0939*

Key words: lipoxygenase, leukotriene, Vilsmeier-Haack reaction

## Introduction

Ever since their structural identification[1], leukotrienes have been implicated in a number of disease states. One important inflammatory state in which their powerful physiological action is manifested is asthma. Based on this recognition, a number of pharmaceutical companies have been working towards the discovery of substances that either neutralize the physiological action or prevent the biosynthesis of leukotrienes. 5-Lipoxygenase is one of the critical enzymes in the leukotriene biosynthetic pathway that transforms arachidonic acid to $LTA_4$ and further to $LTD_4$. Therefore, inhibition of this enzyme has been the target of extensive research efforts at a number of pharmaceutical companies. This effort has already culminated in success inasmuch that the 5-Lipoxygenase inhibitor compound Zileuton,[1d] marketed by Abbott Laboratories, was approved in 1996 by the FDA for the treatment of asthmatic disorders.

We at SmithKline Beecham Pharmaceuticals have also been involved in the discovery and development of 5-Lipoxygenase inhibitors.[2,3] Towards this goal several series of hydroxyurea containing structures have been synthesized. Two members of this class, SB 202235 and 210661 (**9a** and **b**), possessed sufficient pharmacological potency to be approved for clinical development. The chemical development of these two drug candidates will be the subject of this chapter.

## General considerations

The responsibility of supplying clinical development with drug substance of a potential candidate is shouldered by the chemical development group. This task is carried out in three sequential phases:

(i) The initial phase of the work incorporates much of the Medicinal Chemistry devised route to the synthesis of the drug. Modifications are made to this initial supply route to ensure its safety and amenability to scale-up to multi-kg quantities. Commercial intermediates are identified and bought. There are no cost considerations in this effort since time is of the essence.

(ii) Once an initial supply route is established, attention is focused on the discovery of a cost efficient synthesis that will serve as the long-term manufacturing route. This effort starts in the laboratory but evolves into a large scale pilot plant demonstration of the suitability of the route.

(iii)  Finally, the route is transferred to the organization mandated with the commercial production of the drug substance.

During all of this effort, significant consideration is given to the safety and environmental friendliness of the chemistry being developed. Every chemical process which is run in the pilot plant, or has the future potential of being performed in the pilot plant, is subjected to safety evaluation. Initially, the thermal stability of the compounds (starting materials, isolated intermediates etc.) are screened by Differential Scanning Calorimetry (DSC), Thermal Gravimetric Analysis (TGA) and adiabatic calorimetry.  The first method is to detect *endo-* or *exothermic* behaviors while the second method measures the thermal stability of the reactants or products.  The reaction is then run in a calorimeter in order to measure decomposition and off-gassing. A process can not be performed in the pilot plant before passing all three of these safety assessments.

Scheme 1.  Initial Synthesis

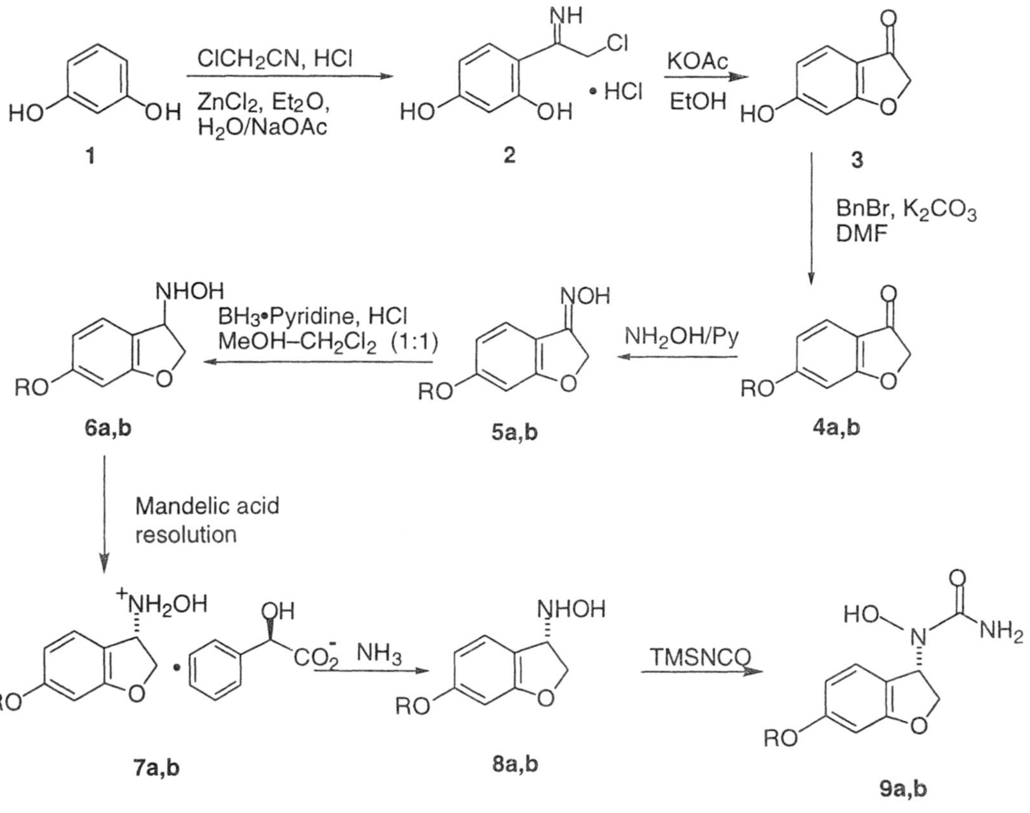

a, R=Bn
b,  =2,6-di-F-Bn

### Phase I of SB 202235/210661 Chemical Development

Upon approval of the compounds **9ab** for clinical development, synthetic procedures for the initial Medicinal Chemistry Route (Route A) were transferred to us for further development. This chemical synthesis is shown on Scheme **1**.

The chemistry starts from resorcinol (**1**) which was chloroacetylated in a Friedel-Crafts type of reaction with chloroacetonitrile and ZnCl$_2$. The resultant imino adduct (**2**) was isolated and hydrolyzed to the chloro-acetophenone and cyclized to the hydroxybenzofuranone (**3**).[4] Benzylation and oximation delivered the required oxime **5** which was reduced to the hydroxylamine by applying the procedure of Kigugawa.[5] This latter transformation, employing an excess of the borane:pyridine complex in the presence of hydrochloric acid, afforded the required hydroxylamine in respectable 80% yields. No over-reduction to the amine stage was observed.

Resolution of the hydroxylamine intermediates (**6**) presented a special problem. Because of their much diminished basicities, as far as we know, classical resolution of chiral hydroxylamines has not been documented in the literature. This technical difficulty prompted the medicinal chemists first

Scheme 2

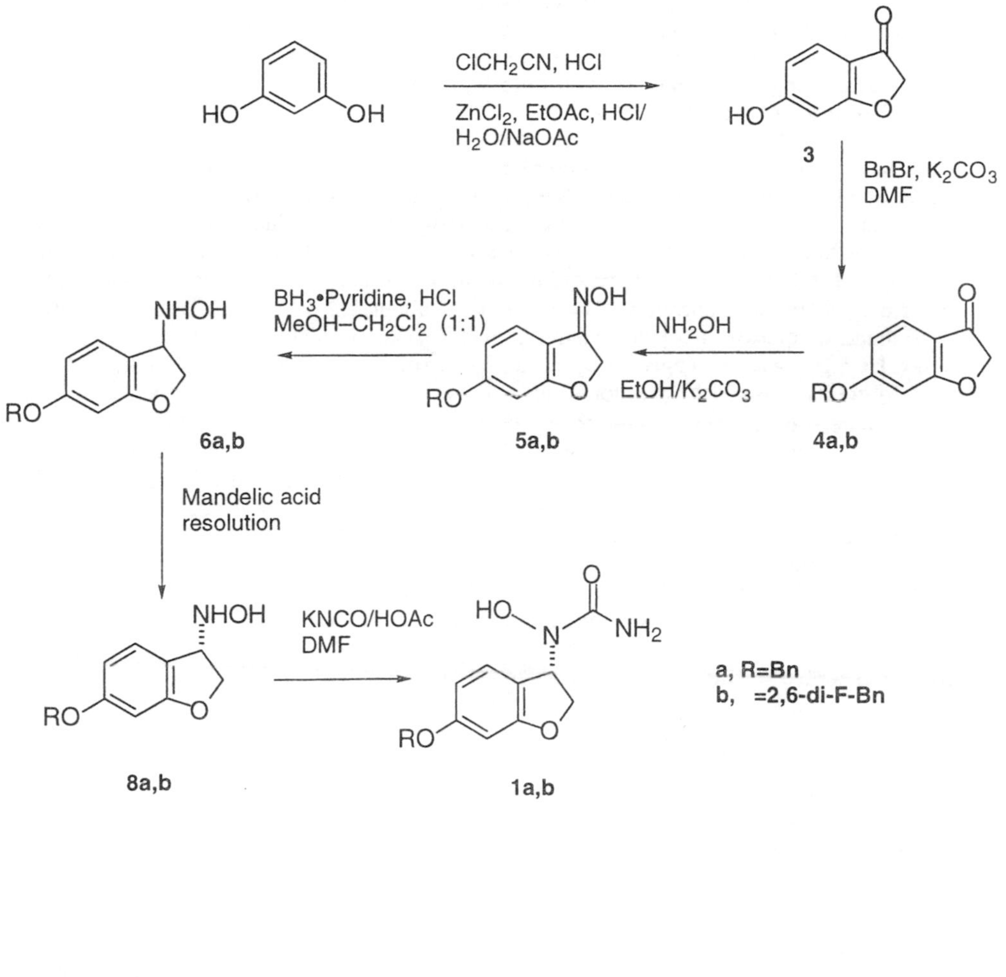

to devise a derivatization procedure using a chiral oxazolidone as the reagent.[3] The diastereomers that resulted were then separated by preparative HPLC and reconverted to the pure enantiomers. This approach, naturally, could not be satisfactorily applied to the production of multi-kg quantities. Medicinal Chemistry therefore devised a classical resolution method employing mandelic acid affording the required S-hydroxylamines **7** in about 30% yield (50% maximum). Final conversion to the urea **9a** or **b** was accomplished with trimethysilyl-cyanate.

When this process entered the Chemical Development department, we immediately turned our attention to several stages that needed further development. In the first stage, a new procedure had to be devised that would avoid isolation of the noxious chloroimidate intermediate **2** and the use of diethyl ether as the solvent needed to be eliminated. The pyridine solvent had to be replaced in the oximation stage and finally, a more commercially available reagent needed to be found to install the urea moiety. To avoid the isolation of **2** we designed a single-pot procedure conducting the Friedel-Crafts reaction in ethyl acetate. Once the reaction was deemed complete, the solvent was slowly removed via vacuum distillation and replaced with water. In this way, the adduct was *in-situ* hydrolyzed. Cyclization of the intermediate was then accomplished by basification of the solution with potassium acetate. The pyridine solvent was also successfully eliminated by switching to a well established classical oximation procedure using hydroxylamine, ethanol and potassium acetate for the reaction.

Much effort was expended in optimizing the resolution. Crystallization of the mixed diastereomeric salts turned out to be fickle. Whereas **6a** was found to reproducibly produce the desired diastereomer **7a** in > 98% e.e., **6b** afforded **7b** in variable optical purity. Although these variations were over the range of 60–95% e.e. one recrystallization of the final product invariably provided better than 99% e.e. Finally, we worked out a conversion of the enatiomerically pure hydroxylamines to the required hydroxyureas using potassium cyanate/HOAc in DMF solution. Scheme **2** represents the process that resulted from these modifications.

With these results in hand we were now able to take the process into the pilot plant and prepare both **9a** and **b** in multi-kg quantities. The following describes the process for **9a**: identical procedures were applied to the preparation of **9b**. The initial phase of this program dealt with a thorough hazard evaluation of each stage of the synthesis. In Stage 1 of the process, the addition of $ZnCl_2$ to a slurry of resorcinol and chloroacetonitrile in ethyl acetate was found to be slightly exothermic yielding at room temperature a heat output of about 2.2 kJ/mol. This heat output, however, was found to quickly dissipate without further heat accumulation. On the other hand, addition of the HCl gas was accompanied by a strongly exothermic event with 59.5 kJ/mol heat output. To perform this phase of the reaction safely, the HCl gas was added over the period of 2 hrs while the temperature was maintained between 20–25° C by occasional cooling. Hydrolysis of the initial adduct via distillation of the solvent at reduced pressure and addition of water did not result in significant heat evolution and was inherently safe. Finally, cyclization to the benzofuranone by treatment of the intermediate chloro-ketone with potassium acetate in ethanol could be carried out safely since the reaction was run at reflux which dissipated the heat released from the reaction. The DSC of **3** revealed a strong endotherm at 251 °C and an exotherm at 265 °C, respectively; the high temperatures indicated a high degree of thermal stability. The second stage of the process was found to be slightly exothermic during the benzyl bromide addition to the mixture of **3** and $K_2CO_3$ in DMF solvent. This heat release could be controlled, however, by the slow addition of the reagent. Thus, the process was safe. Thermal stability of intermediate **4a** was indicated by a slow endotherm at 98 °C in its DSC. The oximation, Stage 3, was conducted in refluxing ethanol. Although the chemistry was found mildly exothermic (24 kJ/mol), no safety hazard existed since the heat was absorbed by the refluxing solvent. The product **5a** however was found to be thermally unstable undergoing rapid and violent decomposition at above 130 °C. Great care had to be taken in the drying of this intermediate. The reduction stage, Stage 4, was the most delicate phase of the entire process. It involved the addition of 6 N HCl to a cooled solution of **5a** containing 5 equiv. of borane-pyridine complex. In addition to substantial heat evolution, copious quantities of gases were released. To conduct this reaction safely in the plant, HCl addition was carried out over a three hr period while the temperature was carefully maintained at 5–10 °C. The temperature was maintained for an additional hr even after the addition was complete and then it was slowly raised to ambient. Thus, by careful control of the addition rate we performed the decomposition of the boronate complexed product and conversion to the HCl salt safely. We did observe however that product **6a** thermally decomposed at 111 °C. This decomposition is no doubt due to the desire of the compound to eliminate hydroxylamine and convert to the highly stabilized benzofuran structure.

The classical resolution was carried out in methanolic solutions using (S)-mandelic acid. The compound was dissolved in methanol by careful heating and a methanolic solution of a stoichometric amount of the acid was added. Upon slow cooling, crystalline diastereomeric salt separated out. No thermal hazards were associated with this reaction, and the temperature was not allowed to approach the decomposition point of the hydroxylamine. The product was collected in a centrifuge, washed with ethyl acetate and carefully dried at reduced pressure around 45–50 °C. Reconversion to the base was effected by partitioning the compound between ethyl acetate and 1 N ammonia. Final conversion to **9a** was performed in cold DMF solution with acetic acid and potassium cyanate. Once all the reagents had been added, the solution was gently warmed to ambient temperature and was maintained there for 30 min. A large excess of water was added and the precipitated product was isolated by centrifugation. The crude product was recrystallized from DMF/TBME mixtures affording **9a** in about 70% yield and 99% e.e. We demonstrated the viability of this process by preparing drug substance in 4 and 40 kg batches for toxicity and clinical studies. Even though this process served us well in generating large quantities of the compounds for our clinical and toxicological studies, the cost of the drug substance was prohibitively high, approximately $4300 per kg. The large cost was mainly the result of the expensive reduction and resolution stages. To solve the cost problem a synthesis was required that would afford the drug substance at below $1000 per Kg. In our opinion, this was best achieved by the discovery of an enantioselective synthesis.

*Attempted Enantioselective Synthesis via Chiral Reductions:* Our first attempts of an enantioselective synthesis explored the feasibility of generating chiral hydroxylamines via the enantioselective reduction of oxime ethers. Although highly effective asymmetric reductions of oxime ethers to optically active amines have been reported in recent years,[6] useful enantioselective conversions of oxime ethers into O-substituted hydroxylamines have been relatively neglected. As far as we know no such successful transformations have been described.[7] A method reported by Burk *et. al.*[8] involving a highly efficient and enantioselective reduction of benzoylhydrazones to hydrazines by

Scheme 3. Attempted Catalytic Oxime Reductions

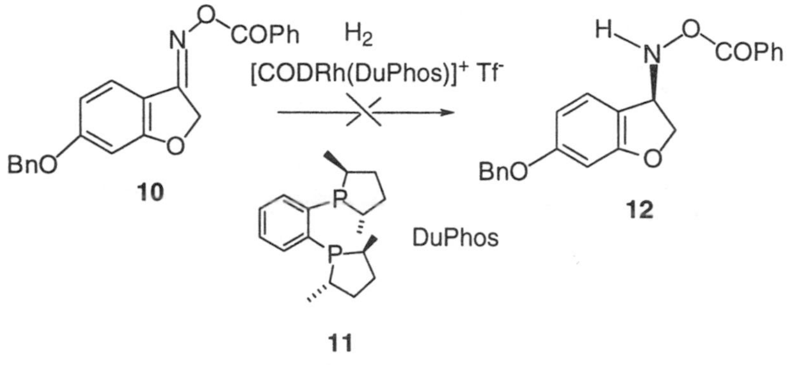

analogy appeared to have potential for the reduction of oximes to hydroxylamines and we decided to investigate it.

The analogous benzoyl-oxime **10** was readily prepared; however, the reduction was not successful. Even though we were able to fully reproduce Burk's reported results on his hydrazone substrates, the oximes totally inhibited the reaction.

After surveying several other chiral reducing agents for the asymmetric reduction of various substituted 6-benzyloxy-2,3-dihydrobenzofuran-3-oxime, we found that reagent **13**, first reported by Sakito[7] and prepared from norephedrine and borane, reduced the O-(*o*-nitrobenzyl) ether **14b** in respectable yields (55%) to the corresponding optically active O-(*o*-nitrobenzyl) hydroxylamine and with *excellent enantioselectivity (99% e.e.)*. Similarly, reductive conversion of the O-benzyloxime **14a** was also successful with 89.2% e.e. and in 63% yield.[9] It is noteworthy that neither the reagent of Corey nor several related reagents were able to carry out these reductions under identical conditions.

In contrast, the asymmetric reductions of ketoxime O-(*p*-methoxyl-benzyl) ether **14c** and ketoxime O-methyl ether **14d** with **13** were less efficient, affording **15c** and **15d** with 64% and 56% e.e., and in 54.2% and 67% yields, respectively. In the course of this study, we found that the chemical yield of the desired O-substituted hydroxylamines depended critically upon the ratio [borane] : [norephedrine]. With the reagent prepared from 1:1 molar ratio, for example, almost no reduction occurred and the starting oxime was recovered. On the other hand, the reagent prepared from 2 mols of borane and 1 mol of norephedrine gave the highest chemical yield of the O-substituted hydroxylamine; a larger excess of the borane decreased the yield of O-substituted hydroxylamine and increased the yield of the amine, clearly the product of over-reduction.

Unfortunately, our success in finding conditions for the enantioselective formation of substituted hydroxylamines did not lead to an efficient solution to the problem of chiral synthesis. Numerous attempts at the selective removal of hydroxylamino substituents resulted in removal of both

Scheme 4

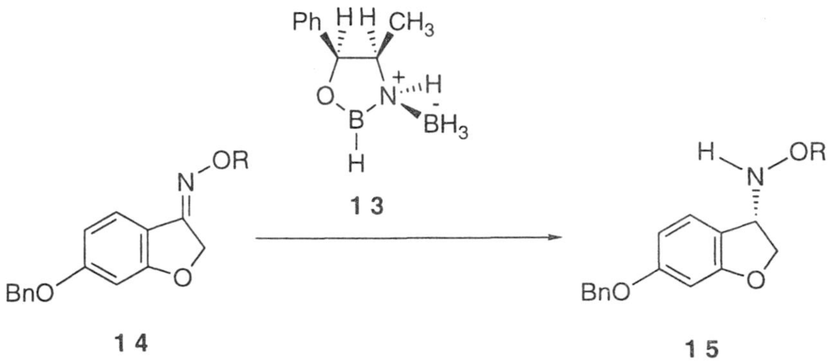

a  R=Bn
b      =o-nitroBn
c      =p-methoxyBn
d      =Me

benzyl substituents in the molecule. Surprisingly, attempted re-benzylation at the phenolic hydroxyl could also not be accomplished. These results were found not only with the penultimate intermediates **15** but also with their corresponding hydroxylurea derivatives. These results forced us to pursue our goal in a different direction.

## Enantioselective Synthesis via Chiral Nitrone:[10]

The transformation of nitrones bearing a chiral auxiliary to chiral hydroxylamines was first noted by Chang and Coates.[11] These workers explored the addition of organometallic reagents to *inter alia* chiral β-methoxy-α-phenylethyl- substituted nitrones. Impressive diasteroselectivities were observed. This approach was further progressed by Hu and Schwartz[12] in their enantioselective synthesis of chiral hydroxylamines via the addition of Grignard reagents to nitrones bearing gulofuranosyl derived chiral auxiliaries. The very high diastereoselectivities obtained and the mild conditions required for auxiliary removal (the highly oxygen-rich character of the aryl moiety in the dihydrobenzofurans makes these molecules inherently acid labile) prompted us to initiate an investigation into applying Schwartz's method to our problem (interestingly it now appears from the literature[13] that three research groups independently came to this same conclusion as witnessed by subsequent publications). A retrosynthetic analysis of the route we have envisaged is shown in equation 1.

Equation 1

It is based on the tandem nucleophilic addition of a suitable nuleophile to the benzaldehyde based nitrones and cyclization of the *in situ* generated intermediates. An essential condition for the proposal was that the anion stabilizing group would in itself be a good leaving group. The commercially readily available mannose bis-acetonide was chosen as the chiral auxiliary and the success of the sulfur ylide addition was anticipated based on the work of Pyne and Hajipour.[14] We were also hoping that the inherently exceptional leaving group ability of dimethyl sulfoxide would allow the initial adduct to undergo spontaneous cyclization.

Before our proposed synthesis could be attempted in practice, the required bezaldehydes and nitrones had to be synthesized. The synthesis of the former has been published.[15a,b] The synthesis begins with the formylation of resorcinol via the application of the Vilsmeyer-Haack reaction and subsequent regioselective alkylation of the *para-* hydroxyl group as in Scheme 5. Vilsmeier-Haack reaction was run in acetonitrile in which the oxalyl chloride and DMF forms a sparingly soluble imino-chloride complex which dissolves upon addition of the resorcinol at -15 °C. Aqueous workup of the initially formed formamidinium chloride salt affords the desired 2,4-dihydroxybenzaldehyde. The regioselective benzylation also had its aspects of interest.[15b] The combination of several bases, solvents and alkylating agents were examined in order to optimize the ratio of 4-substituted vs. 2,4-disubstituted products. The best and most economical conditions turned out to be using the weak base of sodium bicarbonate in acetonitrile solvent with benzyl chloride reagent. After optimization of the reaction conditions, a 60–70% yield of **17a,b** could be isolated.

Scheme 5

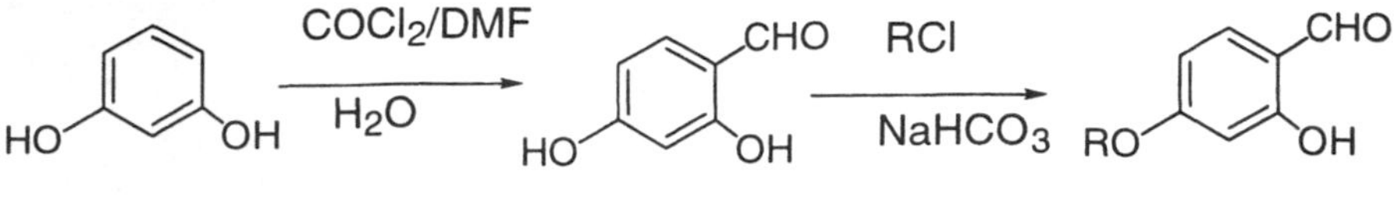

**17a,b**

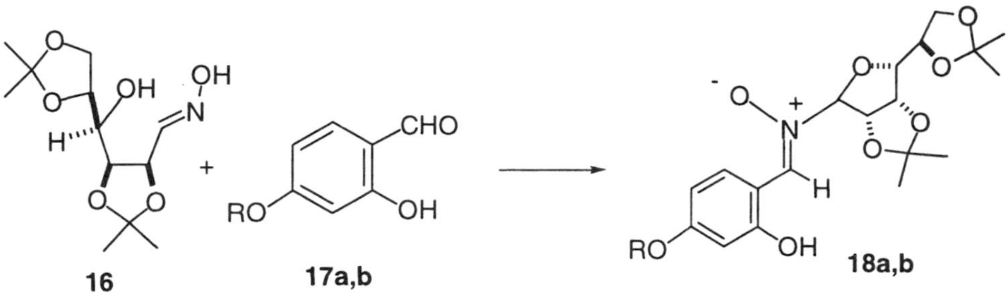

**16**          **17a,b**                                        **18a,b**

When the process was scaled up, the benzylation reaction was combined with the formation of nitrones **18a,b** as in Scheme 5. Thus, once the alkylation was deemed complete by assaying for less than 3% of the dihydroxy starting material by HPLC, water was added and the organic substances were extracted with toluene. The acetonitrile and water were removed by distillation and when the pot temperature reached about 110 °C the solution was cooled and commercially obtained mannose-bis-acetonide oxime (an approximately 1:1 mixture of oxime **16** and the cyclic hydroxylamine) was added. Refluxing the solution for 6 hrs usually completed the condensation and after cooling crystalline nitrones precipitated out upon addition of hexanes. The yield of the nitrones **18 a** and **b** were about 80%.

Having secured the required nitrones we were about to test our proposal of the tandem nucleophilic addition-cyclization reaction for the formation of the chiral hydroxylamines. We were gratified to find that when a slurry of nitrones **18a** in toluene was treated with a solution of dimethylsulfoxonium methylide[16] in THF at -10 °C and the resulting solution was then warmed to 10 °C a mixture of diastereomeric *N*-glycosides, **19a** and **20a** were formed in relative ratios of about 7.5–8.0 to 1 (isolated yields ca 65–75%; solution yields >90%). Identical reaction was carried out with **18b** in THF affording a similar ratio of diastereomers (Scheme 6). The reaction with nitrone **18b** was successfully scaled up in a 50-gallon reactor (THF; -10 °C to +10 °, 1 h; +10 °C, 22 h) to afford 15.91 kg (c.a. 90% solution yield) of *N*-glycoside **19b**.

With both substrates, the diastereoselectivity of the reaction was found to be temperature and concentration dependent; the lower the temperature and the less concentrated the reaction mixture, the higher diastereoselectivity was obtained as shown in Table 1 and Table 2.

**Table 1.** Effects of Temperature on Diastereoselectivity*

| Entry | Temperature | Diastereoselectivity (**19a** to **20a**) |
|---|---|---|
| 1 | 50 °C | 4–5 to 1 |
| 2 | 22 °C | 7 to 1 |
| 3 | 0 °C | 8.5 to 1 |
| 4 | -30 °C | no reaction |

* All reactions were carried out at 0.1 M reaction concentration in THF

Scheme 6

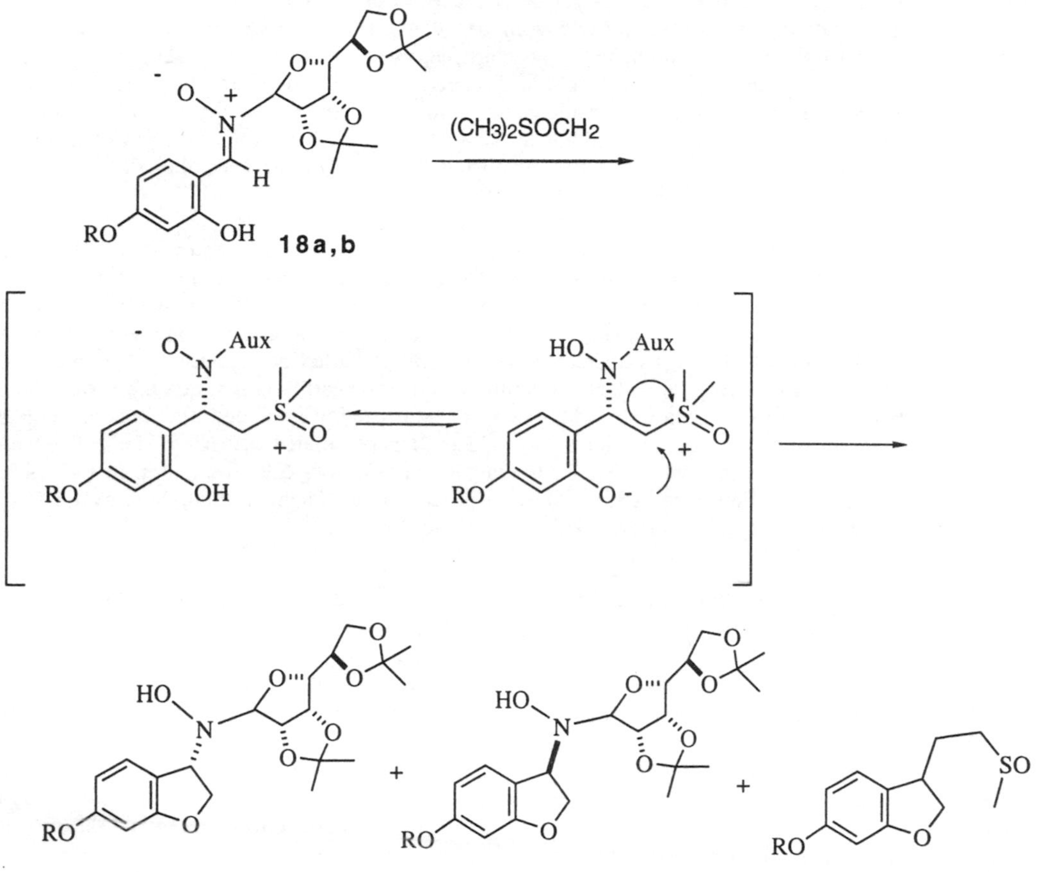


**19a,b** **20a,b** **21a,b**

**a, R=Bn**
**b, =2,6-difluoroBn**

**Table 2.** Effects of Reaction Concentration on Diastereoselectivity**

| Entry | Concentration | Diastereoselectivity (**19b** to **20b**) | Sulfoxide formation **21** |
|-------|---------------|-------------------------------------------|-----------------------------|
| 1 | 0.5 M | 5.6 to 1 | 2–5% |
| 2 | 0.4 M | 5.8 to 1 | 2–5% |
| 3 | 0.2 M | 6.1 to 1 | <1% |
| 4 | 0.1 M | 6.6 to 1 | <1% |

** All reactions were carried out at room temperature in THF

The phenol moiety of *N*-glycosyl nitrones **18** is not deprotonated by the methylide during the addition-cyclization reaction because only one equivalent of methylide is needed for the reaction to take place. If the *N*-glycosyl nitrone is pretreated with one equivalent of potassium *tert*-butoxide before methylide addition, the subsequent addition-cyclization reaction is shut down completely. On the other hand, if one pre-treats the *N*-glycosyl nitrone with one equivalent of potassium *tert*-butoxide and subsequently with trimethylsulfoxonium iodide at temperatures above 40 °C, the addition-cyclization reaction proceeds, albeit with much lower levels of diastereoselectivity due to the rather high reaction temperature required. Based on these observations, we currently believe that the mechanism of the reaction involves initial addition of the methylide to the nitrone, tautomerization of the proton from the phenol to the hydroxylamine moiety, and five-membered ring cyclization with the elimination of dimethylsulfoxide (as shown in Scheme **6**).

The addition-cyclization reaction is extremely sluggish at temperatures below 0 °C. Only one major side-product is formed in these reactions, sulfoxides **21a,b** which could be minimized to < 3% by treating the nitrones with no more than one equivalent of dimethylsulfoxonium methylide at relatively low temperatures (ca less than 10 °C). Addition of more than one equivalent of dimethylsulfoxonium methylide greatly increases the amount of sulfoxide produced. For example, treatment of nitrone **18a** with two equivalents of methylide affords nearly eight times more sulfoxide. Sulfoxide formation is enhanced when the addition-cyclization reaction is carried out at temperatures above 10 °C. The formation of the sulfoxide side-products **21** is postulated as arising from a five-step reaction sequence as outlined in Scheme **8**. Following initial addition, α,β-elimination of the chiral auxiliary occurs. This is followed by a 1,4-addition of a second equivalent of methylide, cyclization, and Stevens' rearrangement to afford the sulfoxide side-product.

The diastereoselectivity seen in the initial addition step of dimethylsulfoxonium methylide to the various *N*-glycosyl nitrones was rationalized via the Vasella transition state model[17] known as the "kinetic anomeric effect". Vasella has used this model to describe both nucleophilic and 1,3-dipolar cycloadditions to *N*-glycosyl nitrones and observed selectivities similar to what we observed in the methylide addition. The argument is schematically represented in Scheme **7**. The sterically less congested 'O-*endo*' conformer of the nitrone reacts preferentially over the 'O-*exo*' conformer. The 'O-*endo*' conformer can be attacked by the methylide from either side of the C(1)–O bond: either '*syn*' to the C(1)–O bond (from the same side) or '*anti*' to the C(1)-O bond (from the opposite side). '*Syn*' attack of the methylide is preferred because of the stabilizing anomeric effect present in the product between the C(1)–O bond in the glycosyl moiety and the lone-pair of the N-atom.

For the purpose of scaling up this chemistry, a practical and economical synthesis of dimethylsulfoxonium methylide in THF from trimethylsulfoxonium iodide was developed. A variety of methods for the generation of dimethylsulfoxonium methylide in THF from trimethylsulfoxonium chloride exist. The chloride salt, however, is much more expensive than the iodide salt. For a large scale reaction using dimethylsulfoxonium methylide, the iodide salt has to be the source. The most common procedure for generating the ylide from the iodide salt involves the use of sodium hydride as base in DMSO at room temperature.[16] These conditions proved undesirable because the presence of DMSO reduces the diastereoselectivity in the addition portion of the one-pot, two-step addition-cyclization reaction. In addition, the combination of sodium hydride and DMSO has been reported to be hazardous on large scale preparations.[18] Mild conditions have been reported for the generation of dimethylsulfoxonium methylide *in situ* from trimethylsulfoxonium iodide consisting of the use of potassium *tert*-butoxide as base in DMSO at room temperature.[18] These reported conditions were slightly modified and optimized to allow the use of THF as the solvent.

Scheme 7

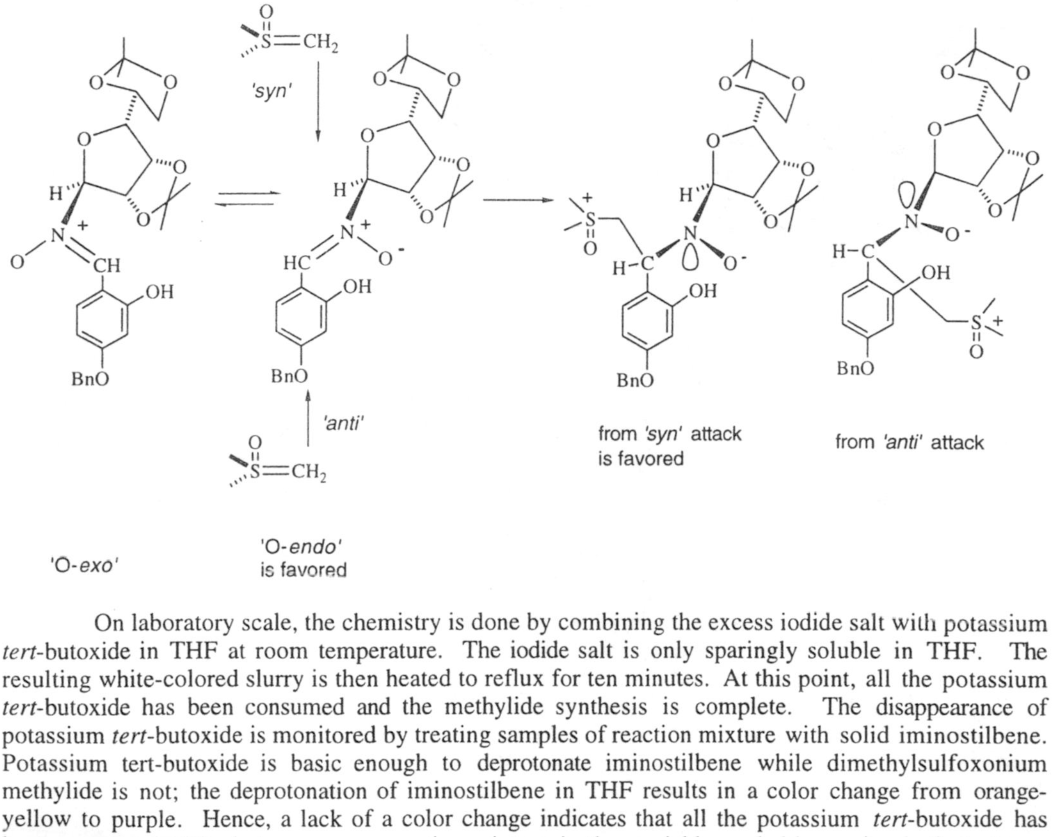

'O-exo'

'O-endo'
is favored

from 'syn' attack
is favored

from 'anti' attack

On laboratory scale, the chemistry is done by combining the excess iodide salt with potassium *tert*-butoxide in THF at room temperature. The iodide salt is only sparingly soluble in THF. The resulting white-colored slurry is then heated to reflux for ten minutes. At this point, all the potassium *tert*-butoxide has been consumed and the methylide synthesis is complete. The disappearance of potassium *tert*-butoxide is monitored by treating samples of reaction mixture with solid iminostilbene. Potassium tert-butoxide is basic enough to deprotonate iminostilbene while dimethylsulfoxonium methylide is not; the deprotonation of iminostilbene in THF results in a color change from orange-yellow to purple. Hence, a lack of a color change indicates that all the potassium *tert*-butoxide has been consumed. The heterogeneous reaction mixture is then quickly cooled in an ice bath to room temperature; the methylide is somewhat unstable at elevated temperatures. The reaction mixture appears as a white-colored slurry because the potassium iodide that is generated in the reaction is only sparingly soluble in THF. The cooled reaction mixture is filtered to remove the potassium iodide and the excess trimethylsulfoxonium iodide. The yellowish-gold tinted filtrate containing the methylide is titrated following the method of Corey[17a] with hydrochloric acid to phenolphthalein endpoint, to determine its molarity. Estimated yields based upon molarity and total volume of solution are routinely greater than 80%. While the filtration of the solution of methylide in THF to remove the potassium iodide could theoretically be carried out at a variety of temperatures, we found that the methylide surprisingly began to oil out in the potassium iodide cake if the filtration was carried out at temperatures at or below 0 °C.

On plant scale, this general procedure was modified primarily because of the time needed to cool a refluxing THF reaction to room temperature. For example, twenty to thirty minutes are generally needed to perform this cooling operation in a typical 50-gallon reactor. We were worried that some of the desired methylide would decompose during this extended cooling down period. We found that the reaction successfully takes place at 40 °C. After about twenty minutes at 40 °C, all the potassium *tert*-butoxide is consumed and the methylide synthesis is complete. This methylide generating reaction is sluggish at reaction temperatures below 35 °C. We successfully ran this reaction in a 50-gallon reactor starting with twenty-five kilograms of trimethylsulfoxonium iodide.

Scheme 8

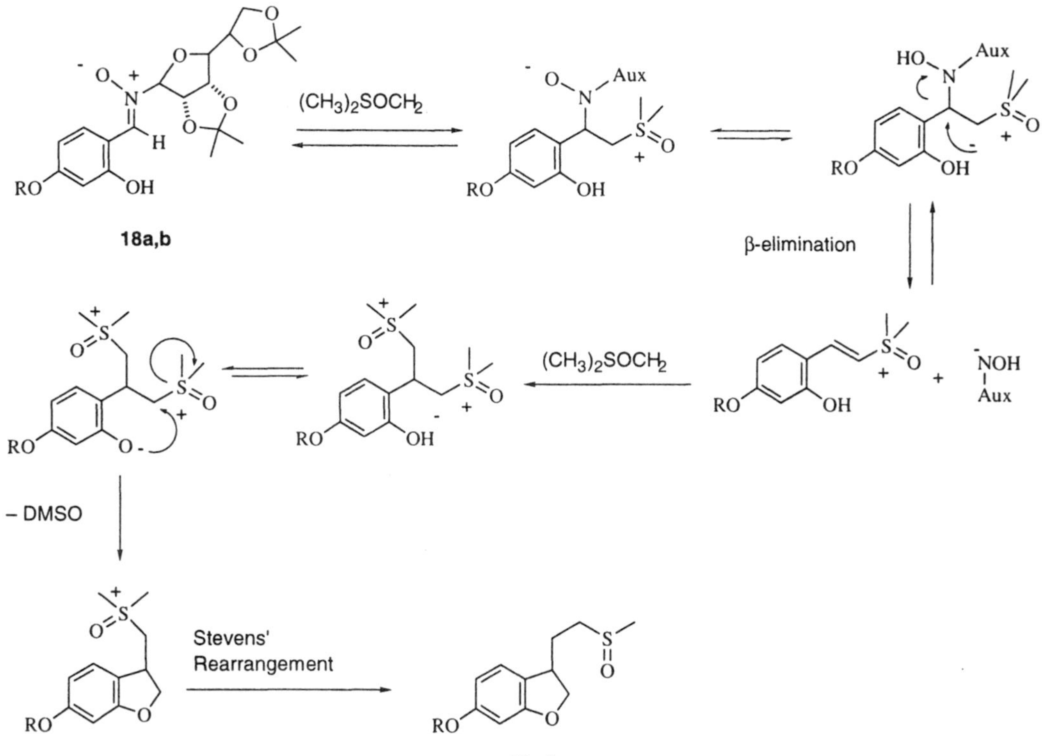

**18a,b**

β-elimination

− DMSO

Stevens'
Rearrangement

**21a,b**

During our initial laboratory experiments, the diastereomeric mixtures of **19** and **20** were isolated by crystallization from hexane/THF solutions. These solid mixtures were then taken on further in the synthesis to **8** and finally to **9**. The possibility of performing the conversion in a single step operation did not escape our attention and attempts were made to convert the diastereomeric mixture to the desired hydroxyureas **9a** and **b** in a single-pot operation as shown in Scheme 9. To achieve this a DMF solution of the mixture was treated with dilute HCl, to remove the auxiliary mannose, and the resulting hydroxylamines were subsequently converted to the ureas by adding dilute NaOH to pH 4.0 followed by potassium cyanate.

Surprisingly, this one-pot procedure afforded, in addition to 65–70% yields of the target compounds, a few percent of the benzyloxy-benzofurans and 5–10 percent of an isomer **22** resulting from hydroxycarbamoylation. When the same isomeric mixture of **19a** and **20a** was subjected to the former two stage procedure none of the by-products resulted. Since we had never observed the appearance of these by-products previously, a fairly extensive investigation was undertaken to find a mechanistic rationale for their formation. It appeared that the strength of the HCl greatly influenced the degree to which **22** was generated. Unexpectedly, higher acid concentration resulted in lower levels of **22**, but more of the benzofuran **23** was formed. It appears that there is a fine balance between the rate of mannoside hydrolysis and loss of hydroxylamine. Thus, 6 N HCl resulted in slower hydrolysis rates but promoted faster conversion to the benzofuran.

Scheme 9

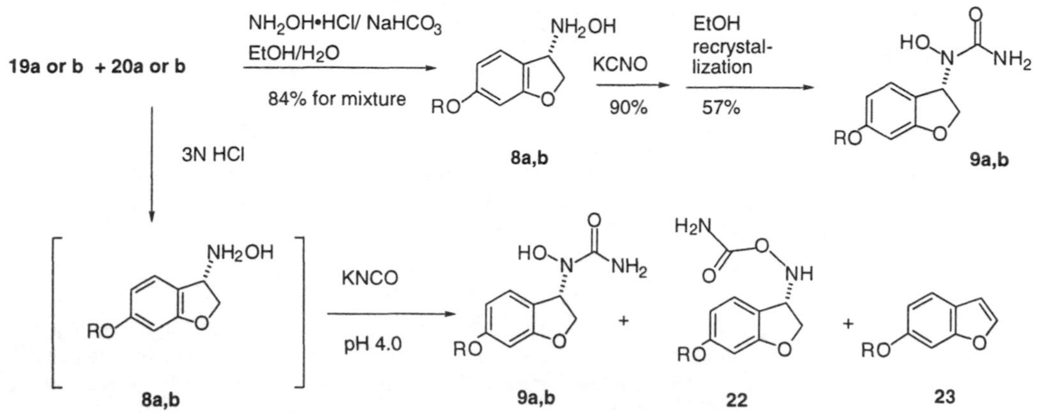

Our continued search for conditions that would alleviate the formation of both **22** and **23** led us to consider, at the suggestion of one of our consultants, the use of hydroxylamine hydrochloride as the reagent which could also potentially regenerate the mannose oxime in the reaction. Our first attempt at employing this reagent led to a much more efficient conversion of **19a** and **20a** to hydroxylamine **8a**. A mixture of the diastereomers was refluxed in 1:1 ethanol-water to which NH$_2$OH•HCl and NaHCO$_3$ were added. The reaction was followed by HPLC and after 1.5 hr removal of the mannoside was found complete. The yellow-cloudy solution was allowed to cool, additional volume of water was added and the enantio- enriched **8a** was collected at 0 °C. The yield of the hydroxylamine was 70–75% from the corresponding nitrones. Furthermore, concentration of the mother liquors enabled us to recover the chiral auxiliary in about 80% yield suitable for reuse. Later, we found isolation of the solid diastereomers was not necessary for the success of the conversion. Simply concentrating the THF solution from the cyclization reaction was sufficient to provide the cyclized intermediates **19** and **20** as starting materials for the hydroxylamine hydrochloride treatment. Both **8a** and **b** could be prepared this way.

Conversion to **9a** (**SB 202235**) and **9b** (**SB 210661**) was carried out using the original protocol, treatment with KOCN in the presence of acetic acid (Scheme 2). Using this protocol no isomer or benzofuran was generated and the crude products were obtained in 96% and 85% yields, respectively, and 77% e.e. Since the enantiomeric mixture was still 8:1 entering the reaction, we were fortunate to find that the enantiopurity of both compounds could be improved to 99% by one recrystallization from isopropanol.

The ultimate success of this route was judged by our ability to perform it in the pilot plant and to deliver the drug substance under $1000 per Kg costs. We demonstrated the large scale amenability of this new route by preparing over 4 kg of **9b.** Thus 6.4 kg (12.22 mols) of the nitrone **18b** were reacted with an equimolar solution of the dimethylsulfoxonium methylide in THF. The reaction afforded 6.4 kg of an 8.3 to 1 mixture of **19b** and **20b** thus fully reproducing the results obtained from the laboratory experiments. The solvent was concentrated and the oily residue containing 6.4 kg of the mixture was dissolved in 50 L of aqueous ethanol and 4 molar equivalents of hydroxylamine hydrochloride and $NaHCO_3$ were added. The mixture was heated at 80 °C for an hr by which time all the starting material was consumed. Upon cooling the solution the product crystallized as an 8 to 1 mixture of **8b** enantiomers. The reaction was run in two identical batches reproducibly affording 80–85% product yields. Final conversion to **9b** was uneventful, 9.15 kg of **8b**, using the previously established protocol, afforded 9.6 kg of a 9:1 enantiomeric mixture of **9b**. One recrystallization from i-PrOH delivered 6.0 kg of the final product in >99% e.e. with 73% recovery of the pure enantiomer. The cost of the drug substance was also reduced to 25% of that obtained by the original route.

## Conclusion

In summary, an efficient 6-step enantioselective synthesis of two related hydroxyurea lipoxygenase inhibitors **1a,b** has been devised using the readily available protected mannose oxime as chiral auxiliary. The synthesis is based on a novel nucleophilic addition-intramolecular cyclization reaction, proceeds in 30% overall yield and >99% e.e. An additional feature is the almost complete recoverability of the auxiliary.

**Acknowledgment.** The authors hereby express their sincere gratitudes to the many colleagues whose hard work and dedication led to the success of this program. Their names appear in references 9 and 10. Special thanks are due to Ms. Kimberly Parker for typing, and Vance Novack for reading the chapter.

## References

1a. Brooks, C. D.; Summers, J. B. *J. Med. Chem.* **1996**, *39*, 2629; b. Stinson, S. C. *Chem. Eng. News* **1997**, *75(1)*, 25; c. Samuelsson, B. *Science* **1983**, *220*, 568; d. Brooks, D. W.; Carter, G. W. in *The Search for Antiinflammatory Drugs*; Merluzzi, V. J.; Adams, J. Eds; Burkhauser: Boston **1995**; Chap. 5.

2. Adams, J. L.; Garigipati R. S.; Sorenson, M.; Schmidt, S. J.; Brian, W. R.; Newton, J. F.; Tyrrell, K. A.; Garver, E.; Yodis, L. A.; Chabot-Fletcher, M.; Tzimas, M.; Webb, E. F.; Breton, J. J.; Griswold, D. E. *J. Med. Chem.* **1996**, *39*, 5035.

3. Garigipati, R. S.; Sorenson, M. E.; Erhardt, K. F. Adams, J. L. *Tetrahedron Lett.* **1993** *35*, 5537.

4. Shriner, R. L.; Grosser, F. J. *Am. Chem. Soc.* **1942**, *64*, 382.

5. Kawase, M.; Kikugawa, Y.; *J. Chem. Soc. Perkins I Trans.* **1979**, 643.

6 a. Itsuno, S.; Nakano, M.; Miyazaki, K.; Masuda, H.; Ito, K. *J. Chem. Soc. Perkin Trans. I* **1985**, 2039-2044; b. Itsuno, S.; Sakurai, Y.; Shimizu, K.; Ito, K. *J. Chem. Soc. Perkin Trans. I* **1989**, 1548-1549; c. Itsuno, S.; Sakurai, Y.; Shimizu, K.; Ito, K. *J. Chem. Soc.*

*Perkin Trans. I* **1990**, 1859-1863; d. Itsuno, S.; Nakano, M.; Ito, K. *J. Chem. Soc. Perkin Trans.* **1985**, 2615-2619; e. Itsuno, S.; Tanaka, K.; Ito, K. *Chem. Lett..* **1986**, 1133-1136; f. Gibbs, D. E.; Barnes, D. *Tetrahedron Lett.* **1990**, *31*, 5555-5558.

7.   Suzukamo *et al.* in their description of the asymmetric reduction of oxime ethers to amines alludes (in the notes to Table 1) to the possibility of obtaining methyl ethers of hydroxylamines which were not characterized; Sakito, Y.; Yoneyoshi, Y.; Suzukamo, G. *Tetrahedron Letters* **1988**, 223-224.

8.   Burk, M.; Feaster, J. E. *J. Am. Chem. Soc.* **1992**, *114*, 6266.

9.   A report describing these reactions has been accepted for publication. Dougherty, J. T.; Flisak, J. R.; Liu, L.; Tucker, L. Lantos, I., Hayes, J.*Tetrahedron Asymmetry* **1997**, *8*, XXX.

10.   Flisak, J. R; Lantos, I.; Liu, Li.; Matsuoka, R. T.; Mendelson, W. L.; Tucker, L. M.; Villani, A. J.; Zhang, W-Y. *Tetrahedron Letters* **1996**, *37*, 4639.

11.   Chang, Z-Y.; Coates, R. M. *J. Org. Chem.*, **1990**, *55*, 3464.

12.   Hu, X.; Schwartz, M. A. *Book of Abstracts*, 204th National Meeting of the American; Chemical Society, Washington, D.C., **1992**; ORGN 5.

13.a.   Rohloff, J. C.; Alfredson, T. V.; Schwartz, M. A. *Tetrahedron Lett* **1994**, *35* , 1011.
    b.Basha, A.; Henry, R.; McLaughlin, M. A.; Ratajczyk, J. D.; Wittenberger, S.J. *J.Org. Chem.* **1994**, *59* , 6103.

14.   Pyne, S.G.; Hajipour, A. R. *Tetrahedron Letters* **1992**, *48*, 9385.

15.a.   Mendelson, W. *Syn Com.* **1996**, *26*, 593. b. Mendelson, W. *Syn. Com.* **1996**, *26*, 603.

16a.   Corey, E. J.; Chaykovsky, M. *J. Am. Chem. Soc.* **1962**, *84*, 867; Corey, E. J.; Chaykovsky, M. *J. Am. Chem. Soc.* **1965**, *87*, 1353; b. For a recent review on dimethyl-sulfoxonium methylide, see Gologobov, Y. G.; Nesmeyano, A. N.; Lysenko, V. P.; Boldeskul, I. E. *Tetrahedron* **1987**, *43*, 2609.

17.   Huber, R.; Vasella, A. *Tetrahedron* **1990**, *46*, 33 and references cited therein.

18.   Ng, J. S. *Synthetic Communications* **1990**, *20*, 1193 and references cited therein).

# Chemistry of Vitamin D:
# A Challenging Field for Process Research

**Masami Okabe**
*Process Research, Chemical Synthesis Department, Hoffmann-La Roche Inc., Nutley, New Jersey 07110-1199*

Key words: vitamin D, photolysis

Vitamin $D_2$ (**1**, ergocalciferol) and vitamin $D_3$ (**2**, cholecalciferol), derived from ergosterol and cholesterol, respectively, are secosteroids, a family of compounds with broken steroid skeletons. Since ergosterol is a plant sterol and not produced in animals, the so-called 'natural form of vitamin D' is **2** whose structure was elucidated by Windaus more than six decades ago.[1] However, vitamin $D_3$ (**2**) is also in reality not a 'vitamin' since this compound is produced in the skin:[2] 7-dehydrocholesterol (**3**) derived from cholesterol undergoes a conrotatory opening of the diene ring upon exposure to sunlight (cleaving the C9-C10 single bond) to give previtamin $D_3$ (**4**), followed by thermal conversion of **4** to give **2** (Figure 1).

**Figure 1**

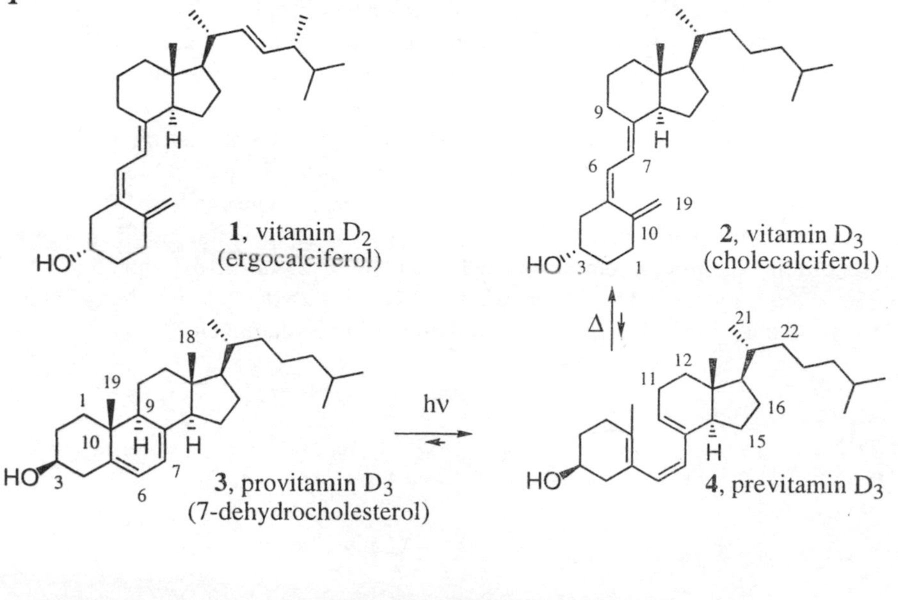

**1**, vitamin $D_2$
(ergocalciferol)

**2**, vitamin $D_3$
(cholecalciferol)

**3**, provitamin $D_3$
(7-dehydrocholesterol)

**4**, previtamin $D_3$

**Figure 2**

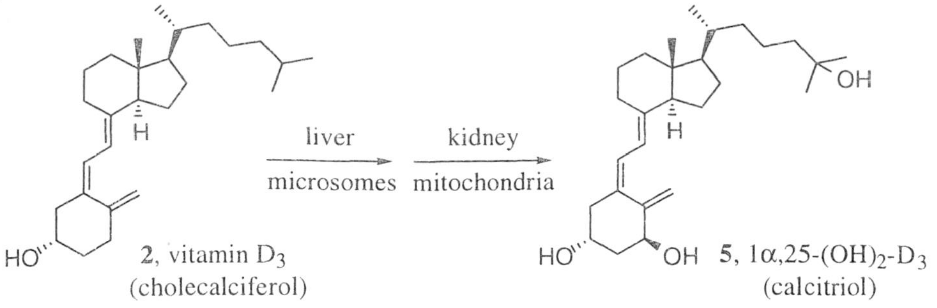

2, vitamin $D_3$
(cholecalciferol)

liver          kidney
─────────→   ─────────→
microsomes  mitochondria

5, $1\alpha,25\text{-}(OH)_2\text{-}D_3$
(calcitriol)

Vitamin $D_3$ (**2**) thus formed in the skin or derived from the diet undergoes several metabolic transformations (Figure 2). In the liver **2** is hydroxylated at C25 to form 25-hydroxyvitamin $D_3$, which is further hydroxylated in the kidney to yield the hormonally active form of vitamin D, $1\alpha,25$-dihydroxyvitamin $D_3$ (**5**, calcitriol).[3] This steroid hormone is one of the most important biological regulators of calcium metabolism and thus is essential for life in higher animals. Certain kidney diseases can cause deficiency of **5** due to the diminished renal function to metabolize the vitamin D and lead to serious health problems. Since 1978 Hoffman-La Roche has provided this important steroid hormone, as Rocaltrol®, for the treatment of osteodystrophy, especially in kidney failure and in hemodialysis patients.[3] Since then it has been found that the same hormone, as well as a number of its analogs, have beneficial therapeutic effects in psoriasis and osteoporosis, and also have high potential for the treatment of certain cancers and as immunomodulators.[4] Thus, the biological activities of the vitamin D derivatives span several therapeutic areas, such as dermatology, metabolic diseases, oncology and autoimmune diseases, and the list is still growing.

These structurally complex compounds pose significant challenges to process research chemists. The inherent instability of the conjugated triene system of vitamin D toward oxygen and the fact that in solution vitamin D exists in a vitamin and previtamin form, which are in thermal equilibrium *via* a [1,7]-sigmatropic hydrogen shift (Figure 1), further complicate the task.[5] Thus, medicinal chemists often obtain them only in milligram quantities as a film or foam after careful HPLC purification. Fortunately for process research chemists, vitamin D and its analogs typically possess very potent activities, thus requiring relatively small quantities (100 g to 5 kg) to be manufactured at peak demand. Thus, even though ease of synthesis is of the upmost importance, cost of goods, environmental impact, etc., are usually not of serious concerns for these products although we, as industrial chemists, are continuously striving to improve these factors. Consequently, a variety of synthetic methodologies and purification methods which otherwise are unsuitable for the manufacture of drug substances can be employed.

In this manuscript I would like to discuss process research efforts at Hoffmann-La Roche in the synthesis of four vitamin D compounds; calcitriol (**5**, $1\alpha,25$-dihydroxy-vitamin $D_3$),[6] Ro 23-8525 (**6**, $1\alpha,25,26$-trihydroxy-22-ene-vitamin $D_3$),[7] Ro 24-2090 (**7**, 25-hydroxy-16-en-23-yne-vitamin $D_3$)[8] and Ro 23-7553 (**8**, 1,25-dihydroxy-16-en-23-yne-vitamin $D_3$).[9]

**Figure 3**

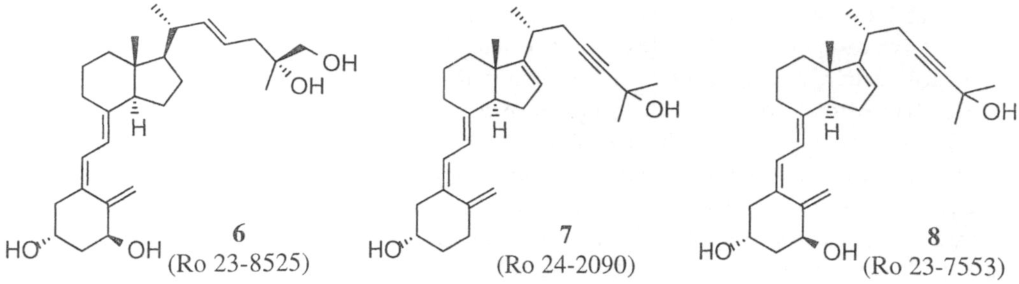

6
(Ro 23-8525)

7
(Ro 24-2090)

8
(Ro 23-7553)

### Synthesis of Calcitriol (5)

Finding efficient syntheses of calcitriol (**5**) has been one of the main topics for many academic and industrial chemists in the past two decades. Through their efforts a number of synthetic routes, which can be categorized in three types, have been developed.[10] The classic Windaus-type approach (Figure 4)[10a] starts from cheap steroid precursors and imitates the biosynthetic route shown in Figure 1. However, problems associated with introduction of the 1α-hydroxy group and the 5,7-diene system and the photochemistry step have made this approach inefficient.

**Figure 4**

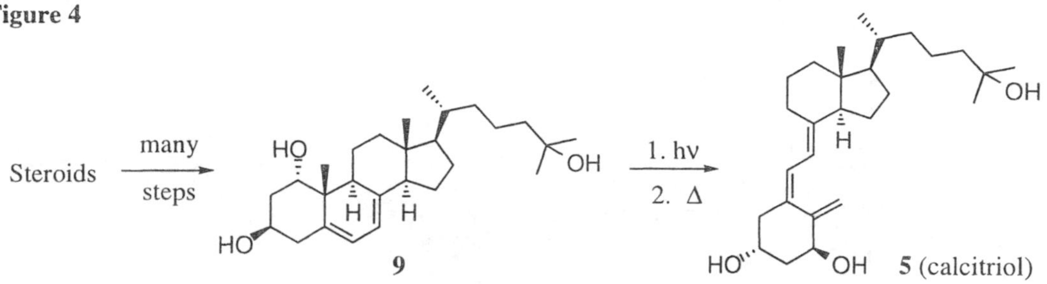

Steroids → many steps → 9 → (1. hv, 2. Δ) → 5 (calcitriol)

A second type of approach is based on coupling of the A-ring and the CD-ring by forming either C5-C6, C6-C7 or C7-C8 bond.[10] Among them the method first developed by Lythgoe,[11] which utilizes the phosphine oxide coupling (Figure 5), has been used extensively, primarily by medicinal chemists. Although quite effective in producing a variety of vitamin D analogs, this approach presents a serious problem for process research chemists--the length. For example, the synthesis of the A-ring **10** alone requires more than 10 steps, rendering this approach undesirable for the large scale (>1 kg) production of calcitriol (**5**).

**Figure 5**

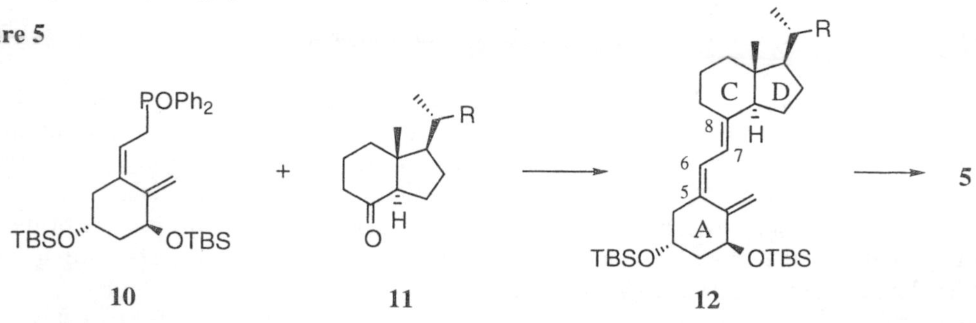

10          11          12          5

The other approach, which we chose to investigate, was first developed by Barton, Hesse, and their collaborators[12] and starts from vitamin D$_2$ (**1**).  The key features of this approach are the regioselective ozonolysis of the side chain to install the 25-hydroxylated side chain and the regio- and stereoselective hydroxylation of 'trans-vitamin D' (5E,7E isomer, such as **16**) to introduce the requisite 1α-hydroxy group.[13]  Our synthesis of **5** basically follows this paragon except for the method of introducing the side chain as shown in Figure 6.  Instead of treating the C22-tosylate with a cuprate reagent prepared from 3-hydroxy-3-methylbutyl bromide,[12] we utilized a nickel-mediated conjugate addition of the C22-iodide to ethyl acrylate.[6]  Using this method we have prepared 100 g quantities of calcitriol (**5**) from vitamin D$_2$ (**1**) in greater than 10% overall yield.  Since the starting material, vitamin D$_2$ is relatively expensive and the product, calcitriol (**5**), is highly potent and valued, several chromatographic purifications were carried out in order to maximize yields in several steps (e.g., after hydroxylation of C1 and photoisomerization).

**Figure 6**

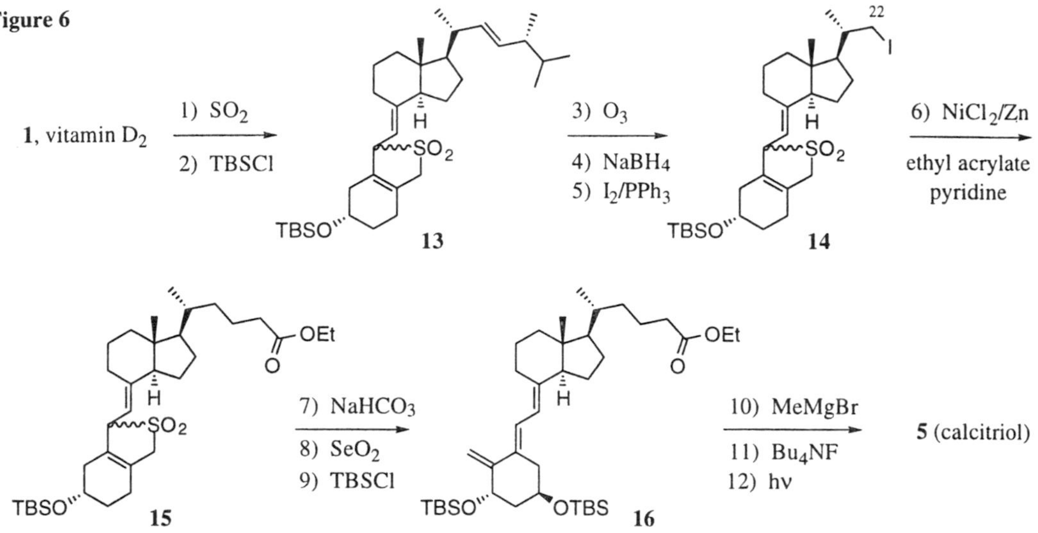

Although this calcitriol preparation[6] was repeated several times without any incident, thereby successfully completing (we thought) our involvement in this project, a new, unspecified impurity (*ca.* 0.2%) unexpectedly arose in the first two batches of calcitriol (**5**) prepared in our production group.  To make matters worse, the final recrystallization procedure used to purify calcitriol was unable to remove this impurity; with only 70% recovery, the recrystallized material still contained 0.15% of the impurity.  Thus, careful HPLC purification of the final product was subsequently carried out to give calcitriol of acceptable quality.

## New calcitriol impurity: What is it? How did it form?

The impurity was analyzed by LC-MS and its molecular weight was found to be 430, 14 mass units higher than that of calcitriol (**5**).  Since it is less polar than **5**, the impurity was assumed to be the 25-*ethyl* analog **19** and this was subsequently confirmed unequivocally by the independent synthesis shown in Figure 7.  The product **19** was obtained as a crystalline material, the melting point of which was found *ca.* 25 °C higher than that of calcitriol (**5**).  It was because of the high melting point of the impurity that the recrystallization procedure previously used to purify calcitriol was rendered ineffective (Murphy's law).

**Figure 7**

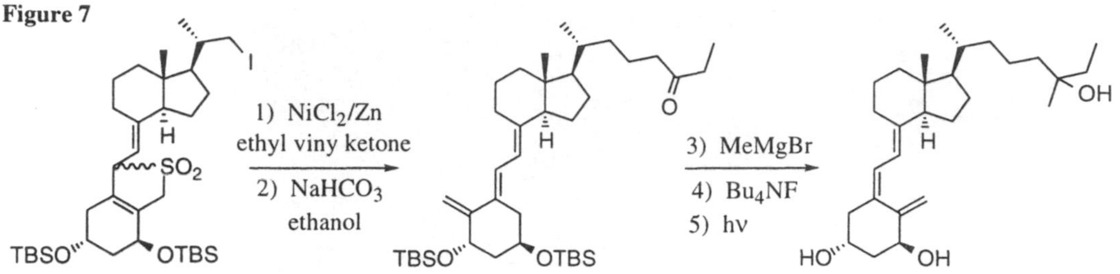

There were two steps where an ethyl group could have been introduced: the nickel-mediated conjugate addition of the C22-iodide to ethyl acrylate and the Grignard reaction. The first scenario, that the product of the nickel reaction **16** could have been contaminated with a low level of the ethyl ketone **18** due to the contamination of the ethyl acrylate with ethyl vinyl ketone, was quickly discarded since no contamination of **18** was found in any retain sample of **16** and ethyl vinyl ketone was not detected in the ethyl acrylate. Thus, the bottle of the Grignard reagent used became the focus of our investigation.

The particular bottle of the Grignard reagent, 3$M$ methylmagnesium bromide in diethyl ether, was from a well-known commercial source and had been in storage at room temperature for more than two months after the reagent was used for the preparation of calcitriol by our production group. Some white sediment, which is not unusual, was evident on the bottom at the time of our investigation. After repeating the Grignard reaction using this bottle of the Grignard reagent and a retain sample of **16**, the silyl groups were removed using tetrabutylammonium fluoride to give trans-calcitriol **20**, which was analyzed for the ethylated impurity **21** (Figure 8). It should be noted that in the production runs in question there were about 0.3% of this impurity present. To our surprise no ethylated impurity **21** was detected in the sample of **20** produced in this experiment! This indicates that the species responsible for the formation of the ethylated impurity may have disappeared or lost its activity after the reagent bottle was opened and stored at room temperature for several months.

**Figure 8**

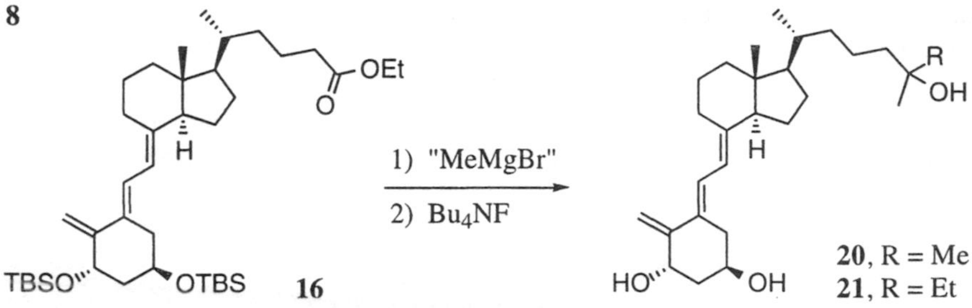

Then methyl bromide, which could be present in the methylmagnesium bromide solution but later would disappear by either evaporation (bp 4 °C) or slow reaction with the Grignard reagent, became a leading suspect. Methyl bromide, if present, would react with the enolate **25** formed in a low level via mono-methylation and enolization of the resulting methyl ketone **23** (Figure 9). Reaction of the ethyl ketone **26** thus formed with methylmagnesium bromide would then furnish the ethylated impurity **27**. To confirm this hypothesis, the Grignard reaction was carried out in the presence of *ca.* 10 equivalents of methyl bromide.

**Figure 9**

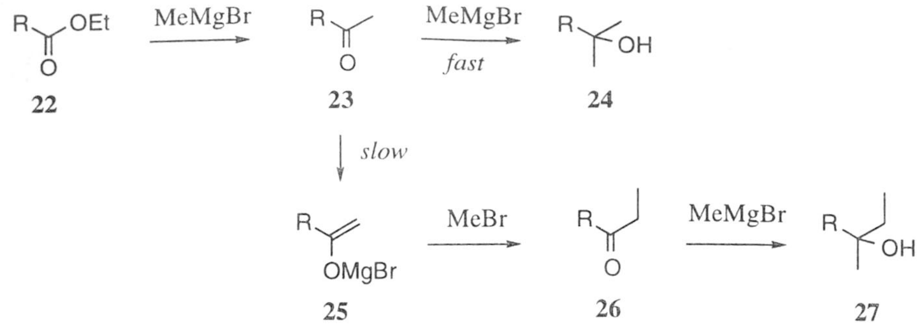

As expected, formation of the ethylated impurity **21** was observed, however, in a rather low level (0.18%) and, with a smaller excess of methyl bromide, a lower amount of the impurity formed. Thus, although the presence of methyl bromide can produce the ethyl impurity, the formation of the impurity to the level (0.3%) observed in the tainted production runs was never attained by adding methyl bromide (up to 10 equivalents) to the Grignard reagent. In addition, it is unlikely that there would ever be such a large quantity of methyl bromide present in commercial Grignard solutions.

An indirect GC method for the evaluation of methylmagnesium bromide solutions was then developed whereby the reaction products (**23**, **24** and **27**, R = *t*-Bu) of the Grignard reagent with ethyl trimethylacetate (**22**, R = *t*-Bu) were analyzed. Using this method, several different lots of unopened bottles of the Grignard reagents from commercial sources were tested. To our surprise, reaction of the majority of the reagent samples resulted in the formation of the ethylated impurity **27** (R = *t*-Bu) albeit in very low levels (*ca.* 0.05%). Having these results in hand, correlation of the methyl bromide level and the amount of the ethylated product **27** formed was investigated. Surprisingly, in this case, addition of methyl bromide did not increase the formation of **27**. Presumably, when R is a bulky *tert*-butyl group, contrast to the ester **16** in Figure 8, the steric effect of the additional two methyl groups sufficiently retarded the alkylation of the corresponding enol **25**, thereby preventing the formation of the ethylated impurity **27** through this mechanism (Figure 9). Nevertheless, the important observation was that *a small amount of the ethylated impurity can form regardless of the amount of methyl bromide.*

The bottle of the Grignard reagent which gave 0.05% of **27** (R = *t*-Bu) was then stored at room temperature for three months and reanalyzed by the GC method. At this point only 0.02% of **27** was detected. Thus, as observed before, the species responsible for the formation of the ethylated impurity slowly loses its activity after the reagent bottle is opened and stored at room temperature.

Could there be ethylmagnesium bromide in the bottle of methylmagnesium bromide? Probably not. If it were ethylmagnesium bromide, it must be at least 1,000 times more unstable than methylmagnesium bromide to selectively decompose in the Grignard solution, and we know that is not the case. Then, what could it be? Unfortunately, we do not know, but we speculate that another metallic species is present in the Grignard reagent since magnesium metal used in the Grignard reagent preparation can not be 100% pure and there are several examples in the literature that impurities in metals are the actual active species in organometallic reactions. Presumably, the Grignard reagent used in the tainted runs was produced from magnesium metal which contained a higher level than usual of the unidentified active metallic species, thus forming *ca.* 0.2% of the ethylated impurity **19** in the presence of suitable amounts of MeBr. In any event, the GC method using ethyl trimethylacetate (**22**, R = *t*-Bu) should be useful in predicting the amount of the ethylated impurity that would be produced in the calcitriol synthesis.

That impurities arise late in the drug development process and even after a product has been launched is not an uncommon occurrence. It is important, however, to react quickly to determine the structure of the impurity, prepare an authentic sample to confirm the structural assignment, determine where and how it is formed, develop a method for preventing its occurrence in the future or to remove it, and, if necessary, qualify the impurity with an appropriate toxicology study.

## Synthesis of Ro 23-8525 (6)

Among a large number of vitamin $D_3$ analogs prepared by the Uskoković group at Roche for a variety of therapeutic indications, (22E,25R)-1α,25,26-trihydroxy-$\Delta^{22}$-vitamin $D_3$ (**6**, Ro 23-8525) was identified as a potential antiosteoporotic agent.[14] The original synthesis of this compound is based on the Lythgoe phosphine oxide approach (Figure 5) and the requisite CD-piece **32** was prepared via a diastereoselective 1,3-dipolar cycloaddition of the nitrone **28** with methyl methacrylate, which gave the desired (R,R)-isomer **30** as a minor product (Figure 10).[15] Since the title compound **6** has similar structural features to calcitriol (**5**), our synthesis of **6** again followed the Barton-Hesse strategy,[12] in which the C22-aldehyde **37** was readily prepared from vitamin $D_2$ using the procedure described by Calverley with some modifications.[6,16] A method for constructing the requisite (22E,25R)-25,26-dihydroxy-$\Delta^{22}$ side chain has been described, which employs a chiral sulfone derived from (R)-citramalic acid.[17] This Julia olefination approach,[18] which has been utilized for the preparation of a variety of vitamin D derivatives,[16b,17,19] requires a multi step synthesis of the sulfone precursor and a reductive elimination step using sodium amalgam to convert the hydroxysulfone intermediate to the desired olefin. Since convergency of process is important, we sought a more direct approach, in which the whole requisite side chain is constructed on the C22-aldehyde **37** in one step (Figure 11).

**Figure 10**

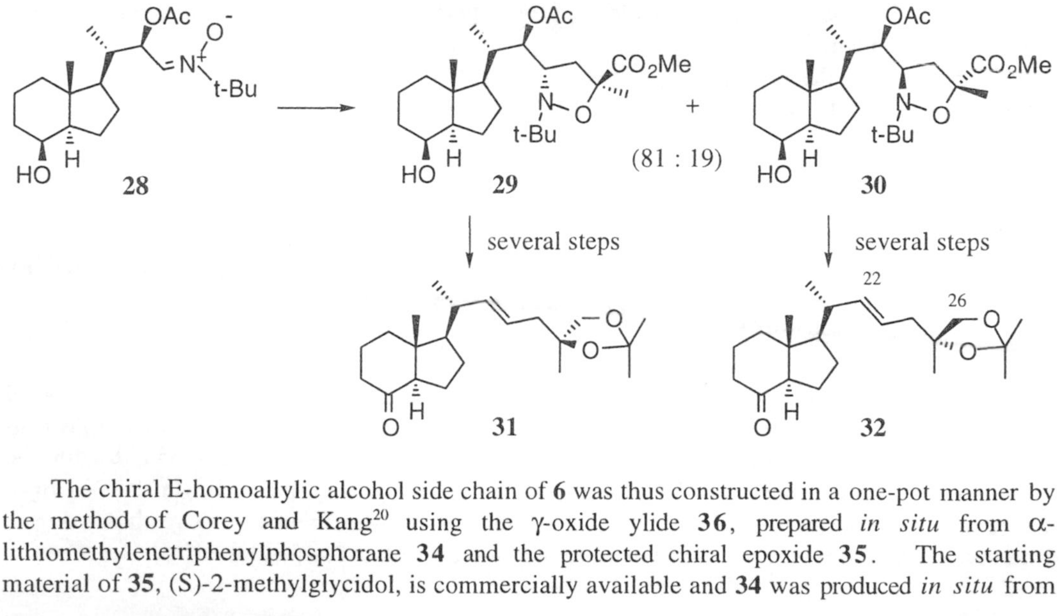

The chiral E-homoallylic alcohol side chain of **6** was thus constructed in a one-pot manner by the method of Corey and Kang[20] using the γ-oxide ylide **36**, prepared *in situ* from α-lithiomethylenetriphenylphosphorane **34** and the protected chiral epoxide **35**. The starting material of **35**, (S)-2-methylglycidol, is commercially available and **34** was produced *in situ* from the phosphonium salt **33** by treating with 2 equivalents of *sec*-butyllithium in anhydrous ether. The product **38**, thus obtained in 71% yield from the aldehyde **37**, was desilylated and photoisomerized using 9-acetylanthracene as triplet sensitizer to give the title compound **6** (Ro 23-8525) in 73% yield.[7]

**Figure 11**

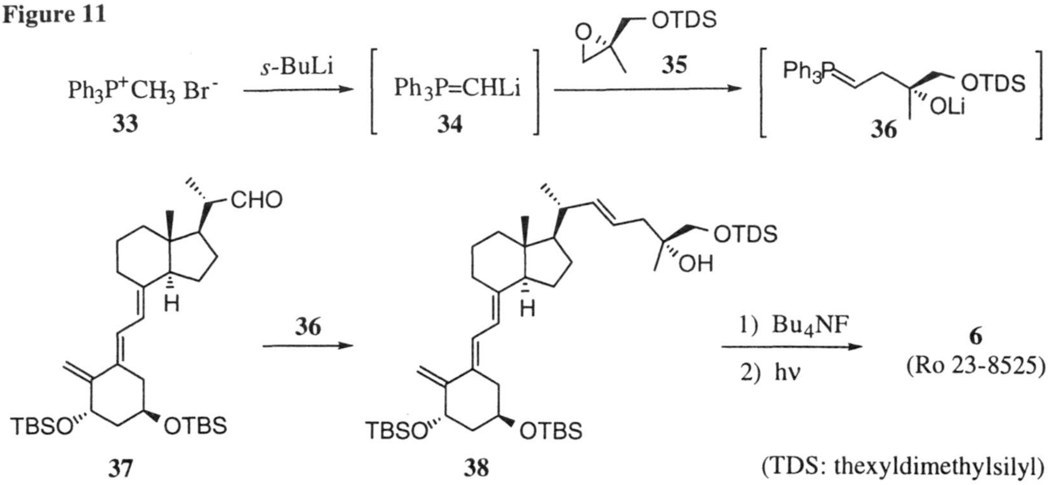

(TDS: thexyldimethylsilyl)

It should be noted that alternative methods involving γ-oxide ylides have been utilized in the preparation of vitamin D derivatives. Salmond reported the reaction of methylene-triphenylphosphorane with isobutylene oxide and subsequent treatment with butyllithium to form the corresponding γ-oxide ylide.[21] However, due to the low reactivity of the methylene ylide toward epoxides, yields of the desired products are often low, and the corresponding methylene compounds are produced as byproducts.[12,20] Others started with the corresponding phosphonium salt having γ-hydroxy group prepared through a series of steps.[22] Thus, the new strategy described here employing the Corey-Kim olefination for the synthesis of vitamin D analogs possessing both C22-23 trans double bond and 25-hydroxy group clearly offers several advantages over the commonly used methods.

Unfortunately, after preparing *ca.* 10 g of Ro 23-8525 (**6**) using this method, which was sufficient to carry out 4-week toxicology studies in two species, the project was terminated.

## Syntheses of the $\Delta^{16}$-Vitamin D Analogs 7 and 8

The advances in understanding the two different mechanisms of calcitriol (**5**) biological activity prompted vitamin D researchers to search for analogs capable of regulating cell proliferation and differentiation with relatively low calcemic effects.[4] A group of analogs with the 16,17-double bond ($\Delta^{16}$) prepared by the Uskoković group at Roche were found to show pronounced separation of these effects.[9] As a consequence, two $\Delta^{16}$-vitamin D analogs were selected for clinical development for two different therapeutic indications; 25-hydroxy-16-en-23-yne-vitamin $D_3$ (Ro 24-2090, **7**) for psoriasis[23] and 1,25-dihydroxy-16-en-23-yne-vitamin $D_3$ (Ro 23-7553, **8**) for the treatment of certain cancers.[24] These compounds were originally prepared via the Lythgoe phosphine oxide approach (Figure 5), in which the requisite CD-fragment **44** was synthesized from 2-methyl-1,3-cyclopentanedione (**39**) in 15 steps (Figure 12); three steps to the Hajos ketone **40**,[25] then seven steps to the Z-olefin **41**,[26] the ene-reaction with the acetylenic aldehyde **42** to give **43**, followed by four more steps.[9] It should be noted that the preparation of the A-ring precursors also required many steps. Thus, this approach was unattractive for the large scale production (>100 g) of these compounds.

**Figure 12**

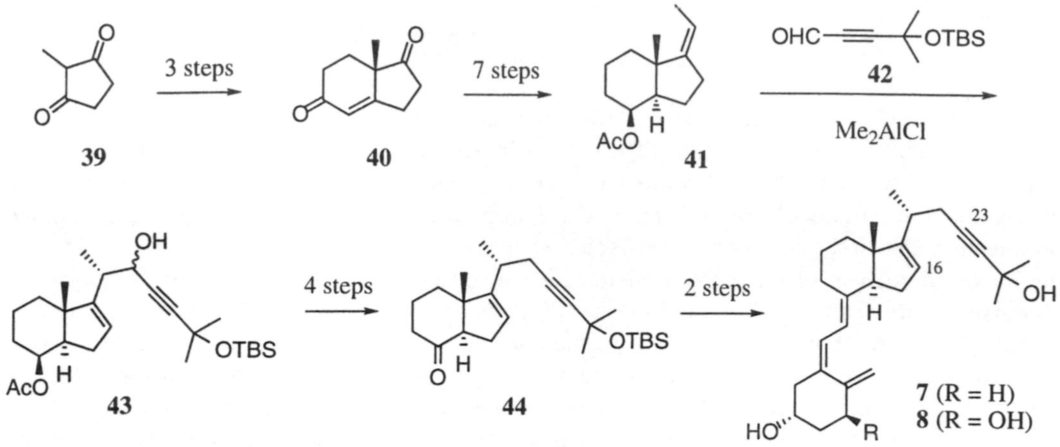

The Barton-Hesse strategy, quite successfully applied to the synthesis of calcitriol (**5**) and Ro 23-8525 (**6**), is also not attractive for the syntheses of these $\Delta^{16}$-vitamin D analogs **7** and **8**. As shown previously, the key intermediates in this approach, such as the iodides **14** and **17** and the aldehyde **37**, do not have the requisite 16,17-double bond and there is no apparent way to introduce this functionality to these compounds. It may be possible to remove the three carbons from the intermediate to give a 17-keto compound, then to reconstruct the whole side chain, but this adds many steps. Thus, we took the classical Windaus-type approach starting from a commercially available steroid, dehydroepiandrosterone (**45**) (Figure 13). For the synthesis of Ro 23-7553 (**8**), which requires a hydroxy group at the C1 position, **45** was subjected to fermentation to give 1α-hydroxy-dehydroepiandrosterone (**46**)[27] and confirmed as described in the next section.

**Figure 13**

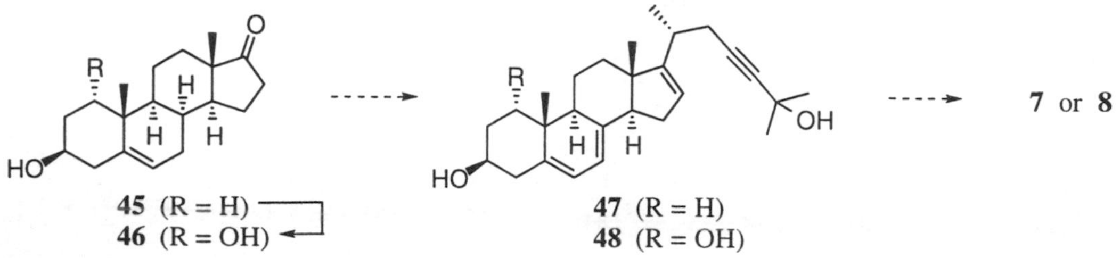

**Synthesis of Ro 24-2090 (7)**[8]

First, the requisite 5,7-diene system was introduced by a method reported by Rappoldt (Figure 14).[28] Thus, **45** was acetylated to give **49** in a quantitative yield. Then, allylic bromination was performed using 1,3-dibromo-5,5-dimethylhydantoin in cyclohexane at reflux. After cooling, the

**Figure 14**

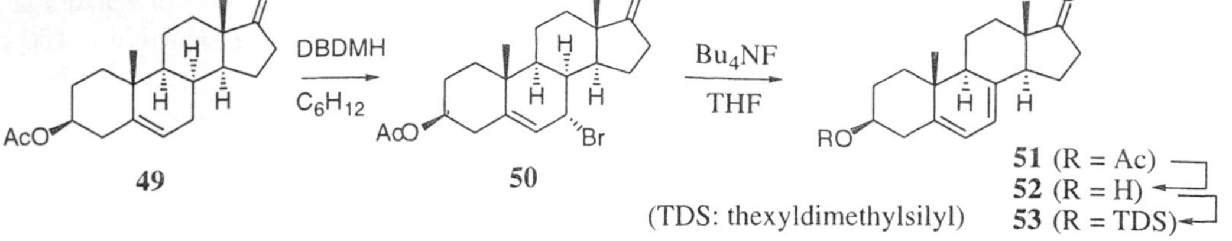

desired 7α-bromide **50** precipitated out from the reaction mixture, then collected by filtration together with 5,5-dimethylhydantoin. The latter was removed by washing with water. The crude bromide thus obtained was dehydrobrominated by treatment with a 1*M* solution of tetra-butylammonium fluoride[28] in THF to give the diene-acetate **51**. It should be noted that the bromination-dehydrobromination is not regiospecific and, although the desired 5,7-diene **51** was the major component of the product mixture, it was quite difficult to purify even by silica gel chromatography. However, after hydrolysis and silylation, pure 5,7-diene **53** was obtained in 44% overall yield from **45** without chromatographic purification. This was possible because steroidal materials tend to precipitate easily, thus allowing easy purification of intermediates. This is clearly one of the advantages of the Windaus-type approach. A more direct approach to **53**, i.e., silylation of **45** followed by bromination-dehydrobromination,[29] was not fruitful as it produced impure **53**, which could not be purified even by silica gel chromatography.

**Figure 15**

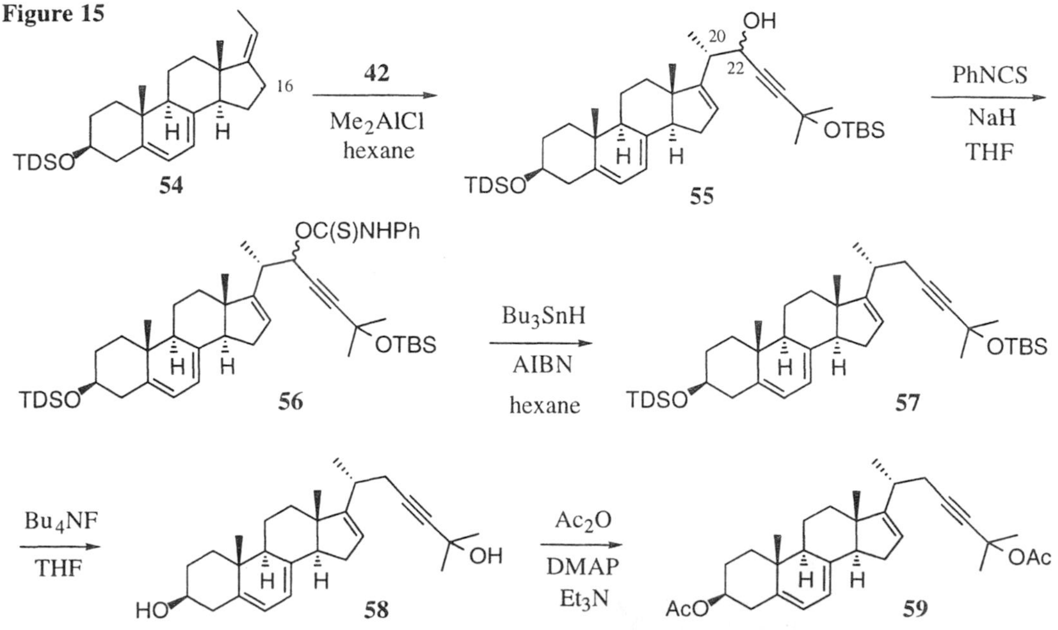

Having established the diene function, the next problem was to construct the requisite side chain with the desired configuration at C20. Granja[30] successfully introduced the propargylic side chain with a hydroxy group at C22 by [2,3]-Wittig rearrangement methodology. However, this approach was precluded for the preparation of Ro 24-2090 (**7**) as it requires a hydroxy group at C16 as a handle. On the other hand, the procedure described by Uskoković et al,[9] in which the ene-reaction of the Z-olefin **41** with acetylenic aldehyde **42** in the presence of $Me_2AlCl$,[31] followed by Barton deoxygenation[32] provided the complete side chain (Figure 12), is quite attractive. However, it was uncertain whether the cis-diene of **53** would tolerate these reaction conditions. Wittig reaction of **53** with ethylidenetriphenylphosphorane gave the requisite Z-olefin **54** in 93.7% yield with high stereoselectivity.[33] The ene-reaction proceeded well with complete stereocontrol at C20 but with moderate selectivity at C22, giving a 5:1 mixture of the C22 epimeric alcohols **55** (Figure 15). This is, however, of no consequence since the hydroxy group is removed in the next two steps to produce the deoxygenated product **57**. Thus, the epimers were not separated but subjected as a mixture to Barton deoxygenation[32]. The phenoxythiocarbonyl and imidazolylcarbonyl derivative worked well for this transformation, despite the presence of the diene function which is potentially sensitive under such conditions.[34] However, the reagents required, phenyl chlorothionoformate and (thiocarbonyl)diimidazole, are too expensive for large scale application. Thus, alternative derivatives that can be prepared using relatively cheap reagents, including the corresponding xanthate and dimethyl thiocarbamate, have been studied. To our pleasant surprise, the thionocarbamate **56** prepared from phenyl isothiocyanate, the cheapest reagent we studied, was found to be equally efficient as more expensive derivatives, such as thiocarbonyl imidazole, to give the deoxygenation product **57** after treatment with 3.5 equivalents of $Bu_3SnH$ in the presence of 0.5 equivalents of AIBN. This relatively large amount of AIBN was necessary for smooth reduction of the carbamate in hexane at reflux. At a higher temperature, the thiocarbamate rearranged to give byproducts. This new deoxygenation procedure, however, may be limited to the system in which relatively stable radicals are produced, since it was reported that 2-decyl thionocarbamates prepared from 2-decanol and isothiocyanates, when treated with $Bu_3SnH$, gave decane only in 7% yield together with 2-decanol (49%).[35] The crude product **57**, after washing with methanol to remove excess $Bu_3SnH$, was deprotected. Diol **58** (ca. 200 g) was thus obtained as a white solid in 47.5% overall yield from the Z-olefin **54**. It should be noted that no chromatographic purification was required to produce **58** from the starting material, 3-dehydroepiandrosterone (**45**). Alcohol **58** was then converted to acetate **59** in 89.8% yield.

As mentioned previously, the key step of the Windaus-type approach is photolytic opening of the diene ring of provitamins (such as **58**) to previtamins (such as **61**), which are then thermally converted to vitamin D derivatives (such as **63**) via a formal [1,7]-sigmatropic hydrogen shift (Figure 16). The overall yields of vitamin D derivatives via the photolyses of provitamins are, in general, only 10-30% (after HPLC purification, typically on a few milligram scale).[36] Difficulties in obtaining higher yields arise from the fact that the photolysis of provitamin gives a photostationary state wherein the distribution of the products (pro-, pre-, lumi, and tachy-isomers, **58, 61, 60** and **62** in Figure 16, respectively) depends on the photolyzing wavelength of light;[37] a shorter wavelength (<270 nm) leads to a predominance of tachy-isomer and a longer wavelength (>305 nm), on the other hand, promotes the ring-closure reaction of previtamin to form pro- and lumi-isomers. Thus, when photolysis is carried out in an optically inert solvent, such as ether, ethanol or THF, using a mercury lamp with a quartz immersion well, tachy-isomer (such as **62**) is the major product and the yield of vitamin D derivative (such as **63**), after thermal isomerization of pre-isomer (such as **61**), is often less than 15%.[17b,29,38] When benzene is used as a solvent, the yield seems to be generally higher (15-40%),[17a,19b-d,34] presumably because benzene acts as a filter to <280 nm light which leads to the tachy-isomer as the major component of the photostationary state (Figure 16).[37] However, photolysis at >290 nm light has a drawback, as 305-320 nm light promotes the ring closure of previtamin to form pro- and lumi-isomers as the major components.[37] Higher yield of previtamin has been achieved by irradiating provitamin with a narrow band of ca. 250 nm light (using a low pressure mercury lamp or a laser) and then selective isomerization of

**Figure 16**

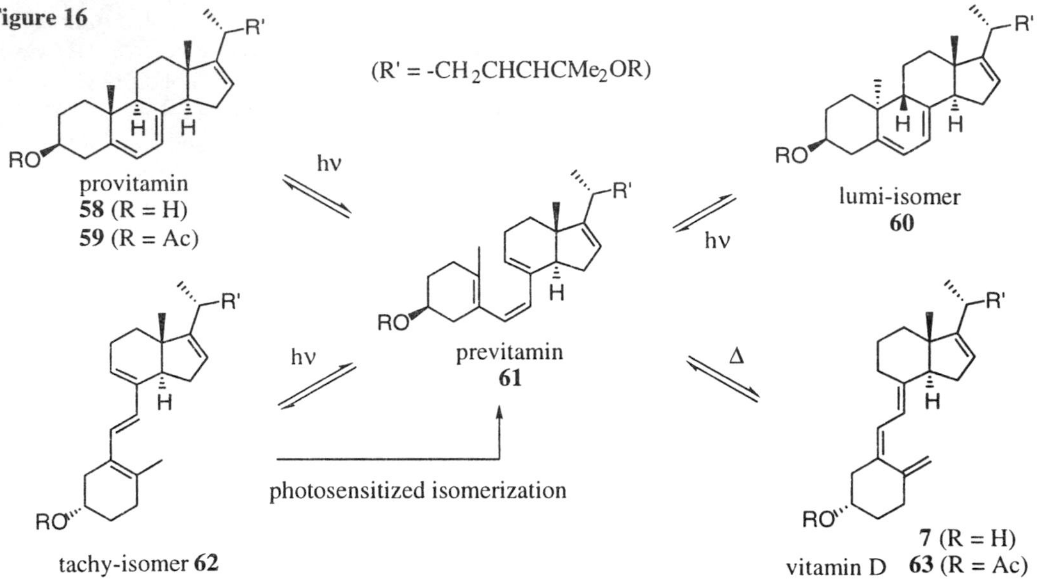

$(R' = -CH_2CHCHCMe_2OR)$

provitamin
**58** (R = H)
**59** (R = Ac)

lumi-isomer
**60**

previtamin
**61**

photosensitized isomerization

tachy-isomer **62**

**7** (R = H)
vitamin D    **63** (R = Ac)

the thus formed tachy-isomer into previtamin by irradiation at ca. 350 nm[37c,d] or by photosensitized isomerization.[39] This approach is unattractive for scale up due to the specialized equipment set up and cost of light sources that have narrow spectral widths and high intensity.

In order to achieve a practical synthesis which eventually can be used to produce several hundreds of grams of the vitamin D analog, we decided to use a medium pressure mercury lamp, a cheap, high-intensity light-source which is commonly used in a synthetic laboratory. Since this lamp emits a wide spectral range of ultra-violet light, it is essential to filter out the 305-320 nm light, which promotes the ring-closure reaction of the previtamin, in order to increase the efficiency of the photolysis process. For this purpose, a small amount of ethyl 4-dimethylaminobenzoate (**64**) was added to the photolyzing mixture. The benzoate **64** has a strong, relatively sharp absorption at 305 nm (ε 32,500), thus effectively filtering out 305-320 nm light, without cutting off 240-270 nm and above 340 nm of light, so the ring opening of the provitamin **58** to the previtamin **61** and photosensitized isomerization of the tachy-isomer **62** to the previtamin **61** can be carried out efficiently. This new protocol worked very well in the conversion of **58** to give, after photosensitized isomerization, the previtamin **61** as the major component of the photolyzed product mixture, which was then subjected to thermal isomerization to the title vitamin D derivative **7**. It should be noted that the previtamin **61** and the vitamin D derivative **7** are in thermal equilibrium in which the ratio is dependent on the temperature.[40] At room temperature, the desired vitamin form is predominant in the thermal equilibrium (>90%). Thus, after photostationary state was established between **58**, **60**, **61** and **62** (in which **61** and **62** are the major component), photosensitized isomerization set a new ratio between **61** and **62** (in which **61** is predominant), then thermal isomerization established another equilibrium between **7** and **61**. Although the desired product **7** was the major component of the product mixture, it was quite difficult to purify even by HPLC as the mixture contained **58** (ca. 15%), **60** (trace), **61** (ca. 10%) and **62** (trace), which are all isomeric to **7**. To make matters worse, the title compound **7** had not been crystallized before, thus no way existed to purify the final product to a satisfactory level. After purification by flash chromatography, **7** was obtained as a foam in 46% yield with a purity of 97%. However, all attempts to crystallize **7** not only failed but also gave less pure material as the conjugated triene system is relatively easily oxidized by air. Thus, we decided to find a crystalline derivative of **7** and, subsequently, the corresponding diacetate **63** was found to be readily crystallized from alcoholic solvents (mp 97-100 °C) to give a stable derivative of the parent compound **7**.

**Figure 17**

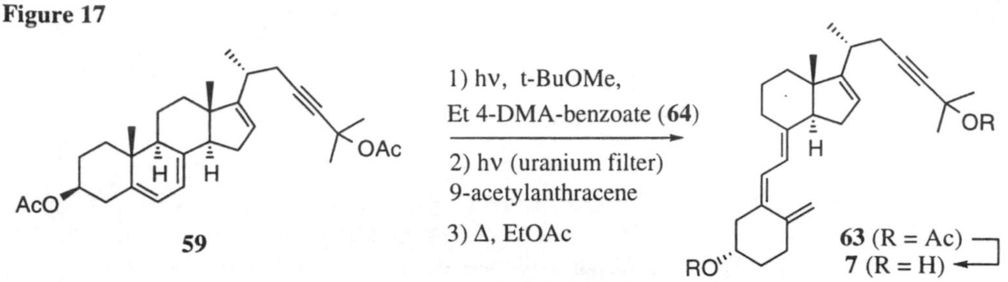

1) hv, t-BuOMe,
Et 4-DMA-benzoate (**64**)
————————————————→
2) hv (uranium filter)
9-acetylanthracene

3) Δ, EtOAc

**59**

**63** (R = Ac)
**7** (R = H)

Consequently, a mixture of **59** (16.4 g) and **64** (1.64 g) in tert-butyl methyl ether (1.7 L) was irradiated for 8 h with a 450 W medium pressure mercury lamp to give a near photostationary mixture (the ratio of **59**, **61** and **62** was about 1:3:2 with a trace amount of **60**). A uranium filter was then inserted to cut off <340 nm light, and the mixture was irradiated with the same lamp in the presence of 9-acetylanthracene as a sensitizer[6-8] to isomerize **62** to **61** (the ratio of **59**, **61** and **62** became about 1:5:<0.1). After a simple chromatography to remove the sensitizer and polar byproducts and then thermal isomerization of **61**,[40] *ca.* 95% pure diacetate **63** (together with *ca.* 5% of **59**) was obtained in 46.7 % yield as a crystalline solid. Fractional crystallization of this crude product gave analytically pure **63** in 38.8% overall yield from **59** (the yield is not corrected for recovered starting material). Hydrolysis of **63** gave **7**, for the first time, as a crystalline solid in 90.9% yield. Subsequent crystallizations required, however, the addition of seed crystals. Crystalline **7** obtained in this study showed unusual chemical properties; the material underwent a facile, autocatalytic oxidative decomposition in the solid state. Apparently, initial degradation products induce an acceleration in the decomposition rate until the crystallinity is completely lost. Hence, as soon as the crystalline material was isolated and dried *in vacuo*, it was immediately dissolved in an appropriate solvent to generate a relatively stable form of the compound.

## Synthesis of Ro 23-7553 (8)

Having completed the synthesis of the closely related analog, Ro 24-2090 (**7**), the preparation of the title compound **8** which possess an additional hydroxy group at C1 was expected to be relatively easy. Thus, the crude fermentation product **46**[27] was acetylated and crystallized to give pure diacetate **65** (Figure 18). Bromination-dehydrobromination[28] of **65** proceeded rather poorly compared with that of **49**, which does not have the 1α-hydroxy function, and **68** was obtained, after hydrolysis and silylation, in only 29% overall yield from **65**, whereas **49** was converted to the corresponding silyl ether **53** in 44%. Since the hindered axial hydroxy group at C1 of **67** did not react with thexyldimethylsilyl chloride, it was protected as a triethylsilyl ether. This allowed the selective deprotection of the latter group later in the synthesis. The requisite side chain was then introduced in 4 steps by the same method described for the preparation of **7** and **69** was obtained in a comparable yield of 47% from **68**.

**Figure 18**

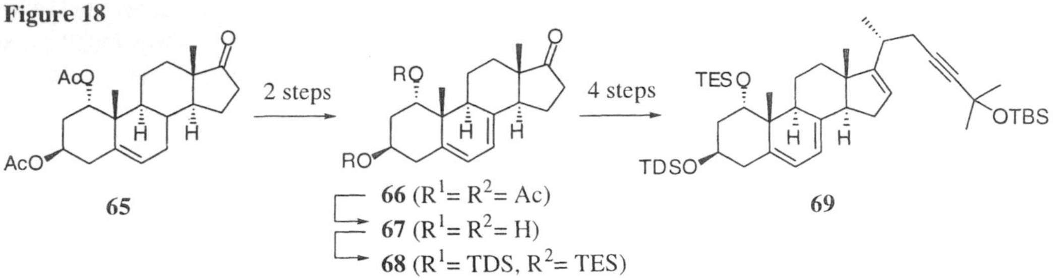

    65          66 (R¹= R²= Ac)                        69
                67 (R¹= R²= H)
                68 (R¹= TDS, R²= TES)

Unexpectedly, the quantum yield of photolytic opening of the ring B diene of **69** was found to be much smaller than that of **58** and **59**, which lack the bulky triethylsilyloxy group at C1. Consequently, extended photolysis was required which led to an increase in the formation of byproducts. In order to clarify this effect, several 1α-trialkylsilyloxy-provitamin D analogs were prepared and the quantum yields of the photochemical ring opening were compared with that of the parent 1α-hydroxy compound. It was found, subsequently, that the relative quantum yield decreases with an increase in the size of the trialkylsilyl protecting group on the 1α-hydroxy group;[41] i.e., conversion of the parent 1α-hydroxy compound to ring-opened photoproducts (pre- and tachy-isomers) is more efficient than that of the 1α-trialkylsilyloxy derivatives. The corresponding triol of **69** was, therefore, subjected to photolysis. Although the reaction proceeded well and the desired product **8** was the major component of the product mixture, it was difficult to purify as in the case of the photolysis of **58** mentioned earlier. Therefore, the partially protected provitamin **70** was employed as the substrate for the photolysis (Figure 19).

**Figure 19**

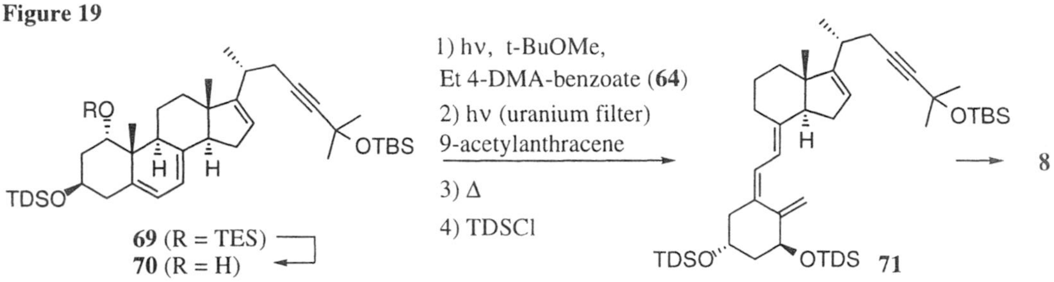

1) hv, t-BuOMe, Et 4-DMA-benzoate (**64**)
2) hv (uranium filter) 9-acetylanthracene
3) Δ
4) TDSCl

69 (R = TES)
70 (R = H)

The triethylsilyl group of **69** was selectively removed by using one equivalent of tetrabutylammonium fluoride to give the corresponding 1α-hydroxy derivative **70**. To a small extent the deprotection of the thexyldimethylsilyl group at C3 also occurred under the conditions, but this has no consequence since the resulting free hydroxy group is easily re-protected using thexyldimethylsilyl chloride in a later step. The crude product **70** obtained after extractive workup was then subjected to the photolysis using the same protocol used for the synthesis of **7** (Figure 20). After thermal isomerization of **73** to **75**, the resulting crude product mixture composed of **70** (*ca.* 15%), **72** (trace), **73** (*ca.* 10%), **74** (trace) and **75** was treated with thexyldimethylsilyl chloride and imidazole in dichloromethane to give the fully protected vitamin D derivative **71**. Since the C1 hydroxy group of the provitamin **70** or lumi-isomer **72** is sterically too hindered to react with this bulky silylation reagent, these isomers were left unprotected, thereby allowing easy removal by silica gel chromatography. It should be noted that provitamins are usually the most difficult ones to remove completely; even if only a few percentage of this compound is present in a recrystallizing mixture, the provitamin tends to co-precipitate with the desired vitamin D derivative. Finally, deprotection of **71** followed by crystallization of the product gave the title compound **8**.

**Figure 20**

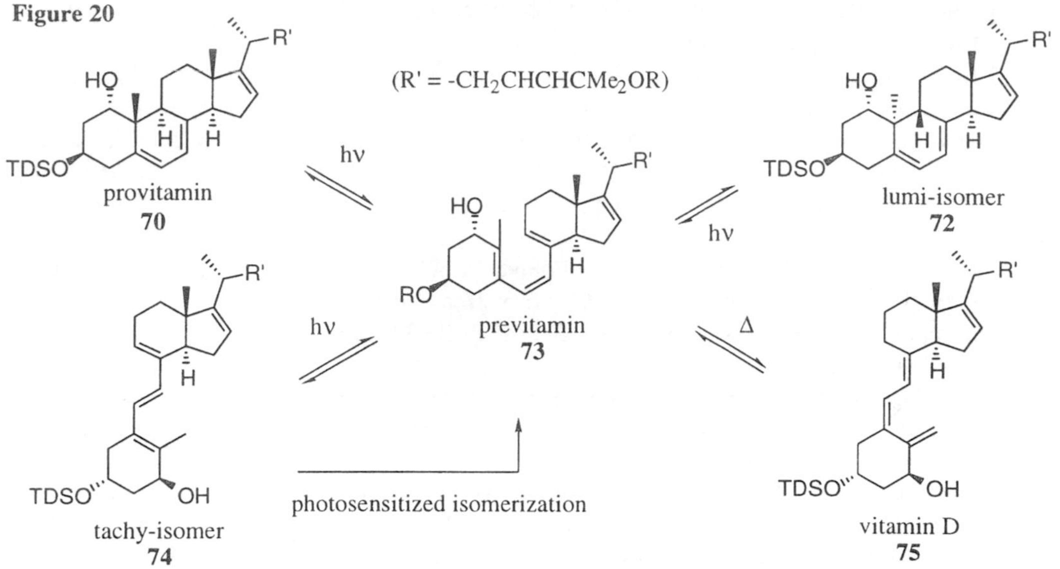

(R' = -CH$_2$CHCHCMe$_2$OR)

provitamin
**70**

lumi-isomer
**72**

previtamin
**73**

tachy-isomer
**74**

photosensitized isomerization

vitamin D
**75**

## Concluding Remarks

Herein, we describe our process research efforts in the synthesis of four vitamin D derivatives, which have been chosen as development candidates at Roche over the past ten years. For side chain modifications, the Barton-Hesse strategy[12] worked well, thus calcitriol (**5**) and Ro 23-8525 (**6**) were prepared based on this paragon. For the $\Delta^{16}$ modification, the Windaus-type approach[10a] was employed and it was quite successful in the synthesis of Ro 24-2090 (**7**), but clearly showed deficiency in the preparation of Ro 23-7553 (**8**), which possesses both the $\Delta^{16}$ modification and a 1α-hydroxy group.

The past several years have witnessed significant new developments in the design[4] and synthesis[10] of vitamin D analogs. As medicinal chemists in search of better therapeutic agents prepare structurally even more complex analogs of vitamin D, it will only become more difficult for process research chemists to come up with efficient syntheses of those drug candidates. Because the additional modifications are often incompatible with the Barton-Hesse and Windaus-type approaches which are based on natural steroids, we may have no choice but to venture into total syntheses of those compounds where more than thirty step synthesis is not uncommon. Combined with increasing interest in vitamin D analogs for a variety of therapeutic indications, undoubtedly, these vitamin D programs will continue to pose significant challenges to process research chemists for many years to come.

## References

1.  Norman, A. W. *Vitamin D: The Calcium Homeostatic Steroid Hormone;* Academic Press: New York, 1979.
2.  Smith, E. L.; Holick, M. F. *Steroids,* **1987,** *49,* 103-131.
3.  DeLuca, H. F. *FASEB J.* **1988,** *2,* 224-236.
4.  a) Bouillon, R.; Okamura, W. H.; Norman, A. W. *Endocrine Rev.* **1995,** *16,* 200-257. b) *Vitamin D, A Pluripotent Steroid Hormone: Structural Studies, Molecular Endocrinology and Clinical Applications;* Norman, A W., Bouillon, R., Thomasset, M., Eds.; Walter de Gruyter: Berlin, 1994. c) *Vitamin D: Gene Regulation, Structure-Function Analysis and Clinical Application;* Norman, A W., Bouillon, R., Thomasset, M., Eds.; Walter de Gruyter: Berlin, 1991.
5.  Okamura, W. H.; Midland, M. M.; Hammond, M. W.; Rahman, N. A.; Dormanen, M. C.; Nemere, I.; Norman, A. W. *J. Steroid Biochem. Molec. Biol.* **1995,** *53,* 603-613.
6.  Manchand, P. S.; Yiannikouros, G. P.; Belica, P. S.; Madan, P. *J. Org. Chem.* **1995,** *60,* 6574-6581.
7.  Okabe, M.; Sun, R.-C. *Tetrahedron Lett.* **1993,** *34,* 6533-6536.
8.  Okabe, M.; Sun, R.-C.; Scalone, M.; Jibilian, C. H.; Hutchings, S. D. *J. Org. Chem.* **1995,** *60,* 767-771.
9.  Uskokovic, M. R.; Baggiolini, E.; Shiuey, S.-J.; Iacobelli, J.; Hennessy, B.; Kiegiel, J.; Danniewski, A. R.; Pizzolato, G.; Courtney, L. F.; Horst, R. L. In *Vitamin D: Gene Regulation, Structure-Function Analysis and Clinical Application;* Norman, A W., Bouillon, R., Thomasset, M., Eds.; Walter de Gruyter: Berlin, 1991; pp 139-145.
10. For reviews, see: a) Zhu, G.-D.; Okamura, W. H. *Chem. Rev.* **1995,** *95,* 1877-1952. b) Dai, H.; Posner, G. H. *Synthesis* **1994,** 1383-1398.
11. Lythgoe, B.; Moran, T. A.; Nambudiry, M. E. N.; Tideswell, J.; Write, P. W. *J. Chem. Soc., Perkin Trans. 1* **1978,** 590-595.
12. Andrews, D. R.; Barton, D. H. R.; Hesse, R. H.; Pechet, M. M. *J. Org. Chem.* **1986,** *51,* 4819-4828.
13. Andrews, D. R.; Barton, D. H. R.; Cheng, K. P.; Finet, J.-P.; Hesse, R. H.; Johnson, G.; Pechet, M. M. *J. Org. Chem.* **1986,** *51,* 1637-1638.
14. a) Drezner, M. K.; Nesbitt, T. *Metabolism* **1990,** *39 (suppl 1),* 18-23. b) Drezner, M. K.; Nesbitt, T.; Quarles, L. D. In *Vitamin D: Gene Regulation, Structure-Function Analysis and Clinical Application;* Norman, A W., Bouillon, R., Thomasset, M., Eds.; Walter de Gruyter: Berlin, 1991; pp 816-822.
15. Wovkulich, P. M.; Barcelos, F.; Batcho, A. D.; Sereno, J. F.; Baggiolini, E. G.; Hennessy, B. M.; Uskokovic, M. R. *Tetrahedron* **1984,** *40,* 2283-2296.
16. a) Calverley, M. J. *Tetrahedron* **1987,** *43,* 4609-4619. b) Choudhry, S. C.; Belica, P. S.; Coffen, D. L.; Focella, A.; Maehr, H.; Manchand, P. S.; Serico, L.; Yang, R. T. *J. Org. Chem.* **1993,** *58,* 1496-1500.
17. a) Yamamoto, K.; Shimizu, M.; Yamada, S.; Iwata, S.; Hoshino, O. *J. Org. Chem.* **1992,** *57,* 33-39. b) Yamada, S.; Nakayama, K.; Takayama, H. *Tetrahedron Lett.* **1981,** *22,* 2591-2594.
18. Julia, M.; Paris, J.-M. *Tetrahedron Lett.* **1973,** 4833-4836.
19. a) Perlman, K. L.; DeLuca, H. F. *Tetrahedron Lett.* **1992,** *33,* 2937-2940. b) Moriarty, R. M.; Kim, J.; Penmasta, R. *Tetrahedron Lett.* **1992,** *33,* 3741-3744. c) Taguchi, T.; Namba, R.; Nakazawa, M.; Nakajima, M.; Nakama, Y.; Kobayashi, Y.; Hara, N.; Ikekawa, N. *Tetrahedron Lett.* **1988,** *29,* 227-230. d) Kutner, A.; Perlman, K. L.; Lago, A.; Sicinski, R. R.; Schnoes, H. K.; DeLuca, H. F. *J. Org. Chem.* **1988,** *53,* 3450-3457.

20. Corey, E. J.; Kang, J. *J. Am. Chem. Soc.* **1982**, *104*, 4724-4725. Also see, Julia, M. *Pure & Appl. Chem.* **1985**, *57*, 763-768.

21. Salmond, W. G.; Barta, M. A.; Havens, J. L. *J. Org. Chem.* **1978**, *43*, 790-792.

22. Hanekamp, J. C.; Rookhuizen, R. B.; van der Heuvel, H. L. A.; Bos, H. J. T.; Brandsma, L. *Tetrahedron Lett.* **1991**, *32*, 5397-5400.

23. Chen, T. C.; Persons, K.; Uskokovic, M. R.; Horst, R. L.; Holick, M. F. *J. Nutr. Biochem.* **1993**, *4*, 49-57.

24. Zhou, J.-Y.; Norman, A. W.; Chen, D.-L.; Sun, G.-W.; Uskokovic, M.; Koeffler, H. P. *Proc. Natl. Acad. Sci. USA* **1990**, *87*, 3929-3932.

25. Hajos, Z. G.; Parrish, D. R. *Org. Synth.* **1985**, *63*, 26-36.

26. Daniewski, A. R.; Kiegiel, J. *J. Org. Chem.* **1988**, *53*, 5534-5535.

27. Dodson, R. M.; Goldkamp, A. H.; Muir, R. D. *J. Am. Chem. Soc.* **1960**, *82*, 4026-4033.

28. Rappoldt, M. P.; Hoogendoorn, J.; Pauli, L. F. In *Vitamin D: Chemical, Biochemical and Clinical Endocrinology of Calcium Metabolism;* Norman, A W., Schaefer, K., Herrath, D., Grigoleit, H.-G., Eds.; Walter de Gruyter: Berlin, 1982; pp 1133-1135.

29. a) Konno, K.; Ojima, K.; Hayashi, T.; Takayama, H. *Chem. Pharm. Bull.* **1992**, *40*, 1120-1124. b) Kubodera, N.; Miyamoto, K.; Akiyama, M.; Matsumoto, M.; Mori, T. *Chem. Pharm. Bull.* **1991**, *39*, 3221-3224.

30. Granja, J. R. *Synth. Commun.* **1991**, *21*, 2033-2038.

31. Mikami, K.; Loh, T.-P.; Nakai, T. *J. Chem. Soc. Chem. Commun.* **1988**, 1430-1431.

32. For reviews, see: a) Ramaiah, M. *Tetrahedron* **1987**, *43*, 3541-3676. b) Hartwig, W. *Tetrahedron* **1983**, *39*, 2609-2645.

33. Batcho, A. D.; Berger, D. E.; Davoust, S. G.; Wovkulich, P. M.; Uskokovic, M. R. *Helv. Chim. Acta* **1981**, *64*, 1682-1687

34. After completion of this study, an additional example of the successful Barton deoxygenation in the presence of the ring B diene was reported. See: a) Yamamoto, K.; Takahashi, J.; Hamano, K.; Yamada, S.; Yamaguchi, K.; DeLuca, H. F. *J. Org. Chem.* **1993**, *58*, 2530-2537. b) Yamamoto, K.; Sun, W. Y.; Ohta, M.; Hamada, K.; DeLuca, H. F.; Yamada, S. *J. Med. Chem.* **1996**, *39*, 2727-2737.

35. Nishiyama, K.; Oba, M. *Tetrahedron Lett.* **1993**, *34*, 3745-3748.

36. For reviews, see: Havinga, E. *Experientia* **1973**, *29*, 1181-1316 and reference 1.

37. a) Dauben, W. G.; Share, P. E.; Ollmann, Jr., R. R.; *J. Am. Chem. Soc.* **1988**, *110*, 2548-2554. b) Dauben, W. G.; Phillips, R. B. *J. Am. Chem. Soc.* **1982**, *104*, 5780-5781. c) Dauben, W. G.; Phillips, R. B. *J. Am. Chem. Soc.* **1982**, *104*, 355-356. d) Malatesta, V.; Willis, C.; Hackett, P. A. *J. Am. Chem. Soc.* **1981**, *103*, 6781-6783. e) Havinga, E.; de Kock, R. J.; Rappoldt, M. R. *Tetrahedron* **1960**, *11*, 276-284.

38. a) Ishida, H.; Shimizu, M.; Yamamoto, K.; Iwasaki, Y.; Yamada, S.; Yamaguchi, K. *J. Org. Chem.* **1995**, *60*, 1828-1833. b) Kobayashi, N.; Hisada, A.; Shimada, K. *J. Chem. Soc. Perkin Trans. I* **1993**, 31-37. c) Hayashi, T.; Ojima, K.; Konno, K.; Manaka, A.; Yamaguchi, K.; Yamada, S.; Takayama, H. *Chem. Pharm. Bull.* **1992**, *40*, 2932-2936. d) Kubodera, N.; Watanabe, H.; Kawanishi, T.; Matsumoto, M. *Chem. Pharm. Bull.* **1992**, *40*, 1494-1499.

39. a) Stevens, R. D. S. US Patent 4,686,023, 1987; *Chem. Abstr.* **1987**, *107*, 237124. b) Eyley, S. C.; Williams, D. H. *J. Chem. Soc. Chem. Commun.* **1975**, 858.

40. a) Okamura, W. H.; Elnagar, H. Y.; Ruther, M.; Dobreff, S. *J. Org. Chem.* **1993**, *58*, 600-610. b) Hanewald, K. H.; Rappoldt, M. P.; Roborgh, J. R. *Recl. Trav. Chim. Pays-Bas.* **1961**, *80*, 1003-1004.

41. Okabe, M.; Sun, R.-C.; Wolff, S. *Tetrahedron Lett.* **1994**, *35*, 2865-2868.

# A Highly Optimized and Concise Large Scale Synthesis of a Purine Bronchodilator[§]

**Hans-Jürgen Federsel[*] and Edib Jakupovic**
*Chemical Process Development Laboratory, Astra Production Chemicals AB, 151 85 Södertälje, Sweden*

Key words: enprofylline; purines; xanthines; asthma; bronchodilation; nitrosation; catalytic transfer hydrogenation; scale up

## Introduction: Pharmacological Concept & Identification of Target Molecule

Bronchial asthma, today an already common but still constantly increasing respiratory tract disease, is characterized by a complex interplay of a number of factors. The most prominent of these is the intermittent constriction of the airways causing an acute phase of severe obstruction to normal breathing, however, also including states of inflammation and other abnormalities. [1,2] Presently, the prevailing medical therapy forsees the usage of selective $\beta_2$-receptor stimulants for the treatment of the former condition, while the latter requires anti-inflammatory agents (e. g. of the steroid class). An additional and complementary possibility is displayed by certain xanthine derivatives (i. e. dioxy purines), notably theophylline, which possess the property of effecting smooth muscle relaxation.[3, 4] Actually, the indirect finding that compounds of this type have a marked asthma curing effect was reported by Salter[5] as long ago as in 1859 (!), when he claimed that coffee (containing multi mg quantities of caffeine per cup) rapidly offers the desired relief.

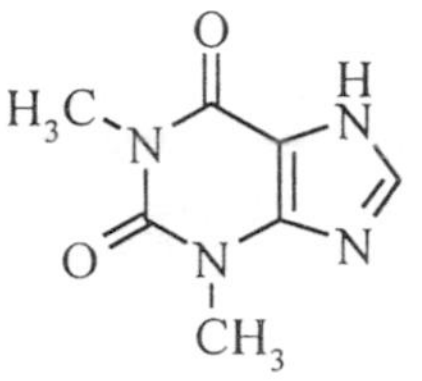

Theofylline

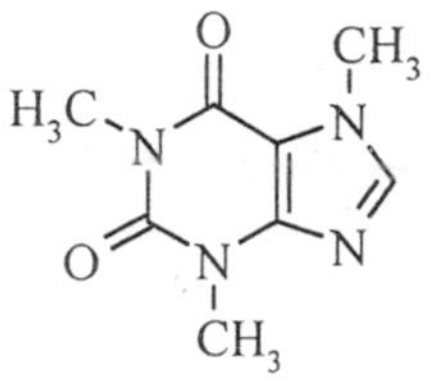

Caffeine

It was this largely unnoticed, by the scientific community, paper that, at least in a formal sense, triggered the start of the current drug development project at Astra Draco around 1980 aiming at designing an analogue to the previously mentioned purine natural products with enhanced efficiency and at the same time a considerably decreased level of side effects. Variation of substituents around the core nucleus gradually generated an understanding of the subtlety and complexity by which changes affected the pharmacological profile.

The targeted structure-activity optimization encompassing the synthesis of virtually hundreds of compounds eventually led to the selection of enprofylline as the candidate drug.[6] As a comparison, a number of structurally similar compounds that have also been selected for clinical evaluation in the same therapeutic area are shown.

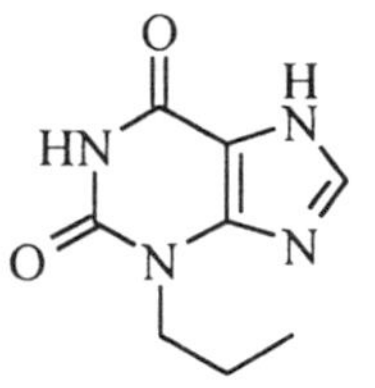

Enprofylline (Astra)

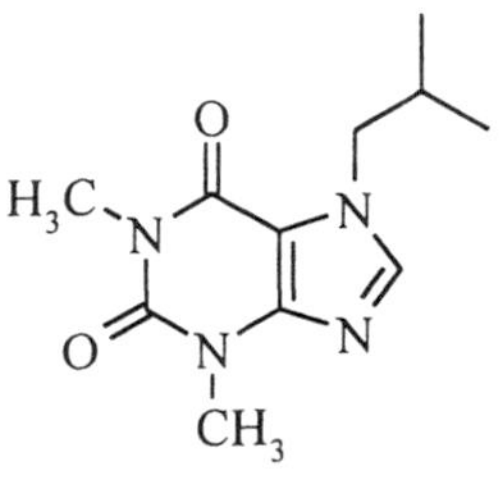

Isbufylline (Malesci) 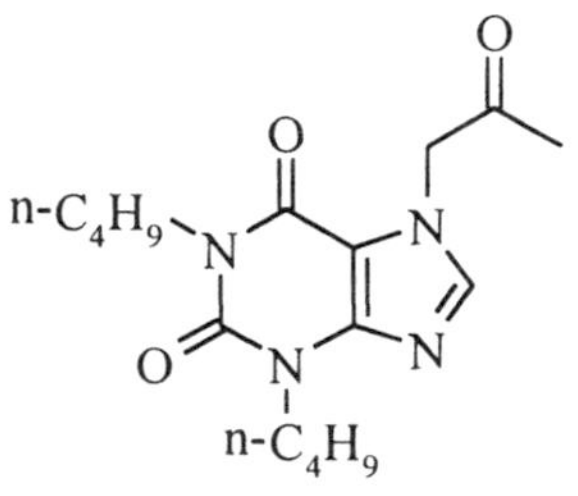          Denbufylline          SK&F-96231
                               (BRL-30892)

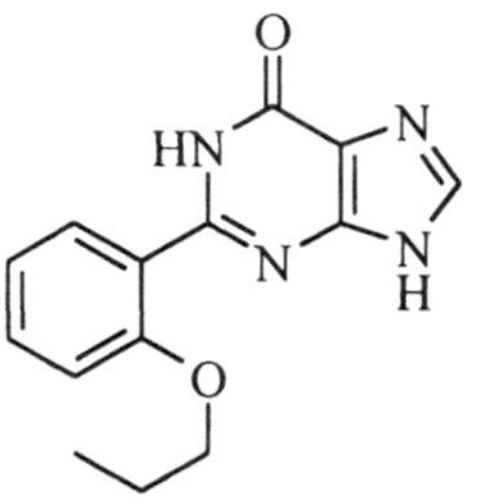

## Starting Point for Process R&D: The Medicinal Chemistry Route

The very onset of the enprofylline project at the Chemical Process Development Laboratory was the presentation of the synthetic method used by the medicinal chemists to prepare initial (smaller) quantities of material. It consisted of a linear 5-step sequence displaying within itself one of the two conceptually available "classical" approaches to assemble the fused imidazo[4,5-*d*]pyrimidine core common to all purines.[7] An outline of these synthetic options is given in scheme 1.

Scheme 1. Classical modes of assembling purines

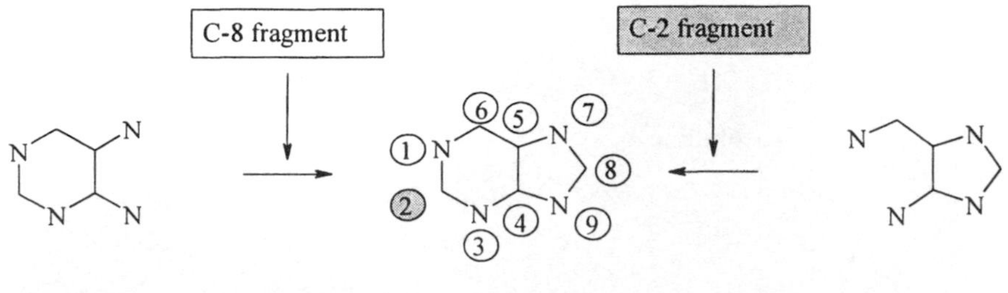

In the present case the attachment of the C-8 fragment had been selected mostly due to the greater ease with which the pyrimidine moiety could be obtained as compared to the imidazole counterpart. Thus, starting from propylamine the isolation of four intermediates before arriving at the desired product was required; see scheme 2 for layout and details of the reactions.

Scheme 2. Original synthesis of enprofylline designed by the Medicinal Chemistry Department

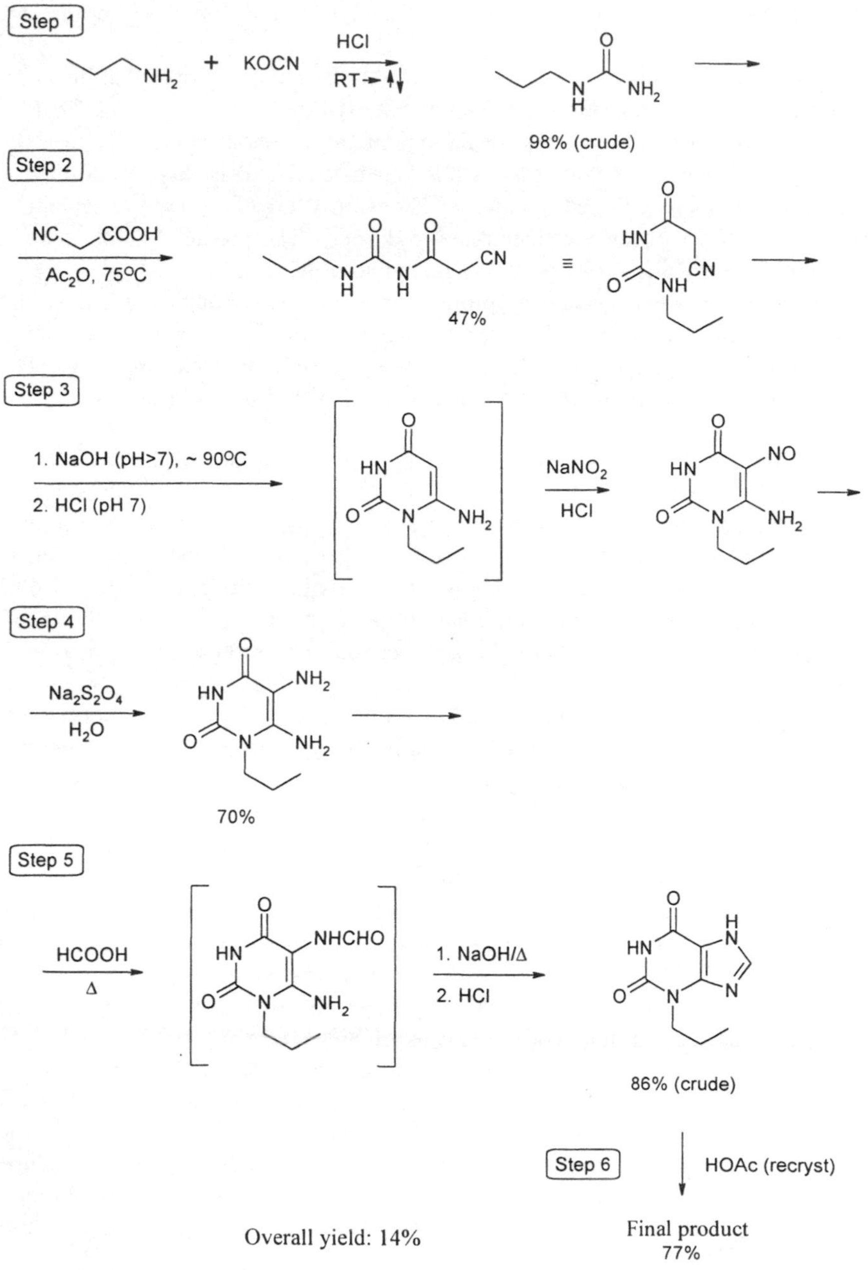

Step 1

+ KOCN $\xrightarrow[\text{RT} \to \Downarrow]{\text{HCl}}$

98% (crude)

Step 2

$\xrightarrow[\text{Ac}_2\text{O, 75°C}]{\text{NC}\diagup\text{COOH}}$

47%

Step 3

1. NaOH (pH>7), ~ 90°C
2. HCl (pH 7)

$\xrightarrow[\text{HCl}]{\text{NaNO}_2}$

Step 4

$\xrightarrow[\text{H}_2\text{O}]{\text{Na}_2\text{S}_2\text{O}_4}$

70%

Step 5

$\xrightarrow[\Delta]{\text{HCOOH}}$

1. NaOH/Δ
2. HCl

86% (crude)

Step 6 | HOAc (recryst)

Overall yield: 14%

Final product
77%

From a structural point of view, the first two steps lead to an acyclic intermediate, i e *N*-cyanoacetyl-*N'*-propylurea which is set up to be cyclised to form the pyrimidine nucleus, followed by a few transformations (nitrosation, reduction, and formylation) to ultimately reach the second cyclisation (creating the imidazole part) which also constitutes the last chemical step, succeeded only by a final recrystallization. It is rather evident from the above scheme that this sequence of reactions involves quite a number of manipulations, some of which deserve being characterized as at least nasty or even environmentally hazardous. This statement and the obvious overall labor intensity of the method as such is further emphasized in the following brief, step-by-step description.

Step 1: propylamine + cyanate → propylurea. To a flask containing propylamine submerged in an ice-bath is added a slight molar excess (10%) of 5M HCl (*aq*) followed by KOCN (1.1 mol eq) dissolved in $H_2O$. The reaction mixture is stirred at ambient temperature for 3 h and is then heated to gentle boiling during ½ h. The solution is allowed to cool and is then evaporated to offer a crystalline material, which is refluxed in ethanol. The warm slurry is filtered to separate inorganic salts (KCl) and the mother liquor is concentrated to dryness. The product is collected and solvent residues are allowed to dry off, affording an isolated yield of 98%.

Step 2: Formation of *N*-cyanoacetyl-*N'*-propylurea. Cyanoacetic acid (1.1 eq) is dissolved in $Ac_2O$ at 60°C in an Erlenmeyer flask. The step 1 product is added at a temperature ≤75°C (yellow discoloration formed at higher levels). During cooling, crystals begin to appear at 60°C, which, once room temperature has eventually been reached, are isolated by filtration and washed giving a yield of 47%.

Step 3: One-pot ring closure to pyrimidine nucleus & nitrosation. A slurry of the substituted urea compound from the previous step in boiling $H_2O$ is treated portionwise with NaOH until basic conditions are retained. The pH is then brought back to neutral by adding some dilute HCl and cooling down to 25°C. An aqueous solution of sodium nitrite (ca 1.09 eq) is then added together with HCl (conc) in small increments. The reaction mixture displayed strong frothing from generating nitrous gases and the temperature should never be allowed to rise above 40°C. After stirring for some hours and leaving overnight, a red crystalline product was isolated in 45% yield after repeated washings with $H_2O$.

Step 4: Dithionite reduction to diaminouracil. To a mixture of the nitroso intermediate in $H_2O$, sodium dithionite was added portionwise while continuously monitoring the disappearance of the red color. The product was obtained as light green crystals and washing with water resulted in a 70% yield.

Step 5: Formylation, ring closure, and final purification. The uracil compound thus prepared was added in portions to warm formic acid and the mixture obtained was then refluxed for 2h. After a quick filtration and subsequent cooling, crystals were precipitated by adding diethyl ether or ethanol affording the $N^5$-formyl derivative, which, without further treatment, was submitted to 2h of boiling in dilute NaOH (*aq*). A polishing filtration was conducted to eliminate extraneous materials followed by a neutralization with HCl (*aq*) to precipitate the final product in a crystalline state, which after repeated washing with $H_2O$ was obtained in 86% yield. Finally, an optional recrystallization of this material from acetic acid could be performed in 77% yield.

In total then, this reaction sequence managed to produce the desired enprofylline in a somewhat mediocre overall yield of 14% with the bottleneck yieldwise being the cyanoacetylation in the second step (running at only 47%). However, a deeper stepwise analysis of the procedure in terms of crucial parameters such as reaction conditions, technical and chemical operability, potential hazards, capacity of the process, and product quality revealed a number of inherent shortcomings and drawbacks:

• The essential insolubility of the intermediates in solvents of conventional type (which in this context means those solvents used on a regular basis in our commercial production).

• A severely negative environmental impact due to the formation of highly toxic fumes (nitrous gases in step 3).

• Difficult isolations of several of the solid products handled leading to centrifugation times of considerable length.

• A pronounced quality problem clearly visible in the high coloration of the end product.

At this point it became quite obvious that considerable development efforts would have to be invested in order to overcome the problems thus identified in the existing route or eventually, in the case they could not be solved satisfactorily, move back to the stage of discovery trying to design a new synthetic sequence.

### Initial Development Leading to Identification of a Viable Synthetic Route

In order to increase the level of understanding and to identify critical parameters, series of experiments in the laboratory scale were conducted on each of the transformations which altogether constitute the whole process. This approach creates a broad set of basic data from which a number of conclusions can be drawn regarding for example, attainable yields, stability properties of process streams and intermediates, process break points, materials output capacity in weight per volume, specific as well as overall productivity in an assumed manufacturing environment, work-up procedures, and purification efforts to reach the desired quality.

Most fortunate it was during this procedure that we managed to isolate the hitherto non-isolated 6-amino-1-propyluracil, an intermediate in the step 3 ring-closure/nitrosation "cascade", which, once its formation and properties had been studied in greater detail, attracted our interest more and more as being a potentially quite ideal pivotal starting material for the whole synthesis.

Aminopropyluracil

This rather unexpected finding enabled us to reduce the number of steps to be performed in-house by 50%, but, even more importantly, the environmentally cumbersome nitrosation could be separated from the rest of the sequence allowing the development of a tailor-made and optimized procedure for this particular transformation. Also as a result of this, safety and environmental issues could be controlled and handled considerably easier. At this point of time it was decided, following our outsourcing policy, to present the key features of the preliminary 6-amino-1-propyluracil process to an external supplier, which eventually led to an exclusive rights agreement on long term production and delivery of the material under toll manufacturing conditions. The full scale process for this important building block is depicted in the following scheme.

Scheme 3. Commercial manufacturing of 6-amino-1-propyluracil

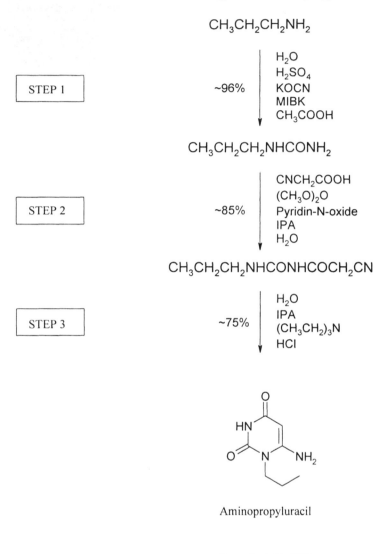

Aminopropyluracil

Overall yield: 61%

A comparison with the original procedure (scheme 2) clearly shows that the major improvement (yield increase from 47 to 85%) has occurred in the second step forming the cyanoacetyl propyl urea. Essentially two modifications were introduced to effect this change: (*i*) the addition of catalytic quantities of pyridine-*N*-oxide and (*ii*) a product wash with isopropanol. A brief summary highlighting the main differences/adjustments in the final version of the method versus the original one is given in table 1.

**Table 1**. Process Characteristics & Developmental Conclusions

| Step no. | Action taken/Parameter changed | Effect observed/Advantage |
|---|---|---|
| 1 | Addition of $H_2SO_4$ instead of HCl | Allows potassium salts ($K_2SO_4$) formed to be filtered off |
| | pH$\geq$9 | Minimizes formation of impurities |
| | Addition of HOAc | Reaction mixture is stirrable at 60°C |
| 2 | Use of pyridine-*N*-oxide | Reaction time reduced by 50% |
| | Washing of product with 2-propanol | Elimination of by-products |
| 3 | High pH | Minimization of hydrolysis which is a dominating side reaction at pH<7.5 |

Having determined reaction conditions including reactants, solvents, and reagents as well as the set-points of key parameters, the process was now ready for the final scale up to the ultimate production size of 4000 L. The flow scheme of the procedure resulting in a batch capacity of some 828 kg, indicating also the maximum volume in each step is depicted below (scheme 4).

Scheme 4. Full scale (=4000 L) batch data for the commercial production of aminopropyluracil

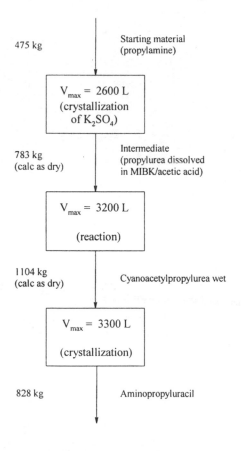

## Design of Final In-House Process & Scale-Up

With the establishment of the aminopropyluracil as the future raw material to be delivered on a toll manufacturing basis, it now remained to direct our attention to the latter half of the original procedure with the aim to design a robust method which could safely and efficiently provide the large quantities (=several tens of tons per annum) that had been forecast as the requirement once the product had been successfully introduced on the market. Thus, the transformations that required careful scrutiny and development to put them into the overall context of a production process were C-5 nitrosation, reduction, *N*-formylation, and ring closure. On top of this, and of extreme importance in the pharmaceutical area for regulatory reasons, the final purification had to be fully defined and locked as soon as possible to allow critical toxicity studies (mainly the long-term cancer types) and human clinical trials to be performed with material produced at a consistent quality level displaying by-product profiles and content levels without any appreciable batch-to-batch variation. From a structural point of view, the chain of events that will now receive full attention are outlined in scheme 5 (see also scheme 2 above for a more detailed display of the reactions involved).

Scheme 5. Conversion of aminopropyluracil to target enprofylline

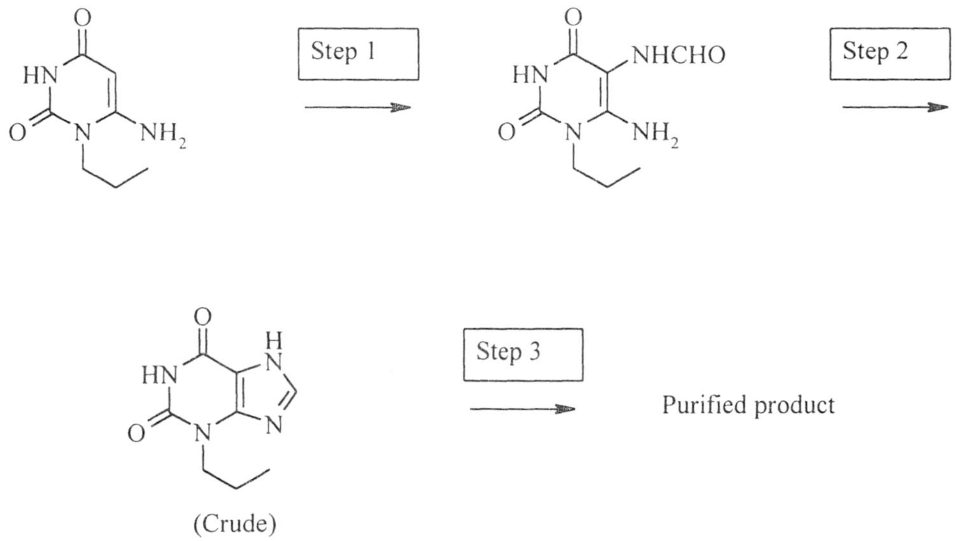

(Crude)

Again, extended series of experiments had to be performed over a long period of time, where, besides considering chemical, technical, environmental, and quality factors, the cost of goods situation also had to be sharply focused due to the indeed very tough competition from rather cheap products already existing on the market and used therapeutically (e. g. caffeine). This proved, no doubt, to be quite a challenge on its own! The findings resulting from our endeavors have been compiled and summarized in table 2.

**Table 2.** Results from Developmental Studies on the Three-Step In-House Sequence

| Step no. | Action taken/Parameter changed or modified | Effect observed/Advantage |
|---|---|---|
| 1 | Nitrosation in HOAc instead of HCl | Eliminates problems with foaming and evolution of nitrous gases |
| | Nitrosation performed with $NaNO_2$ (*aq*) in HCOOH followed by *in situ* reduction ($PtO_2$) and formylation | Combining three chemical transformations into a one-pot reaction |
| | Switch to Pt/C (5%) catalyst | Increased efficiency and lowered catalyst requirement |
| 2 | Ring-closure effected by charging to NaOH at 80°C | Decreases hydrolysis to diamine |
| 3 | Two reprecipitations in NaOH/HCl instead of recrystallization from HOAc | Yield increase from 77→83%; eliminates formation of the undesired 5-acetylamino analogue |
| | Addition of HCl at 80-90°C | Elimination of 2nd reprecipitation |

As is fairly evident from the table, with these improvements in place the process began to assume the contours of quite a promising production method. As a matter of fact, even the overall yield at this point (=84%) was at least somewhere in the neighborhood of the target level to be reached in order to achieve the required cost efficiency. The combined one-pot reaction of step 1 would most probably be characterized as the foremost advancement from a methods perspective, albeit the less "spectacular" changes also deserve a great deal of credit in giving a high quality label to the development work. Nevertheless, in this first step formamidoylation chemistry, formic acid plays a rather unique triplicate role of being both reactant (formyl-group donor), solvent, and reducing agent (hydrogen equivalent). Thus, after forming the nitroso intermediate a transfer hydrogenation takes place where a precious metal catalyst (i. e. Pt, but also others such as Pd, Rh, and Ir are reported in the literature[8]) provides the necessary reduction power in the form of $H_2$ or $H_2$-equivalents (= atomic species acting as elemental hydrogen), whereas the remainder of the HCOOH is discarded as inert gaseous $CO_2$. An indeed environmentally friendly and cheap methodology to effect these kinds of transformations! As a matter of fact, literature reports describing this procedure especially from a preparative perspective are relatively scarce (one of the few available examples is the deprotection of benzyl substituted amino acids and peptides[9]) and virtually nothing has been published on the nitroso→amine transformation.

Secondly, in order to keep up with the demand for a high productivity process, HCOOH is ideally suited to dissolve the rather insoluble starting material to obtain elevated concentrations (≈20% w/w) at moderate temperatures. Finally, HCOOH is an excellent formylating agent not requiring any activation to the mixed anhydride (or similar reactive derivatives) to perform its "duties" in the current case. A detailed reaction scheme specifying the materials used is presented below.

Scheme 6. Full scale synthesis of enprofylline

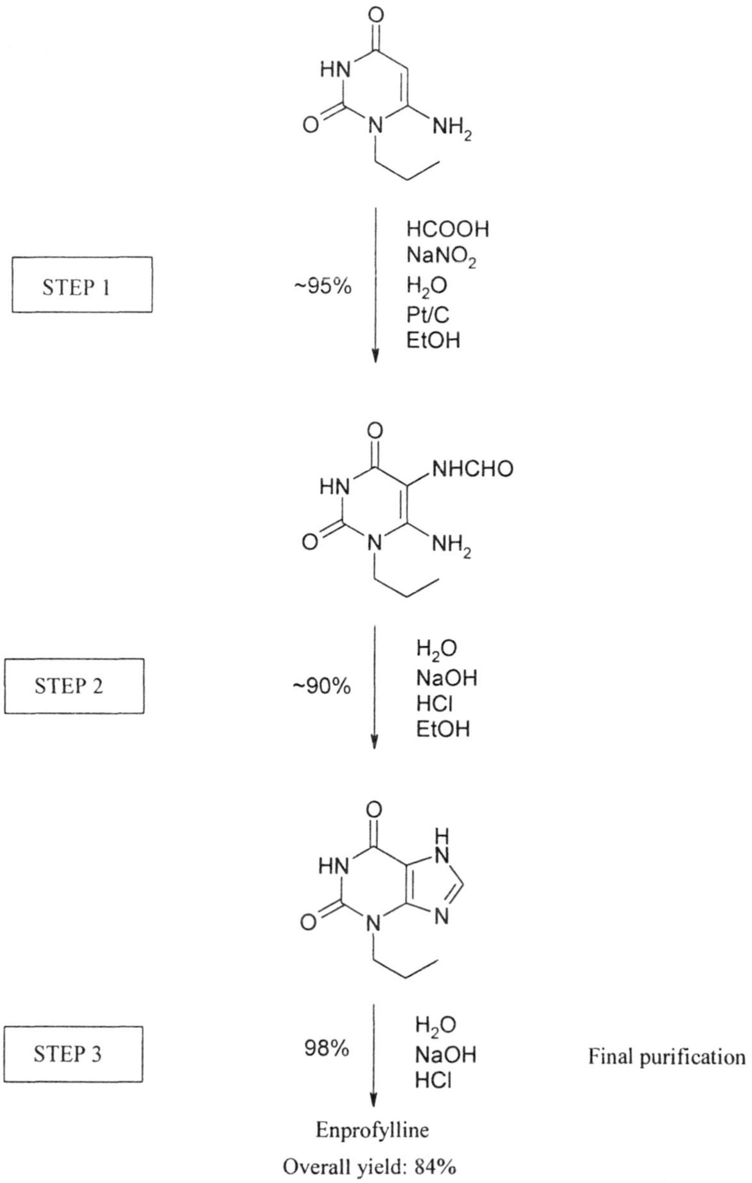

As a complement to this entirely structural and formula-centered scheme and to explain the process flow in considerably more detail, a step-by-step presentation is offered covering the specific conditions and procedures including the amounts of materials actually applied on the commercial (=2 kmol) scale.

Step 1: Sodium nitrite (163 kg, 2.4 kmol) is dissolved in $H_2O$ (240 L) and allowed to react for 2 h at 10°C with a suspension of aminopropyluracil (400 kg, 2.4 kmol) in HCOOH (1871 kg, ≈ 1534 L). The mixture is then heated to 40°C to obtain a homogeneous solution, which is subsequently added slowly (safety precautions!) to a slurry of Pt/C, 5% (3.2 kg as dry weight) in HCOOH (407 kg, ≈ 334 L) and $H_2O$ (96 L) at 25°C. After completing the reaction at 45°C, the catalyst is filtered off at 60°C followed by distilling off $HCOOH/H_2O$ (880 L) and adding ethanol (1470 L). Cooling the temperature to 2-5°C precipitates the product which is isolated and washed with cold ethanol and $H_2O$.

Step 2: The wet product (formamidouracil) used as is from step 1 (477 kg dry weight, 2.25 kmol) is allowed to react with conc NaOH (aq) (244 kg) in $H_2O$ (2400 L) containing $Na_2SO_3$ (4.8 kg) for 2 h at ≈100°C. After a slight decrease of the temperature to 95°C, conc HCl (230 L) and $H_2O$ (268 L) are added which renders a final pH of 2, however, the addition is discontinued at the point of crystallization onset (pH≈9) in order to allow for some crystal growth to occur before proceeding to the end value. The product (enprofylline crude) is then isolated and washed with cold water, ethanol, and finally water again.

Step 3: Crude enprofylline (≈393 kg dry weight, 2.03 kmol) is dissolved in $H_2O$ (2250 L) and conc NaOH (*l*) (176 kg) at 90°C and then reprecipitated by adding conc HCl (201 kg) and $H_2O$ (172 L). After cooling to ≈0°C, the final product (enprofylline) is isolated and washed with cold water yielding 385 kg (1.98 kmol) of material.

This detailed methods description is translated into a commercial process flow chart (see following Scheme) which highlights the, from a production point of view, crucial maximum obtainable volume per step in the given batch size (= 2.4 kmol) also specifying the critical unit operation where this occurs.

Scheme 7. Process layout (4000 L production size) in quantities and $V_{max}$

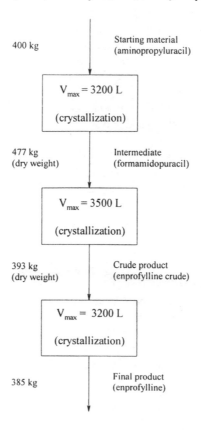

400 kg — Starting material (aminopropyluracil)

$V_{max}$ = 3200 L (crystallization)

477 kg (dry weight) — Intermediate (formamidopuracil)

$V_{max}$ = 3500 L (crystallization)

393 kg (dry weight) — Crude product (enprofylline crude)

$V_{max}$ = 3200 L (crystallization)

385 kg — Final product (enprofylline)

### Characteristic Process Features & Summary of Developmental Conclusions

In order to rationalize the chosen process design (e g in the form of stoichiometries applied and types of reactants), which has been outlined down to quite some detail above, as well as the selection of set points for critical parameters (pH-values, temperatures), the following step-by-step overview is provided. Special emphasis is also given to a mechanistic explanation of the formation of two rather peculiar by-products. A brief addendum finally summarizes the results from the final optimizations established by laboratory experiments but which have not yet been implemented in full scale production.

### Step 1

• The starting material (=aminopropyluracil) should be charged to formic acid at <10°C in order to depress unwanted $N^5$-formyluracil impurity.

• Effective stirring is essential to prevent the precipitation of a highly insoluble solvate complex (with HCOOH).

• The addition of $NaNO_2$ should be balanced at exactly 1.00 mol eq (compared to aminouracil) to reduce the formation of impurities.

• Reduction of the nitroso intermediate should be initiated on only a minor part (≈5%) of the full batch due to safety reasons. This operational procedure causes the evolution of 2.4 $m^3$ of $CO_2$ (g) and a temperature increase of 10°C within 1-2 min.

• A full scale batch generates ≈103 $m^3$ (!) of $CO_2$ which has to be handled by the gas exhaust system.

• Evaporation of $H_2O$/formic acid from the reaction mixture to the correct (predetermined) concentration is crucial for both isolable yield and stirring capability.

• $C_2H_5OH$ should be added at the point where complete crystallization has occurred to avoid isolation problems related to small particles clogging the centrifuge cloth and therefore preventing, or severely retarding, the mother liquor from being efficiently withdrawn.

• The work-up procedure designed guarantees that the residual amounts of Pt-catalyst <1 ppm in the end product.

• Set the temperature during the reduction phase at <35°C to suppress formation of pyrimidopteridine type by-products. Their origin is, in both cases, an over-reaction of the desired step 1 formamidouracil product to the diamine (via formyl hydrolysis), which then undergoes self-condensation to two compounds having almost identical molecular mass and whose formation, most interestingly, is controlled by geometrical and sterical factors expressed in the way the two reactants are aligning themselves (=their relative orientation); see scheme 8 for details. Under ordinary reaction conditions, the amounts formed of these products were in the range of 0.5-1% each.

Scheme 8. Proposed mechanism for generation of pseudo-isomeric pyrimidopteridines

I. Michael-type addition
  (syn-alignment)

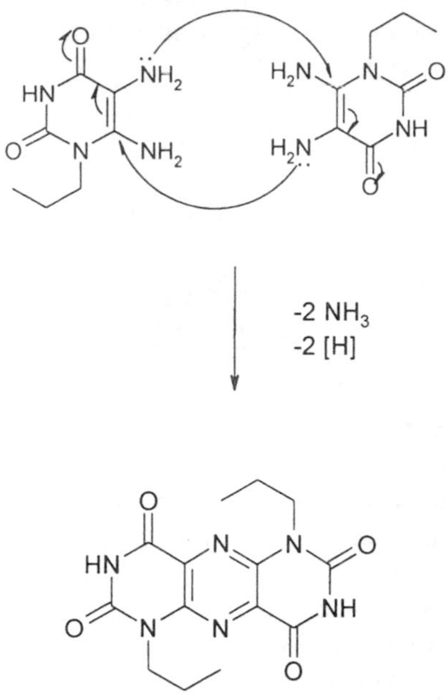

$$-2\ NH_3$$
$$-2\ [H]$$

$(C_{14}H_{16}N_6O_4;\ M=332.320)$

II. Amino-carbonyl condensation
  (anti-alignment)

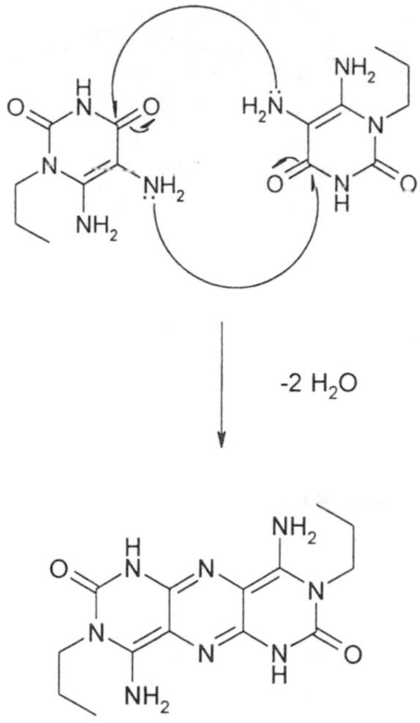

$$-2\ H_2O$$

$(C_{14}H_{20}N_8O_2;\ M=332.368)$

**Step 2**

• The cyclization rate shows a strong pH and temperature dependence: at 95°C and pH≈9.3 the full reaction time is 20 min whereas at pH≈8.3, maintaining the same temperature, it is prolonged to 150 min. A preferential minimization of the reaction time is counteracted by the finding that the extent of hydrolysis becomes appreciable at pH>9.

• Under the given procedure ≈10% of diamine is formed.

• Addition of catalytic amounts of $Na_2SO_3$ prevents discoloration of the reaction mixture.

• The loss of enprofylline in the mother liquor is <1%.

**Step 3**

• The addition of HCl (*aq*) is discontinued at the point of onset of product precipitation (=pH 9).

At this stage of the process development material could be produced in bulk on the large scale and many of the controlling parameters had been studied in great detail which also included the determination of their specific value or interval under running conditions. Accumulating experience eventually allowed us to conduct some further optimizations which, however, have not yet been transferred to full scale production. With these modifications in place we strongly believe that the goals originally set up for the chemical process part of the project have been met in all aspects. These ultimate changes are summarized in table 3.

**Table 3.** Summary of final optimizations in the lab scale

| Step no | Action taken/change conducted | Advantage obtained/Effect observed |
|---|---|---|
| 1 | Using formic acid of recovered quality (containing 10% $H_2O$) | Environmental (considerable reduction in effluent streams) |
| | The ratio of formic acid/$H_2O$ increased by 7.5% | The reaction mixture after the nitrosation turns into a homogeneous solution at 30 instead of 40°C |
| | $NaNO_2$ is added in the solid state (crystalline form) directly to the reaction mixture | The process occupies one vessel less (capacity benefit) |
| | Formic acid/$H_2O$ is distilled off in two steps: firstly 900 L, then allow for precipitation followed by further 600 L | The centrifugation is facilitated (shorter process cycle time) |
| | EtOH is replaced by acetone and product is isolated at -5°C and washed with acetone | A higher yield is obtained: 98% (instead of 95%) |
| 2 | The intermediate from step 1 is charged to $H_2O$ at ambient temperature followed by heating which causes residual acetone to be simultaneously distilled off | Charging of crystalline material is easier from a handling point of view |

Table 3, continued

| Step no | Action taken/change conducted | Advantage obtained/Effect observed |
|---------|-------------------------------|-------------------------------------|
| | The cyclization reaction is performed at 80 instead of 90°C | Hydrolysis to unwanted diamine is prevented causing the yield to increase from 90 to 98% |

## Concluding Statements

The achievements described above can in essence be summarized in the following five points, which one by one represents key areas in process development and whose fulfillment of preset targets was one of the most important prerequisites for a successful termination of the whole project.

• The synthesis designed was short (only three steps), straightforward, and economically feasible and provided the enprofylline product in an 84% overall yield.

• The final product was obtained in excellent quality (>99.9% purity on HPLC) as a white to off-white crystalline material.

• The upscaled process offered a good weight per volume output ($\approx$400 kg/4000 L).

• Safety issues are kept under full control (especially the strong exothermicity in the nitrosation and reduction steps).

• The process displays an acceptably low environmental impact (E-factor describing the required amount of material input per kg of final compound output $\approx$50).

## Acknowledgement

The authors want to express their sincere gratitude to all members of the project team, process chemists, analytical chemists, and chemical engineers to mention only some of the foremost categories, without whose full dedication, skill, and efforts the accomplishments described herein would not have been possible. Special thanks go to Dr Gunnar Kjellin and his staff in Medicinal Chemistry at Astra Draco, the research company where the original project was first laid out and where the target molecule was designed under his supervision, for the stimulating and fruitful collaboration throughout the entire time spent completing this work.

## References

§   This work has in part been presented orally by H.-J. Federsel at a) *The 11th Process Development Symposium*, Cambridge, UK, December 16-17, 1993. b) *The 10th IUPAC Conference on Organic Synthesis*, Bangalore, India, December 11-16,1994.

*   To whom correspondence should be addressed.

1.  Holgate, S.T. *Postgrad. Med. J.* **1988**, *64*, 82.
2.  Reed, C.E. *Chest* **1988**, *94*, 175.
3.  Persson, C.G.A. *Eur. J. Respir. Dis.* **1980**, *61*(suppl 109), 7.

4. Persson, C.G.A. *Trends Pharmacol. Sci.* **1982**, *3*, 312.
5. Salter, H. *Edinb. Med. J.* **1859**, 1109.
6. Persson, C.G.A.; Kjellin, G. *Acta Pharmacol. Toxicol.* **1981**, *49*, 313.
7. (*a*) Shaw, G. In *Comprehensive Heterocyclic Chemistry*; Katritzky, A.R.; Rees, C.W.; Potts, K.T., Eds., Pergamon: Oxford, 1984; Vol 5, pp 567-598 and references therein. (*b*) Shaw, G. In *Comprehensive Heterocyclic Chemistry II*; Katritzky, A.R.; Rees, C.W.; Scriven E.F.V., Eds., Pergamon: Oxford, 1996; Vol 7, pp 419-422 and references therein. (*c*) For a recent example of a synthetic approach, see also: Müller, C.E.; Sandoval-Ramírez, J. *Synthesis* **1995**, 1295.
8. Fahrenfort, J.; van Reyen, L.L.; Sachtler, W.M.H. In *The Mechanism of Heterogeneous Catalysis*; de Boer, J.H., Ed., Elsevier: Amsterdam, 1960; pp 23-48.
9. Babiker, EA.; Anantharamaiah, G.M.; Royer, G.P.; Means, G.E. *J. Org. Chem.* **1979**, *44*, 3442.
10. A rigorous definition of the E-factor can be given as the weight-based ratio between chemicals input in a given process (including starting materials and building blocks, reagents, catalysts, solvents and various inorganic materials such as acids, bases and salts) and the obtained amount of a desired product. The figure reported in the present case (ca 50) is on the lower end of the interval indicated for chemical processes in the pharmaceutical industry (25- >100); cf. Sheldon, R.A. Chem. Ind. (London), **1992**, (December 7), 903.

# Ontazolast: The Evolution of a Process

**Vittorio Farina,**[†] **Karl Grozinger,**[†] **Hermann Müller-Bötticher**[¶] **and Gregory P. Roth**[‡]

† *Department of Chemical Development, Boehringer Ingelheim Pharmaceuticals, 900 Ridgebury Rd, Ridgefield CT 06877 USA*
‡ *Department of Inflammatory Diseases and Medicinal Chemistry, Boehringer Ingelheim Pharmaceuticals, 900 Ridgebury Rd, Ridgefield CT 06877 USA*
¶ *Geschäftseinheit Feinchemikalien, Boehringer Ingelheim KG, D-55216 Ingelheim, Germany*

## INTRODUCTION

Ontazolast (BIRM270), **1**, a novel inhibitor of Leukotriene B$_4$ biosynthesis in human neutrophils,[1] was discovered in 1992 after optimization of a lead from high-throughput screening. The compound possesses one chiral center, of the (S) configuration. Initially, *in vitro* activity was optimized using racemates, but very soon enantiomeric selectivity was demonstrated by acylation of amine **2** with Boc-L-valine, separation of the diastereomers by chromatography, followed by amide hydrolysis and final conversion to **1**.

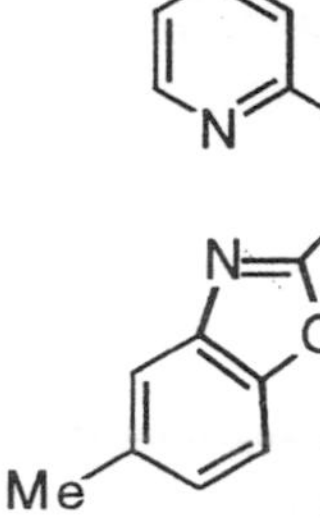

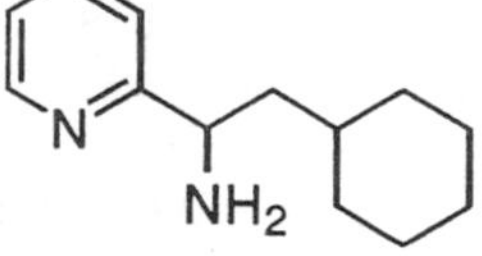

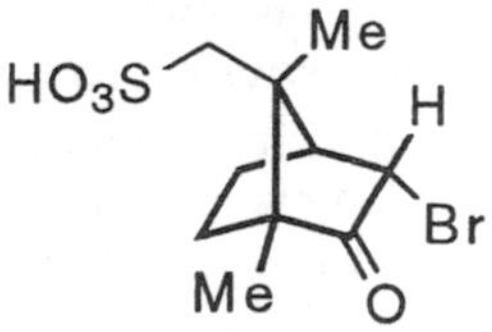

In order to complete the necessary pre-development activities, a rather large batch of the drug (500 g) was requested. As the chemistry to prepare racemic **2** was in place and was deemed suitable for scale-up, the primary challenge at this stage was to address in a rapid and effective manner the problem of preparing a large supply of enantiomerically pure drug. Because chromatographic separation on such a scale was impractical, we launched on a two-prong campaign: a team of scientists proceeded to screen a variety of commercially available chiral carboxylic acids in an attempt at separating the (S) enantiomer of amine **2** by resolution. From this early screening activity, [(1R)-*endo, anti*)]-(+)-3-bromocamphor-8-sulfonic acid, **3**, emerged as a practical resolving agent. At the same time, a second team began studying asymmetric approaches to **2**: these capitalized on a vast array of known methods, with the ultimate goal of defining the commercial process to Ontazolast. We will first describe our initial synthesis of **1**, which was used in the initial stages of the development of the drug. We will then describe the details of the synthesis that we decided to use for phase IIb-III supplies, and we will conclude with a description of the "runner-up" synthesis, *i.e.* the one that, among the many investigated, underwent a preliminary scale-up and was at least considered for the preparation of clinical supplies. The practical issues that guided our "synthesis-switch" decision and our route selection will be discussed.

## BACKGROUND AND SYNTHETIC STRATEGY

Among the possible asymmetric approaches that we considered, the first is illustrated in Scheme 1: formation of an anion α to a nitrogen atom is well precedented and there are asymmetric versions of this operation that employ a chiral template at N.[2,3] Alternatively, a riskier but potentially more rewarding approach may be to use an achiral alkylation substrate in the presence of a chiral non-racemic activator or catalyst.[4,5] All these possibilities were explored to some extent.

**Scheme 1**

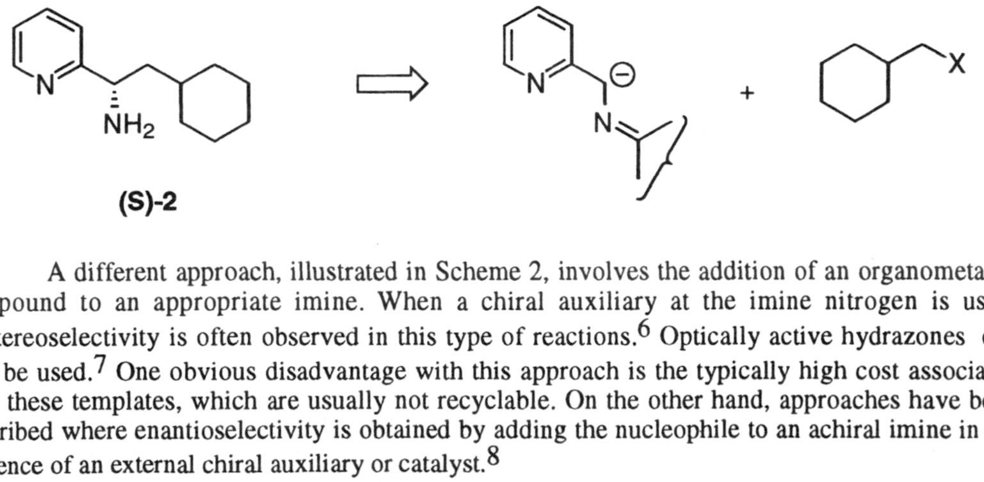

**(S)-2**

A different approach, illustrated in Scheme 2, involves the addition of an organometallic compound to an appropriate imine. When a chiral auxiliary at the imine nitrogen is used, diastereoselectivity is often observed in this type of reactions.[6] Optically active hydrazones can also be used.[7] One obvious disadvantage with this approach is the typically high cost associated with these templates, which are usually not recyclable. On the other hand, approaches have been described where enantioselectivity is obtained by adding the nucleophile to an achiral imine in the presence of an external chiral auxiliary or catalyst.[8]

**Scheme 2**

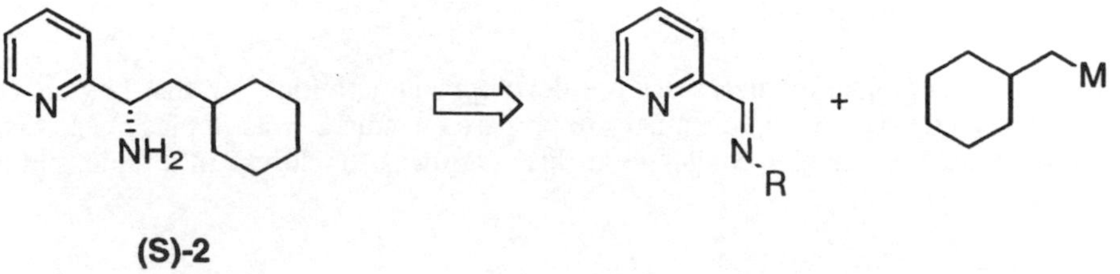

**(S)-2**

These approaches have better potential for recycling the chiral material. Finally, reduction of an imine or enamine represents a potentially simple approach to (S)-**2**. Approaches that utilize chirally substituted imines or their synthetic equivalents have been described,[9] as well as approaches that employ reduction of a prochiral imine in the presence of a chiral catalyst (Scheme 3).[10] The catalytic method is obviously preferred unless the chiral template used (*i.e.* the amine) is very cheap.

**Scheme 3**

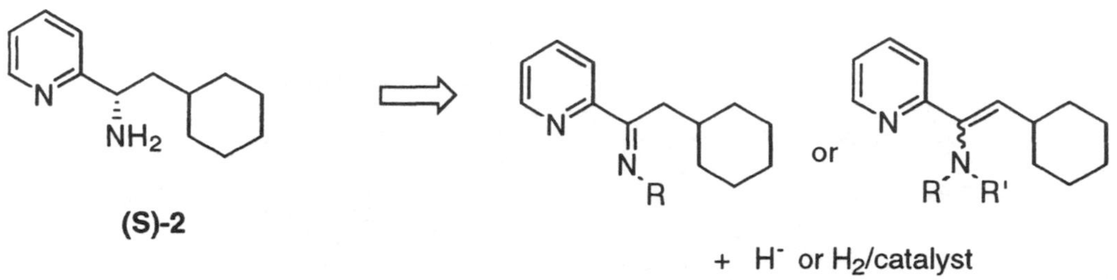

+ H⁻ or H₂/catalyst

This approach was investigated aggressively. Unfortunately, the enantioselectivities observed in the asymmetric reductions of prochiral imines are not uniformly high or require extensive manipulations after the reduction step; since we had found early on in our synthetic work that the use of cheap, readily available chiral amines afforded useful diastereoselectivities, the chiral catalyst approach was not extensively studied.

## FIRST SYNTHESIS OF ONTAZOLAST VIA RESOLUTION

**Racemic synthesis and resolution strategy.**

The original racemic synthesis of Ontazolast is quite straighforward and is shown in Scheme 4: the Grignard reagent from cyclohexylmethyl bromide was prepared in a mixture of THF and methyl *t*-butyl ether (MTBE) and slowly treated with a solution of 2-cyanopyridine in MTBE. The crude imine salt **6** was directly reduced with sodium borohydride in methanol at room temperature. Workup afforded the crude racemic **2**, which was coupled with commercially available **7** to afford racemic Ontazolast after crystallization from ethanol/water. The racemate was typically obtained as a hydrate.

In order to prepare optically pure **1** by resolution in a practical fashion with commercialization potential, the following requirements were set: a) The resolving agent must be available in bulk at a reasonable price. b) Recovery of the resolving agent must be facile and high yielding. c) The fractional crystallization has to produce enantiomerically pure material (typically >90% ee for the initial toxicology studies and higher later on) in a small number of crystallizations. d) A practical and high-yielding recycling procedure must be found for the unwanted (R)-enantiomer of **2**. After screening a number of optically active organic acids which are available in bulk at reasonable prices, [(1R)-*endo, anti*)]-(+)-3-bromocamphor-8-sulfonic acid (**3**) emerged as the most efficient one. This resolving agent is available from Calaire Chimie as an ammonium salt at an attractive price (*ca.* $200/Kg). Conversion to the free acid needed for the resolution is achieved by passing an aqueous solution of the ammonium salt through a strongly acidic cation exchange resin.[11]

Scheme 4

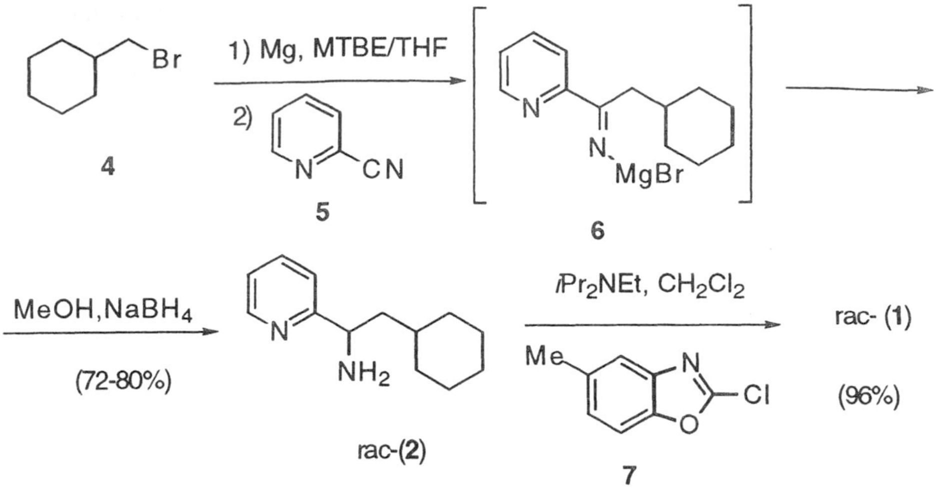

In order to avoid the cumbersome ion-exchange column, we explored alternative procedures, and found that the free acid can be prepared in a two-step protocol which consists of first forming the hemi calcium salt by adding $Ca(OH)_2$ to the ammonium salt of **3** and then concentrating *in vacuo* until all the ammonia has been removed. The solution of the Ca salt is then treated with sulfuric acid in the appropriate stoichiometric amount. Filtration of $CaSO_4$ and evaporation yielded a syrup of **3** free acid, which was dissolved in ethyl acetate for the resolution step.

## Pilot plant-scale resolution.

In a typical resolution procedure, a solution of 66.9 Kg (215 mole) of free (+)-**3** in 160 L ethyl acetate was added to a stirred solution of racemic amine **2** (87.5 Kg, 428 mole) and stirred at room temperature for 12-18 h. Centrifugation and drying gave material typically showing 50% ee. Further recrystallization from isopropanol yielded ee's of 74%, 92%, 96% and finally, after the fourth recrystallization, 99%.

For the racemization, the ethyl acetate filtrate from the above resolution is treated with one equivalent HCl and the resulting amine hydrochloride is racemized simply by refluxing with 10 mol % salicylaldehyde[12] over several days. The racemized lots are combined as appropriate, neutralized with NaOH and the ethyl acetate layer, containing racemic **2**, is reprocessed as above. The isopropanol solutions of (R)-enriched **2** are evaporated and racemized in the same fashion.

## Conclusion of the synthesis.

After the fractional crystallization, free (S)-**2** is isolated by adding ammonium hydroxide to an aqueous solution of the bromocamphorsulfonate of **2** and extracting the free base into dichloromethane; typically one obtains enantiomerically pure (>99%) **2** in 33-55% yield, depending on the scale and on how far the various filtrates are pursued after each recrystallization. From the aqueous layer the bromocamphorsulfonic acid is recovered in *ca.* 95% yield as an ammonium salt by simply evaporating the water; it can be used directly in the resolution without further purification. The dichloromethane solution of (S)-**2** is treated with **7** in the presence of diisopropylethyl amine, to obtain crude Ontazolast in 96% yield (Scheme 4). Final purification is achieved by recrystallization from ethanol/water.

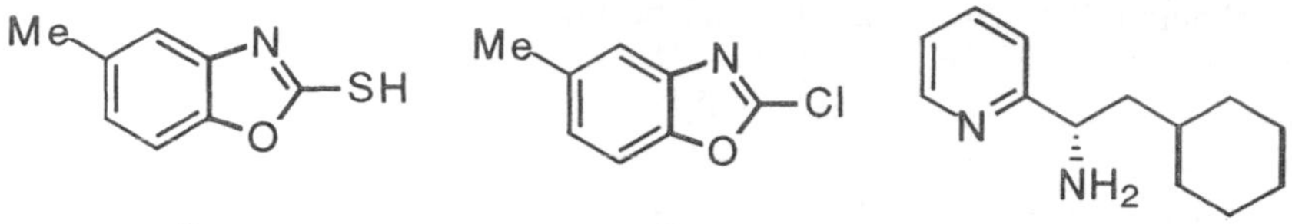

**Scheme 5**: Possible impurities in clinical batches of Ontazolast (in addition to [R]-1).

Three large batches of the drug substance, ranging from 7.6 to 43.1 Kg, were prepared by this method and used mainly for toxicology and phase I and IIa clinical studies. Validated HPLC assays showed that these lots easily met the 98% purity specification (total organic impurities, including the ones identified in Scheme 5, were in the range 0.08-0.19%) and had enantiomeric excesses that ranged from 96% for the first toxicology batch to 98.4-99.2% in subsequent ones, fully meeting our specifications.

**Commercialization potential.**

The resolution process outlined above was never seriously considered as a candidate for commercialization. Although resolution is a straightforward approach to the synthesis of enantiomerically pure organic acids and bases and many pharmaceutical products have been and still are produced in this way,[13] the present resolution requires five crystallization steps to produce **2** of acceptable enantiomeric purity, and is, therefore, too labor-intensive and uses too large an amount of solvents; these two factors rendered its use in a large plant quite unattractive, especially when viewed against the two asymmetric syntheses that were developed in parallel. These new approaches are described in the next sections.

## SECOND SYNTHESIS OF ONTAZOLAST BY REDUCTION OF A CHIRAL IMINE

**Initial considerations and laboratory-scale asymmetric synthesis of (S)-2.[14]**

The approach in Scheme 3, where R is a chiral auxiliary, presupposes that the chiral amine template is cleaved after the reduction step and cannot therefore be recycled. The success of this route depends therefore on choosing a cheap optically active auxiliary. Our first choice was phenethylamine, whose enantiomers are both commercially available in large quantities at a very economic price because of a new and efficient biocatalytic process developed by BASF.[15] In relation to the original synthesis illustrated in Scheme 4, we felt that the initial step was quite practical and could be used in our new approach. Specifically (Scheme 6), the Grignard reagent from **4** was treated with 2-cyanopyridine and the resulting product subjected to acidic hydrolysis, which yielded ketone **9** in very good yield.

From literature studies aimed at the enantioselective synthesis of arylalkylamines,[9f] it was apparent that (S)-phenethylamine **10** would be required in order to obtain the required stereochemistry at the prochiral center in the asymmetric reduction. Imine formation between ketone **9** and amine **10** turned out to be a non-trivial task, because standard conditions (toluene, *p*-toluenesulfonic acid) did not effect the desired conversion. The reaction responded much better to heterogeneous acidic catalysts and initially we employed suspended silica gel as a promoter.

Complete conversion still needed a slight excess (10-20%) of amine **10**. The resulting **11** was a mixture (NMR analysis) of E/Z isomers (*ca.* 9:1), whose composition does not change over time. Given that the conversion is carried out in refluxing toluene (temperature around 110-120°C), this ratio almost certainly reflects the equilibrium composition.

**Scheme 6**

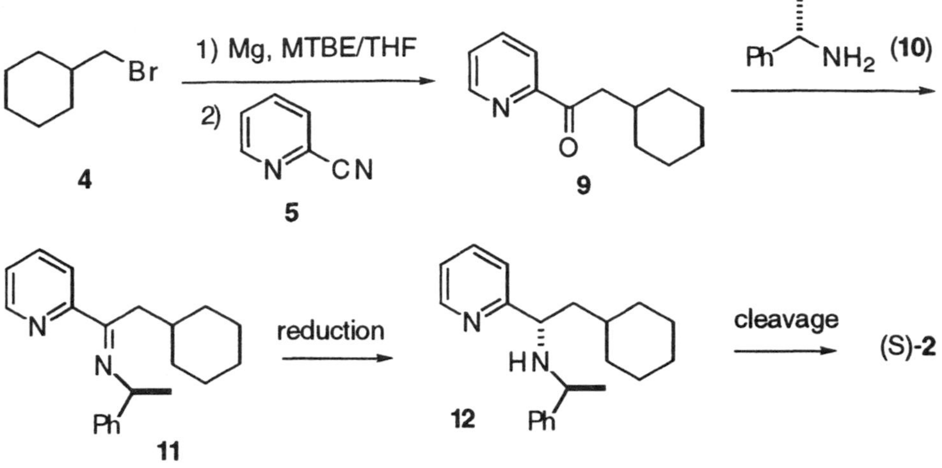

Many reducing agents were screened in order to optimize the diastereoselectivity of the reduction of **11** to **12**. A few selected examples are shown in Table 1. Very good selectivities were achieved with bulky reagents, such as L-selectride and sodium triacetoxyborohydride; selectivities with simple borohydride and catalytic hydrogenation were also acceptable.

**Table 1:** Diastereoselective Reduction of Imine **11**.

| Reducing Agent/Solvent | Temperature (°C) Time | Diastereomeric excess (S,S:R,S) | Yield (%)[a] |
|---|---|---|---|
| L-Selectride/ THF | 60, 4h or 20, 24h | 95:5 | 43 |
| NaBH(OAc)$_3$/PhMe-THF | 60, 24h | 94:6 | 42 |
| NaBH$_4$/PhMe-EtOH | -10, 40h | 89:11 | n.d. |
| NaBH$_4$/PhMe-EtOH | 20, 4h | 88:12 | 60-63 |
| NaBH$_4$/PhMe-EtOH | 60, 2h | 86:14 | n.d. |
| Ra Ni/H$_2$/ EtOH | 30, 5h | 86:14 | 52 |

a: Isolated yields of pure (S,S)-diastereomer **12** as the tartrate salt (*vide infra*).

The sterically demanding reducing agents, although slightly more selective than NaBH$_4$, are also more expensive and wasteful since they contain only one equivalent of active hydride per molecule. In addition, these reductions require extensive periods at room temperature to reach completion. Catalytic hydrogenation works well in EtOH, but is sluggish in the presence of toluene. Since we wanted to carry out the reduction directly on the imine solution without isolation, the presence of toluene is hardly avoidable. Sodium borohydride, on the other hand, performs well in toluene/ethanol mixtures and at room temperature. The selectivity depends only slightly on the temperature (see Table 1), and it was decided to carry out the reduction at room temperature. Moreover, the overall yields of **12** are consistently 10-20% higher with this reagent *vs.* the others, in spite of the lower selectivity. For this reason, NaBH$_4$ was selected as the reducing agent for the process.

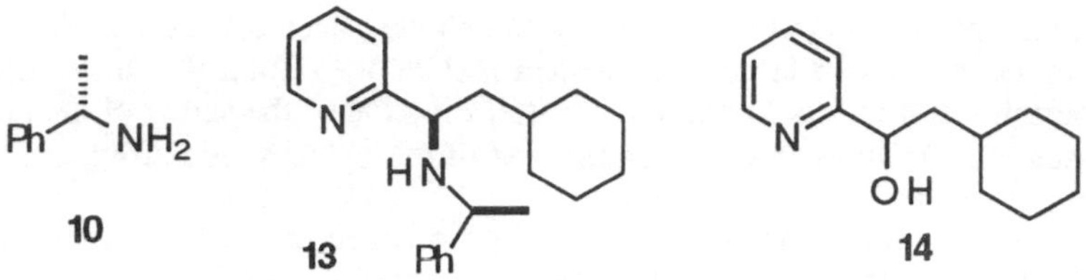

**Scheme 7**: Major impurities contained in crude **12**.

Several side products are detected in crude preparations of **12** after the reduction (Scheme 7). In addition to the excess **10** (up to 20%), the (R,S) diastereomer **13** was detected in 12% proportion and had to be removed; also, carbinol **14** was isolated in 5% yield, most likely the product of *in situ* imine hydrolysis followed by reduction.    At this point, we hoped that salt formation would facilitate purification of **12**, and we found that organic diacids (*e.g.* fumaric, oxalic) readily form crystalline salts with **12**. We also found that (L)-tartaric acid was very effective in removing the unwanted diastereomeric impurity **13**: the tartrate of **12** was obtained in high yield and was essentially free of **13** in one single crystallization.

The necessary excess of phenethylamine turned out to be the most serious problem, since this chiral auxiliary, still present in crude **12**, also crystallized as a tartrate salt. In order to resolve this problem, we came up with a solution that hinged on the higher reactivity of primary amine **10** *vs.* secondary amine **12**: a competition study of the acylation reaction of **10** and **12** with pivaloyl and isobutyryl chloride showed that the reactivity difference was substantial (>100 fold), whereas the reactivity difference with acetic anhydride was too small to be practical. In the end, propionic anhydride was selected as the scavenger of choice. Addition of enough propionic anhydride (typically 16% mol) to consume all the excess **10**, followed by addition to a suspension of (L)-tartaric acid in acetone yielded crystalline **12** in high diastereomeric purity as the tartrate salt. The acylated phenethyl amine and unreacted propionic anhydride remained in solution and were removed by filtration, together with **13** and **14**. The typical yield obtained was 60-63% based on 2-cyanopyridine, over the two chemical steps plus the crystallization/diastereomeric enrichment step, and this was quite satisfactory for our purposes.

The next step of the sequence is the reductive cleavage of the auxiliary to generate (S)-**2**. A wide variety of methods exist in the literature for this type of operation, but our initial experiences were disappointing, since competitive pyridine hydrogenation occurs, accompanied by the product of C-N bond cleavage α to the pyridine ring. In addition, racemization of the stereocenter, probably by dehydrogenation/ rehydrogenation[16] was encountered in initial runs. Catalytic transfer hydrogenation using Pd/C and cyclohexene[17] was successful in the laboratory, but the use of cyclohexene is not acceptable in the plant, because benzene formation by dehydrogenation is unavoidable, which is not environmentally acceptable. We found that catalyst type was the key to avoiding racemization, and of the ones tested the best results were obtained with a Pd/C catalyst obtained from Degussa.[18] This breakthrough in experimental conditions allowed us to obtain directly (S)-**2** as the tartrate salt. The yield was reproducibly 87% on a 1Kg scale. At this stage we were able to verify that indeed (S)-phenethylamine induces the S configuration at the newly formed chiral center in **2**. The overall yield of the process was, up to this point, 52-55%, and provided our key intermediate as an optically pure (>99.8% ee) material.

**Design of the manufacturing process for Ontazolast.**

The synthesis described above provides optically pure (S)-**2** in three simple chemical steps using relatively inexpensive raw materials. In addition, it uses the same first step as the original synthesis, and some optimization of the chemistry had already been carried out. Some of the issues to be addressed in the design of a large-scale process included: 1) Control of the Grignard initiation and of the exothermic nature of its formation. 2) Determination of the safety of the $NaBH_4$ process on a very large scale. 3) Evaluation of the throughput of the process, and efficient use of time and reactor capacity.

In addition, we were faced with a classical problem in process development: switch to a new synthesis had the potential of producing a "new impurity" that had not been present in the lots used for the toxicological studies in animals and the safety studies in humans. We were therefore required to control the levels of this impurity to very low levels or prove, through further preclinical and clinical studies, that lots containing this impurity were also safe. The preferred approach was, of course, to set very tight specifications for allowable content of **15** in the drug substance (0.1%) and endeavor to design a manufacturing process that reproducibly met such specification.

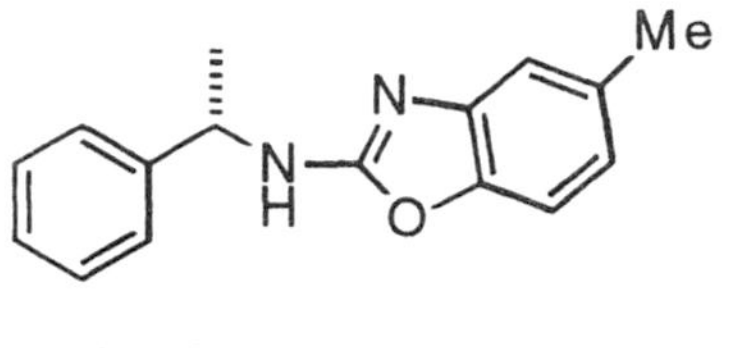

**15, New impurity in drug substance**

Our optimized synthesis of ketone **9** is shown in Scheme 8. Several minor but important changes were made with respect to the laboratory synthesis. First of all, the Grignard reagent was prepared not in MTBE but in THF. The higher boiling point of this solvent allowed us to reach a wider safety margin due to the higher boiling point of THF. The exotherm of this process is quite high, *i.e.* about 250°C adiabatic temperature rise ($\Delta H$= 580 KJ/Kg). Furthermore, the Grignard reagent is extremely soluble in THF, and this allowed us to carry out the reaction at high concentration (*e.g.* 150 Kg of cyclohexylmethyl bromide, **4**, can be reacted with Mg in a total of 300 L THF), in this way facilitating a high throughput. Most critical in the lab had been initiation of the Grignard in a reproducible fashion and control of the reaction during the addition of the bromide. Of all methods tried, the reductive activation of Mg with Vitride (Red-Al) was the most reliable and did not result in any failures.[19]

**Scheme 8**

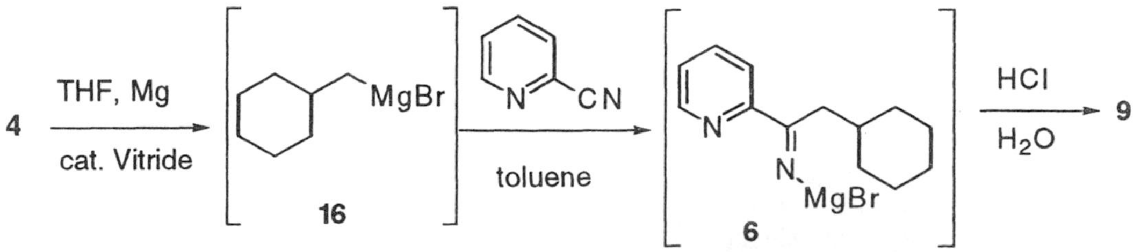

The reaction is easily controlled by the rate of addition of **4**. Once the formation of the Grignard is complete, addition of 2-cyanopyridine was carried out. Addition as a THF solution led to a very thick slurry that was essentially impossible to stir. We eventually found that adding the nitrile in toluene produced a much thinner slurry that could be easily handled on a large scale. For complete conversion, we settled on using a 25% excess of **4** *vs.* **5**. A separate study showed that lower levels left unreacted nitrile. The need for an excess of **4** is easily explained by the finding that considerable amounts (*ca.* 8%) of Wurtz product (1,2-dicyclohexyl ethane) were present in crude preparations of **9**. Using these conditions, we were able to carry out formation of **9** in a 600 L reactor on a 940 mole scale (based on input of **4**). Quenching of the slurry containing **6** was carried out by its controlled addition to a cooled solution of 1:1 concentrated hydrochloric acid-water, in such a way as to keep the internal temperature below 20°C. After the quench was complete, the pH was adjusted to 1.9-2.5 and then extraction with toluene gave the crude product. The pH has to be carefully controlled, because a pH lower than 1.9 leads to losses of **9** to the aqueous phase, and a pH above 2.5 yields gelatinous precipitation of **Mg** salts, which render the phase separation very difficult. The above toluene solution was used directly in the next step: the choice of toluene in the extraction allows us to eliminate a time-consuming evaporation step.

Imine formation had to be modified with respect to the laboratory experiments. It was found that solid catalysts, such as silica gel or montmorillonite, were effective but hard to remove because filtration was exceedingly slow in a variety of attempted protocols. Eventually we found that a small amount (0.5% mol) of thionyl chloride was effective in catalyzing Schiff base formation (Scheme 9).

**Scheme 9**

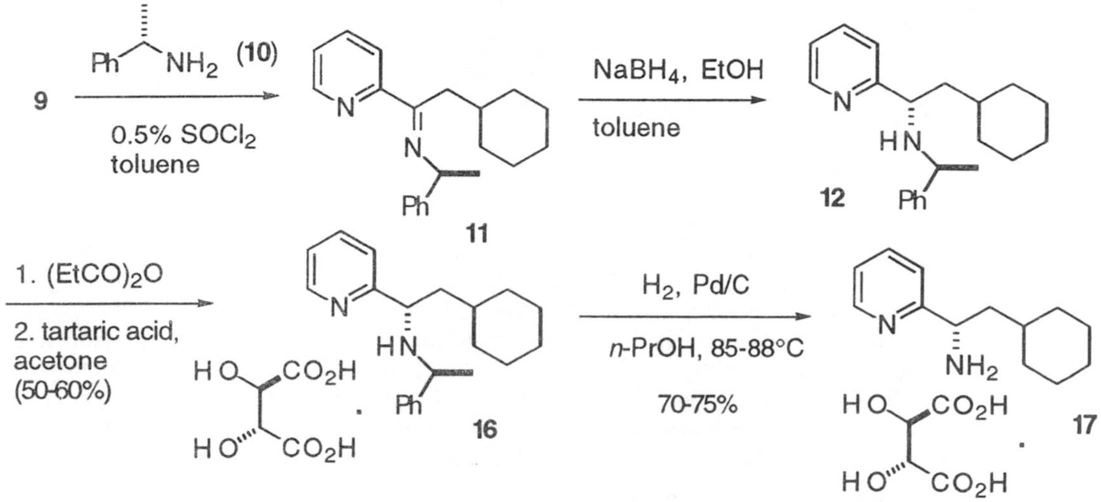

The diastereoselective reduction of **11** was carried out by slow addition of crude imine in toluene to a suspension of granular sodium borohydride in ethanol. Safety is an important issue in this step: sodium borohydride decomposes in ethanol with evolution of hydrogen. At 10°C, 1% of the borohydride decomposes within 6h, whereas at 40°C 52% decomposes in the same time frame. To avoid massive decomposition and extensive foaming, the addition was done at such a rate as to keep the internal temperature between 10°C and 20°C. At the end of the addition, water and more toluene are added until a clearcut phase separation is obtained. The organic phase is then treated with propionic anhydride; this time 0.32 equivalents (twice the relative amount used in the lab) are needed to remove all the excess phenethyl amine, probably a reflection of a lower yield in the imine formation. The solution is then added to a suspension of tartaric acid in acetone kept at 45°C-55°C and the crystalline precipitate of **16** is collected by filtration (Scheme 9).

The process can be run very efficiently in a set of six reactors. In the first two reactors (600 and 1,200 L) the Grignard reaction is carried out and the workup is performed. Imine formation and reduction are done in the next set (500 and 1,200 L) and finally propionic anhydride acylation and product crystallization are performed in another 500 L/1,200 L setup. From 166Kg of cyclohexyl bromide **4**, in three days 193 Kg of crystalline **16** was obtained. The process avoids isolation of intermediates and is very practical from the standpoint of throughput.

The quality of the product was quite acceptable: the only major impurity is phenethylamine tartrate. We set a specification of 2% maximum of this impurity (quantitative HPLC assay). It was found that intermediate **16** with impurity content under this limit gave rise to a final product meeting the specification for the maximum allowed level of **15** (0.1%). Some early batches in which insufficient amounts of propionic anhydride was used did not meet this specification, and had to be reprocessed. For all the three "registration" batches the optimum amount of 0.32 equivalents of propionic anhydride was used, and the impurity specifications for **16** were consistently met. The excess unreacted phenethylamine correlated with a lower yield in the pilot plant *vs.* the lab (ca 50% vs. 60-63%, respectively). The optical purity was consistently better than 99% over ten batches on a 150-200 Kg scale.

The scale-up of the hydrogenation for the cleavage of the chiral auxiliary group to the 100 Kg level brought additional problems: whereas uptake of 120% hydrogen (optimum uptake in terms of yield) took only 20 minutes on a 1 Kg scale, on the 110 Kg scale it took 2-4 hours and process monitoring by GC showed the presence of various side products derived from partial hydrogenation of the pyridine ring. Reoptimization of the $H_2$ uptake (135%) and temperature (85-88°C, internal) led to a consistent yield of 70-75% on the 110 Kg scale. These more drastic conditions led to some esterification of tartaric acid to propyl tartrate, and a slight excess of the acid had to be added for complete crystallization.

With these changes, excellent product quality was obtained: **17** was >99.5% pure with >99.6% ee. The only measurable impurity was phenethylamine tartrate, usually in the 0.2-0.4% range, and well within our specifications. In contrast to our original process, we define now **17** and not the free base **2** as our key intermediate. The salt **17** is crystalline and much easier to analyze and characterize. This illustrates a general strategy when optimizing processes that proceed through basic or acidic intermediates.

The final coupling step (Scheme 10) had been originally carried out in dichloromethane using Hünig's base as HCl scavenger. Although the yields and product quality were good, we had to find a replacement for the environmentally unacceptable dichloromethane and introduce an inorganic base in place of the relatively expensive tertiary amine. Since we did not want to isolate the free amine **2**, we looked for a process that could be performed in a single solvent, beginning from liberation of the free base, coupling with **6**, work-up and crystallization. This would simplify the scale-up tremendously and streamline it in terms of time and labor. Methylcyclohexane was chosen after extensive screening. Partitioning between aqueous NaOH and methylcyclohexane at 50°C (to maximize solubility of **2** in the organic phase) left the free base (S)-**2** in the organic layer; addition of **6**, a catalytic amount of sodium acetate, and an excess sodium carbonate[20] gave a suspension containing **1**. Filtration to remove the inorganic salts gave better results than aqueous workup. From the filtrate, Ontazolast crystallized upon concentration and cooling.

A new problem arose: it was not possible to dry **1** from methylcyclohexane because the "wet" crystals melt around 30°C in the drying oven. The solvent was then removed azeotropically upon adding ethanol and water. The ternary mixture has the lowest b.p. (69.6°C, for 61% methylcyclohexane, 32% ethanol, 7% water[21]). In the condensed phase, methylcyclohexane separates, and the ethanol/water phase is drained back into the reaction mixture. This guarantees that the residual content of methylcyclohexane in the drug substance is below 10 ppm, according to specifications. Further addition of water leads to crystallization. After validated drying, **1** is isolated as a monohydrate in 87% yield meeting all the predetermined specifications. The ee was higher than 99.8%. The impurity profile is essentially identical to the one found in the drug manufactured by the resolution process. The only new impurity, **15**, can be kept below the required 0.1% if the precursors contains less than 2% phenethylamine tartrate.

The process described above is quite adequate in producing the drug substance at an acceptable cost. Most importantly, the efficient throughput and minimum solvent change/manipulations ensures a rapid and effective synthesis of Ontazolast.

**Scheme 10**

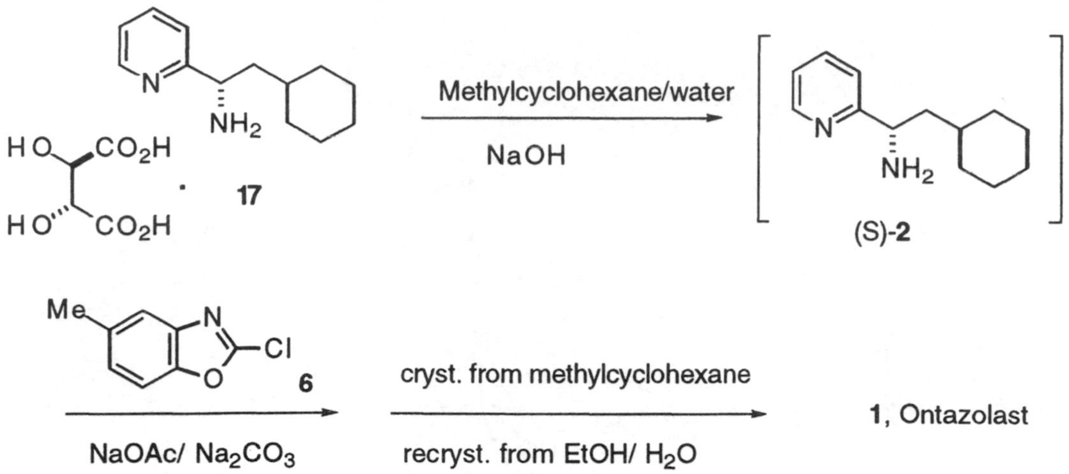

ALTERNATIVE ASYMMETRIC SYNTHESIS OF (S)-2 BY
DIASTEREOSELECTIVE ALKYLATION

**Initial studies: development of a diastereoselective alkylation.**

In recent years there have been a number of reports describing the diastereoselective α-alkylation of Schiff bases obtained from chiral carbonyl compounds to yield, after hydrolysis, amino acid derivatives as well as substituted benzylamines in variable optical purity (Scheme 11). Among the several naturally-derived auxiliaries that have been investigated are camphor (20),[22] atrolactic acid derivative 21,[23] ketopinic acid derivative 22,[24] and 2-hydroxy-3-pinanone (23).[25]

It was hoped that such an alkylation strategy, with the proper auxiliary, could be applied to the synthesis of the key pyridylamine building block (S)-2. Preliminary studies, using camphor (20), only provided modest results (Scheme 12). We found that deprotonation with either *n*-BuLi or lithium diisopropylamide, at low temperature, followed by quenching the resulting anion with a two-fold excess of cyclohexylmethyl bromide, afforded the intermediate alkylated product 25. After Schiff base hydrolysis, chiral HPLC analysis indicated the alkylation product was formed in a 35% diastereomeric excess. A variety of conditions were explored without yielding further improvement in stereoselectivity.

Next surveyed was the commercially available, although expensive (*ca.* $10/g), auxiliary 1-(R)-(+)-2-hydroxy-3-pinanone 23. Although catalog reagent suppliers quoted extremely high prices for >98% enantiomerically lots, literature precedents compelled us to evaluate its utility in the process under consideration.

Initial experiments, following the literature,[25a,b] and using cyclohexylmethyl bromide as the alkylating agent, indicated that alkylation could be achieved with complete diastereoselectivity, but the isolated yield of product was low due to incomplete alkylation (Eq. 2).

The starting imine 26 could also be recovered from the reaction mixture. The experimental conditions shown above (-78°C) are clearly not suitable for an industrial process and, although the stereoselectivity is high, yields are poor. In light of this, our effort was focused on optimization of this step. Parameters that were optimized during the course of the study were: time, temperature, stoichiometry, and electrophile. The results are shown in Table 2.

Scheme 11

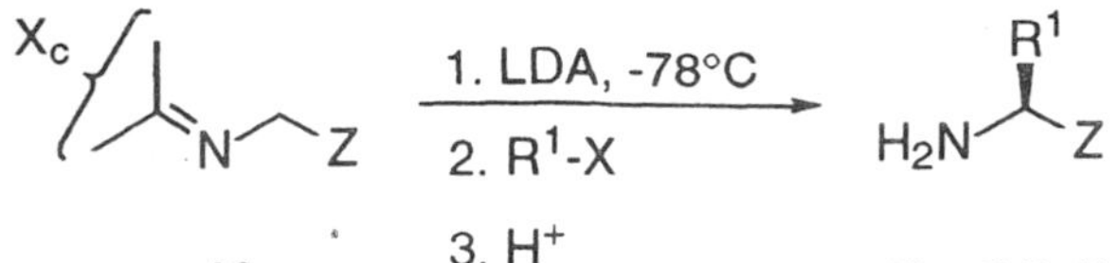

**18**

**19a**: Z = CO$_2$R   0-98% e.e.

**19b**: Z = Ph       36-90% e.e.

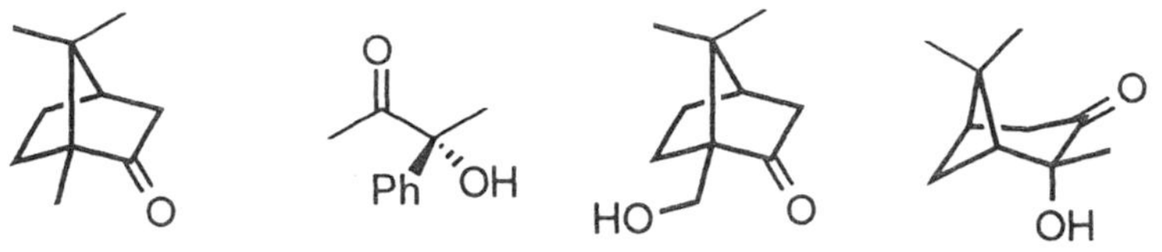

**20**          **21**          **22**          **23**

Examination of the results from the optimization study shows several trends. The initial experiments using lithium diisopropylamide (LDA) as the base were abandoned due to potential downstream problems of separating the diisopropylamine from the desired (S)-pyridylamine **2**, as well as for cost reasons. It was discovered that a reaction temperature of -78°C was not necessary for efficient stereocontrol of the alkylation and that -10 to 0°C was sufficient. A slight excess of butyllithium is necessary to quench any adventitious water present in the reaction since technical grade solvents were used. The desired dianion is formed immediately on addition of *n*-BuLi and extended reaction periods are not necessary. Additives such as TMEDA had no beneficial effect and the reaction can be run in THF or MTBE with similar results. Potassium *t*-butoxide was basic enough to deprotonate the imine, but the alkylated product was obtained as a 2:1 mixture of diastereomers. Thus, the Li counterion seems to play a key role in affecting the diastereoselectivity.

Scheme 12

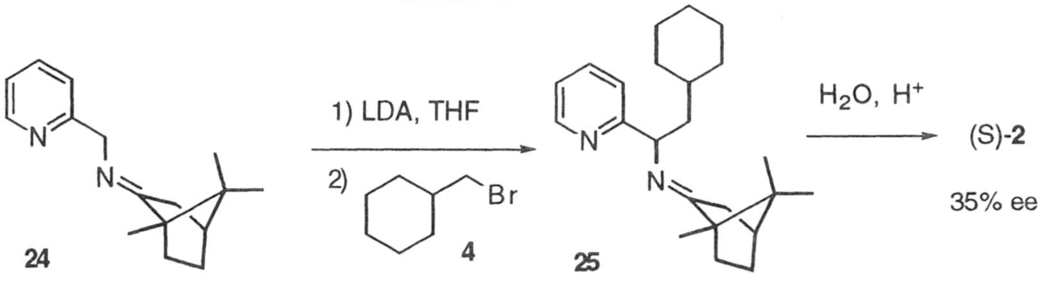

**24**                                        **25**                    (S)-**2**

35% ee

Choice of leaving group on the electrophilic component appeared to be critical to the success of the alkylation. If less than two equivalents of alkyl bromide are used, incomplete conversion is seen. The use of a mesylate as a leaving group does not give complete conversion even if used in excess. The alkyl iodide is the reagent of choice, with only a slight excess being necessary for complete reaction. Since we already had multikilogram quantities of the alkyl bromide in house, we opted to use two equivalents of this over the more efficient iodide. Cost estimates indicated that the cost of raw materials price for either process is about the same.

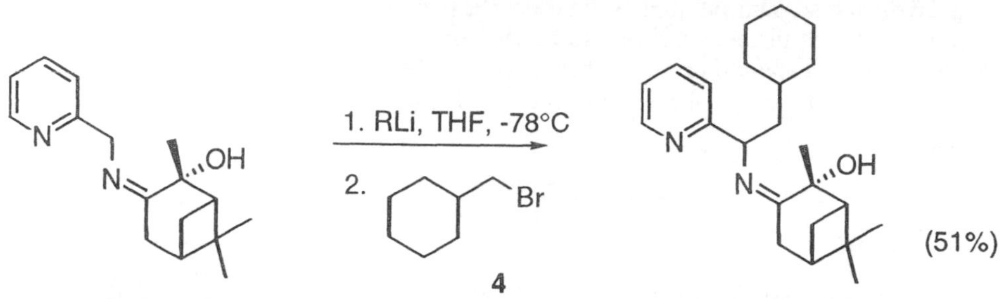

$$ (2) $$

**26** >99% e.e.    **27** >99% d.e.    (51%)

**Table 2**: Optimization of the Alkylation Reaction (Eq. 2, 1 mmol scale)

| Base | Equiv | Temp. (°C) | $T_1{}^a$ | RX | Equiv RX | $T_2{}^b$ | % conv$^c$ |
|------|-------|-----------|-----------|-----|----------|-----------|-----------|
| LDA | 2.1 | -60 | 3h | RBr | 2.2 | 2h(RT) | 56 |
| n-BuLi | 3.0 | 0 | 3h | RBr | 1.1 | 4h | 70 |
| n-BuLi | 3.0 | 0 | 3h | RBr | 2.1 | 4h | 100 |
| n-BuLi | 2.1 | 0 | 3h | RBr | 1.1 | 4h | 70 |
| n-BuLi | 2.1 | 0 | 3h | RBr | 2.1 | 4h | 90 |
| n-BuLi | 2.1 | 0 | 10 min | RBr | 2.1 | 16h | 100 |
| n-BuLi | 2.1 | 0 | 10 min | RBr | 1.5 | 16h | 73-81 |
| n-BuLi | 2.1 | 0 | 10 min | RBr | 1.0 | 16h | 40 |
| n-BuLi | 2.1 | 0 | 10 min | ROMs | 2.1 | 16h | 50 |
| n-BuLi/ TMEDA$^d$ | 2.1 | 0 | 10 min | RBr | 2.1 | 3h | 60-80 |
| n-BuLi/ TMEDA$^d$ | 2.1 | 0 | 10 min | RBr | 2.1 | 16h | 100 |
| n-BuLi | 2.1 | -10$^e$ | 10 min | RI | 1.0 | 16h | 70-95 |
| n-BuLi | 2.1 | -10 | 10 min | RI | 1.15 | 7h | 100 |
| **n-BuLi** | **2.1** | **-10** | **10 min** | **RI** | **1.15** | **16h$^f$** | **100** |
| n-BuLi$^g$ | 2.1 | -10 | 10 min | RI | 1.15 | 16h | 100 |
| t-BuOK | 2.1 | -10 | 10 min | RI | 1.15 | 16h | 100$^h$ |

a) Time allowed for anion formation after addition of base; b) Time allowed for quenching after addition of RX; c) % Conversion as determined by NMR; d) ratio = 1:1; e) This temperature on larger scale is used to control the exothermic nature of the deprotonation reaction; f) Reaction was complete after 7h but was allowed to stir overnight for convenience; g) The reaction was run in MTBE dried over molecular sieves; h) diastereoselectivity was low (2:1).

**Development of the manufacturing process.**

With proven stereoselectivity in the key alkylation step, we next turned our attention towards developing this chemistry into a manufacturing process. The remaining steps to be examined were: economic synthesis of the auxiliary, imine formation, imine hydrolysis, and final bulk product formation and isolation. In addition, recycle of the auxiliary had to be studied.

α-Pinene (**28**) is commercially available in bulk lots of different enantiomeric purity and, since several methods are available for its oxidation to **23**, we decided to study a number of methods for producing **23** in house. Initial studies involving a two-step sequence involving Sharpless dihydroxylation to **29** with further oxidation to **23** (Scheme 13): this proved fruitful and offered high yields, but it was not explored on larger than a 500 g scale due to the cost of $K_2OS_4 \cdot H_2O$ and disposal issues concerning the use of a toxic heavy metal reagent. The details of this study have been reported elsewhere.[26]

**Scheme 13**

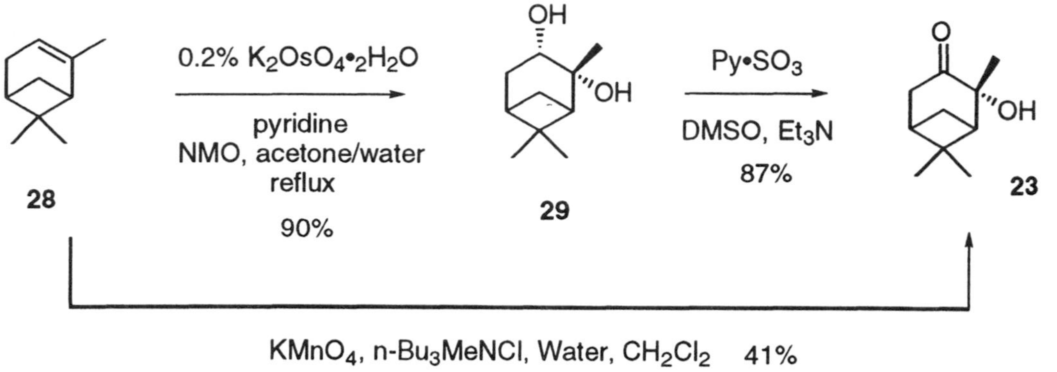

KMnO$_4$, n-Bu$_3$MeNCl, Water, CH$_2$Cl$_2$    41%

The published procedure by Carson and Pierce[27] calls for direct oxidation of **28** using KMnO$_4$ in acetone and provides the desired product in a reproducible manner. In addition, although the catalog prices for high e.e. **28** are quite high, we were able to locate a source of 93.5% e.e. **28** at a very attractive price.[28]

During the safety evaluation for this step, a report was found describing an explosion hazard when an alkylamine was oxidized to the corresponding nitroalkane in an identical permanganate/acetone system.[29] Although no problems were encountered during the initial 500 g run, it was felt that an alternative procedure should be devised. Since conventional methods for using permanganate as an oxidant for organic substrates is severely limited by the lack of compatible solvents, it was thought that use of phase transfer catalysis might be advantageous. It was found that **28** could be safely and effectively oxidized with permanganate in the presence of a catalytic amount of a quaternary ammonium salt using a water/dichloromethane solvent mixture (Scheme 13).

After several successful laboratory experiments, the reaction was scaled to two 11.0 Kg runs (based on charged **28**) in a 50 gallon glass-lined reactor. Due to the exotherm on charging aliquots of solid potassium permanganate to the reactor, it was found that several initial charges of 1 kg each were necessary in order to keep the internal reaction temperature at 2°-5°C. As the reaction proceeded, 2.0 Kg charges were suitable, although toward the end of the reaction, where heat transfer in the thick slurry was marginal, charges of 1.0 Kg only were again necessary. When addition was complete, the reaction was stirred overnight. A thick slurry formed consisting of manganese oxides and water. The mixture was too thick to be drained through the bottom valve, and therefore the supernatant was removed using suction. The remaining slurry was rinsed several times with toluene. We were ultimately able to demonstrate this reaction, with no loss of stereochemistry, on a 20 Kg scale.

The best laboratory procedure for the formation of the chiral imine **26** involved condensation of 2-(aminomethyl)pyridine with the (R)-hydroxyketone **23** in refluxing benzene using 10 mole% of BF$_3$•Et$_2$O as the catalyst. In this way, reproducible yields of 98% were obtained. Since benzene is not acceptable in a manufacturing process and BF$_3$ is not compatible with glass-lined reactors, an alternative procedure was required. *p*-Toluenesulfonic acid was not an effective catalyst. It was found that titanium isopropoxide was an effective promoter if at least 10 mole% was used. Lower amounts (1-5%) increased reaction time and decreased the quality of **26**. The resulting titanium solids had a tendency to form gels, and therefore addition of silica gel was found necessary to aid filtration. The crude filtrate and washes were then concentrated to completely remove the toluene and the crude ketimine was used directly in the next step. Although this protocol was effective for laboratory scale synthesis (<500 g), the filtration proved problematic on scale up. Initial pilot plant studies showed that the finely dispersed solids would not settle even after 14h and the resulting suspension had to be filtered through a pad of celite. Additional laboratory work was carried out and it was found that 5 mole% of thionyl chloride effectively catalyzes the condensation reaction (Eq.3).

The critical alkylation step proved uneventful using the optimized conditions highlighted in Table 2. The reactions were run at a 0.4M concentration (76L THF) in a glass-lined 50 gallon reactor under a N$_2$ atmosphere. No major exotherm was noted during the addition step and *n*-BuLi could be charged at a fairly rapid rate (26 L within 40 min.). After addition was complete, the reaction was held at -10°C and the deep red solution was allowed to stir for 1 hour. Neat cyclohexylmethyl bromide (2 equiv) was added through the dispenser at a rate such that the internal temperature was maintained at -10°C. Once the addition was complete, the reaction was allowed to stir overnight at -10°C. The resulting mixture was quenched by addition of 20% aqueous ammonium chloride solution while keeping the temperature below 5°C. The crude **27** was directly subjected to hydrolysis (Eq. 4).

Some problems were encountered in the hydrolysis step. Ketimines, in general, can be hydrolyzed either by acids or bases. Acidic hydrolysis, the method of choice for amino acid synthesis using this auxiliary, was not suitable for our application. Attempts at hydrolysis using 1 to 4M citric acid solutions were unsuccessful, whereas HCl (1- 2M) proved to be too harsh and led to decomposition as well as racemization of the imine. Imine exchange is also useful for cleavage of ketimines and the literature reports that hydroxylamine acetate is an effective reagent for this cleavage.[30] It was found that treatment of the imine **27** from the crude reaction mixture led directly in good yield to amine **2**. Therefore, after quenching the alkylation reaction, agitation was stopped and the reactor contents were allowed to settle for 30 minutes. The aqueous phase was removed through the bottom valve. The THF solution was treated with a small amount of a 2:1 mixture of ethanol/water to enhance solubility and then with hydroxylamine hydrochloride.

After stirring at 20°C for a 20 hour period, the mixture was acidified by addition of 1.5M hydrochloric acid solution and extracted with ethyl acetate in order to remove the hydrolyzed auxiliary, excess cyclohexylmethyl bromide, and *n*-pentylcyclohexane, a by-product. It is worth noting that the extra equivalent of cyclohexylmethyl bromide could be recovered by distillation in 85% efficiency. The amine-rich aqueous fraction was made basic by the addition of concentrated ammonium hydroxide. Cooling was necessary in order to maintain a reaction temperature of less than 20°C. The mixture was extracted with dichloromethane. After extraction, the organics were charged back to the reactor and dried using sodium sulfate. This formed a hard cake and it was possible to drain the reactor through the bottom valve without filtration. The dichloromethane was then distilled off while adding isopropanol to the reactor. The isopropanol solution of (S)-**2** was further distilled to a small volume. Using a measured aliquot, it was estimated by weight that the 22.95 Kg amine/isopropanol solution contained 3.30 Kg (54%) of the desired product.

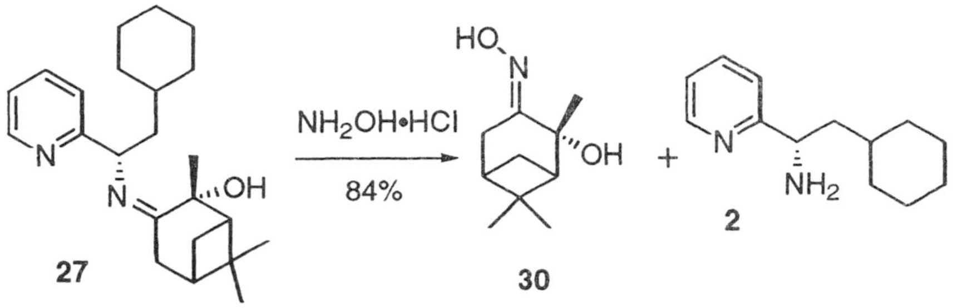

Scaling the reaction from millimole to up to a 30 mole scale proceeded smoothly and imine formation, alkylation, and hydrolysis sequence to provide **2** were carried out without the need for isolation of any intermediates. A second alkylation run was conducted in an identical manner. The only modification was to eliminate the use of sodium sulfate for the drying step. Instead, the residual water was removed via azeotropic distillation with the added isopropanol. The yield (56.5%) was estimated in a manner identical to that described for the first run.

Chiral HPLC analysis of the crude (S)-**2** from each run showed the material to be 91.6% and 89.9% e.e. respectively. This indicates that there was about a 1-2% erosion of e.e. from auxiliary **23**. At this point, the product solutions from both runs were combined into a single lot for the enrichment step.

Obviously, since **28** was only available in 93.5% e.e., the resulting (S)-pyridylamine **2** required an enantiomeric enrichment step. Several commercially available chiral acids were surveyed, and it was found that L-tartaric acid worked best. As shown in Eq. 5, treatment of a wet isopropanol solution of crude **2** with one equivalent of L-tartaric acid, with slow stirring overnight, furnished the corresponding salt **17**, which could be isolated as a stable white solid. This convenient method not only allowed for the isolation of the key intermediate as a stable free-flowing solid but also increased the e.e. to >99%.

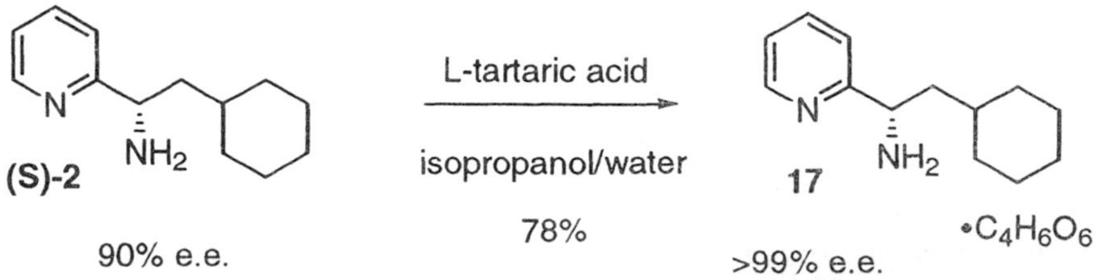

Overall, our campaign in fixed equipment was successful and suggested that this is a viable process for the synthesis of the (S)-pyridylamine building block for Ontazolast. In addition to demonstrating a feasible process, a reproducible, safe oxidation procedure was developed for the synthesis of the chiral auxiliary from α-pinene. Although any efficient diastereoselective process must address the recovery of the auxiliary at the end of the synthesis, this operation was not studied extensively. Even without this recycle, the process was found to be roughly comparable from the viewpoint of cost and efficiency to the one discussed in the previous section. The ultimate choice was made in favor of process #2 (Scheme 6) because - at a comparable cost and throughput - it was at least one year ahead in development and, given our extensive experience in scaling this route, we were very confident that >100 Kg quantities of Ontazolast meeting specifications could be safely and rapidly obtained.

## REFERENCES

1 Farina, P. R., Graham, A.G.; Hoffman, A.F.; Watrous, J.M.; Borgeat, P.; Nadeau, M.; Hansen, G.; Dinallo, R.M.; Adams, J.; Miao, C.K.; Lazer, E.S.; Parks, T.P.; Homon, C.A. *J. Pharmacol. Exp. Ther.* **1994**, *27*, 1418.

2 Beak, P.; Zaidel, W.J.; Reitz, D.B. *Chem. Rev.* **1984**, *84*, 471.

3 Gawley, R.E.; Rein, K in *Comprehensive Organic Synthesis*; Trost, B.M., Ed.; Pergamon Press: Oxford, 1990; Vol.1, Chapter 2.1.

4 Dolling, U.H.; Davis, P.; Grabowski, E.J.J. *J. Am. Chem. Soc.* **1986**, *106*, 446; Hughes, D.L.; Dolling, U.H.; Ryan. M.; Schoenewaldt, E.F.; Grabowski, E.J.J. *J. Org. Chem.* **1987**, *52*, 4745.

5 O'Donnell, M.J.; Bennett, W.D.; Wu, S. *J. Am. Chem. Soc.* **1989**, *111*, 2353; O'Donnell, M.J.; Wu, S.; Huffman, J.C. *Tetrahedron* **1994**, *50*, 4507.

6 Annunziata, R.; Cinquini, M.; Cozzi, F. *J. Chem. Soc. Perkin Trans. I* **1982**, 339; Takahashi, H.; Chida, Y.; Yoshii, T.; Suzuki, T.; Yanaura, S. *Chem. Pharm. Bull.* **1986**, *34*, 2071; Takahashi, H.; Hsieh, B.C.; Higashiyama, K. *Chem. Pharm. Bull.* **1990**, *38*, 2429; Alberola, A.; Andres, C.; Pedrosa, R. *Synlett* **1990**, 763; Wu, M.-J.; Pridgen, L.N. *J. Org. Chem.* **1991**, *56*, 1340; Miao, C.K.; Sorcek, R.; Jones, P.-J. *Tetrahedron Lett.* **1993**, *34*, 2259.

7 Denmark, S.E.; Nicaise, O. *Synlett* **1993**, 359.

8 Tomioka, K.; Inoue, I.; Shindo, M.; Koga, K. *Tetrahedron Lett.* **1991**, *32*, 3095; Inoue, I.; Shindo, M.; Koga, K.; Tomioka, K. *Tetrahedron* **1994**, *50*, 4429.

9 a) Demailly, G.; Solladie, G. *Tetrahedron Lett.* **1975**, 2471; b) Pirkle, W.H.; Hauska, J.R. *J. Org. Chem.* **1977**, *42*, 2436; c) Wiehl, W.; Frahm, A.W. *Chem. Ber.* **1986**, *119*, 2668; d) Bringmann, G.; Geisler, J.-P. *Tetrahedron Lett.* **1989**, *30*, 317; *Synthesis* **1989**, 608; e) Periasamy, M.; Devasagayaraj, A.; Satyanarayana, N.; Narayana, C. *Synth. Commun.* **1989**, *19*, 565; f) Fuller, J.C.; Belisle, C.M.; Goralski, C.T.; Singaram, B. *Tetrahedron Lett.* **1994**, *30*, 5389; g) Moss, N.; Gauthier, J.; Ferland, J.-M. *Synlett* **1995**, 142; h) Kowalczyk, B.A.; Rohloff, J.C.; Dvorak, C.A.; Gardner, J.O. *Synth. Commun.* **1996**, *26*, 2009.

10 Kang, G.-J.; Cullen, W.R.; Fryzuk, M.D.; James, B.R.; Kutney, J.P. *J. Chem. Soc. Chem. Commun.* **1988**, 1467; Spindler, F.; Pugin, B.; Blaser, H.-U. *Angew. Chem. Int. Ed. Engl.* **1990**, *29*, 558; Burk, M.J.; Feaster, J.E. *J. Am. Chem. Soc.* **1992**, *114*, 6266; Cho, B.T.; Chun, Y.S. *Tetrahedron: Asymmetry* **1992**, *3*, 1583; Burk, M.J.; Martinez, J.P.; Feaster, J.E.; Cosford, N. *Tetrahedron* **1994**, *50*, 4399; Murahashi, S.-I.; Watanabe, S.; Shiota, T. *J. Chem. Soc. Chem. Commun.* **1994**, 725.

11 Kaufmann, G.B. *J. Prakt. Chem.* **1966**, *33*, 295.

12 Reider, P.J.; Davis, P.; Hughes, D.L.; Grabowski, E.J.J. *J. Org. Chem.* **1987**, *52*, 955.

13 (a) Roth, H.J.; Kleeman, A.; Beisswenger, T. *Pharmaceutical Chemistry*; Ellis Horwood, Chichester, 1988; vol.1; (b) Wilen, S.H.; Collet, A.; Jacques, J. *Tetrahedron* **1977**, *33*, 2725.

14 Langbein, A.; Schneider, H.; Bressler, G.-R. DE 4428531 A1 (1994)

15 Balkenhol, F.; Hauer, B.; Landner, W.; Pressler, U. DE 4332738 (1993).

16 Murahashi, S.-I.; Yoshimura, N.; Tsumiyama, T.; Kojima, T. *J. Am. Chem. Soc.* **1983**, *105*, 5002.

17 a) Farkas, E.; Sunman, C. *J. Org. Chem.* **1985**, *50*, 1110; b) Rico, J.G.; Lindmark, R.J.; Rogers, T.E.; Bovy, P.R. *J. Org. Chem.* **1994**, *58*, 7948; c) Rajagopal, S.; Spatola, A.F. *J. Org. Chem.* **1995**, *60*, 1347.

18 The catalyst was developed for the debenzylation of amines; it is designed as 10% Pd/C E106XNO/W and contains 50% water.

19 Simanek, V.; Klasek, K. *Collect. Czech. Chem. Commun.* **1973**, *38*, 1614.

20 Stoichiometric sodium acetate may not be a strong enough base to promote the process (indeed it worked poorly), but sodium carbonate alone leads to very long reaction times. Probably sodium acetate, being more soluble in methyl cyclohexane, serves to transfer protons to the  much less soluble carbonate.

21 Swietoslawski, W.; Zieborak, K.; Galska-Krajewska, A. *Bull. Acad. Pol. Sciences, Ser. Sci. Chim. Geol. et Geogr.* **1959**, *7*, 43.

22 a) Yaozhong, J.; Guilan, L.; Changyou, Z. *Synthetic Commun.* **1987**, *17*, 1545-1548; b) McIntosh, J.M.; Mishra, P. *Can. J. Chem.* **1986**, *64*, 726-731.

23 Nakatsuka, T.; Miwa, T.; Mukaiyama, T. *Chem. Lett.* **1981**, 279-282.

24 Ikota, N.; Sakai, H,; Shibata, H.; Koga, K. *Chem. Pharm. Bull.* **1986**, *34*, 1050-1055.

25 a) Yuanwei, C.; Aiqiao, M.; Xun, X.; Yaozhong, J. *Synthetic Commun.* **1989**, *19*, 1423-1430; b) Aiqiao, M.; Xun, X.; Lanjun, W.; Yaozhong, J. *Synthetic Commun.* **1991**, *21*, 2207-2212; c) El Achqar, A.; Boumzebra, M.; Roumestant, M.L.; Viallefont, P. *Tetrahedron* **1988**, *44*, 5319-5332; d) Solladié-Cavallo, A.; Simon, M.C. *Tetrahedron Lett.* **1989**, *29*, 2411-2414.

26 Krishnamurthy, V.; Landi, J.J.; Roth, G.P. *Synth. Commun.* **1997**, *27*, 853.

27 Carlson, R.G.; Pierce, J.K. *J. Org. Chem.* **1971**, *36*, 2319.

28 We purchased ca. 2,000 lbs. of **28** (93.5% ee, determined by chiral GC) from SCM-Glidco Organics (Jacksonville, FL) for $ 2.5/lb.

29 *Bretherick's Handbook of Reactive Chemical Hazards*; Urben, P.G., Ed.; 5th Ed., Vol. 1; Butterworth-Heinemann: London, 1995, 1627.

30 Yamada, S.I.; Oguri, T.; Shioiri, T. *J. Chem. Soc. Chem. Commun.* **1976**, 136.

# Dilevalol Hydrochloride:
# Development of a Commercial Process

**Derek Walker**

*Chemical Development, Schering-Plough Research Institute, Union, N.J. 07083*

Key words: dilevalol; optical resolution; chiral synthesis; outsourcing; process plant; regulatory

## 1. Background

The turbulent nature of New Chemical Entity (NCE) development is widely recognized. Only 1 in 10 or so new entities identified for development will reach the market place[1]. Pharmacological, Toxicological and Clinical findings can halt development at any time. The rate of progression of a new entity can be greatly changed by toxicology issues, by metabolic findings, formulation difficulties, the availability of bulk supplies, readouts from both clinical studies and the FDA, changing market conditions and so on. In short, priorities are constantly shifting. The process of NCE development also runs so fast that, from a Chemical Development standpoint, conflicts between keeping the IND synthesis vs. improving or changing to a better synthesis provide endless challenge and, frequently, much frustration. In an NCE development respect, Chemical Development organizations mostly act in a service role, providing high quality NCE's for toxicology, clinical, pharmaceutical development and analytical programs in a timely manner. Beyond the service role, Chemical Development's contribution to the identification and development of a commercial process is generally crucial. In this area, involvement with Manufacturing organizations is essential — indeed the most successful Chemical Development organizations have a close link with Manufacturing.

In achieving its mission a Chemical Development organization relies upon Chemists, Engineers and Analysts as the primary professionals and works to forge important collaborations with other groups. Above all, Chemical Development ensures the safety of its chemical processes from the beginning and increasingly concerns itself with creating as environmentally sound a process as possible. Beyond Safety the groups of most importance, especially at the start of a project, are Quality Control and Regulatory Affairs — a continuous dialogue addresses issues and expedites the filing of IND's, IND updates and NDAs.

The development of a commercial process for the manufacture of dilevalol hydrochloride was not much affected by toxicology, pharmacology, or formulation considerations since experience with labetalol hydrochloride had provided a knowledge base on which to build.  The greatest challenges were provided by the need to produce large quantities very quickly, by the cost-of-goods (COG) target and by chiral synthesis requirements — we assumed that a chiral synthesis would lead to the lowest COG.

## 2. Introduction

The development of dilevalol hydrochloride[2] for use as a vasodilator and competitive antagonist at ß-andrenergic receptor sites was an outgrowth of efforts in Schering-Plough, initiated in the late 1970's, to determine whether a single enantiomer of the racemic $\alpha$ - and ß-antagonist labetalol would offer an advantageous marketing situation.  It was speculated that one of the four enantiomers which comprise labetalol would carry enhanced ß -andrenergic receptor blocking activity and fewer side effects.

Testing the concept required the preparation of pure samples of each of the four enantiomers.  These were prepared by Gold et al.[3] and Chemical Development using classical resolution and chiral synthesis methods.  It was quickly found that the R R enantiomer, later named dilevalol, was virtually free of $\alpha$-andrenergic receptor blocking activity, and also possessed superior vasodilator properties vs. labetalol.

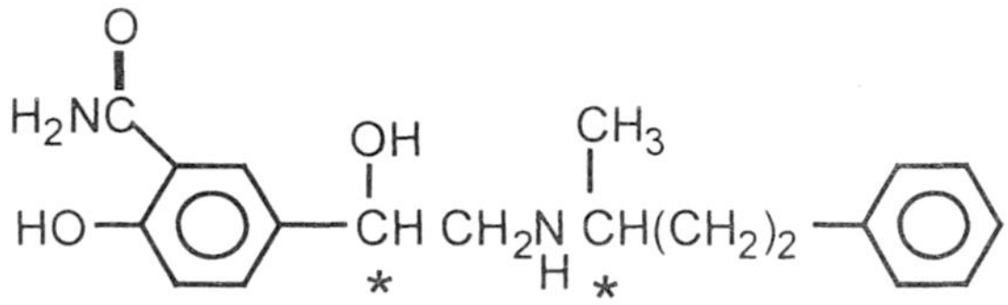

| labetalol | RS | RS |
|-----------|----|----|
| dilevalol | R  | R  |

The receptor blocking properties of the four enantiomers were published by Gold et al.[3] and Hartley[4].  A summary of Gold's figures is presented in Table 1.

Table 1. Summary of Comparative Cardiovascular Effects of
Labetalol and its Stereoisomers Relative Potencies*

| Compound | ß$_1$ Receptor Blockade | $\alpha$ Receptor Blockade | Vasodilation |
|----------|------------------------|----------------------------|--------------|
| labetalol | 1 | 1 | 1 |
| RR - Isomer | 3.5 | < 0.2 | 7 |
| RS - Isomer | < 0.06 | 0 | - |
| SR - Isomer | < 0.05 | 5.1 | - |
| SS - Isomer | 0 | 1.5 | - |

*Potencies normalized to labetalol = 1. ß$_1$ - Blockade and $\alpha$-blockade are on different absolute scales — see Gold et al.[3] for detail and qualifications.

These results demonstrate that the RR and SR enantiomers are most responsible for the ß and α- blocking activities, respectively. Hartley[4] also showed that a 1:1 mixture of RR and SR enantiomers was twice as active a ß-blocker as labetalol and about 1.3 times as potent an α-blocker.

Based on knowledge of the activity of the RR enantiomer and other Marketing considerations, the Schering Cardiovascular Therapy Team decided to pursue the development of a 100-200 mg maintenance dose twice daily. This situation raised cost-of-goods (COG) questions which were addressed to Chemical Development through the Marketing representative in the Cardiovascular Therapy Team. By the time of the COG request, Chemical Development had taken over the implementation and development of the chiral synthesis of dilevalol outlined by the Schering-Plough Research organization. Although this chiral synthesis was improved and shown to be workable from an initial supply standpoint (indeed it was scaled up to meet urgent bulk drug supply needs) it was not considered a good candidate for commercial operation. Nevertheless, in the spirit of reaching to achieve the lowest COG, we in Chemical Development projected that considerable simplification of the Research chiral synthesis should be possible (see later). Since both the research chiral synthesis and the projected simplified synthesis of dilevalol were based on the original labetalol synthesis, we "guesstimated," following discussion with Schering Manufacturing who produced labetalol, that a fully absorbed manufacturing cost (raw materials, labor and overhead) at a 50 tonne/annum scale (a figure provided by Marketing for 3 years after launch) should be in a range as follows:

Cost of Dilevalol = [3x cost of Labetalol] ± 25%

Not surprisingly, given these figures, Marketing promptly set the COG target at [3x Cost of Labetalol] **minus** 25%!

It is worth adding a cautionary note in regard to COG projections. In the early phases of a project, such as was the case with dilevalol, there is a danger that COG projections might be used to justify termination of a project, rather than serve to challenge the creativity of process R&D Chemists to invent a better synthesis. Fortunately, in the dilevalol case, the Cardiovascular Therapy Team and particularly Marketing and Manufacturing, aggressively supported the Development Chemists and Engineers in their efforts to create a simpler synthesis. COG projections were used, as the project developed, to validate that the core simpler, lower-cost synthesis strategy was viable and to identify those features and components of the synthesis most in need of improvement.

### 3. Early considerations in Selecting a Synthesis Route for Further Development

The possibility of separating dilevalol from labetalol was considered as an option at the commencement of the dilevalol project. However, it quickly became clear, from work carried out in both Research and Development, that this approach might not be a viable option. Although the racemic pairs (RR + SS and RS + SR) were separable by crystallization, and although the optical resolution of the RR and SS enantiomers could be achieved through salt formation with a chiral acid, the direct yield of dilevalol was less than 20%. Nevertheless, it was recognized that, if the recovery and recycling of the waste streams from the physical and optical resolutions could be carried out efficiently, considerable economies would be obtained (Scheme 1).

**Scheme 1**

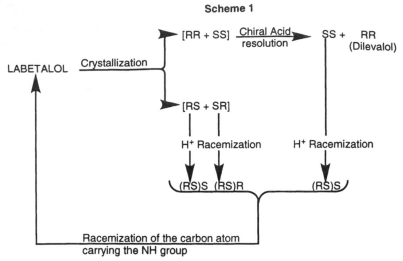

Acid catalyzed racemization of the benzylic alcohol in the waste was shown to be quite straightforward. However, racemization of the carbon carrying the amino group appeared likely, from probing experiments, to prove difficult especially on a manufacturing scale. Moreover, COG concerns haunted the prospects of creating an efficient separation and recycling process especially one starting with the most expensive molecule in the labetalol synthesis, labetalol itself. In summary, labetalol was considered unsatisfactory as a starting material on account of the logistics of the initial separations, the low one-pass yield of dilevalol , the need for two steps to racemize the very large quantity of waste and the excessive solids handling requirements. Simple calculations showed that the Marketing COG target for dilevalol would not be attainable starting with labetalol. Furthermore it was quickly evident that a large labor-intensive manufacturing plant would be needed to develop a dilevalol process based on labetalol.

The above realties, and the realization that large quantities of dilevalol would be needed quickly for the Toxicology, Clinical and Pharmaceutical Development programs, led Research to propose a chiral synthesis for the initial supplies. The process identified was based on the use of a labetalol intermediate and analogous chemistry to that used in the subsequent manufacturing steps for labetalol.

## 4.  Synthesis of Initial Supplies of Dilevalol for Cardiovascular Therapy Team Programs

The process for the manufacture of labetalol is outlined in Scheme 2.

**Scheme 2**

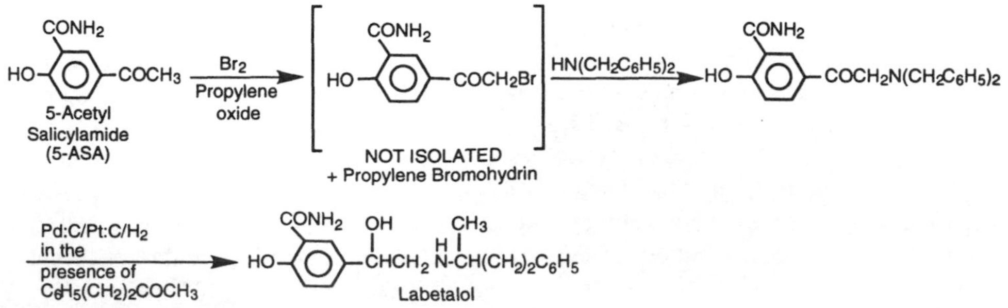

It was clear that a dilevalol synthesis strategy based on Scheme 2 would be advantageous. Use of the same intermediates and synthesis scheme as used for labetalol introduces operating economies. In addition faster implementation and lower costs were anticipated by building on existing operations. Furthermore, although it was recognized that dibenzylamine in the labetalol synthesis was an expensive way of introducing the $NH_2$ group needed in labetalol, it was reasoned that use of a secondary amine, with the desired chirality already built in, may lead to induction of chirality in the subsequent reduction step. The Research synthesis was based on the O-benzyl derivative of 5-ASA (I), and is outlined in Scheme 3.

**Scheme 3**

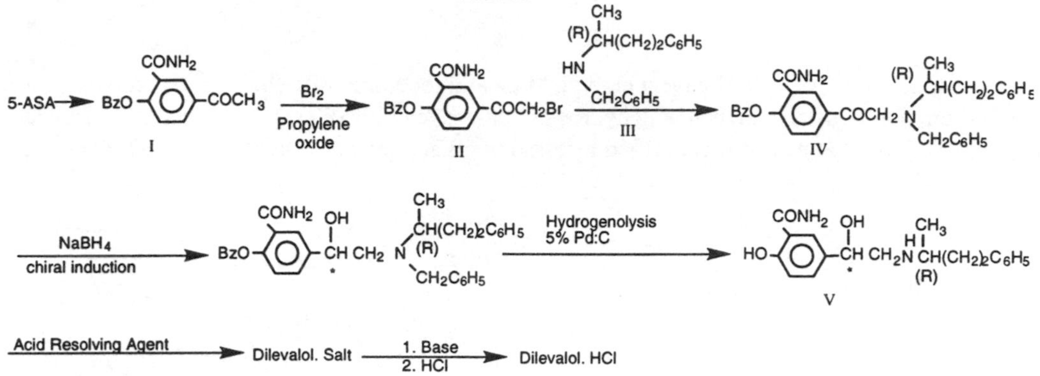

5-ASA → BzO—⟨○⟩(CONH₂)—COCH₃  (I)  —Br₂, Propylene oxide→  BzO—⟨○⟩(CONH₂)—COCH₂Br  (II)  —III→  BzO—⟨○⟩(CONH₂)—COCH₂N((R)CH(CH₂)₂C₆H₅)(CH₂C₆H₅)  (IV)

III: HN((R)CH(CH₃)(CH₂)₂C₆H₅)(CH₂C₆H₅)

—NaBH₄ chiral induction→  BzO—⟨○⟩(CONH₂)—CH(OH)CH₂N*((R)CH(CH₃)(CH₂)₂C₆H₅)(CH₂C₆H₅)  —Hydrogenolysis 5% Pd:C→  HO—⟨○⟩(CONH₂)—CH(OH)CH₂N H CH(CH₃)(CH₂)₂C₆H₅ (R)  (V)

—Acid Resolving Agent→ Dilevalol. Salt —1. Base, 2. HCl→ Dilevalol. HCl

*Enriched in (R)

In this approach the R-amine moiety in III was considered likely, in view of the work of Koga and Yamada[5] and later Kametani et al.[6], to provide some inductive control in the sodium borohydride reduction of IV. Moreover, the R-amine moiety is a necessary component of dilevalol. Desired inductive control was quickly demonstrated by Gold et al. (internal communication, October 25, 1979). However, a broad study of process conditions, particularly of solvent and temperature effects, only gave, at best, a ratio of RR to SR of

| RR | SR |
|----|----|
| 75% | 25% |

A later publication by Hartley[4] validated these figures.

The above result provided a basis for the idea that if both alkyl substituents on the amine moiety were of R-configuration inductive control in the sodium borohydride reduction of the keto group might be greatly increased. Since α-methylbenzylamines are known to hydrogenolyse relatively easily (cf benzylamine itself) α-methylbenzyl substitution was considered a good choice. The ready availability of both RS-and R-α-methylbenzylamine prompted investigation of this proposal.

The RR-secondary amine (VI) was synthesized as follows:

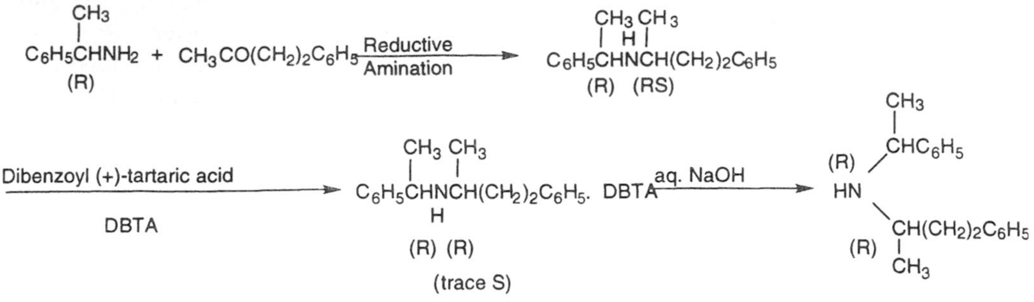

CH₃
|
C₆H₅CHNH₂ + CH₃CO(CH₂)₂C₆H₅ →(Reductive Amination)→ 
(R)

CH₃ CH₃
| H |
C₆H₅CHNCH(CH₂)₂C₆H₅
(R) (RS)

Dibenzoyl (+)-tartaric acid ─(DBTA)→
CH₃ CH₃
| |
C₆H₅CHNCH(CH₂)₂C₆H₅ · DBTA ─(aq. NaOH)→
H
(R) (R)
(trace S)

CH₃
|
(R) CHC₆H₅
HN
(R) CH(CH₂)₂C₆H₅
|
CH₃

VI

Reaction of VI with II gave a high yield of aminoketone VII which when reduced first with sodium borohydride and then with hydrogen Pd/C gave crude dilevalol isolated as its acetate (Scheme 4). The acetate salt was purified by dissolution and crystallization as its DBTA salt.

**Scheme 4**

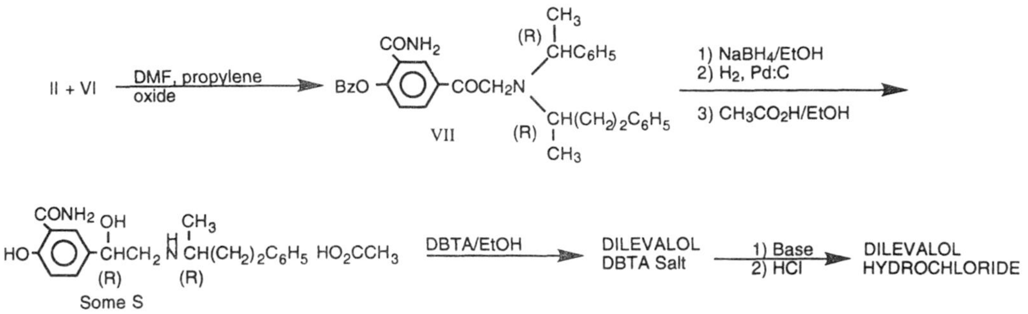

II + VI ─(DMF, propylene oxide)→ 
CONH₂
BzO─⟨⟩─COCH₂N
(R) CHC₆H₅ / CH₃
(R) CH(CH₂)₂C₆H₅ / CH₃
VII
─(1) NaBH₄/EtOH; 2) H₂, Pd:C; 3) CH₃CO₂H/EtOH)→

CONH₂ OH CH₃
| | | |
HO─⟨⟩─CHCH₂ N CH(CH₂)₂C₆H₅ HO₂CCH₃ ─(DBTA/EtOH)→ DILEVALOL DBTA Salt ─(1) Base; 2) HCl)→ DILEVALOL HYDROCHLORIDE
(R) (R)
Some S

In order to determine whether or not added steric effects in using α-methylbenzyl contributed to the induction of R configuration in the sodium borohydride reduction, the same sequence of reactions was carried out employing RS-α-methylbenzylamine. Analysis of the products from both reaction sequences gave the results summarized in Table II.

Table II

| Conformation of 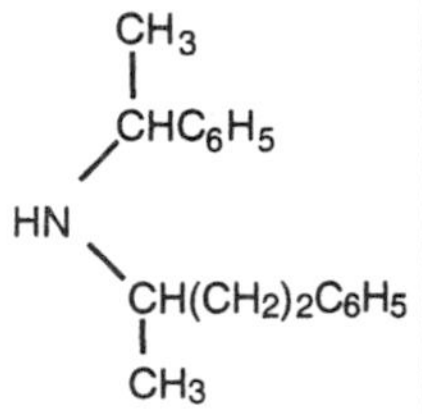 (CH₃ CHC₆H₅ HN CH(CH₂)₂C₆H₅ CH₃) | Stereoisomer Analysis of Crude Dilevalol Acetate | | | | Stereoisomer Analysis of Dilevalol DBTA Salt | | | | Yield of Dilevalol DBTA Salt based on Starting RR Sec. Amine |
|---|---|---|---|---|---|---|---|---|---|
|  | RS | SR | RR | SS | RS | SR | kR | SS |  |
| RR:RS*=52:48 | — | 21.3 | 77.5 | 1.2 | — | 4.3 | 94.6 | 1.2 | 50.5% |
| RR > 98% | 0.7 | 10.2 | 86.9 | 2.1 | 0.2 | 2.3 | 96.2 | 1.3 | 59.0% |

*S component derives from RS-α-methylbenzylamine.

Table II indicates that there is little difference between benzyl and RS-α-methyl benzyl in terms of inductive effect in the ketone reduction. Furthermore, as expected, the use of the secondary amine VI, in which both amine substituents were R in conformation did give a desirable increase in the yield of dilevalol. As an aside it is known that hydrogenation of the keto group in compound IV gives a 1:1 mixture of the RR and SR enantiomers[4]. From this it is clear that the most important factors in the induction of maximum chirality in the ketone reduction step are the complexation of borohydride with the amine function[5] and the like chirality of the N-alkyl substituents.

The above Scheme 4 process utilizing RR-amine (VI), referred to as the "RR-amine process", became the IND process for producing hundreds of kilos of dilevalol hydrochloride needed for the early clinical, toxicological and pharmaceutical sciences work. The Scheme 4 process was patented[7]. It should be noted that, despite a great deal of work on process conditions, solvents and reducing agents the ratio of RR to SR could not be improved. Thus, in a practical sense, it proved impossible to eliminate the DBTA resolution as a step for creating desired chiral purity. Minor changes in the process and improving the optical purity of VI gave an acceptable quality dilevalol with an RR assay of generally greater than 97%. Process changes were filed, as IND updates, with the FDA as they were validated. The full analytical specification set for dilevalol, as its hydrochloride salt, was

- Description: White to off-white powder.
- Identification:
    A. I.R.- Agrees with reference standard specification.
    B. Chloride - Responds to test.
    C. TLC - Sample spot migrates at the same rate ($R_f$) as the reference standard spot.
- Related Compounds: Maximum 1% total with not more than 0.5% of any one substance.
- Stereoisomer Content: Maximum 3% total of other stereoisomers (SS + RS + SR).
- Specific Rotation $[\propto]^{26}_D$ = -26.5° to -30.5°
- Loss on Drying: Maximum 0.5%.
- Residue on Ignition: Maximum 0.1%.
- Heavy Metals: Maximum 0.002%.
- Assay (HPLC): 97-102% (calc. on dry basis).

The impurity profile (synthesis related impurities) was actually better for dilevalol than for the parent compound, labetalol, reflecting the benefits of further purification during the DBTA resolution step. Thus, although dilevalol hydrochloride contained traces of DBTA itself, no tertiary amine impurities or brominated dilevalol could be detected — both types of impurity are present at very low levels in labetalol.

## 5. Selecting the NDA Process and Reducing the Cost-Of-Goods

The above "RR-amine process", based on RR-amine (VI), was used for the first approximately two year production program supplying most of the early requirements of bulk drug for the Clinical, Toxicology and Pharmaceutical Development programs. During this time efforts were undertaken to improve the process, and assess its commercial potential. At the same time process research was going on in Chemical Development to evaluate the simpler synthesis, referenced earlier, which was projected as likely, if successfully developed, to meet the Marketing COG targets. Other ideas for improving the simpler synthesis, as well as ideas for radically different syntheses, were "championed" during this period.

Initially, only a small effort was disposed to assess the feasibility of the simpler synthesis. This grew at the expense of the "RR-amine process" as it became evident that this process was, like its forerunner (the separation of labetalol isomers), unlikely to achieve the Marketing COG target. Ideas for radically different syntheses were given an even lower priority and were often left to "bootleg efforts" by the originators.

The simpler synthesis was built on the same premise as the "RR-amine process", namely that it should be based on the existing labetalol process, particularly in terms of using the same or similar raw materials and intermediates, wherever possible, and also using similar plant equipment. The simpler synthesis grew out of a critique of the disadvantages of the "RR-amine process" (Scheme 4).

| Main Disadvantages of Scheme 4 | Improvements |
|---|---|
| 1. Too Many Steps. | Avoid steps or combine them. |
| • Is benzylation of the phenol necessary. | Test elimination of Bz protection. |
| • RR-Amine is a new compound — low cost source needed. | Go to third party - minimize investment. |
| • Chiral reduction not 100% — DBTA resolution unavoidable. | Minimum is to recycle DBTA. |
| • Two reductions necessary ($BH_4^{\ominus}$ and $H_2$/Pd:C) | One reduction if Bz eliminated |
| • High solvent and reagent usage. | Increase reaction concentration/Recycle. |
| 2. Costs are high. | |
| • Expensive chiral α-methylbenzylamine is lost as $C_2H_5C_6H_5$. | Since DBTA resolution is unavoidable, eliminate enhancement of chiral induction. |
| • Recovering/recycling wastes adds costs. | Minimize wastes. |
| • Considerable capital investment needed — high depreciation | ⎫ |
| • Labor intensive | ⎬ Simply process to reduce/avoid these costs. ⎭ |

Based on the above critique the simpler process was defined as follows (Scheme 5):

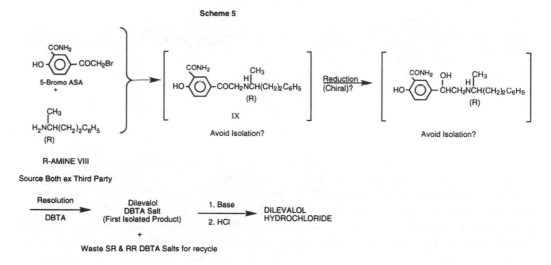

Scheme 5

The simplified process concept was itself the subject of criticism and doubt:

- Could R-1-methyl-3-phenylpropylamine (R-amine, VIII) be sourced at low enough cost?
- Would the likely dialkylation of R-Amine (VIII) introduce new impurities which are difficult to remove?
- Would R-amino ketone (IX) be isolable relatively free of dialkylated impurities, thereby serving as a purification step if needed?
- Would the DBTA resolution of an expected 1:1 SR : RR mixture be efficient?
- Would the dilevalol hydrochloride obtained by this Scheme contain any new impurities which would complicate the Regulatory registration process?
- Could the work needed to demonstrate and prove that the simplified process gives dilevalol hydrochloride acceptable to Regulatory Affairs and the FDA be done in the time frame needed to update the IND prior to NDA filing?

## 5.1 Raw Materials Sourcing

There was relatively little problem in sourcing 5-bromoacetylsalicylamide (in-house) or dibenzoyl-(+)-tartaric acid (large tonnage Italian source). Although RS-1-methyl-3-phenylpropylamine was available at low cost ($10-12/Kg) in tonnage quantities (Germany and Holland) no supplier of the R-amine VIII was known.

### 5.1.1 R-1-Methyl-3-phenylpropylamine

Many chiral acids were evaluated, with water as the solvent, before N-formyl-L-phenylalanine (FPA) was selected as the best acid for resolving RS-1-methyl-3-phenylpropylamine. The resolving acid and process were patented[8]. An outline of the commercial process implemented in Germany is given in Scheme 6.

Scheme 6

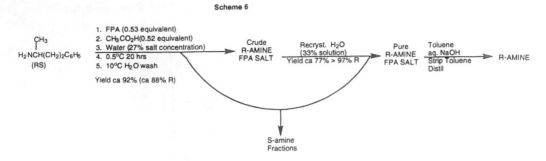

The resolution process step worked very well on a commercial scale. In an early version, the R-amine was obtained as a methylene chloride solution. This had not posed a problem in the Chemical Development plant. Methylene chloride was rapidly distilled using a condenser system which efficiently liquefied the distilled methylene chloride. When this process was transferred to Germany the rate of distillation of the methylene chloride in the equipment available was greatly extended (from 3-4 hours to > 30 hours) to enable the German plant to stay within its environmental emission permit for methylene chloride. Under these conditions the R-amine was alkylated

$$CH_3$$
$$H_2NCH(CH_2)_2C_6H_5$$
(R)

$$\longrightarrow$$

$$Y-N\phantom{xx}N-Y$$
with N–Y at top

$$CH_3$$
$$Y= CH(CH_2)_2C_6H_5$$
(R)

In hindsight, the effects of using methylene chloride as a solvent for a primary amine should have been predictable. The reality is that the possibility of adverse reactions occurring was lost with repeated successful use of methylene chloride under the rapid distillation conditions prevailing in the Chemical Development plant.

Various solvents were evaluated as methylene chloride replacements. Toluene was selected as best meeting all the needs. Thus, the R-amine was sufficiently soluble, water phases were readily separable at 25°C and toluene contamination was of no consequence since the reaction of R-amine and 5-bromoacetylsalicylamide was already conducted in the presence of traces of toluene.

The chemical resolution process of Scheme 6 was developed into an economically favorable one by the ready recycle of both the FPA resolving agent and the combined (mostly S) amine fractions. FPA, which was prepared by methyl formate reaction with L-phenylalanine, was shown to be stable (no racemization and no hydrolysis) when the resolution and work-up processes were conducted within the pH range of 2-12 at temperatures below 25°C (higher temperatures were not studied). FPA meeting specification was isolated in yields of ca 95% by simple acidification of its aqueous salt solutions and filtration.

The 1-methyl-3-phenylpropylamine containing fractions of largely S conformation were found to be readily racemized without degradation by heating at 150 C and 150 psi hydrogen in the presence of Raney Nickel[9]. Thus the manufacturers of RS-1-methyl-3-phenylpropylamine, who produced this compound by the reductive amination of benzylacetone, were well able to racemize the byproduct S-containing fractions, thereby providing some additional cost reduction and also avoiding a waste disposal problem.

Although the above commercial process succeeded in providing R-1-methyl-3-phenylpropylamine (VIII) for significantly less than $100/Kg several other companies carried out research to find even lower cost processes based on the RS raw material. A few of these processes will be described later.

## 6. Development of the New NDA Process to a Commercial Scale

### 6.1 Early Considerations

A degree of nervousness developed in the early stages of evaluation and promotion of the simpler synthesis. Many were concerned by the risks associated with process change. In particular QC and Regulatory Affairs raised questions on quality, especially the impurity profile, and the equivalence of dilevalol hydrochloride from the simpler process. By this time the 'biobatch' had already been produced via the "RR amine process" for toxicological and pharmacological work. Understandably Pharmaceutical Development, in harmony with QC and Regulatory Affairs, wanted assurance that such parameters as crystal size, bulk density, particle size distribution and tablet dissolution rates would not change and that tablets from the simpler synthesis would be bioequivalent to the dilevalol hydrochloride from the "RR-amine process" already registered in the IND application. In short, urgent evaluations of the product of the simpler synthesis were needed in order to gather the data required for an IND update. Regulatory Affairs proposed that, since the NDA filing was only about 2 years away, the data needed to validate the simpler synthesis should be obtained as quickly as possible and presented to the FDA Cardiovascular Division reviewer, as much in advance of the NDA filing as we could manage.

In defence of the process change, it was pointed out that the R-amine and RR-amine processes were based on the same chemistry. Nevertheless, we still needed to accommodate the views of those who queried whether the hydrogenolysis step in the "RR-amine process" was somehow also introducing a purge of "something" during the debenzylation step in Scheme 4! Those with concerns did, however, concede that by intersecting the "RR amine process" before the DBTA resolution and purification steps, the simpler R-amine process did maintain considerable "clean-up" capability.

The Schering Manufacture Division was, at the same time, also looking to the future by initiating outside evaluation of the "RR-amine process". This effort was intended to determine whether others, with under-used manufacturing plant capacity, could take on the production of dilevalol hydrochloride, or a late intermediate, and generate cost projections, at a 50 tonne/ annum production rate, which would be advantageous vs. in-house projections. The primary objective was to source a late intermediate thereby allowing Schering Manufacturing to undertake the final steps itself under the strictest GMP control. Another scenario was also initiated, with both Chemical Development and Manufacturing determining the capital investments which would be required to manufacture dilevalol hydrochloride in-house. The figure for a 50 tonne/annum manufacturing plant on the Schering manufacturing site in Ireland via the "RR-amine process", assuming raw material outsourcing, was estimated at $48-50 million. Chemical Development proposed that since it had much unused plant equipment in Union, New Jersey (bequeathed by Manufacturing when it moved operations to Puerto Rico), that an alternative would be to carry out early launch manufacture of a late stage intermediate in Union by using the simpler synthesis. Ireland would then take on the manufacture of the dilevalol hydrochloride in mostly existing plant. The reasoning was that it would be advantageous to limit capital spending and delay major investment in manufacturing plant until the process was better defined and the market needs were better known. Since the Union capital investment was projected at only $5-6 million, this strategy was adopted. This approach also made best use of Chemical Development's chemical engineers who were closely involved in the design of the process as well as in the testing and selection of process equipment.

## 6.2   Process Development Leading to FDA Review

One of the core premises in the simpler process as outlined in Scheme 5 was that it would be highly desirable, especially from a cost reduction standpoint, to avoid isolating solids. In this way all the equipment and labor needs associated with solids handling would be avoided, thereby reducing the product cost by minimizing labor and overhead costs and also by boosting the process throughput and plant capacity. The main disadvantage of such a strategy is that the process loses an outlet for byproducts (impurities) and generally requires that high reaction yields are obtained to minimize the need for purging reaction byproducts. The following describes the successful efforts to combine process steps such that the first isolated product from the reaction sequence is dilevalol as its DBTA salt.

### 6.2.1 From Raw Materials to Dilevalol DBTA Salt

In order to provide the best chance of success in the Scheme 5 sequence, great emphasis was placed on starting with the highest quality raw materials.

$CONH_2$

$HO$ —⟨ ⟩— $COCH_2Br$

5-Bromoacetyl Salicylamide
(5-Br ASA)

Appearance: White to cream solid.
Identity: IR agrees with standard.
Purity (HPLC): > 95%
The major impurity is 5-acetylsalicylamide (5-ASA). Small amounts (< 0.3%) of ring brominated impurity are also present.

$CH_3$
|
$H_2NCH(CH_2)_2C_6H_5$
(R)
R-1-Methyl-3-phenylpropylamine

Appearance: Clear, colorless to light yellow liquid.
Identification: IR agrees with standard.
Specific Rotation: $[\alpha]_D^{20}$ = -19.0 ± 1.5° (c=5 in cyclohexane).
Enantiomeric Purity: Minimum 98% R (Moshers Acid method)
Residual Solvent: ≤ 0.5% (gc).
Related Compounds: ≤ 1% (gc).
Assay: 98-102% ($HClO_4$ titration)

Quality criteria and analytical release specifications were set for all the solvents and reagents used in the process. Operating conditions for the process were established to ensure the minimum decomposition of intermediates and reagents, such as DBTA, at the same time as maximizing product yield.

In order to minimize dialkylation of the amino group, 5-BrASA was added to a large excess of the R-amine. This necessitated the engineering of a simple countercurrent toluene extraction system for recycle of the excess R-amine. This first reaction step was studied at a variety of temperatures, concentrations, excesses of R-amine and extraction conditions. The process was monitored by HPLC. In summary the following outlines the optimum process to the sodium salt of R-aminoketone (X)-Scheme 7. The plant equipment layout is shown in Figure 1.

Scheme 7

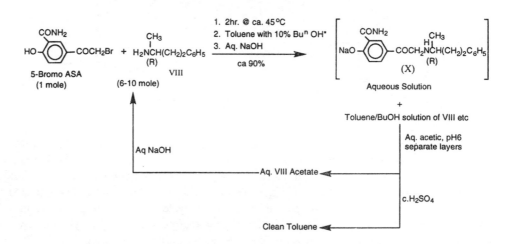

* $CH_2Cl_2$ used in an early version of the process

Typical HPLC Chromatogram
Column: $C_{18}$ Novapak (or equivalent)
Element: 0.004 M 1-DSA Na in. $CH_3OH:H_2O$
(60:40) with 1% HOAc
Detection: UV at 254 nm

FIG 1. Plant Equipment for R-Amine reaction with 5-Bromo ASA (Scheme 7)

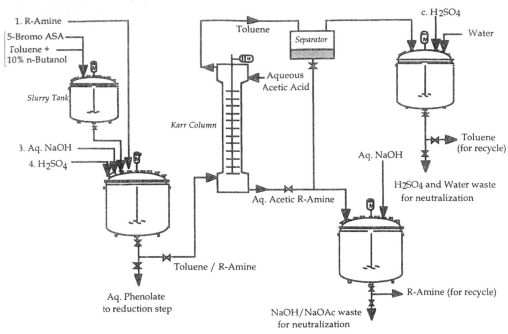

Although R-amine (VIII) forms a phenolate salt with phenols, no evidence of phenolate reaction with the bromoketone was found. As expected the only byproduct of consequence was the dialkylated R-amine (up to ca. 2%). The toluene solution containing the unreacted R-amine was extracted with aqueous acetic acid and the aqueous layer treated with aqueous sodium hydroxide to give neat R-amine for recycle. The color of the R-amine did increase with multiple recycles but this did not appear to affect the quality or yield of IX. As a precaution however, the R-amine was distilled after about every tenth batch. Although toluene could be recovered for re-use by distillation it was found that simple washing with c. sulfuric acid, separating the layers and washing with water gave toluene suitable for re-use.

In a further investigation, it was found that the aqueous solution of the sodium salt of R-aminoketone (X) could be acidified, the R-aminoketone extracted into a solvent and precipitated as a pure hydrochloride. Although this step never proved necessary in commercial operation substantial quantities of the hydrochloride of R-aminoketone IX were produced for work on alternative routes to dilevalol (qv).

The optimum process for converting aqueous solutions of the sodium salt of R-aminoketone (X) to dilevalol DBTA salt resulted from a study of such parameters as reaction solvent, temperature, mole equivalents of sodium borohydride, mole equivalents of DBTA and crystallization conditions. The reduction process, which essentially yielded a 1:1 mixture of RR and SR compounds, was monitored by HPLC (Scheme 8). The plant equipment layout is shown in Figure 2. The process containment equipment (Krauss Maffei Titus system) used for the filtration, washing and drying steps is shown in Figure 3.

**Scheme 8**

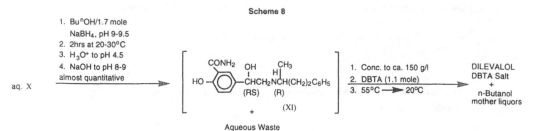

aq. X

Overall yield 5-Bromo ASA to Dilevalol DBTA salt = 30-40%

Typical HPLC Chromatogram
Column: $C_{18}$ Novapak (or equivalent).
Element: 0.004 M 1-DSA Na in. $CH_3OH:H_2O$
(60:40) with 1% HOAc
Detection: UV at 254 nm

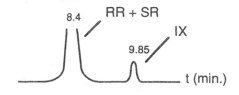

**FIG 2. Plant Equipment for Reduction/Resolution Step (Scheme 8)**

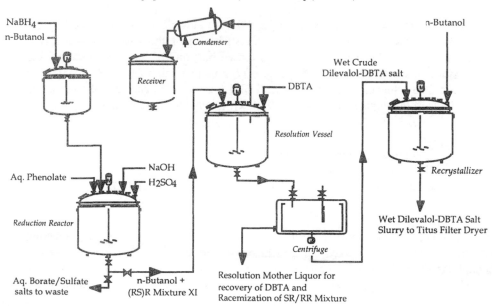

FIG 3. **Plant Equipment for Filtration, Washing and Drying of Dilevalol DBTA salt**

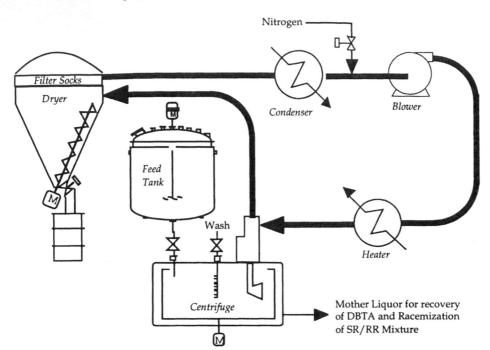

The purpose of the acidification step (to pH 4.5 with sulfuric acid) was to destroy the borate esters and complexes which compromised the distribution of the RR/SR mixture (XI) into the n-butanol layer. Considerable work was carried out on the crystallization of the dilevalol DBTA salt in an effort to avoid oiling and to crystallize a salt with an RR content of ca. 97%. This could be achieved by dissolution with DBTA at ca. 55°C and crystallizing at ca. 45°C. Higher yields of lower purity product (ca. 95% RR to 5% SR) were obtained by cooling to 0°C.

For day-to-day process monitoring and assay of the synthesis related impurity levels it proved more convenient to use a thin layer chromatographic assay [with an elution system comprising ethyl acetate (100), isopropanol (60), water (32) and ammonia (8)] than to use HPLC.

For assaying the enantiomeric purity of dilevalol in the DBTA salt, the Schering Research Analytical Department worked out an efficient glc procedure, utilizing methylboronic acid. Although the method did not separate RR and SS enantiomers or RS and SR enantiomers it served to quickly indicate the efficiency of the resolution process since no racemization of the R-amine moiety was ever found. A typical glc trace was as follows:

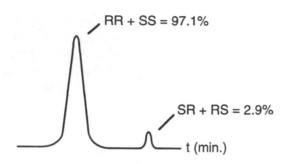

The specification set for the dilevalol DBTA Salt was

      Appearance:  White to off-white solid

      Chemical Purity (tlc):  < 3% related substances

      Enantiomeric Purity:  RR + SS > 97%, SR + RS < 3%

As indicated by the above enantiomeric purity assay the quality of the DBTA salt prepared according to Scheme 8 was often borderline (ca. 97%).  A process was registered (as an IND update) for occasional use in which the salt was split back to the base (in n-butanol) and the DBTA salt formation step repeated.  As time passed it became apparent that it would be better, from an operations and economics standpoint, to seek a higher first crop yield (by cooling to ca. 0°C) and to recrystallize the wet first crop routinely.  This position became the subject of criticism during the FDA's pre-approval inspection (q.v.).

### 6.2.2  Racemization and Recycle of the n-Butanol Mother Liquors from the DBTA Resolution

The aqueous base extraction of DBTA followed by acidification of the aqueous layer and filtration of the DBTA proved relatively straight forward.  On the other hand much work was needed to identify and develop the most cost effective system for racemizing the SR/RR mixture (approximately 70:30 respectively in composition) and to recycle the racemate produced.

Initially the SR/RR mixture was precipitated as an oxalate salt, with azeotropic drying of the n-butanol to enhance the yield.  This oxalate salt was filtered and the washed crystals racemized by heating with aqueous sulfuric acid at 40-60°C.  The racemate was extracted into n-butanol at pH 8.3 to 8.8 for recycle to the resolution step.  This process, although workable, proved cumbersome, necessitating the handling of large quantities of the oxalate salt, adding to equipment and labor requirements and also waste disposal problems.  An elegant solution to the problem was generated by the finding that water wet n-butanol solutions of the 70:30/SR:RR mixture, as obtained after extracting the DBTA from the resolution mother liquors, could be racemized utilizing a strong cation exchange resin — Dowex XFS 43279 ($H^+$) was particularly effective.  This operation was conducted in a batch fashion using a 2000 gallon jacketed glass-lined vessel equipped with a filter in the bottom valve for drawing off liquids.  After loading the SR/RR mixture, the n-butanol was sucked away, the loaded resin washed successively with one bed volume of n-butanol, followed by water, aq 5% sulfuric acid and water.  The racemization was carried out by heating the water resin slurry at ca. 90°C for 2 hours.  The racemized SR/RR mixture was removed from the resin by treatment with aq. sodium hydroxide/n-butanol.

The pH was adjusted to 8.5 to retain carboxylic acid (ca. 7% hydrolysis of the amide) in the aqueous layer. The n-butanol layer containing the 1:1/SR:RR mixture (88-90% recovery) was recycled to the resolution step. The resin was regenerated via sulfuric acid treatment. This process was patented[10]. The equipment layout for this step is shown in Figure 4.

**FIG 4. Equipment for Racemization of n-Butanol Solution of Mostly SR Waste**

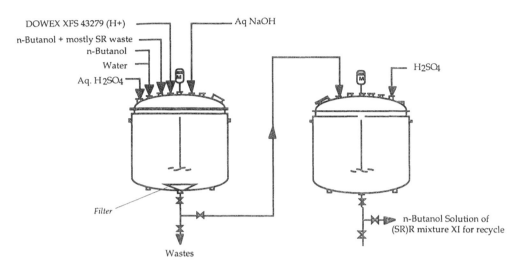

### 6.2.3 Dilevalol DBTA Salt to Dilevalol Hydrochloride:

Very little improvement in the RR composition results from the transformation of dilevalol DBTA salt to dilevalol hydrochloride. Ethyl acetate was used as the solvent vehicle in early work. In a search for a more stable solvent methyl isobutyl ketone (MIBK) was selected as the best alternative. DBTA was removed by extraction into water with sodium hydroxide (DBTA of excellent quality was recovered from the aqueous phase in high yield > 90%). The MIBK solution of dilevalol was then treated with hydrochloride acid to precipitate dilevalol hydrochloride (Scheme 9). The pH needed for maximum efficiency in the crystallization of dilevalol hydrochloride was 0.5 (this is in sharp contrast to the pH required for maximum efficiency in the crystallization of labetalol hydrochloride — pH 3.0). It should also be noted that dilevalol hydrochloride could not be handled in stainless steel equipment. Hastelloy, plastic or ceramic equipment was employed to eliminate the risk of coloration of dilevalol hydrochloride by traces of iron compounds. The plant equipment layout for this step is shown in Figure 5.

**Scheme 9**

1. $H_2O$/ aq. NaOH to pH 8.4 to 9.8
2. MIBK (3 volumes) and separate layers
3. $H_2O$ (2 volumes deionized and trace citric acid)
4. Add c.HCl, 50-60°C to pH ca. 0.5
5. Continue stir (45-50°C) to complete cryst.
6. Cool to below 20°C. Hold several hours
7. Filter, wash DI $H_2O$. Dry below 75°C

DILEVALOL · DBTA $\longrightarrow$ DILEVALOL · HCl

Yield ca 90%

FIG 5. Equipment for Conversion of Dilevalol DBTA Salt to Dilevalol Hydrochloride (Scheme 9)

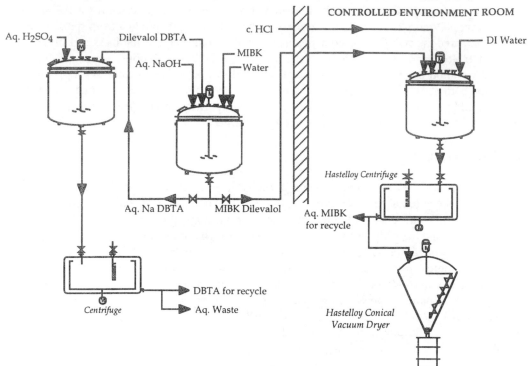

The conditions used in the hydrochloride formation/crystallization step were somewhat different when MIBK/water was used in place of ethyl acetate. In particular hydrochloride formation needed to be carried out by adding concentrated hydrochloric acid to the MIBK solution of dilevalol base at ca. 55°C (vs. ca. 25°C for ethyl acetate). In this way oiling out of the hydrochloride salt was avoided. A small amount of citric acid was included in the crystallization system to chelate any traces of iron which may be introduced. The amount of water in the system is more than sufficient to dissolve the small amount of citric acid — in early versions of the process, using much less water, precipitation of some citric acid caused a slight discoloration of the dilevalol hydrochloride. The crystallization conditions were carefully chosen to produce a crystal which filtered and washed well, which dried well (to MIBK < 0.05%) and which gave a bulk density (ca. 0.3 g/ml) which met Pharmaceutical Development's criteria for operation of their tabletting process.

The work carried out on the dilevalol hydrochloride step was undertaken in close collaboration with Schering Manufacturing in Ireland who contributed greatly to the establishment of the IND/NDA process conditions for final NCE preparation. The more aggressive process conditions of temperature, coupled with the use of a pH < 1, as employed in the MIBK-based production process, were examined in depth. It was shown that the 50-60°C process condition in the crystallization step at low pH did not lead to detectable racemization at the carbinol center.

The product of the simpler synthesis was compared in detail with the product of the "RR-amine process". In particular the Research Quality Control Unit searched for the presence of different polymorphs and new impurities (e.g. the dialkylation byproduct from the first step). They compared the stabilities of both products and also compared the hardness and dissolution rates of tablets made from both products. Since the DBTA resolution, crystallization and product isolation steps, as well as the final dilevalol hydrochloride preparation step, were the same for both the "RR-amine process" and the simpler synthesis it was anticipated that these steps should protect against the introduction of new impurities or changed physical parameters in the final crystalline product. Such proved to be the case.

## 6.2.4 Process Engineering

Chemical Development's chemical engineers worked closely with the chemists and analysts in the internal team created to progress the dilevalol hydrochloride project. The engineering input and sharing of points of view contributed greatly to the speedy simplification of the process and the early focus on cost reduction through minimizing isolations and recycling solvents, as well as utilizing waste streams. Safety issues were identified and overcome. Emissions control needs were met. Existing plant was adapted to the requirements of the simpler synthesis. Additional needed process equipment was evaluated, selected, purchased and set up. Automation opportunities were defined and process control instruments tested, purchased and installed. An existing clean (HEPA-filtered) area was upgraded for the final isolation of the dilevalol hydrochloride made in New Jersey — this was needed to serve the requirements for the parenteral dosage form.

The equipment flow sheets (Figures 1-5) outline vessel needs for the process. Equipment is mostly conventional. The only equipment purchases were a 2000 gallon glass-lined vessel for the racemization of fractions containing S-carbinol, a counter-current extraction column (Karr Column) and the automated Krauss Maffei Titus system. This latter piece of equipment (Figure 3) is designed for closed system crystallization, filtration, washing and drying. It provides nitrogen blanketing, solvent capture and drying capabilities under totally contained conditions. The only exposure of operators to the hypotensive dilevalol DBTA salt is during the step of offloading the dry powder — protective clothing is worn during this procedure. It is pertinent to add that process containment equipment of the type of the Titus system is invaluable in the processing of solids where dusts have explosion potential — dilevalol hydrochloride dust, for example, was found to be more explosive than coal dust.

## 6.2.5 FDA Review and Compliance Activities

A package of information detailing the above simpler synthesis and the definitive work carried out in Chemical Development, Research QC and Pharmaceutical Development to show equivalence vs. the original "RR-amine process" was approved by the Review Branch of the FDA's Cardiovascular Division at a meeting in Rockville. This package provided the basis for the NDA filing.

Approval of several additional process changes was sought post the NDA filing. Although it is considered risky to request FDA approval of process changes after NDA filing (because of the potential that changes may set back the NDA review) Schering-Plough Regulatory Affairs was able to review additional changes, and all the supporting data, with the

Cardiovascular Division and gain agreement that the changes were of a non-critical nature such that there was no risk of compromising the quality of dilevalol hydrochloride. The NDA chemistry section was updated without penalty and the changes adopted. The changes were:

1. The registration of toluene or n-butanol as the slurry solvent for adding 5-Bromo ASA to the R-amine in the first step — minimizes solids handling.
2. The substitution of sulfuric acid for hydrochloric acid to reduce the pH to 4.5 after the borohydride reduction step — reduces the risk of plant corrosion and product contamination by iron.
3. The use of 55-60°C in the DBTA crystallization step with cooling to 25-45°C to replace the 50°C and cooling to 0-5°C — gave more consistent enantiomeric purity results (and slightly lower yields). It should be noted that we later reverted to 55-60°C in the crystallization steps with cooling to 0-5°C, followed by routine recrystallization.
4. The use of a wet n-butanol recrystallization for reprocessing out-of-specification DBTA salt to replace the original split back to the base and repeating the DBTA salt formation. The recrystallization process gave a product with higher enantiomeric purity.
5. Detail of the 50-60°C crystallization of dilevalol hydrochloride from MIBK/water. A comprehensive comparison report with the earlier process using 25°C hydrochloride crystallization temperatures was provided.

In today's more formalized review climate, it seems unlikely that such initiatives would be attempted. As a result many companies in the Pharmaceutical Industry have taken on the challenge of accelerating definition of the NDA process, and today essentially freeze the process by the start of the Phase III program.

In regard to compliance with FDA Regulations for bulk drug substance manufacture the Pharmaceutical Industry has, over the years, built a strong formalized program to meet GMP requirements. The industry continues an energetic dialogue with Regulatory Administrations around the world, primarily with United States, European and Japanese Agencies. Harmonization of Regulatory guidelines is a major interest at this time.

For the manufacture of dilevalol hydrochloride (and other NCE's) Schering created needed process and control documentation and set up formal compliance programs to ensure GMP guidelines were met. Major programs for ensuring GMP compliance include:

> Operator and Management Training
> Batch Sheet Preparation and Change Control
> Materials Management and Control
> Process Operation and Control
> Equipment and Instrument Calibration
> Equipment Monitoring and Maintenance
> Validation (process chemistry, plant operation and cleaning)
> Facility and Equipment Cleaning Program
> Quality Assurance Auditing and Continued Monitoring

These programs are also matched in the areas of Safety and Industrial Hygiene as well as Environmental Compliance.

At the time of the FDA's pre-approval inspection (our first) of the Chemical Development dilevalol DBTA manufacturing operation (dilevalol hydrochloride itself was manufactured in Schering's Ireland facility) Chemical Development received an FDA 483 notification stating the full time use of the butanol/water recrystallization process for dilevalol DBTA salt was in violation of the NDA. The NDA stated the recrystallization process was registered for use only when the first crystallization of the DBTA salt gave a product outside specification — the FDA interpreted this to mean no more than about 10% of the time. Since the process had evolved to taking a higher first crop yield of lower enantiomeric purity, followed by routine recrystallization, the FDA criticism was justified. The fact that higher purity dilevalol DBTA was being produced by the change was subordinate to the wording of the NDA for which approval was given. An NDA supplement providing detail of the reasoning for the change, and analytical comparison of batches made before and after the change, was filed with the FDA and approved.

## 7.  Ongoing Process Development and Alternative Routes to dilevalol

Once the recycle operations for R-amine, for DBTA, for the racemization of SR byproduct and for solvent recovery were in place, the simpler synthesis as described above met the Marketing COG targets [3X cost of Labetalol minus 25%). Several additional cost reduction programs were in hand at the time of the NDA filing. The ones which were significant in terms of laboratory and pilot plant effort are worthy of brief reviews. Efforts on these programs illustrate the diversity of ideas and individual endeavor which flourished in the challenging climate created to solve the cost of goods problem.

Cost reduction efforts were undertaken both inside and outside the company. They covered the preparation of the raw materials, particularly R-1-methyl-3-phenylpropylamine and derivatives, and also several exciting programs for the direct preparation of dilevalol from chiral intermediates.

### 7.1  Raw Materials

The Celgene Company, Warren, New Jersey, building on the knowledge that RS-1-methyl-3-phenylpropylamine costs only $12/kg, proposed an enantiomeric enrichment process. This process, utilizing Celgene technology, is based on the ability of omega-amino acid transaminases to preferentially convert one of the two chiral forms of the racemic amine, in our case the S-enantiomer, to a ketone[11]. In this approach the S-enantiomer acts as the preferred nitrogen source (Scheme 10).

**Scheme 10**

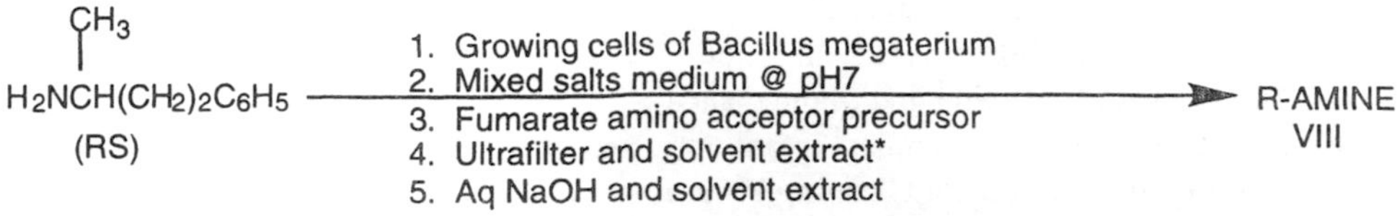

```
  CH3
   |
H2NCH(CH2)2C6H5    1. Growing cells of Bacillus megaterium
   (RS)            2. Mixed salts medium @ pH7              ───►  R-AMINE
                   3. Fumarate amino acceptor precursor            VIII
                   4. Ultrafilter and solvent extract*
                   5. Aq NaOH and solvent extract
```

* Removes benzylacetone produced

The process evolved to one in which the converting enzyme was isolated and used in a batch process with a small amount of pyridoxal 5-phosphate as co-factor and pyruvate as the amine acceptor[12]. This process was in a pilot plant phase when dilevalol hydrochloride was withdrawn from the market.

Another initiative in the Schering Manufacturing Division was based on the idea the N-benzylated RS-1-methyl-3-phenylpropylamine may be readily resolved, and used advantageously in a process analogous to Scheme 3 without the phenol blocking group. This process did indeed work well giving a high RR enantiomer yield (ca. 80-85%) in the borohydride reduction step. The cost of the benzylation/debenzylation steps was not worked out (for comparison with the NDA process costs) by the time dilevalol hydrochloride was withdrawn.

## 7.2 Alternative Routes to Dilevalol Hydrochloride

The major disadvantage of the NDA process lies in the need for the classical resolution of the RS-carbinols XI using DBTA. Although the recycle of DBTA and the waste SR/RR mixture (ca 70:30) did enable the COG target to be met it would be far more elegant and potentially lower in cost if a more direct process could be found which would eliminate the classical resolution and recycling operations. The following proposals were evaluated.

**Scheme 11**

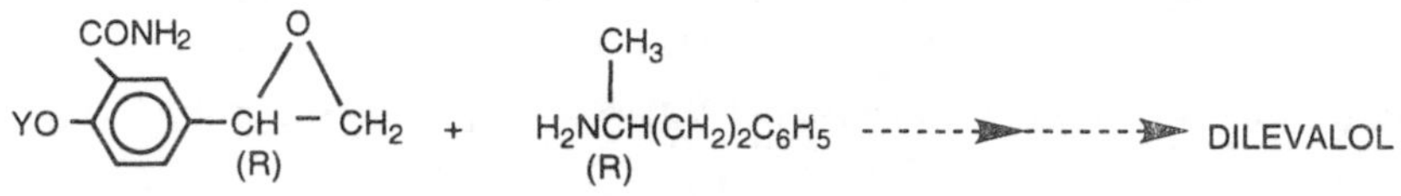

**Scheme 12**

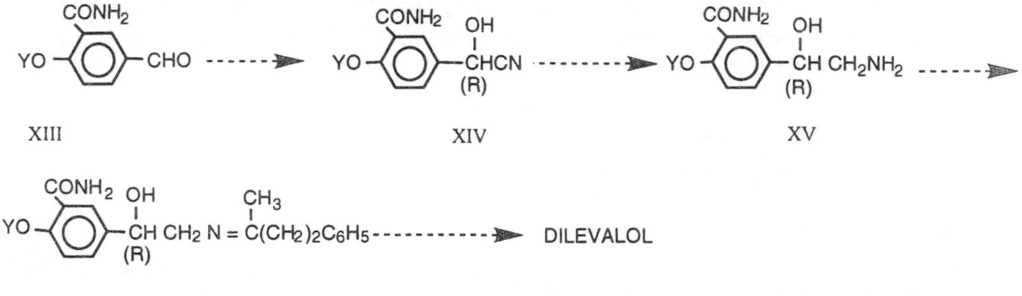

**Scheme 13**

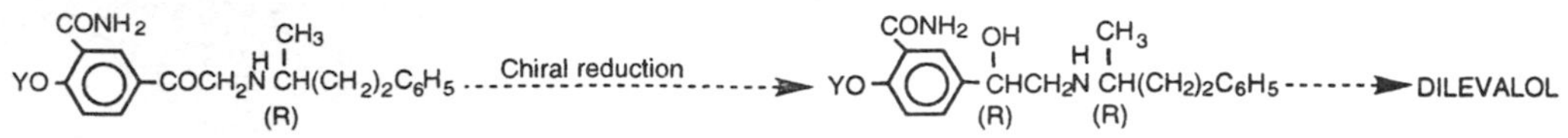

IX   (Y=H)
XVII (Y=C₆H₅CH₂)

Dilevalol (Y=H)
XVIII (Y=C₆H₅CH₂)

Y can be H or a blocking group

### 7.2.1 Scheme 11 Option

A great deal of work was carried out on the preparation of R-epoxides (XII) and their reaction with R-amine (VIII). It has long been known that styrene oxides react with primary amines at either of the epoxide carbons, and also that neat amines appear to favor the desired reaction, attack at the methylene carbon atom of XII[13].

The epoxide (XII, Y=C₆H₅CH₂) was prepared in high yield and high ee(>98%) by the enantio-selective reduction of bromoketone (II), using Itsuno chemistry[14] [with R-diphenylvalinol borane complex], followed by cyclization of the bromohydrin. Epoxide XII readily formed the desired aminocarbinol XVIII (Y=C₆H₅CH₂) with R-amine VIII which yielded dilevalol after hydrogenolysis of the benzyl group.

XII (Y=C₆H₅CH₂) + VIII ────────────→ XVIII ──H₂/Pd:C──→ DILEVALOL

The corresponding series with the free phenol (XII, Y=H) gave a poor result.

The greatest problem with the epoxide sequence, apart from the extra blocking and deblocking steps, lies in the need for a large excess of the expensive R-diphenylvalinol borane complex. The Corey modification[15] of Itsuno's method, in which only catalytic amounts of a chiral auxiliary are needed, failed probably owing to complexation of borane with the ring amide and phenol (or protected phenol) groups, leading to chirally uncontrolled reduction of the keto group. Several other routes for preparing the chiral epoxide were pursued without success. These included biological approaches to the reduction of bromoketone (II) as well enzyme mediated selective hydrolysis of RS halohydrin esters.

The epoxide route was eventually abandoned on the grounds that cost reduction vs the NDA process did not appear to be attainable.

### 7.2.2 Scheme 12 Option

The oxynitrilase catalyzed HCN addition to the aldehyde XIII appeared to offer an attractive prospect presuming the R-cyanohydrin (XIV) could be formed and this then converted to dilevalol via intermediates XV and XVI. Although the oxynitrilase catalyzed formation of chiral aromatic and aliphatic cyanohydrins and their reduction to chiral aminoalcohols has been known for some time[16], the selective reduction of XIV to XV and the likelihood of 100% induction in the reduction of the Schiff base XVI raised many questions.

Work by our Swiss Chemical Development group demonstrated that when the hydroxyl group of the racemic form of XV (Y=$C_6H_5CH_2$) was blocked by t-butyldimethylsilyl, selective reduction of the nitrile to $CH_2NH_2$ could be achieved (NaBH$_4$/CoCl$_2$/CH$_3$OH). However when the free cyanohydrin was reduced with the same reagent only the hydroxymethyl compound could be obtained, presumably owing to cyanohydrin conversion to the aldehyde prior to the reduction step.

In addition our Swiss group found, in probing experiments, that the racemic form of XV (Y=$C_6H_5CH_2$) did not give the Schiff base corresponding to XVI.

Since the Scheme 12 option appeared likely to require the extensive use of blocking groups it lost its simple appeal and was abandoned.

7.2.3 Scheme 13 Option

A great deal of work was carried out to find a reduction procedure for the chiral reduction of the readily available R-aminoketone (IX). In addition, despite the benzyl blocking group, the R-aminoketone (XVII) was also the subject of chiral reduction work.

It quickly became clear that the use of Itsuno chemistry for reducing the carbonyl group of XVII [with R-diphenylvalinol borane complex] would not be economic, again owing to the need for excesses of the borane complex. The catalytic elaborations of Itsuno's chemistry also failed.

Biotransformations have attracted increasing attention as more chiral NCE's are being created by Pharmaceutical Companies. The use of microorganisms and enzymes is particularly attractive in that systems can often be "engineered" to achieve desirable goals. Moreover, biotransformations are generally carried out in water. It was logical therefore to screen the microorganisms in ATCC banks which are known to reduce ketones to carbinols. Some 50 microorganisms were screened including bacterial, such as Schizomycetes and fungi such as Ascomycetes, Basidiomycetes and Phycomycetes. Unfortunately, none of these was active in reducing the keto group of R-aminoketone (IX) to the desired R-carbinol, dilevalol. A major breakthrough occurred when Dr. William Charney of Schering's Biotechnology Development group observed a large underground oil storage tank being removed near his laboratory. Dr. Charney, who was about 70 years old at the time, clambered to the bottom of the approximately 15 foot deep pit for a soil sample. He isolated a novel fungus from this sample which rapidly carried out the desired transformation of IX to dilevalol. The organism was separated by the soil enrichment method, wherein the soil sample is mixed with a compound which restricts the growth to those organisms which can use that compound. In this case the compound was 5-methoxyacetyl-2-hydroxybenzamide. Incubation was carried out for several days and the mixture sampled using standard microbiological techniques and plated out. The active pure culture was a white mould, characterized as belonging to the genus Aspergillus and was further identified as Aspergillus niveus. An investigation of other members of the family, Aspergillus niger (ATCC 11488) Aspergillus orxyae (ATCC 1454) and Aspergillus oryzae (ATCC 11488) failed to provide chiral reduction of the keto group of R-aminoketone (IX). Aspergillus niveus, ATCC 20922, was the subject of patent claims[17].

An outline of the process for the biotransformation of R-aminoketone (IX) to dilevalol at the point the project was cancelled is as follows:

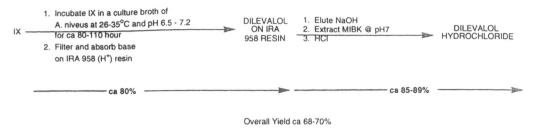

Since this process involved an Aspergillus fermentation as the last chemical transformation step, considerable concern was expressed concerning possible contamination of the dilevalol hydrochloride with such as citrinin or allergenic proteins. All test carried out by the time dilevalol hydrochloride was withdrawn were negative. However, a full testing program had not been completed.

It is clear from the conversion yield that the biological process was worthy of further development for potential use in the longer term. Work was especially needed to improve the concentration (8g/l at the time) and to deal with the slightly different impurity profile (total 0.3% with 0.1% identified as the R-amine VIII). Also technologies, e.g. ultrafiltration, needed to be evaluated to ensure that proteinaceous material did not contaminate the product.

## 8.   Withdrawal of Dilevalol Hydrochloride from the Market

Dilevalol hydrochloride was on the market in Japan and Portugal when it was withdrawn. Approximately 34 hepatic events were recorded in a population of 176,000 patients taking dilevalol hydrochloride. Most of the hepatic events were reversible but there were two deaths. During an extensive clinical research and development program (ca. 10,000 patients), there had been no significant evidence to indicate that hepatotoxicity would become a problem once the drug was marketed. Since labetalol hydrochloride (which contains about 25% dilevalol hydrochloride) was being continually compared with dilevalol hydrochloride more detailed evaluation and comparison of the hepatic data accumulated on both compounds was undertaken. Interestingly, in the first year of marketing labetalol hydrochloride, 15 hepatic events were recorded/million prescriptions. This rate dropped such that in the 7 years of marketing labetalol hydrochloride to the time of dilevalol hydrochloride withdrawal, only 80 reports of hepatic reactions were recorded for labetalol hydrochloride. In line with this data, it was quickly shown that the patterns of hepatic injury associated with the two medications were not similar. In the case of dilevalol hydrochloride those patients affected demonstrated a fairly rapid onset of hepatic events, expressed as showing jaundice, dark urine, nausea, vomiting and fatigue.

The withdrawal of dilevalol hydrochloride represents a unique milestone. As far as the author is aware, this is the first case wherein a deliberately produced chiral NCE may have demonstrated more toxic liability than the racemic mixture.

## 9. Acknowledgments

I am indebted to the many Chemists, Engineers, Analysts and Regulatory people whose dedication and hard work led to the technical success of the dilevalol hydrochloride project. In particular special thanks go to those who worked tirelessly to produce the needed supplies and to engineer the technology to meet the COG target (Messrs. Bruce Shutts, Raymond Werner, and their staffs), to those who "imagined" and carried out the exploratory work on future options [Dr. Richard Draper (chiral epoxides), Dr. Ingrid Mergelsberg (chiral cyanohydrins), the late Dr. William Charney (biological reductions), Dr. Maurice Fitzgerald (N-benzyl R-amine approach)] to senior management for their support and encouragement, to those who made suggestions to improve the manuscript and to Lavonne Wheeler who did all the typing.

## 10. Bibliography

1. Success Rates for New Drugs Entering Clinical Testing in the United States, J. A. DiMasi, Pharmacology and Therapeutics, July 1995, **58**, 1. See also Scrip 1995, August 18, page 18. Overall the significance of prior testing of an NCE outside the U.S. is apparent: for licenced-in compounds the success rate was 29.4%; for compounds first tested abroad, 14.6%; for those originated and first tested in the US 10.4%.

2. U.S. Patent 4,619,919 to Schering Corporation, Nov. 14, 1986; U.S. Patent 4,950,783 to Schering Corporation, Aug. 21, 1990.

3. Gold, E. H.; Chang, W.; Cohen, M.; Baum, T.; Ehrreich, S.; Johnson, G.; Prioli, N. and Sybertz, E. J. , J. Med. Chem., 1982, **25**, 1363.

4. Hartley, D., Chem. and Industry, 1981, 551.

5. Yamada, S. and Koga, K., Tetrahedron Letters, 1967, No. 18, 1711; Koga, K. and Yamada, S. Chem. Pharm. Bull., 1972, **20**, 526.

6. Kametani, T.; Kigasawa, K.; Hiiragi, M.; Wagatsuma, N.; Kohagizawa, T.; and Inoue, H., J. Pharm. Soc. Japan, 1980, **100**, 839.

7. U.S. Patent 4,658,060 to Schering Corporation, Apr. 26, 1982.

8. European Patent 320898 to Schering Corporation, June 21, 1989.

9. Finding by Dr. N. Carruthers and R. DeVelde in our laboratories.

10. International Patent Application, WO 91/08196 to Schering Corporation, June 13, 1991.

11. U.S. Patent 4,950,606 to Celgene Corporation, Aug. 21, 1990.

12. U.S. Patent 5,300,437 to Celgene Corporation, Apr. 5, 1994.

13. Parker, R. E, and Isaacs, N. S. Chem. Rev., 1959, 59, 737.

14. Itsuno, S.; Hirao, A.; Nakahama, S. and Yamazaki, N.; J. Chem. Soc., Perkin Trans.I, 1983, (8), 1673. Itsuno, S.; Ito, K.; Hirao, A. and Nakahama, S., J. Chem. Soc. Chem. Commun., 1983 (8), 469. Itsuno, S.; Ito, K.; Hirao, A. and Nakahama, S., J. Org. Chem., 1984, **49**, 555.

15. Corey, E. J.; Bakshi, R. K. and Shibata, S., J. Am. Chem. Soc., 1987, **109**, 5551. In this paper Corey et al. used S-diphenylprolinol as the chiral auxiliary.

16. Becker, W.; Freund, H. and Pfeil, E., Angew. Chem. Int. Ed., 1965, **4**, 1079.

17. U.S. Patent 4,948,732 to Schering Corporation, Nov. 7, 1989.

# Design and Development of Practical Syntheses of LY228729, a Potent 5HT$_{1a}$ Receptor Agonist

**Michael J. Martinelli** and **David L. Varie**
*Chemical Process R&D, Lilly Research Laboratories, A Division of Eli Lilly and Company, Lilly Corporate Center, Indianapolis, Indiana 46285-4813*

Key words: LY228729, epoxidation, 5HT1a, tryptophan, aziridine formation

## Introduction

The interface between medicinal chemistry and process research groups is critical to the rapid evaluation and development of new clinical candidates. The discovery of drug molecules is often followed with structural optimization for potency and selectivity by diverse and thorough structure-activity-relationship studies. These studies rely upon versatile and general synthetic techniques and strategies that are often not the most pragmatic approaches to specific congeners. As a result, the biological activity of a family of molecules can be investigated in a timely fashion to reveal the most potent drug substance with the least side effects. Notwithstanding, the resulting chemical synthesis that allowed the initial construction of the drug candidate is frequently inappropriate for scale-up beyond several milligrams of material. The design of a synthesis for a general class of compounds, therefore, rarely considers the same criteria as that for a specific target compound.

Once selection of a clinical candidate has been made, the drug substance is required for a number of biological and supporting studies. Due to time constraints, early stage route selection is often defined by whatever technology is available to enable the production at the 100+ gram level. With current industry standards, the attrition rate for projects at this stage of development is quite high. Thus, adopting a manufacturing strategy to define the best possible synthesis so early is not always the most prudent strategy. Typically, it is possible to answer many of these crucial questions with small quantities of material obtained through scale-up of an early phase route.

Route selection becomes an important activity once the safety assessment in a toxicology laboratory has been accomplished and the compound appears more viable as a clinical candidate. The final sequence for a synthesis often dictates where critical purification steps occur for quality control. The new synthetic target, therefore, may become an advanced intermediate that could be prepared by a variety of strategies. This tactic affords the most optimal and versatile approach to the synthesis of drug candidates without compromising quality. In the discussion of LY228729 (**3**), a common intermediate was crucial to two of the best synthetic routes.

Serotonin is an important neurotransmitter, responsible for signal transduction and ultimately for example, muscle control. The receptors for serotonin are accountable for a variety of functions and have thus been the target for a number of therapeutic reagents. The serotonin (5-HT) receptors are divided into families and subtypes, including $5HT_1$, $5HT_2$, $5HT_3$, $5HT_4$, $5HT_5$, $5HT_6$, and $5HT_7$. Both agonists and antagonists for many serotonin receptors have been identified and developed into valuable therapeutic reagents. LSD (**2**) is an agonist for serotonin receptors, but is not very specific at any given receptor. Although the activity may not be surprising, the lack of specificity is intriguing based upon its conformational rigidity. The minor structural similarities of an indole tethered with an amine moiety suggested that the conformationally more fixed compound **3** should provide unique activity. SAR revealed the optimal structure with a di-*n*-propylamino substituent and the 5-carbamoyl moiety in LY228729 (**3**).[1b] Thus, this molecule became the target drug candidate for development in our chemical process facility.

5 HT (**1**)                     LSD (**2**)                     LY228729 (**3**)

This chapter outlines the evaluation of the original chemical synthesis for LY228729, including method optimization. The second and third generation syntheses are also discussed in detail, with particular focus on some of the key transformations that enabled the synthesis of LY228729 for clinical trials. These synthetic approaches are then summarized and compared in the conclusion of this chapter. Based upon the Jacobsen catalytic asymmetric epoxidation, it was possible to achieve kinetic resolution in the epoxidation step and hence afford a fourth generation synthetic option. However, competing aromatization proved to be a difficult problem in this asymmetric epoxidation and will not be discussed further.

### Original Synthetic Route for LY228729

The first gram quantities of enantiomerically pure LY228729 were prepared using a strategy such as that shown in Scheme 1.[1] This route and its variations were studied with two goals in mind: 1) determine whether the route could supply 100 g quantities of bulk drug for early biological evaluation, and 2) evaluate this route strategically (overall yield, number of steps, costs of reagents and processing) as compared to alternative synthetic routes which were being considered concurrently. This work began with optimization studies which focused on steps that most critically affected throughput or required particularly hazardous or expensive reagents. Only selected parts of the optimization studies which contributed to achieving the above goals, will be discussed in this section.

The synthesis started with construction of the AC (tetralin) ring of LY228729. Commercially available 2-bromophenylacetic acid (**4**) was converted to the acid chloride and treated with $AlCl_3$ and ethylene gas to give 8-bromo-2-tetralone (**5**). Reductive amination of the tetralone with $NaCNBH_3$ and (*R*)-4-nitrophenethylamine yielded a 50:50 mixture of diastereomeric

2-aminotetralins **6a** and **6b**.[2a,b] The desired 2-(*S*) diastereomer was obtained in 22% overall yield by fractional crystallization of the ditoluoyltartaric acid salts. The HCl salt of the tetralin **6a** was formed and the phenethyl chiral auxiliary cleaved by hydrogenolysis using a sulfided platinum catalyst.[2c] This intentionally poisoned catalyst allowed the nitro group to be reduced without cleavage of the sensitive carbon-bromine bond. Once the nitro group was reduced, solvolysis of the auxiliary occurred to provide the HCl salt of the enantiomerically pure 2-amino-8-bromotetralin (**7**). The dipropyl group on the nitrogen was introduced by acylation with propionyl chloride, followed by borane reduction and another acylation with propionyl chloride to give tertiary amide **8**. To introduce the B-ring of the target, amide **8** was nitrated to provide a 60:40 ratio of the desired 5-nitro amide **9a** and the 7-nitroamide, **9b**. The regioisomers were chromatographically separated and the desired 5-nitro isomer was formylated with a dimethylformamide acetal to give nitro aldehyde **10**. Nitro group reduction with $TiCl_3$ and subsequent cyclization afforded indole **11**. The amide was then reduced with borane to give the tertiary amine **12**. The synthesis was completed by introducing the carboxamide in a two step sequence: CuCN displacement of the bromide followed by hydrolysis of the nitrile with neat polyphosphoric acid (PPA).

From the onset, we believed the most difficult process chemistry problems were contained in the end of this synthesis, especially the non-selective nitration and the installation of the carboxamide group. In order to prepare sufficient material to study these steps, we anticipated performing the early steps of the synthesis on multi kilogram scale, possibly in a pilot plant. Thus we first studied the preparation of 8-bromo-2-tetralone (**5**). The conversion of phenylacetic acid chlorides to 2-tetralones with ethylene under Friedel-Crafts conditions was well known.[3] Initial lots of this compound were prepared by conditions derived from the preparation of 2-tetralone. Ethylene gas was added to mixture of 2-bromophenylacetyl chloride and $AlCl_3$ at -20 to 0 °C. Tetralone **5** was obtained as an oily semi-solid in less than 50% yield. After considerable study, the temperature at which ethylene was added was found to be the key to producing crystalline **5** with acceptable purity. In this instance, if the ethylene was added to the acid chloride/$AlCl_3$ at 10-20 °C, the desired cyclization reaction was competitive with polyethylene formation. Under these conditions, less than 2 equivalents of ethylene was required, and tetralone **5** was isolated in 65-75% crystallized yield on 20 kg scale.

Having pure 8-bromo-2-tetralone, we next studied the reductive amination reaction with the goal of eliminating sodium cyanoborohydride (not available in bulk quantities at the time), simplifying the product isolation, and possibly improving the diastereoselectivity of the reduction. Our initial study of reaction solvent, hydride sources, and temperature led to no improvement in the diastereoselectivity of the reduction. However we did find that the enamine derived from **5** and (*R*)-4-nitrophenethylamine formed easily in $CH_2Cl_2$ or THF (with azeotropic removal of water) and was rapidly reduced by sequential addition of solid $NaBH_4$ and 4 equivalents of acetic acid. The borate salts precipitated from THF leaving a solution of amine diastereomers **6a** and **6b**. We subsequently found that the trifluoroacetate salt of the desired amine diastereomer (**6a**) selectively precipitated from THF. At this point, we had a streamlined reductive amination/covalent resolution process which provided **6a** in >35% overall yield with >96% de on a multi kg scale. The TFA salt of **6a** was converted to the HCl salt and the chiral auxiliary was removed with sulfided 5% Pt/C under 1 atm of hydrogen to give the 2-aminotetralin HCl salt (**7**). (The HCl salt of the amine **6a** was preferred over the TFA salt because the resulting HCl salt of amine **7** was less soluble and easier to crystallize.) While we were confident that this clever covalent resolution sequence could be scaled up in the short term, we were concerned about being dependent on (*R*)-4-nitrophenethylamine. This chiral auxiliary had to be prepared in four steps from readily available (*R*)-phenethylamine,[4] or it could be purchased in bulk quantities (100-1000 kg) for $500-1000/kg.

Scheme 1. Tetralone-Based Synthesis of LY228729

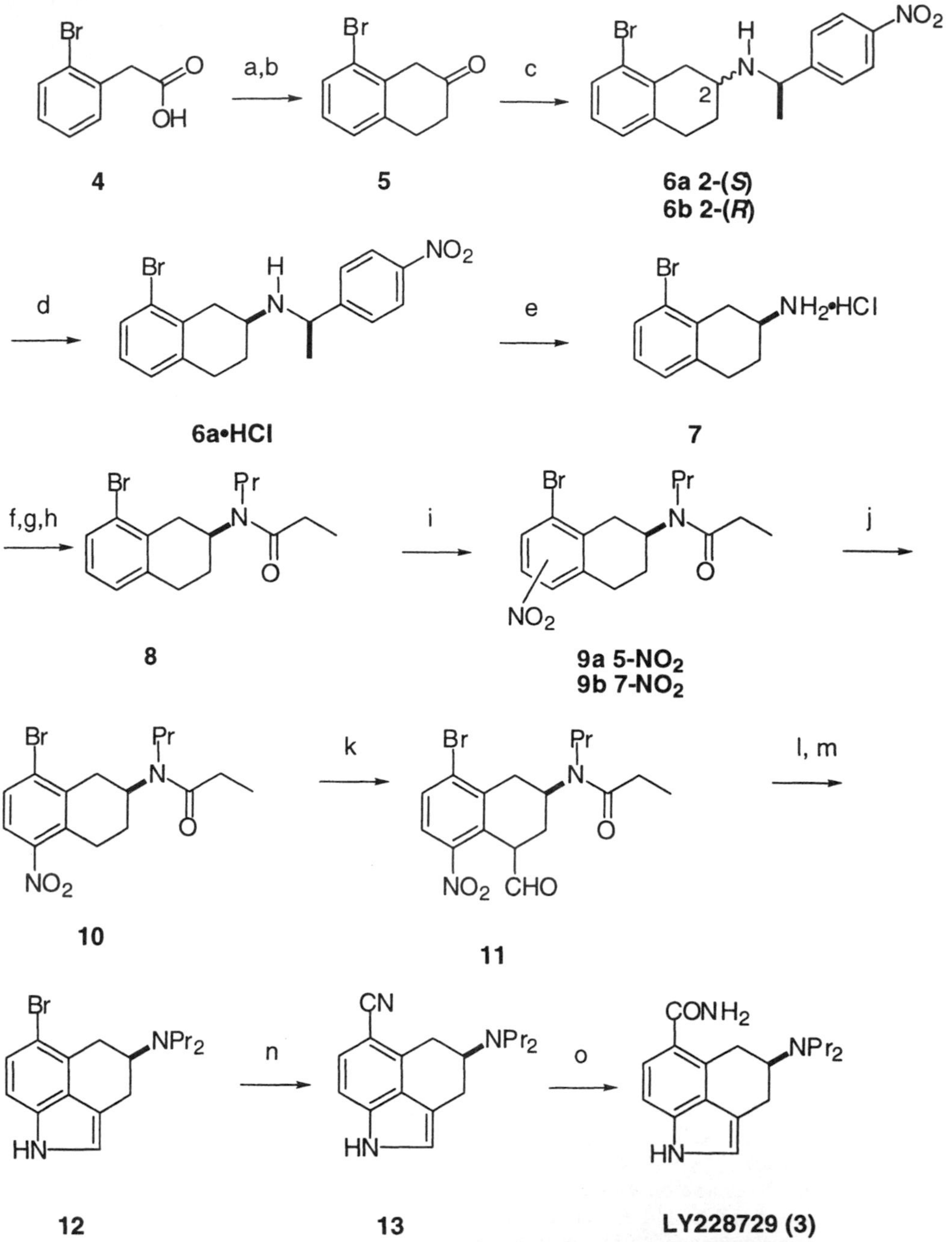

a) SOCl$_2$; b) AlCl$_3$, ethylene; c) 1. (*R*)-4-nitrophenethylamine 2. NaBH$_4$, HOAc;
d) 1. L-DTTA/recrystallize, 2. HCl; e) H$_2$/sulfided Pt-C, MeOH f) EtCOCl; g) BH$_3$
h) EtCOCl i) NH$_4$NO$_3$ TFAA,CHCl$_3$ ; j) chromatography; k) DMF acetal;
l) TiCl$_3$, NH$_4$OAc, MeOH; m) BH$_3$; n) CuCN, CuI, DMF, 140 °C; o) PPA, 80 °C

We next investigated the nitration of amide **8**, and replaced the $NH_4NO_3/CHCl_3$[5] conditions with a safer alternative: 1 equivalent of nitric acid in a mixture of acetic acid and sulfuric acid. As might be expected, the amount of the desired regioisomer in the mixture was only 60%. Realizing this poor regioselectivity was a significant long-term issue that needed to be dealt with strategically, we continued to investigate process issues in the latter steps of the synthesis. The $TiCl_3$ reduction of nitroaldehyde **10** produced copious quantities of titanium waste and a laborious workup was required to isolate the product. The sulfided Pt-C catalyzed hydrogenation seemed to be an ideal replacement for $TiCl_3$. However, when **10** was hydrogenated using this catalyst, the intermediate hydroxylamine was trapped to give the *N*-hydroxy indole **14** (eq 1).

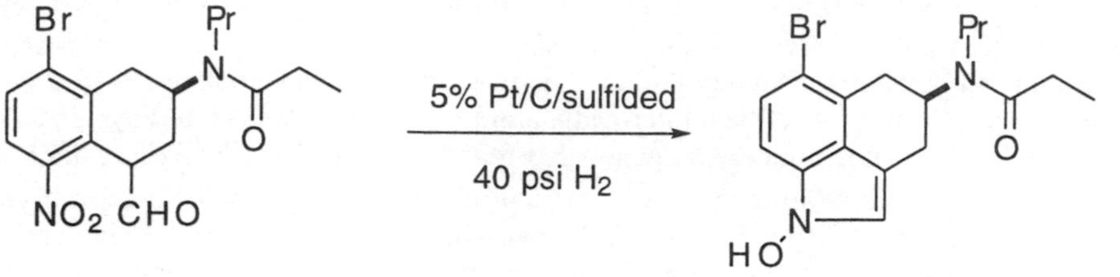

The final two steps of the synthesis also posed a number of concerns for large scale processing. The cyanide displacement reaction of bromide **12** generated cyanide and copper waste and the yield was modest (50%). The cyano group hydrolysis with PPA suffered from handling problems, including the high viscosity of PPA and the need to neutralize all of the acid during the work-up of the reaction. (More conventional nitrile hydrolysis conditions gave no reaction or decomposed the nitrile and/or amide product.) A more streamlined process to introduce the carboxamide group was investigated (eq 2). Bromo indole **12** was treated with KH and then *t*-BuLi to provide a dianion which was quenched with trimethylsilyl isocyanate at -78 °C. The silyl amide obtained readily hydrolyzed to the amide LY228729. This sequence presented numerous obvious process chemistry concerns, including the lack of a bulk supply of trimethylsilyl isocyanate and handling of pyrophoric reagents such as KH and *t*-BuLi. However, we minimally hoped that the reaction might be useful for the preparation of small quantities of bulk drug needed for testing. Unfortunately, the reaction gave only 30-50% yields of LY228729 on a 1 g scale, and gave inconsistent and lower yields on any larger scale. These results, while disappointing, indicated that new strategies were needed in the long-term to introduce the carboxamide moiety.

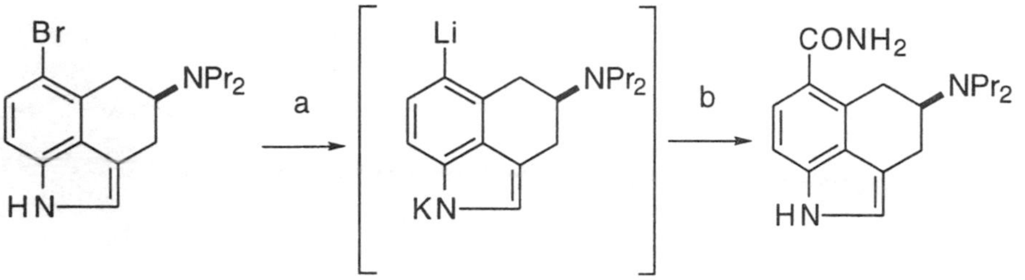

a) 1. KH/THF 2. *t*-BuLi/-78 °C; b) 1. TMS-NCO 2. $H_2O$

The efforts on this route indicated that all of the steps could be performed at least on modest scale and equally important, showed that other approaches must be examined. Using the best yields from the original and modified sequence, the synthesis was 13 steps, contained a covalent resolution, a late stage separation of regioisomers, and an expensive, non-reusable chiral auxiliary. The overall yield was 1%, with each kg of LY228729 requiring 72 kg of 2-bromophenylacetic acid and 41 kg of (*R*)-4-nitrophenethylamine.

### Second Generation Synthetic Route. Diastereoselective Epoxidation

The first total synthesis of Lysergic acid was published by Kornfeld, Woodward and colleagues from the Lilly Research Laboratories in 1954.[6] This elegant campaign was originally meant to address the difficulty in obtaining meaningful quantities from nature, and stood the test of time for several decades which followed. A critical intermediate in the Kornfeld-Woodward synthesis was a compound which became affectionately known as Kornfeld's ketone (**15**). NaBH4 reduction of the ketone and subsequent dehydration afforded a crystalline olefin **16** in excellent overall yield. The epoxidation of this olefin with peracids was known since the early lysergic acid synthesis days of Kornfeld. However, the stereochemical outcome of this epoxidation remained unknown for the three decades that followed the first disclosure.[7] In any event, these molecules were successfully employed as valuable intermediates in a total synthesis of the complex natural product.

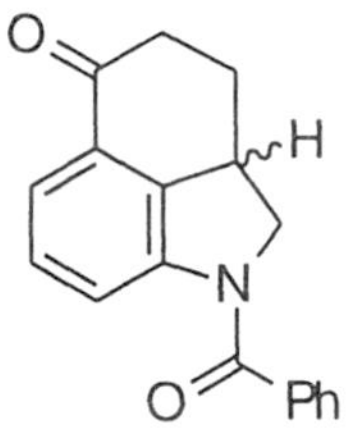

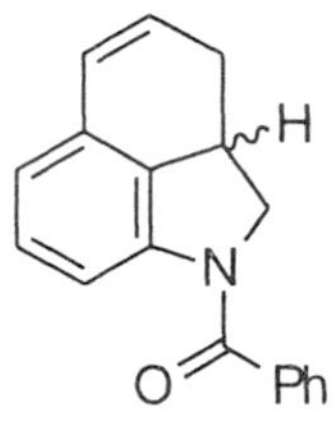

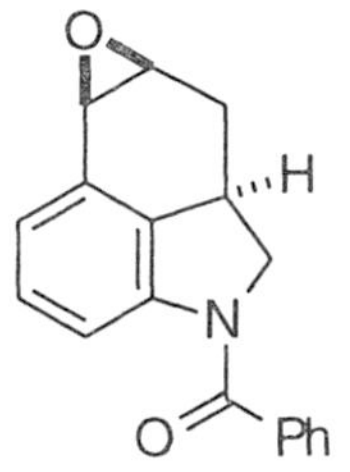

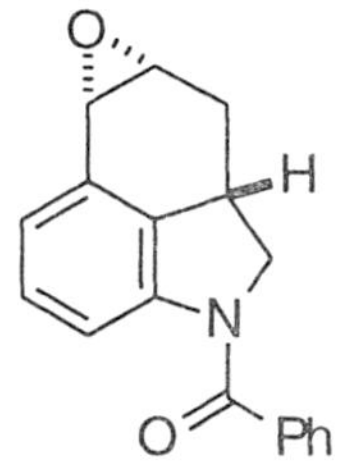

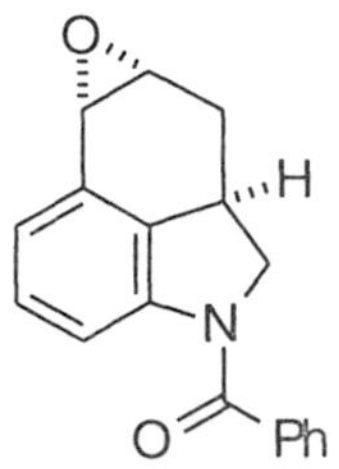

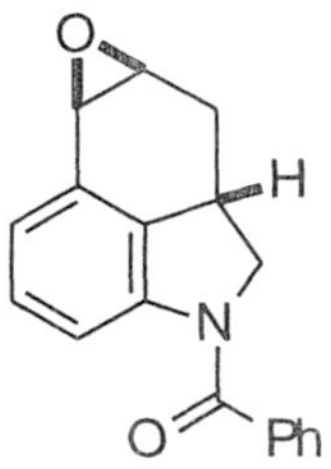

In late 1989, we had the opportunity to reinvestigate the epoxidation of the olefin, believing that a mixture of diastereomeric epoxides would result. To our surprise, we determined that the epoxidation of olefin **16** was highly diastereoselective, affording primarily the anti-epoxides **18** with >96% de.[8] The rationale for this result was based upon a torsional model combined with transition state modeling.[9] The importance of this insight into the chemical process suggested that the epoxidation stereochemistry should be reagent independent. In other words, *m*-CPBA, NBS/$H_2O$, MMPP (mono-magnesiumperoxyphthalate) or $OsO_4$ should provide the same stereofacially selective oxidation with respect to selectivity sense and magnitude (de). That this was indeed the case allowed us the flexibility to choose the optimal oxidant with confidence that the reaction could be reliably scaled up.

Interestingly, *m*-CPBA was employed in the initial epoxidation studies with great success in a variety of solvents including $CH_2Cl_2$, toluene and ethereal solvents. During that time, the availability of relatively pure *m*-CPBA became difficult being replaced with somewhat more stable but less pure grade of the reagent (~50% potency). Also coincident with this transition was the introduction of mono-magnesiumperoxyphthalate (MMPP)[10] as a potential replacement of the former oxidant. Since we believed the epoxidation was reagent and solvent independent, our incorporation of this new reagent into the process met only with the difficulties inherent to using the reagent itself. Namely, the relative insolubility of MMPP limited our solvent choice. Nonetheless, this resulted in the development of 50% aqueous *n*-BuOH biphasic reaction medium. With substrate and reagent soluble under these conditions, the reaction proceeded smoothly. Upon complete reaction, the layers were separated to effect removal of the spent oxidant. The organic phase contained the desired epoxide with an isomer profile identical to *m*-CPBA. The solvent choice resulted partly from a survey, but equally importantly from consideration of how the epoxidation reaction would dovetail into the subsequent process.

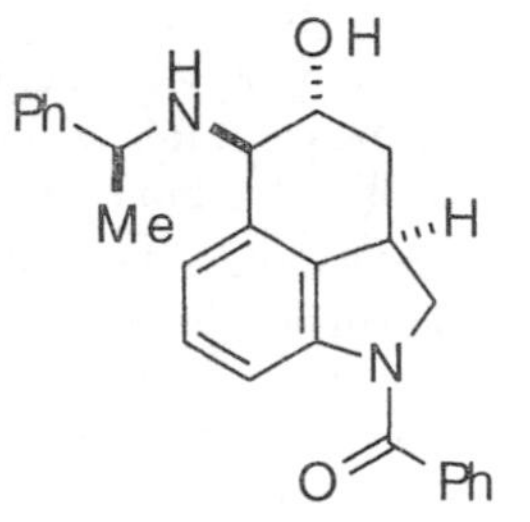

19 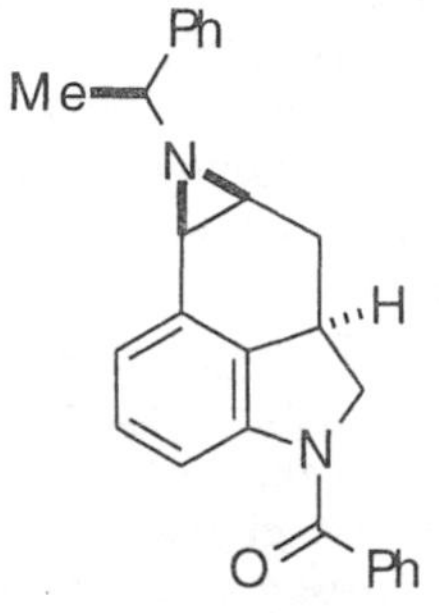          21 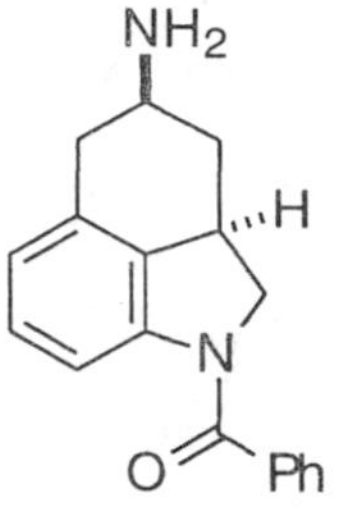          23

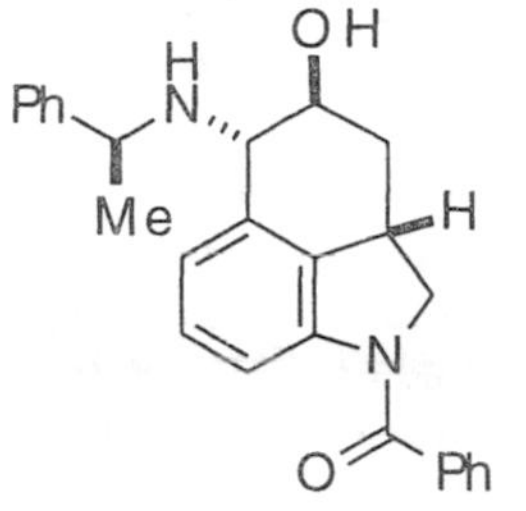

20 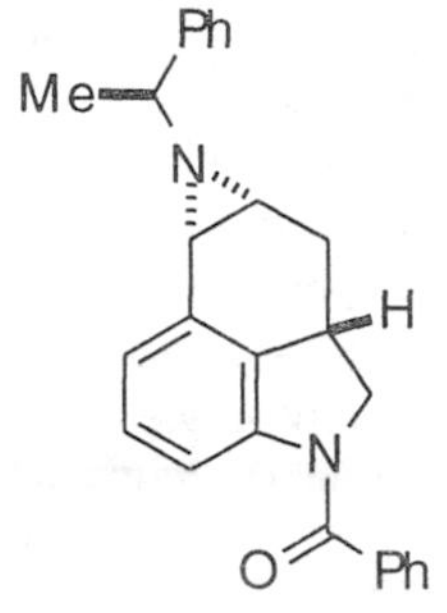          22 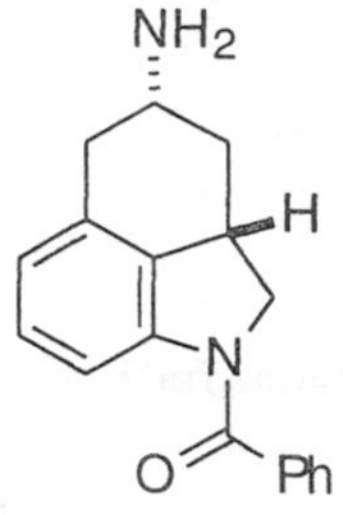          *ent*-23

The epoxide-opening of trisubstituted epoxide substrates with a variety of amine nucleophiles was previously studied by Kornfeld and shown to be selective for attack at the 4-

position.[6] Thus, we concluded that an amine opened the epoxide **18** to afford an amino alcohol, which in principle, could be deoxygenated to provide the correct relative stereochemistry. However, it was disappointing that attack occurred at the more electrophilic (but rational) benzylic position to afford the amino alcohol **19**. A variety of 1° and 2° amines were used to open this racemic epoxide in excellent yield. These epoxide openings were best conducted in *n*-BuOH at 110 °C, thus fitting very well with the epoxide forming step above in the same solvent. Consequently, a solution of the racemic epoxide when reacted with an optically pure amine, such as (*S*)-α-phenethylamine, produced an equal mixture of diastereomers **19** and **20**. This mixture upon cooling provided the single isomer **19** in 43% yield, thus affording an efficient means for separation of optically enriched material. Simple reslurry of this crude material in warm *i*-propanol effectively removed the more soluble less desired isomer (**20**).

As noted above, it was disappointing due to the structural attributes of the target molecule that the epoxide opening had occurred at the benzylic position. Transposition of the nitrogen around the ring through the intermediacy of the aziridine **21** was then considered. Initially, the Mitsunobu reaction furnished this aziridine in high yield, but concomitant with the production of highly crystalline contaminants, triphenylphosphine oxide and the reduced form of DEAD (diethyl-) or DIAD (di-isopropylazodicarboxylate). While chromatographic removal of these impurities was feasible on a small scale, it was less likely at the larger scale. The aziridine could likewise be formed with $Ph_3PCl_2$, $Ph_3PBr_2$ and other similar reagents but with the same purification liabilities. A new procedure employing methanesulfonyl chloride and triethylamine was serendipitously discovered and then implemented to overcome this purification requirement, securing the aziridine **21** as a viable intermediate. That mesyl chloride could be utilized for this transformation is remarkable since *N*-mesylation was the expected primary course of reaction. It was not possible to isolate any sulfonamide derivative, nor mesylate from this reaction. Application of this protocol to substrates without alkyl branching resulted only in *N*-mesylation.

Tandem hydrogenolysis of the aziridine bond and the auxiliary bond were achieved by exposure to hydrogen in the presence of a palladium catalyst. Thus, cleavage of the aziridine bond occurred at 0 °C under 1 atm $H_2$ gas. Upon complete conversion to the 2° amine, the benzylic auxiliary bond was cleaved at 55 °C in the same vessel. The only detectable side product from these reactions was the des-amino compound in which the exocyclic nitrogen was completely lost. This unwanted pathway could be minimized by careful temperature control, i.e., conduct the endocyclic reductive ring opening at low temperatures, followed by auxiliary cleavage at the elevated temperature. To prevent competing solvolytic reactions, nucleophilic solvents such as acetic acid or alcohols were avoided. With those solvents, aziridine opening could be observed to afford the corresponding acetate or methyl ether derivatives. The optimal binary solvent medium for this hydrogenolysis was therefore THF and phosphoric acid. To further prevent hydrolysis of the benzamide protecting group, the pH was adjusted to 3 prior to temperature elevation. Under these conditions, both diastereomeric amino alcohols could be converted to the enantiomers of **23** with equal but opposite signs of rotation ($[\alpha]_D = 59°$, ($c = 1$, THF)).

Regioselective aromatic electrophilic *para*-substitution on an indoline moiety is well precedented,[11] although many reaction conditions were not compatible with the existing functionality. The mildly directing and activating effects of the *p*-benzamide and *o*-alkyl moieties were essential for regiocontrol. The primary amine **23** thus obtained could undergo aromatic electrophilic substitution reaction with bromine in a carefully buffered NaOAc/HOAc reaction medium, thereby preventing oxidation of the amine moiety, to afford the aryl bromide **24a** in 89% yield. Alternatively, iodination with iodine and periodic acid furnished the aryl iodide **24b** in 85% yield. Subsequent alkylation with excess iodopropane and $K_2CO_3$ in acetonitrile at 80 °C afforded the tertiary amine **25b** in 80% yield. Quaternization of the amine group, rather than incomplete alkylation, was the preferred option since the quaternary ammonium salt could be easily removed by aqueous extraction.

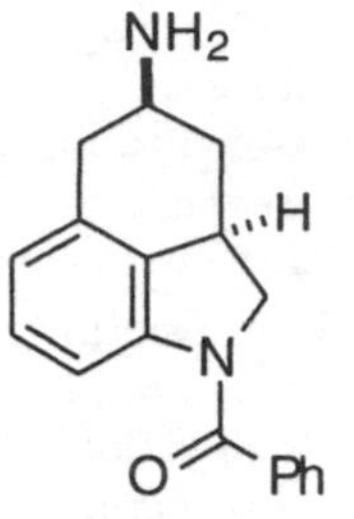

**23**

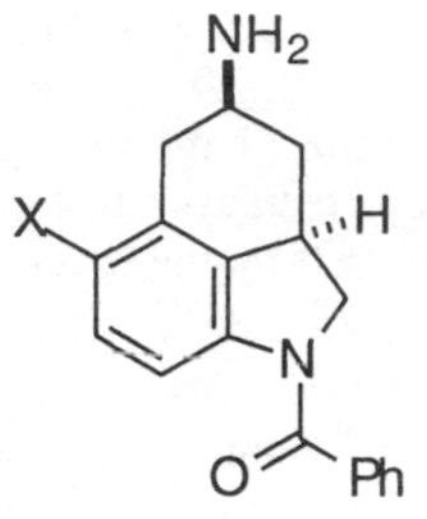

**24a** X=Br
**24b** X = I

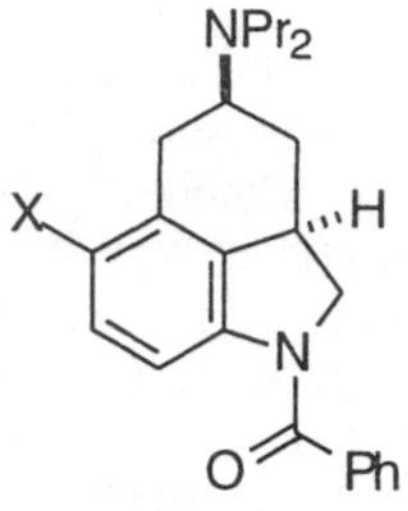

**25a** X=Br
**25b** X = I

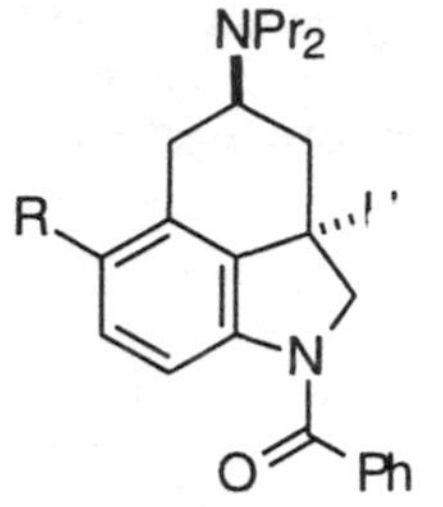

**26a** R = CN
**26b** R = CONH$_2$
**26c** R = CO$_2$Me
**26d** R = CO$_2$Et

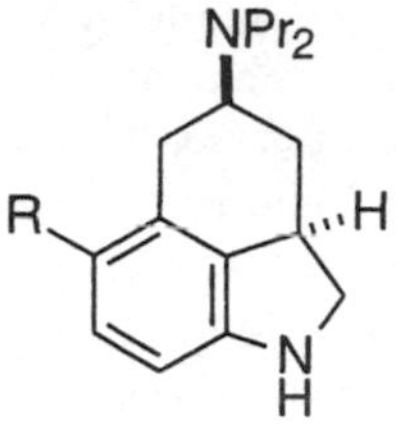

**27a** R = CN
**27b** R = CONH$_2$

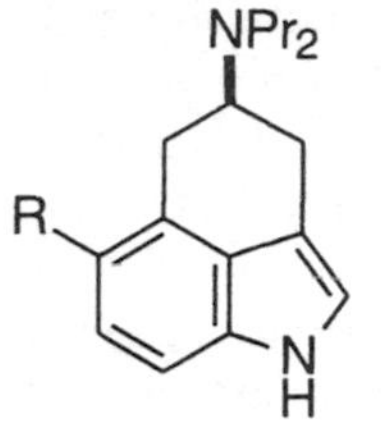

**13** R = CN
**3 (LY228729)** R = CONH$_2$

To introduce the requisite carboxamide moiety, several options were available. It was expected that Friedel-Crafts acylation followed by amide formation would be problematic. However, the bromide moiety offered several possibilities for aromatic substitution or metal catalyzed replacement. Rosemund-von Braun reaction (as described earlier) of the bromide **25a** with CuCN in NMP (N-methyl pyrrolidinone) at 200 °C gave the nitrile **26a** in 76% yield. An alternative to NMP included DMF at reflux, but the reaction was much slower. A competing side reaction was the cleavage of an *N*-propyl group under these extreme conditions. Conversion of the nitrile into the corresponding amide (**26b**) was accomplished quite smoothly with neat polyphosphoric acid (PPA) at 90 °C in excellent yield. This reaction was run quite concentrated, but required large volumes of water in the workup procedure to break the intermediate complex. This latter fact presented us with a throughput issue, suggesting a search for alternatives.

Deprotection of the nitrile **26a** could be accomplished with a variety of reagents. In the early phases of the project, we noted the facile amide cleavage with *n*-BuLi. This observation was made while attempting to transmetalate the aryl bromide. In fact, the cleavage was so efficient that we actually implemented this protocol to prepare initial quantities of the desired indoline. On small scale, however, we also noted the indole as the major byproduct. This could perhaps be formed from autoxidation of a dipole stabilized α-anion and was noted in the Rebek lysergic acid synthesis.[12a] The suspected autoxidation could not be developed into an efficient process for net oxidation. Later, it was found that the more conventional KOH/EtOH conditions were well-suited for benzamide cleavage, and these were the conditions ultimately implemented for our large scale work. The deprotection workup incorporated an acid-base extraction to separate the benzoic acid and neutral impurities. Thus, deprotection could occur at either the nitrile or amide stage with equal efficiency to provide a substrate compatible with subsequent oxidation conditions.

Two alternative strategies for introduction of the primary amide functionality utilized Heck carbonylation technology. Two potential precursors to amide **26b** were envisioned to be esters or secondary amides, which could be prepared by the reaction of an aryl halide with carbon monoxide in the presence of a palladium catalyst.[13] While the aryl bromide **25a** proved to be an unreactive substrate for carbonylation, the iodide **25b** reacted with carbon monoxide and palladium catalysts in alcohol solvents to give methyl and ethyl esters (**26c,d**) in high yields. Both esters were extremely resistant towards amidation with ammonia. Under forcing conditions (NaNH$_2$) only cleavage of the benzoyl group occurred. We were pleased to find however, that iodide **25b** smoothly underwent carboxamidation in the presence of carbon monoxide, ammonia, and palladium catalyst to give the desired primary amide **26b** in excellent yield. We believe this reaction is the first reported example of preparing a primary amide *via* this methodology.[14]

With the optically enriched indoline, our attention was now focused on the indole oxidation. The literature offers many suggestions for this thermodynamically favored conversion. Manganese dioxide in CH$_2$Cl$_2$ was initially quite effective but required a large excess of reagent, often >25 mole equivalents. Solvent change to acetic acid showed a dramatic rate acceleration even at 1.5-2 equivalents of MnO$_2$. This oxidative protocol seemed reasonable at first, but the workup revealed some hidden troubles. Filtration of the manganese waste at the end of the reaction was problematic, due presumably to a myriad of acetate salts. The filtrate obtained after these lengthy separations contained a new species (**28**), derived from *N*-depropylation. Employing Mn(OAc)$_3$ as the primary oxidant also proved successful and avoided dealkylation since it is a much milder oxidant.

28                                            29                                            30

Alternatives to the manganese-based oxidations included DDQ, which was less effective at completing the oxidation. Swern oxidation of the indoline provided the indole very efficiently and concomitantly dehydrated the amide to the nitrile. Palladium on carbon in MeOH at reflux smoothly transformed the indoline to the indole in 75-85% yield. As with the manganese procedure, scale-up proved that *N*-dealkylation was the major by-product. Potentially, coordination with either nitrogen could lead to an oxidative insertion across a carbon-nitrogen bond. Reductive elimination would then provide the desired indole, or the undesired propyl-cleaved secondary amine. New components in the reaction mixture were characterized as the transposed ketone **29** and the saturated compound, **30**. These were rationalized as arising from oxidative addition to form an enamine, which produced the ketone. This ketone, when isolated and resubjected to the reaction conditions afforded the corresponding saturated material. The reaction profile varied with various lots of catalyst. Catalyst water content and age played a significant role in securing the optimal catalytic system. Isolation of LY228729 as the hippurate salt[15] offered a convenient method for purification by recrystallization (Scheme 2).

Scheme 2. Synthesis of LY228729 from Olefin **16** ("Kornfeld Ketone Route")

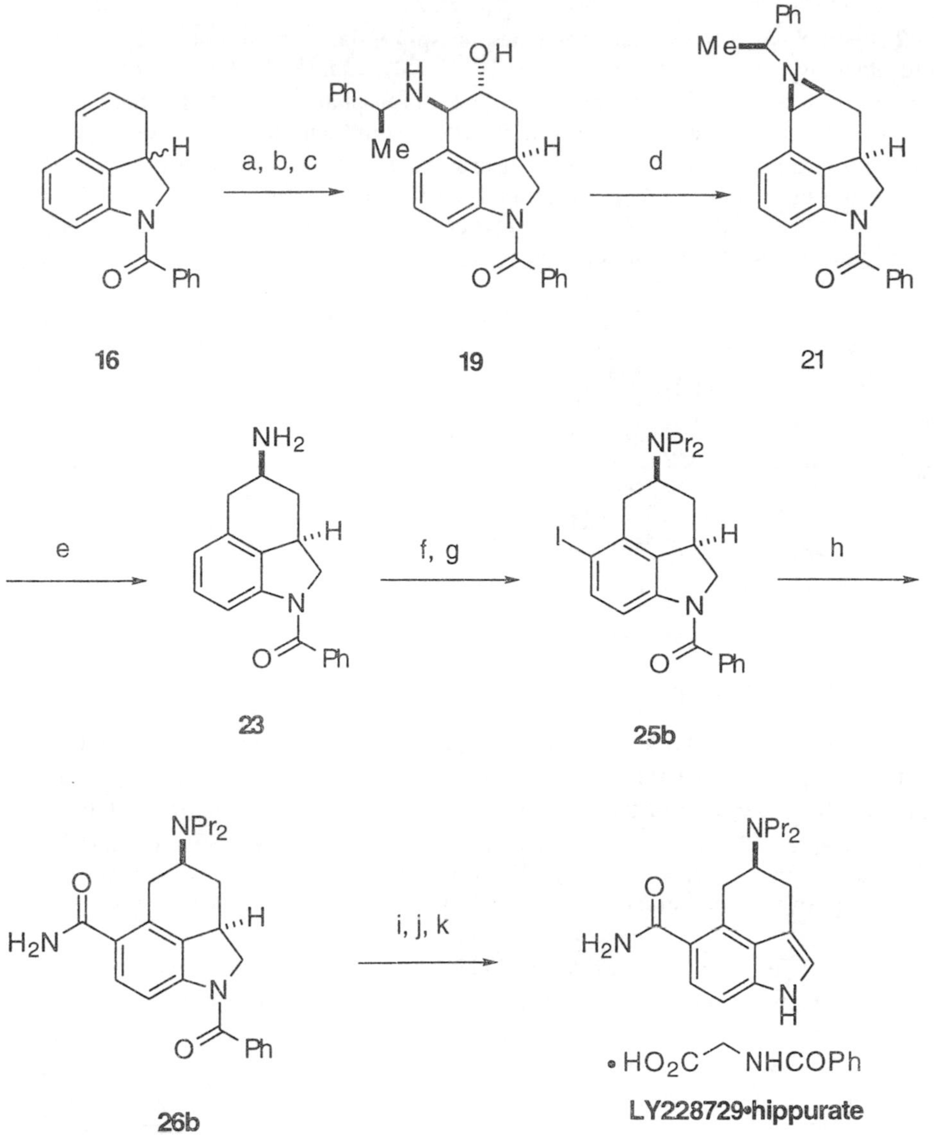

a) MMPP, *n*-BuOH;  b) *(S)*-phenethylamine, *n*-BuOH;  c) crystallize from *i*-PrOH;
d) MsCl, NEt₃;  e) 10% Pd-C, H₂, THF, H₃PO₄;  f) H₅IO₆, I₂;  g) *n*-PrI, K₂CO₃;
h) CO, NH₃, Pd(PPh₃)₂Cl₂;  i) NaOH, EtOH;  j) 10% Pd-C, MeOH;  k) hippuric acid

## Tryptophan Route

L-tryptophan (**32**) contains the carbon framework and an appropriately positioned stereocenter, with the desired absolute configuration needed for LY228729. In 1984, Rebek reported the synthesis of lysergic acid and its C-4 epimer (**33**) from L-tryptophan.[12] These factors and the ready availability of L-tryptophan lead us to pursue it as a starting material for the synthesis of LY228729.[16] Based on Rebek's work, we believed there was a high probability that L-tryptophan could be converted to tricyclic amine **31**, and thus intersect with the synthetic route described in the previous section. As will be discussed, we were less certain this "chiral pool" based synthesis could be developed into an efficient, manufacturable process.

|  |  |  |  |
|---|---|---|---|
| **31** | | **32** | **33** |

(3)

### Reduction and Protection of L-Tryptophan

We planned to form the C-ring of LY228729 using an intramolecular Friedel -Crafts acylation reaction, as had Rebek. The required *N*-protected indoline was prepared by trifluoroacetylation[17] of L-tryptophan, ionic hydrogenation of the C-2 double bond (TFA/Et$_3$SiH), and benzoylation of the indoline nitrogen (Scheme 3). Using this scheme, a 45:55 mixture of diastereomeric indolines (**36a:36b**) were obtained in 67% yield from **34**. Fractional crystallization of the mixture from CHCl$_3$ provided the desired diastereomer (*vide infra*) **36a** with >96% de in 21% overall yield from **34**.

Although convenient for small scale work, the Et$_3$SiH reduction required TFA as solvent. More economical catalytic hydrogenation methods were investigated. Reduction of an indole C-2 double bond under neutral conditions typically requires forcing conditions (e.g. Raney Ni/>1000 psi hydrogen).[6] Lower pressure hydrogenations employ strongly acidic conditions, presumably to form an N1-C2 iminium tautomer which is readily reduced. Optimal catalytic hydrogenation conditions for **34** were found to be over 5% PtO$_2$ catalyst in water with 10-20 equivalents of TFA at 10 psi of hydrogen. Notably, when the hydrogenation was performed in acetic acid or 1N HCl, extensive overreduction of the aromatic ring occurred (eq 4). Even with the TFA/H$_2$O reaction medium, the reaction had to be carefully monitored to prevent overreduction.

Although the catalytic hydrogenation was not diastereoselective, it had processing advantages over the ionic hydrogenation namely, indolines **35a,b** were benzoylated *in situ* without isolation. In the former reaction, when the hydrogenation was complete the PtO$_2$ was filtered, aqueous NaHCO$_3$ was added to the reaction mixture, followed by benzoyl chloride. This process was coupled with the CHCl$_3$ fractional crystallization to provide indoline **36a** in 18% overall yield from **34**. Unfortunately, regardless of which indole reduction and benzoylation sequence used, consistent crystallization of acid **36a** was difficult (even with seeding) and CHCl$_3$ proved to be the optimal crystallization solvent.

Scheme 3. Reduction and Protection of L-Tryptophan

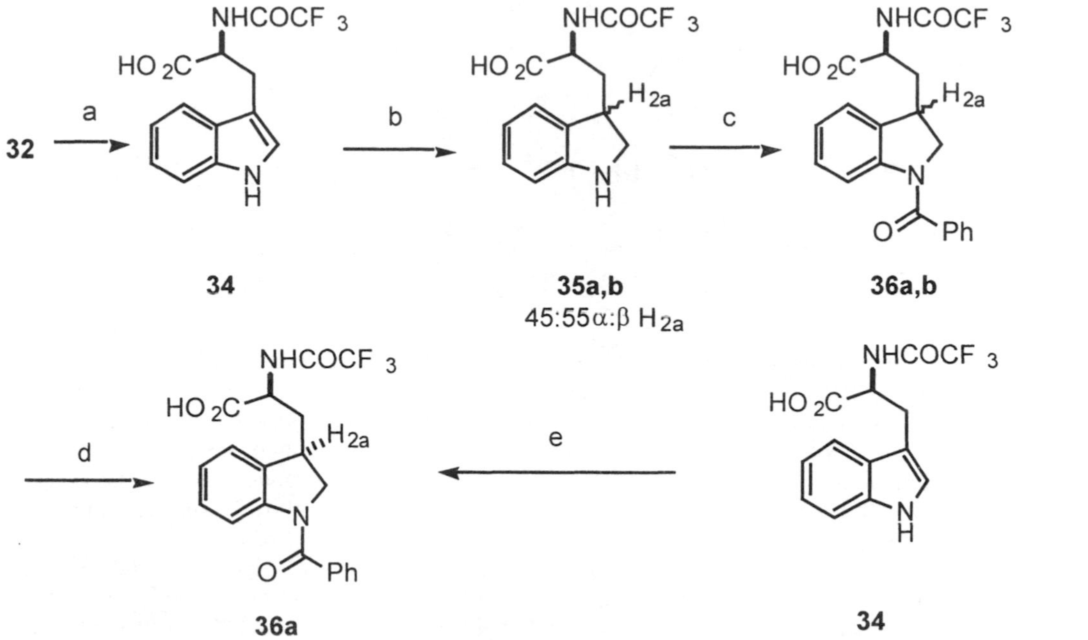

a) CF$_3$CO$_2$Et, MeOH, NEt$_3$; b) Et$_3$SiH, TFA; c) PhCOCl, NEt$_3$, THF; d) CHCl$_3$ cryst.
e) 1. 5% PtO$_2$, H$_2$, H$_2$O, TFA 2. NaHCO$_3$, PhCOCl 3. CHCl$_3$ cryst.

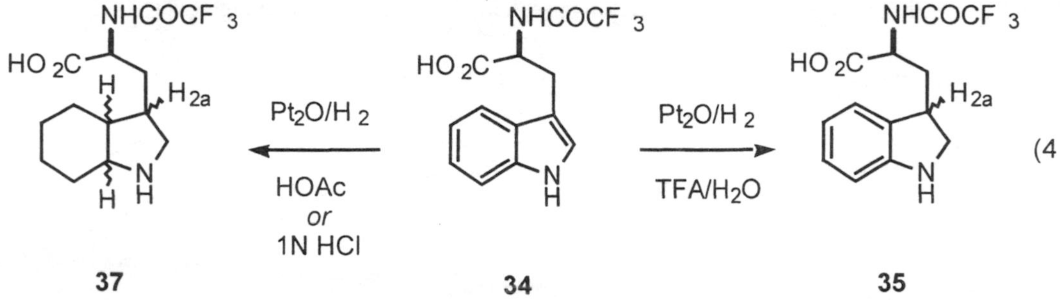

(4

**Friedel-Crafts Acylation Reaction and Stereochemistry**

Preserving the configuration of the C-4 stereocenter during the Friedel Crafts acylation reaction was essential for using this strategy to prepare LY228729. While several enantiomerically pure α-amido 1-tetralones have been prepared *via* Friedel-Crafts cyclizations,[18] Rebek's work suggested the desired tricyclic ketone and/or the carboxylate precursor would be more prone to epimerization at C-4. As shown in eq 5,[12] the azlactones derived from indoline diastereomers **38a** and **38b** cyclized in the presence of AlCl$_3$ at 80 °C to give enantiomers of a single ketone diastereomer **39**, having the C-2a and C-4 hydrogens on the same face of the molecule. As expected, cyclization of 50:50 mixture of acid diastereomers (**38a, 38b**) gave racemic ketone **39**. In short, the configuration at C-2a dictated the configuration at C-4.

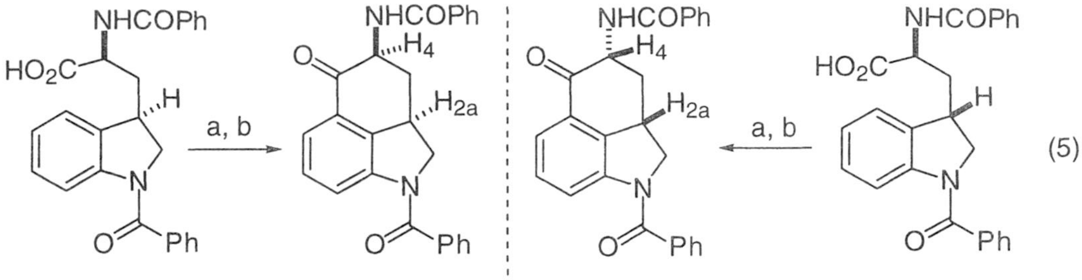

|  38a  |  (+)-39  |  (-)-39  |  38b  |

a)  Ac$_2$O, 100 °C;  b)  4 equiv AlCl$_3$, ClCH$_2$CH$_2$Cl, 80 °C

We set out to determine whether the Friedel-Crafts acylation reaction could be performed under milder conditions that would enable us to determine when C-4 epimerization occurred. In other words, could both diastereomers of a tricyclic ketone be formed and would they be configurationally stable under the reaction conditions? If so, the covalent resolution of acid **36** could be omitted from the process. In this instance optimizing and understanding the reaction parameters for the Friedel-Crafts acylation reaction would not only improve this step, but could have profound impact on the throughput of the synthesis.

Indeed we found that **36a** could be converted to the tricyclic ketone **40a** under relatively mild conditions. The acid was treated with 2 equivalents of oxalyl chloride and catalytic DMF at -5 °C, and the resulting acid chloride was treated with 4 equivalents of AlCl$_3$ at 22 °C for 20-30 h . The crude ketone **40** was obtained as a 98:2 mixture of diastereomers and was recrystallized from *n*-BuOH to provide **40a** in 70% overall yield with >99% ee (eq 6). Other Lewis acids were examined in this reaction; TiCl$_4$ and FeCl$_3$/MeNO$_2$ gave no reaction at ambient temperature. Reactions with AlBr$_3$ gave the desired ketone product, but offered no advantage over AlCl$_3$.

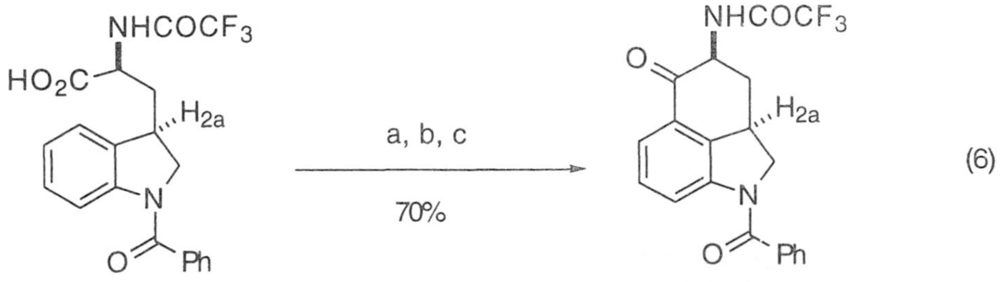

|  36a  |  40a  |

a)  2 equiv (COCl)$_2$, - 5 °C, CH$_2$Cl$_2$ ;  b)  4 equiv AlCl$_3$, 22 °C;  c)  *n*-BuOH cryst.

We then subjected a 45:55 mixture of acid diastereomers **36a:36b** to the above reaction conditions and followed the reaction by HPLC and chiral HPLC. These studies showed:
1) the acid diastereomers were consumed at similar rates at 22 °C
2) the ee of the crude **40a** decreased over time

3) the diastereomer ratio of ketones **40a:40b** increased (relative to the **36a:36b** diastereomer ratio in the starting acids) over time.

Scheme 4. Intramolecular Friedel-Crafts Acylation Reaction: Stereochemistry

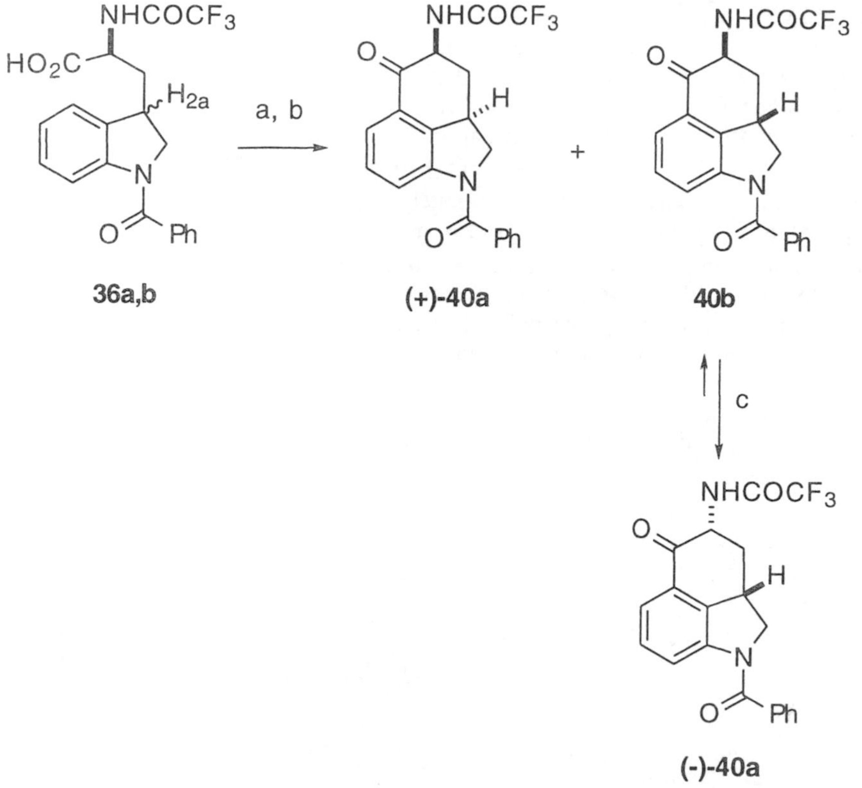

a) 2 equiv (COCl)$_2$, cat DMF, -5 °C; b) 4 equiv AlCl$_3$; c) AlCl$_3$

Table 1. Friedel-Crafts Reaction Data

| Reaction Temp (°C) | Time (h) | Ratio of **40a:40b** (HPLC) | %ee **40a** (HPLC)[a] |
|:---:|:---:|:---:|:---:|
| 0 | 27[b] | 43:57 | 91 |
| 22 | 30 | 52:48 | 82 |

[a]SS Whelk 01 column
[b]Reaction at 50% conversion

As shown in Scheme 4 and Table 1, when the cyclization reaction was performed at 22 °C, the crude ketone was obtained as a 52:48 mixture of diastereomers **40a:b**, respectively. The enantiomeric purity of **40a** was (an unacceptable) 82% ee. When the reaction was repeated at 0 °C, the rate of decline in ee of **40a** was slower, as was the reaction rate. After 24 h at 0 °C, crude **40a** was obtained with 91% ee. Unfortunately the reaction had proceeded to only 50% conversion. In general, cyclization reactions run at 0 °C could not be driven much beyond 50% conversion due to what appeared to be the precipitation of an aluminum complex of the acid chloride after 4-5 h. Interestingly, samples taken from the reaction performed at 0 °C, during the first 4 h, showed that the acid chloride derived from diastereomer **36a** was consumed at approximately twice the rate of the acid chloride derived from diastereomer **36b**. Ketone diastereomer **40a** was correspondingly formed at twice the rate of ketone **40b**.

We concluded from these experiments that the decrease in ee of ketone **40a** during the reaction was primarily due to the epimerization of ketone **40b**. When ketone **40b** (isolated by careful preparative HPLC) was treated with 1 equivalent of $AlCl_3$ in $CH_2Cl_2$ at 22 °C, it epimerized to **40a** with a half-life of approximately 3 h. Ketone **40b** was more rapidly isomerized with 1 equivalent of triethylamine in $CH_2Cl_2$, and was stable in aqueous acid (e.g. acetonitrile containing 1 equivalent of 1N HCl /24 h). We found that any mixture of **40a** and **40b** could be isomerized to an equilibrium 98:2 ratio of diastereomers. The thermodynamic preference for diastereomer **40a** versus **40b** can be rationalized as the difference between an equatorial and an axial C-4 amide substituent, respectively (Scheme 5).

Scheme 5. Possible Conformations of Ketone **40** Diastereomers

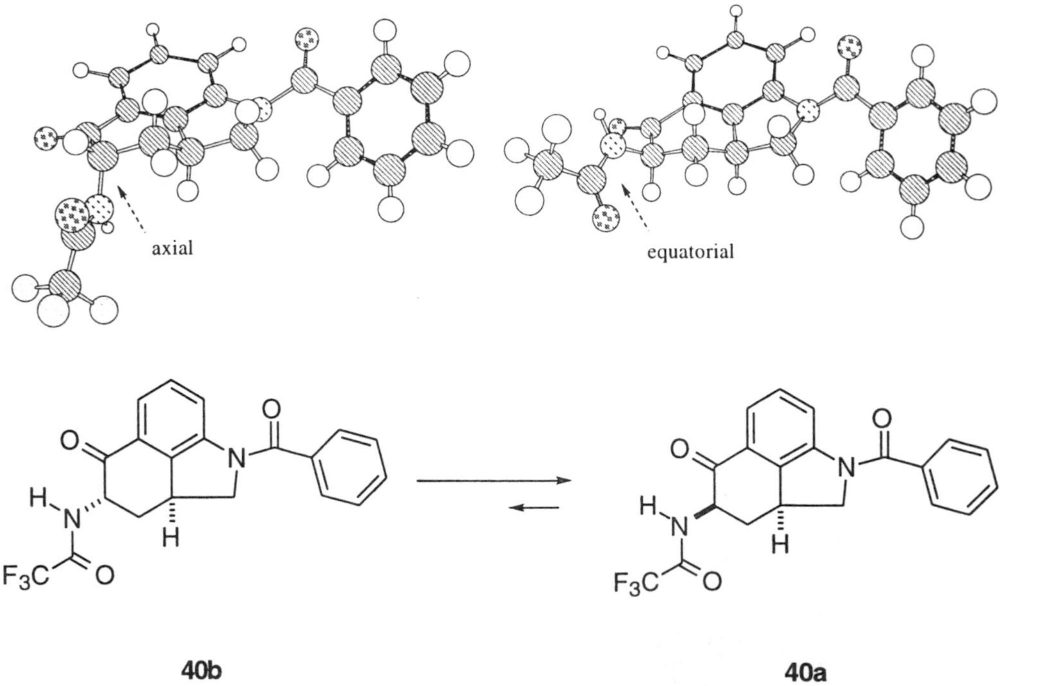

axial                    equatorial

**40b**                                    **40a**

### Intersection with the Kornfeld Ketone Route

With ketone **40a** in hand, all that remained to intersect the Kornfeld ketone route was deoxygenation of C-5 and deprotection of the primary amide. Deoxygenation of the ketone proved to be non-trivial. Catalytic reduction of **40a** with 10% Pd/C gave the fully reduced compound (**42**) in 47% yield, but required TFA as solvent. The most efficient deoxygenation method was a two step protocol (Scheme 6). Ketone **40a** was reduced with NaBH$_4$ to give alcohol **41 as** an 8:1 mixture of epimers at C-5 in 90% yield. Alcohol **41** was then trifluoroacetylated in situ with TFAA under hydrogenolysis conditions (10% Pd-C, 50 psi H$_2$, THF) to give ketone **42** in 86% yield. Simple base hydrolysis of the trifluoroacetamide gave amine **23** identical to that prepared from the Kornfeld-Woodward ketone.

In summary, we demonstrated that a key intermediate in the LY228729 synthesis (amine **23**) could be prepared in seven steps and 8% overall yield from L-tryptophan. However, due to the stereochemical consequences of the Friedel-Crafts acylation reaction, this strategy still required a difficult covalent resolution which detracted from its large scale viability.

Scheme 6. Completion of the Synthesis of Amine **23** from L-Tryptophan

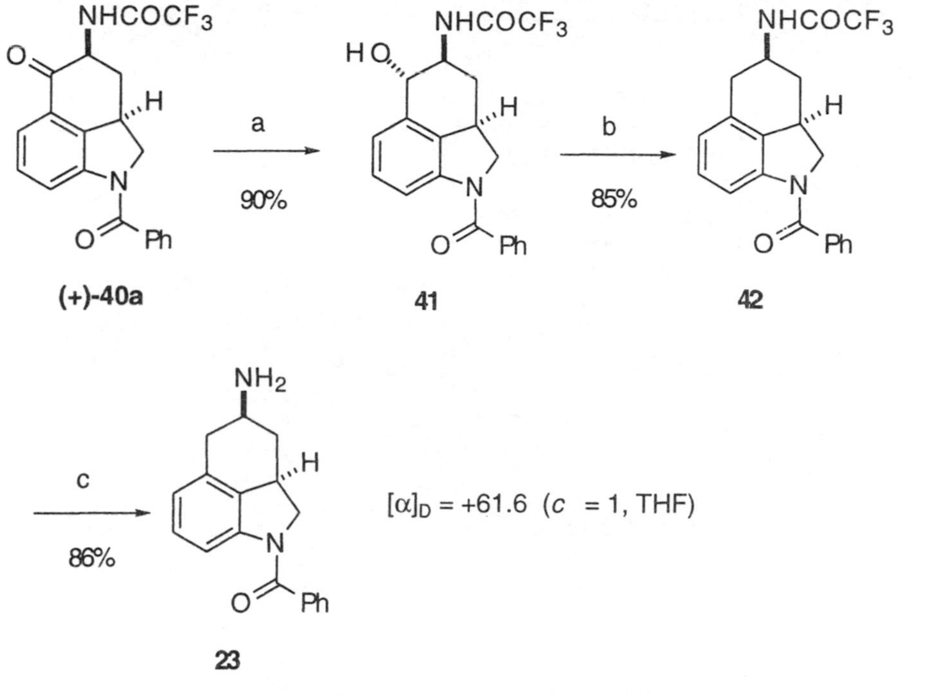

a) NaBH$_4$, MeOH, THF, 0 °C; b) TFAA, THF, H$_2$,10% Pd-C; c) NaOH, THF, H$_2$O

### Comparison of Synthetic Routes

Our route selection was guided by our laboratory and pilot plant experiences on all three synthetic approaches. Strategically, all three routes produced enantiomerically pure product *via* a covalent resolution. The cost of the chiral auxiliary (($R$)-4-nitrophenethylamine), and the low demonstrated throughput of the original strategy, led us to focus on the Kornfeld ketone and tryptophan-based routes. A comparison of the syntheses of the key intermediate, amine **23**, from commercially available materials is shown in Scheme 7. In summary, the synthesis of LY228729 was accomplished in 9 steps in 9% overall yield from commercially available olefin **16**. While L-tryptophan cost is approximately one-tenth that of olefin **16**, the synthesis of LY228729 from L-tryptophan requires 13 steps and proceeds in 3% overall yield. In addition to requiring fewer steps, the conversion of olefin **16** to amine **23** was operationally much simpler and more reliable than the conversion of L-tryptophan to **23**. Notably the fractional crystallization of amino alcohol **19** was much more robust than the fractional crystallization of the tryptophan derived acid **36a**. These factors all pointed to the Kornfeld ketone route as the most viable long-term route of choice.

Scheme 7. Comparison of Synthetic Routes

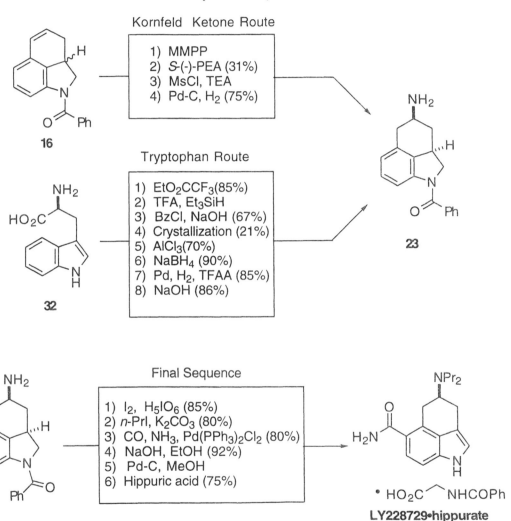

• HO$_2$C $\frown$ NHCOPh

**LY228729•hippurate**

## Acknowledgments

This chapter represents the outstanding contributions of many scientists within Lilly Research Laboratories. Among these, the authors would like to acknowledge the following colleagues for their intellectual and technical contributions to this project: T. J. Kress, M. A. Carr, W. M. Chang, P. E. Creviston, M. E. Flaugh, A. L. Glasson, J. H. Kennedy, V. V. Khau, M. R. Leanna, J. D. Marshall, G. L. Murdoch, B. C. Peterson, R. Waggoner, and J. P. Wepsiec. We are grateful to T. J. Kress for contributing the description of the Heck carboxamidation chemistry to this chapter.

## References

1 (a) Flaugh M. E.; Murdoch G. L. "Total Synthesis of LY228729, a Novel 5-HT$_{1A}$ Agonist". 200th National Meeting of the American Chemical Society. Washington, DC; August 26, 1990. (b) For the synthesis of racemic LY229729 see: Flaugh, M. E.; Mullen, D. L.; Fuller, R. W.; Mason, N. R. *J. Med Chem.* **1988**, *31*, 1746.

2. For use of 4-nitrophenethylamine to prepare enantiomerically pure 8-bromo-2-aminotetralins see: (a) Flaugh, M. E.; Schaus, J. M.; Titus, R. D. European Patent Application 385658, **1990.** (b) Gidda, J. S.; Schaus, J. M. European Patent Application 455510, **1991.** (c) The sulfided Pt-C catalyst was purchased from Engelhard Corp., Seneca, SC.

3. Hunden, D. C. *Org. Prep. Proceed. Int.* **1984**, *16*, 294.

4. Perry, C. W.; Brossi, A.; Deitcher, K. H.; Tautz, W.; Tettel, S. *Synthesis*, **1977**, *7*, 492.

5. Crivello, J. V. *J. Org. Chem.* **1981**, *46*, 3056.

6. Kornfeld, E. C.; Fornfeld, E. J.; Kline, G. B.; Mann, M. J.; Morrison, D. E.; Jones, R. G.; Woodward, R. B. *J. Am. Chem. Soc.* **1956,** *78*, 3087.

7. Nichols, D. E.; Robinson, J. M.; Li, G. S.; Cassady, J. M.; Floss, H. G. *Org. Prep. Proc. Int.* **1977**, *9*, 277.

8. Leanna, M. R.; Martinelli, M. J.; Varie, D. L. Kress, T. J. *Tetrahedron Lett.* **1989**, *30*, 3935.

9. Martinelli, M. J.; Peterson, B. C.; Khau, V. V.; Hutchison, D. R.; Leanna, M. R.; Audia, J. E.; Droste, J. J.; Wu, Y-D.; Houk, K. N. *J. Org. Chem.* **1994**, *59*, 2204.

10. Brougham, P.; Cooper, M. S.; Heaney, H.; Thompson, N. *Synthesis* **1987**, 1015.

11. (a) Russell, H. F.; Harris, B. J.; Hood, D. B.; Thompson, E. G.; Watkins, A. D.; Williams, R. D. *Org. Prep. Proc. Int.* **1985**, *17*, 391. (b) Johnson, H. E.; Crosby, D. G. *J. Org. Chem.* **1963**, *55*, 2794. c) Borror, A. L.; Chinoporos, E.; Filosa, M. P.; Herchen, S. R.; Petersen, C. P.; Stern, C. A.; Onan, K. D. *J. Org. Chem.* **1988**, *53*, 2047.

12. (a) Rebek, J.; Tai, D. F.; Shue, Y. K. *J. Am. Chem. Soc.* **1984,** *106*, 1813-1819. (b) Rebek, J.; Shue, Y. K.; Tai, D. F. *J. Org. Chem.* **1984,** *49*, 3540-3545.

13. a) Schoenberg, A.; Bartoletti, I.; Heck, R. *J. Org. Chem.* **1974**, *39*, 3318. b) Schoenberg, A.; Heck, R. *J. Org. Chem.* **1974**, *39*, 3327.

14.  Kress, T. J.;  Wepsiec, J. P.  US Patent 5039820, **1991.**

15.  Kress, T. J.;  Varie, D. L.  US Patent 5397799, **1995**.

16. For a preliminary account of ths work see:  Varie, D. L.  *Tetrahedron Lett.* **1990**, *31*, 7583.

17. Curphey, T. J.  *J. Org. Chem.* **1979,** *44*, 2805-2807.

18.  Examples of intramolecular acylations of α-amino acid derivatives:  (a) Melillo, D. G.; Larsen, R. D.;  Mathre, D. J.;  Shukis, W. F.;  Wood, A. W.;  Colleluori, J. R.  *J. Org. Chem.* **1987,** *52,* 5143-50.  (b) Buckley, T. F.;  Rapoport, H.  *J. Org.  Chem.* **1983,** *48,* 4222-4232.  (c) McClure, D. E.;  Lumma, P. K.;  Arison, B. H.;  Jones, J. H.;  Baldwin, J. J.  *J. Org. Chem.* **1983,** *48,* 2675-2679.  (d) McClure, D. E.;  Arison, B. H.;  Jones, J. H.;  Baldwin, J. J.  *J. Org. Chem.* **1981,** *46,* 2431-2433.

# Process Research and Development of CGS 19755, an NMDA Antagonist

**Peter Giannousis[1]\* and John Carlson[1⊥]**
*Chemical Development, Novartis Pharmaceuticals Division, [IV] Summit, New Jersey, 07901-1398*

**Marius Leimer**
*Chemical Development, Novartis Pharmaceuticals Division, [IV] Basel SWITZERLAND, CH-4001*

Key words: piperidine, pipecolic acid, phosphonate, radical carbamoylation, reduction

## Introduction

CGS 19755 (**1**, (±)-*cis*-4-phosphonomethyl-2-piperidinecarboxylic acid) is an N-methyl-D-aspartic acid (NMDA) antagonist, that has been investigated as an antiischemic agent in the treatment of stroke and head trauma.[2]  This chapter presents the evolution of the synthetic processes to prepare this compound from the initial research laboratory-bases synthesis to the launch-ready multi kilogram pilot plant campaign.  A brief explanation of the development process at the former Ciba-Geigy during the timeframe of the CGS 19755 project is necessary in order to put the entire effort into perspective.  Medicinal chemistry scientists discovered CGS 19755 in 1985 (Scheme 1; retrosynthetic strategy: Scheme 2).[3]  After passing several preclinical screens, the compound was promoted to early development and as a result it was necessary to prepare very quickly about 300 g of CGS 19755 in a non-GMP Kilo Lab (Scheme 3).  Additional material was subsequently needed to meet the needs of formal development (formulation, toxicology, clinical, etc.).  After some process research, two preparations were carried out under GMP (Good Manufacturing Practice) in a pilot plant, giving 8 kg and 51 kg respectively (Scheme 4).  The project lay dormant for some time, but rekindled interest led us to develop further improvements.  Two similar routes emerged for scale-up evaluation (Schemes 6 and 7).  The selected process was then transferred to a Process Development group for final optimization using statistical methods.  As a result, three parallel campaigns led to the production of a total of 440 kg of CGS 19755 during 1995.  This final process is suitable for further scale-up and registration.

**Structure and Properties of CGS 19755 (1)**

|  | 1 | 2 |
|---|---|---|
|  | CGS 19755 | AP-5 |

CGS 19755 (**1**) is racemic (±)-*cis*-4-phosphonomethyl-2-piperidinecarboxylic acid and was designed as a cyclic analog of 2-amino-5-phosphonopentanoic acid (AP-5, **2**).[4] It features the thermodynamically-favored 2-4-*cis*-relationship of the carboxylic and methylenephosphonic acid moieties. It is a polyionic compound, and exists as a zwitterion at the isoelectric point (pH ca. 2.5), with the amino group protonated and one of the phosphonohydroxyls deprotonated. For the purposes of drug development it is highly desirable to have a purifiable well-characterized Active Ingredient, as well as good analytical methods for determining its impurities. An advantageous property of CGS 19755 is that only one polymorph has been identified. A major technical problem for CGS 19755, however, is that it is not a thermally recrystallizable solid, in other words, no solvent has been found in which the compound could be dissolved with heating and crystallized with cooling. In addition, recrystallization via solvent/antisolvent approach was not feasible due to extremely large volumes involved and no improvement in purity. These characteristics have made purification of a recrystallizable precursor or a salt key in achieving the high purity levels required for clinical development compounds. Finding a workable purification approach proved difficult. Because of the polyionic nature and very low solubility of **1** in organic solvents, its salts with bases tended to be unstable. Attempts to recrystallize them by heating/cooling cycles often resulted in disproportionation such that **1** precipitated from solution and salts of the di-anion remained in solution. In addition, because of its acidic nature, crystallization of the hydrochloride salt from alcohols afforded esterification byproducts. These problems made preparation and isolation of this deceivingly simple compound quite challenging.

## Research Synthesis

CGS 19755 (**1**) was first prepared by Medicinal chemists at the former Ciba-Geigy using the route outlined in Scheme 1. 4-Chloromethylpyridine (**4**) was prepared from the alcohol **3** and thionyl chloride, and reacted with sodium diethyl phosphite to give phosphonate **5**. N-oxidation with *m*-chloroperbenzoic acid, followed by heating with trimethylsilyl cyanide and triethylamine gave the 2-cyanopyridine derivative **7**. Selective hydrolysis of the nitrile with sulfuric acid gave the amide-diester **8**. Hydrogenation with platinum oxide in acetic acid, followed by base treatment gave the *cis*-piperidinecarboxamide **9**. Finally, acidic hydrolysis, followed by precipitation with propylene oxide, gave CGS 19755 (**1**) with predominantly *cis* stereochemistry. The *cis/trans* ratio of the final product was later found to be about 95:5, which generally was not acceptable for clinical development because the *trans*-isomer is considered an impurity. In addition, some of the reagents that were used in this process (e.g., *m*-CPBA, TMS-CN, propylene oxide) presented considerable handling, safety, and/or cost problems.

Scheme 1.  Research Synthesis of CGS 19755 (**1**)

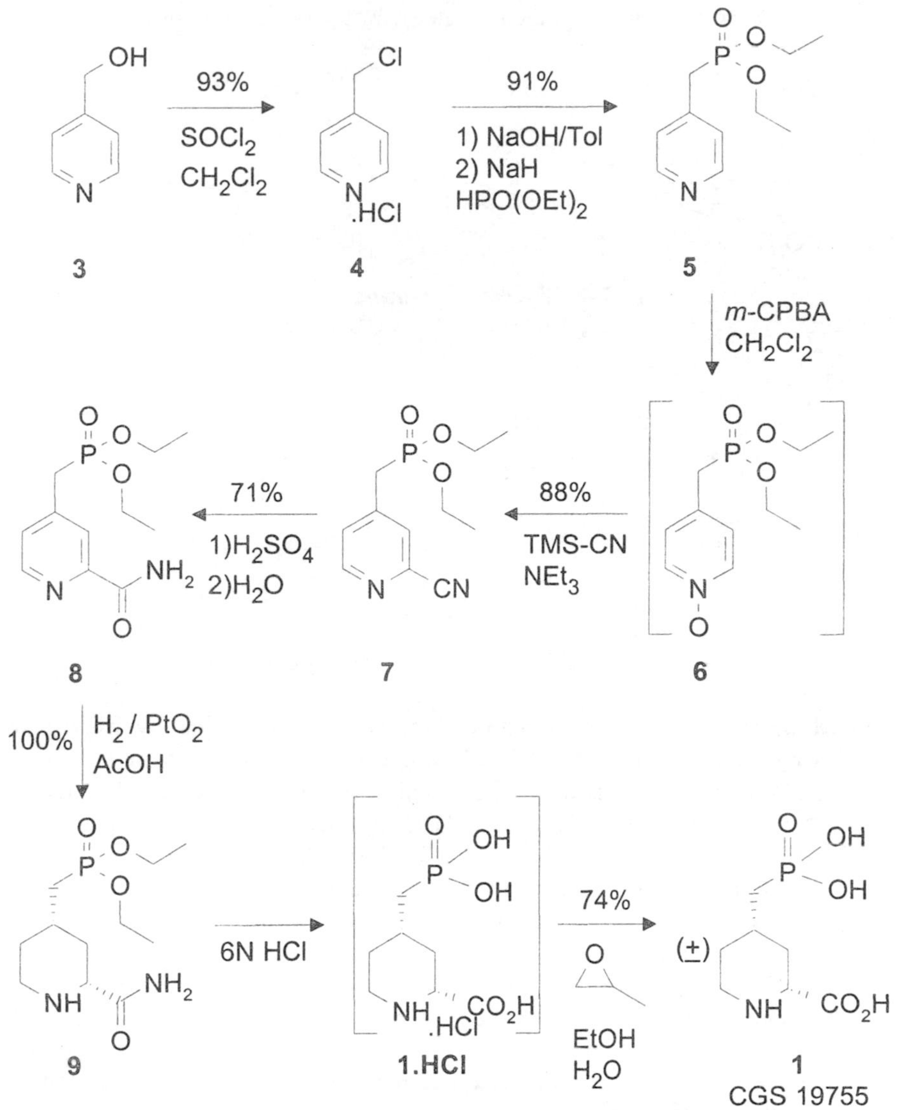

## Retrosynthetic Approach for Development

All of our approaches to **1**, including the route used in Research, have used the retrosynthetic analysis outlined in Scheme 2:

Scheme 2. Retrosynthetic strategy for the preparation of **1**

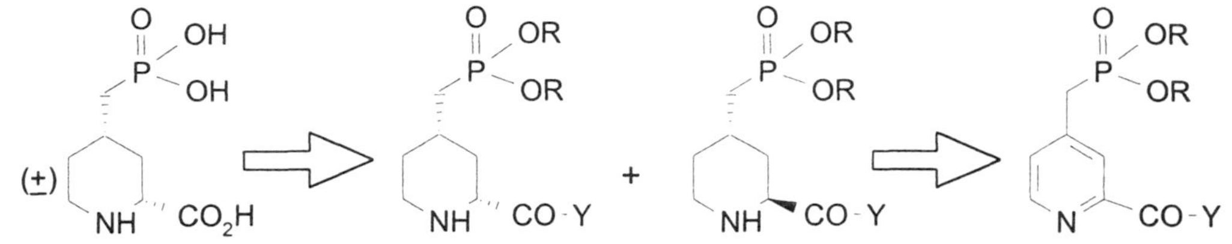

| **1** CGS 19755 | **1**: R=H; Y=OH CGS 19755 | **1-*trans***: R=H; Y=OH CGS 22567 | **10** (8: R=Et, Y=NH₂) |

**10**          **11** (**5**: R=Et)          **12** (**4**: X=Cl)          **13**

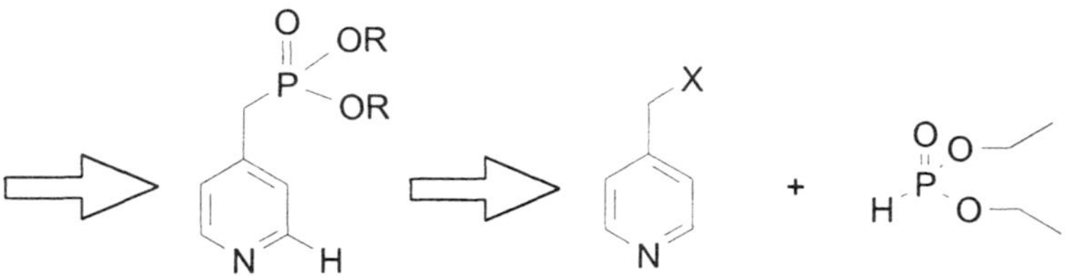

We envisioned obtaining the *cis*-configuration on the piperidine ring by hydrogenating an appropriate 2,4-pyridine intermediate **10**. Of course such a reaction is not known to give exclusively the *cis*-isomer, but a mixture of *cis*- and *trans*-isomers. Removal of the undesired *trans*-isomer ("CGS 22567" if R=H and Y=OH) and hydrolysis of the phosphonate and carboxylate moieties, if necessary, would give CGS 19755. Preparation of **10** was envisioned as arising via an oxidation/substitution reaction sequence on a diester 4-phosphonomethylpyridine **11** (**5**: R=Ethyl), which is available by alkylation of anion of diethyl phosphite (**13**) with 4-X-methylpyridine **12**, where X is a suitable leaving group. The overall approach outlined above has been used in all of the preparations of CGS 19755 within Chemical Development. In order to avoid confusion, some definitions that have been used at the former Ciba-Geigy and by FDA are as follows[5]:

*Final Intermediate*: Last intermediate isolated and controlled during the manufacturing process, before the final step that creates the crude API

*Active Pharmaceutical Ingredient (API)* or *Active Ingredient*: Substance that is represented for use in a drug and that, when used in the manufacturing, processing, or packaging of a drug, becomes an drug substance or a finished dosage form of the drug.

*Batch*: Specific quantity of an intermediate or API intended to have uniform character and quality, within specified limits, and produced according to a single manufacturing order during the same cycle of manufacture. There were seven separate manufacturing cycles of **1**, Batches 0 through 6.

## Batch 0/1: Speed in synthesis

The initial scale-up of the Research synthesis in our Kilo Lab was focused on preparing **1** with higher purity (i.e., >98%) as quickly as possible. As a result, optimization of many of the steps was not performed. Several changes were made as shown in Scheme 3, mainly in order to remove the undesired *trans*-isomer of CGS 19755 in two stages, a problem which was not addressed in the Research synthesis.

Scheme 3. Kilo Lab synthesis of **1** (Batches 0 and 1)

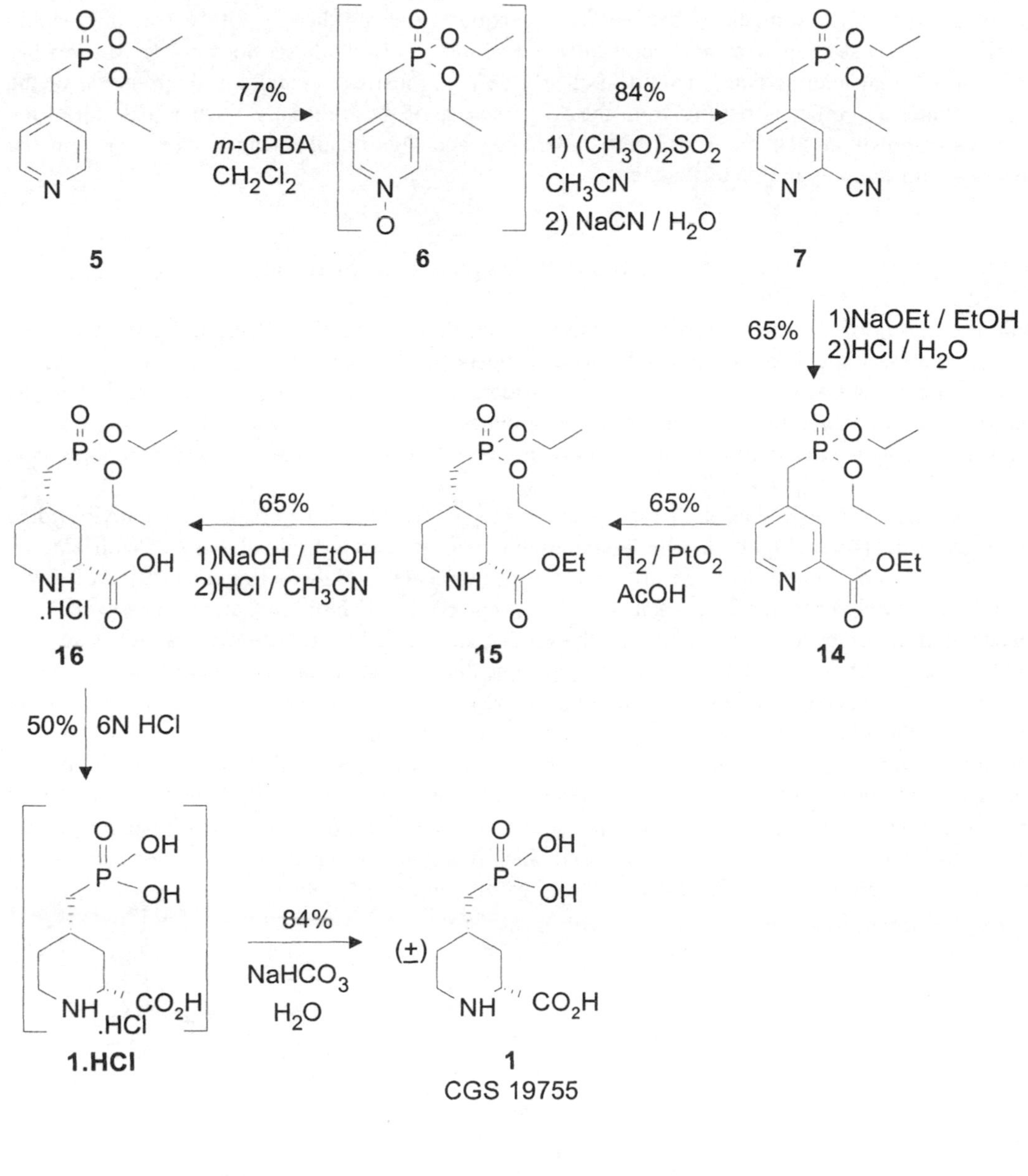

The TMS-CN used in preparing the 2-cyanopyridine was replaced: the N-oxide was reacted with dimethyl sulfate followed by sodium cyanide to give the 2-cyanopyridine **7**. Acidic hydrolysis of the cyano group proved to be extremely exothermic, therefore, the nitrile was converted to an ethyl ester via the iminoether. The pyridine-triester **14** was hydrogenated, giving piperidine **15**, with about a 95:5 *cis:trans*-isomer ratio. Selective hydrolysis of the carboxylate ester, followed by hydrochloride salt formation and several manipulations including a recrystallization, removed most of the impurities. Acidic hydrolysis of the remaining esters of **16**, followed by recrystallization gave the *cis*-isomer of the HCl salt in >99.5% purity. Sodium bicarbonate neutralization of **1.HCl** gave about 300 g of **1** in 6.3% overall yield, starting from 4-pyridinecarbinol. The relatively low overall yield with this route was mainly due to the fact that purification of all intermediates between the N-oxide **6** and the diester salt **16** was not readily attainable because they were not recrystallizable solids. This synthesis does not have a readily purifiable "final intermediate", **16**, nor a clearly defined purification strategy for removal of the *trans*-isomer and other impurities from the AI. Because of these reasons, further scale up of the synthesis shown in Scheme 3 was not undertaken and significant changes were made in the synthesis for the preparation of Batches 2 and 3.

### Batches 2/3: Rapid Development and Scale-up

Initial scale-up of a synthetic process to multi-kg reactions in the Pilot Plant at the former Ciba-Geigy Pharmaceuticals is not allowed unless it meets strict safety, environment, and regulatory requirements. As a process matures, cost considerations become important as well, but in the early stages of development, speed is more important. In order to meet these requirements, substantial changes were made in the synthesis between Batch 1 and Batch 2, as shown in Scheme 4.

The synthesis began in the same way as Scheme 3: 4-Chloromethylpyridine hydrochloride (**4**), a highly irritating compound, was prepared from alcohol **3** with thionyl chloride. Because of worker exposure concerns on this scale (50 kg reactions), isolation of **4** was avoided: the hydrochloride suspension was prepared in dichloromethane, the solvent was exchanged *in situ* to toluene, and the free base of **4** was formed in toluene using aqueous sodium hydroxide. The toluene solution of **4** was azeotropically dried *in vacuo*, and diethyl phosphite was added. In the previous synthesis, sodium hydride was used as the base for forming the sodium salt of diethyl phosphite. This was not deemed suitable for scale-up because it creates a safety hazard (hydrogen gas) and adds mineral oil to the reaction mixture. As an alternative, potassium *tert*-amylate solution in toluene, which can be purchased, was used as base. Thus the base was added slowly to the solution of **4** and diethyl phosphite, giving the anion of diethyl phosphite *in situ*, followed by displacement of the halide to give **5**. A major side-product formed in this reaction (in about 11%) is phosphonate **17**, which is formed by alkylation of **5** with **4**.

**17**

The level of this impurity was insensitive to alterante addition sequences, but was lowered to about 7% when the reaction was run at a lower temperature.

Pyridine **5** is not thermally stable, so it was purified by WFE (wiped-film evaporation) vacuum distillation, a technique with very short residence time under high temperature.

The conversion of **5** to a 2-carboxypyridine according to Scheme 3 involved five chemical steps and some undesirable chemicals (e.g., dimethyl sulfate, sodium cyanide). A literature search showed that the same overall conversion could be achieved by nucleophilic radical substitution, as reported by Minisci and coworkers.[6] A protonated heteroaromatic base can be substituted with nucleophilic carbon-centered radicals in excellent regioselectivity and good yields. These radical substitution reactions are complementary to the Friedel-Crafts (electrophilic) reactions because the resulting products have opposite regioselectivity. A limitation of these reactions is that it is sometimes difficult to stop at monosubstitution, especially when alkyl radicals are used. Using carbonyl radicals, however, greatly reduces the problem of disubstitution because once a carbonyl moiety is attached to the heteroaromatic base, the molecule becomes much less basic and essentially inert to further substitution. Thus it is possible to control the extent of substitution by controlling the pH of the reaction mixture. A major problem seen in the radical substitution of pyridines is that they tend to react in both the 2- and 4-positions. In our case, however, pyridine **5** is already 4-substituted, so we expect only 2-substitution. At first we attempted to prepare the triethyl ester **15**, also prepared in Scheme 3, by reacting protonated **5** under conditions reported by Minisci[7] (ethyl pyruvate, hydrogen peroxide, and ferrous sulfate). Up to a 65% conversion to **15** could be attained, but isolation was difficult by flash chromatography. We noted that the first synthesis of CGS 19755 involved the amide **8**, a crystalline intermediate, and decided to prepare the same compound via a radical coupling and purify by crystallization. A brief search of oxidants indicated that benzoyl peroxide with formamide could be used with good results. It should be noted that in this reaction benzoyl peroxide is not only the initiator but is required stoichiometrically as an oxidizing reagent in rearomatizing the pyridine ring after the nucleophilic radical attack. To avoid possible oxidant accumulation and runaway exotherms, a feed-controlled process was developed where the benzoyl peroxide was added slowly as a formamide slurry to the reaction mixture containing protonated **5** (3 eq. of $H_2SO_4$) at 85-90 °C. The workup involved aqueous base, to remove the byproduct benzoic acid, and solvent exchange from dichloromethane into *n*-butyl acetate, affording crystalline **8** (>97% purity) in 44% yield from **5** (20 kg scale).

In the first route to CGS 19755, hydrogenation of the pyridine nucleus to a piperidine was followed by hydrolysis to remove the amide and esters. This approach was not attractive for scale-up mainly because the intermediate **9** was an oil, not readily purifiable, and susceptible to epimerization. In the timeframe of this work (late 1980s), FDA required that such a "final intermediate" should be well characterized and purifiable. For these reasons, we chose to develop the alternative approach of hydrolyzing all of the protective groups first, and then hydrogenating the pyridine nucleus. An advantage is that the intermediate after hydrolysis is a solid that can be purified by base and acid treatment if necessary. Hydrolysis of the purified **8** in refluxing 6N HCl followed by solvent exchange to water gave the pyridine triacid **18** in 89% yield (>99% purity; 9 kg scale). An unexpected occurrence in the Pilot Plant, however, required a rapid solution. During the hydrolysis of a 9-kg batch in a glass-lined reactor, the tip of one of the stirrer blades came apart, contaminating the batch with glass and metal (>600 ppm Fe, 35 ppm Ni). The removal of the metal ions was accomplished by dissolving the contaminated material in 4 equiv. of aqueous NaOH, thus precipitating the metal salts. Filtration gave an essentially metal-free solution, which was acidified with 5 equiv. of HCl to precipitate **18**, but a modest yield was obtained (46%). This is because **18** is quite soluble in aqueous salt solutions. To remove the

NaCl salt, we took advantage of the common ion effect. Thus the solvent was exchanged to 12N HCl (in which only **18** was soluble), the NaCl precipitate was filtered off, and the solvent was exchanged again to water, giving another 27% of **18** with acceptable purity.

Scheme 4.  Initial Pilot Plant Synthesis of **1** (Batches 2 and 3)

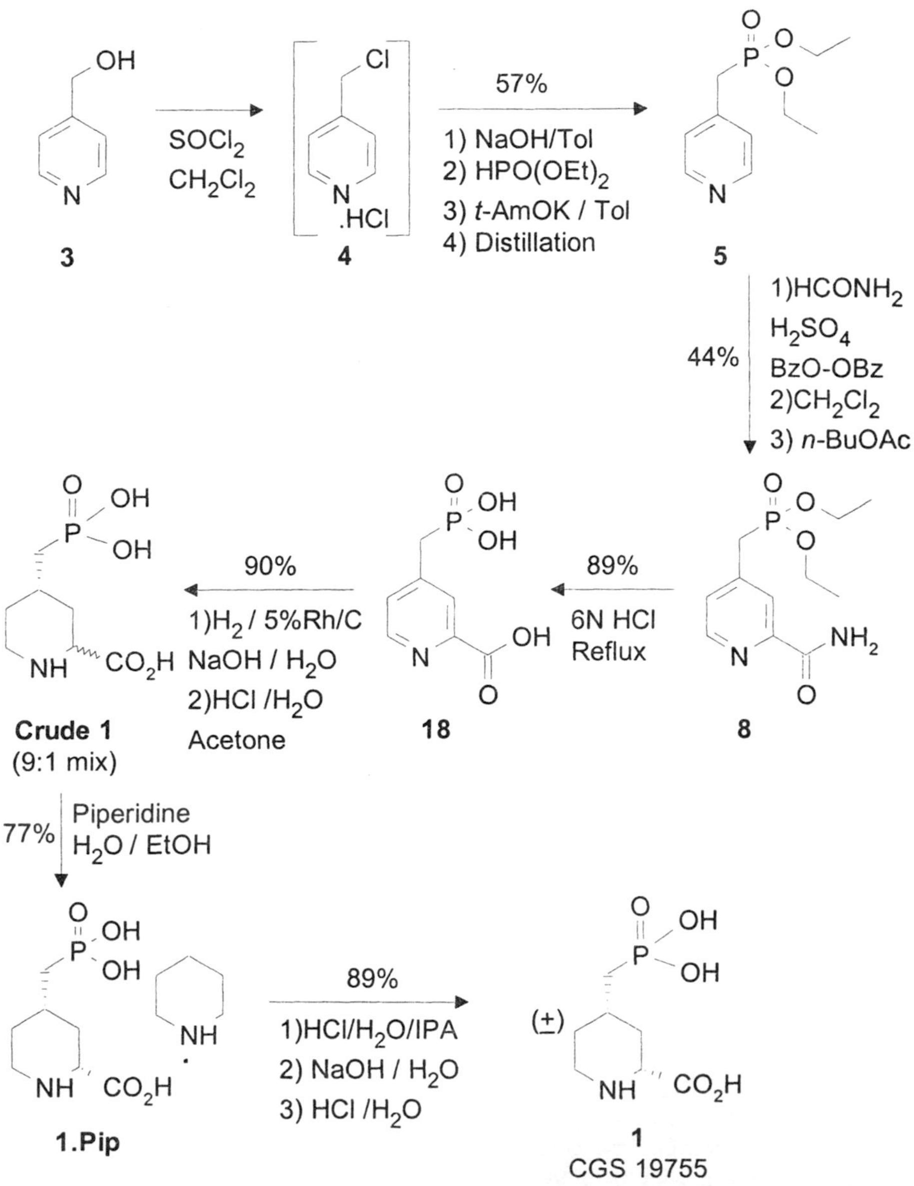

Triacid **18** is essentially insoluble in most organic solvents and only very slightly soluble in water. Therefore it was converted to the monosodium salt (with NaOH) for hydrogenation in water. A brief search of heterogeneous catalysts showed that 5% rhodium on alumina (or carbon) could provide a reasonable reaction rate with good *cis:trans* selectivity. After the hydrogenation was complete, acidification with aqueous HCl provided **Crude 1**, as a 90:10 mixture of *cis:trans*-isomers. Removal of the *trans*-isomer is a difficult problem because CGS 19755 is not a normal recrystallizable solid. Previously the *trans*-isomer was removed by recrystallizing the hydrochloride salt in alcohols in low overall yield. However, because of transesterification problems seen in this approach, we decided to investigate salts of **1** with bases. Even this approach was tricky because in most cases attempts to recrystallize base-salts of **1** by heating/cooling cycles resulted in disproportionation such that **1** precipitated from solution and salts of the di-anion remained in solution. Eventually it was found that the monopiperidine salt **1.Pip** behaved best and the *trans*-isomer level could be reduced to below 1% by recrystallization from aqueous ethanol. We also faced a difficult analytical problem in monitoring the %-*trans*-isomer content during the purifications. The piperidine salt of **1** does not have a UV chromophore, and the HPLC technology available at that time did not provide a very good separation of the isomers. We found that we could use quantitative NMR techniques to obtain accurate information, especially using $^{31}$P NMR. A well-defined sample preparation and spectrum acquisition allowed accurate determination of the *trans*-isomer to the <0.1% level.[8] In proton-decoupled $^{31}$P NMR, the phosphorus-containing compounds appear as singlets. The inherent $^{31}$P-$^{13}$C coupling gives "$^{13}$C satellites", one small peak on either side of the main peak with a relative intensity of 0.55% (due to the 1.1% natural abundance of $^{13}$C). Thus, comparison of peak heights of the P-containing impurities to the peak heights of the $^{13}$C satellites (after simple correction for peak shape) allowed accurate determination in the 0.1% to 5% range.

The zwitterionic CGS 19755 was prepared from **1.Pip** by an acid-base-acid neutralization sequence in aqueous medium in 89% yield. Because CGS 19755 was formulated as a sterile solution, special precautions had to be taken on the quality of the water used in these steps. It was necessary to use water that met water-for-injection limits, especially with respect to biological load and endotoxins.

The advantages of the Scheme 4 synthesis versus the two previous routes are numerous and make it suitable for further scale-up with some additional optimizations. One great advantage is that triacid **18** and its precursor amide **8** are readily recrystallizable, stable solids, whereas ester **15** and its precursors are hydrolytically unstable oils. Another big improvement is the use of piperidine salt **1.Pip** for a well-defined purification of CGS 19755 via recrystallization. In addition, the one-step conversion of the pyridine phosphonate **5** to the 2-carboxamide **8** saves considerable processing time (about 3 steps). The overall yield from 4-pyridinemethanol in the two batches of CGS 19755 obtained following Scheme 3 was 6.3% on 0.3 kg scale. In comparison, the overall yield from <u>2</u> in the two batches of CGS 19755 obtained following Scheme 4 was 8.4% to 13.8% giving 8.1 and 51.0 kg of API with acceptable quality for further clinical and preclinical development. Despite these advantages, several aspects of the process needed improvement. The isolation of pyridine **5** (lengthy distillation) needed improvement, or even finding an altogether new way to make **5**. The yield in the conversion of **5** to **8** was modest. The *cis:trans*-isomer ratio in the hydrogenation of **18** to **Crude 1** could be improved, with the expectation that the yield of CGS 19755 should be higher when there is less of the *trans*-isomer in the crude CGS 19755. These shortcomings were addressed in the subsequent development of the process.

**Batches 4, 5, 6: Final Process Selection and Optimization**

**Batch 4 Synthesis: Optimal Approach**

The shortcomings of the synthesis outlined in Scheme 4 were addressed in further laboratory investigations. It was desired that the isolation of **5** should be simplified by preparation of a crystalline product. It was found the phosphoric acid salt of **5** is a recrystallizable solid. Thus crude **5** could be purified by adding anhydrous phosphoric acid to a solution of **5** and filtering the resulting solid. Furthermore, the product could be used directly in the radical carbonylation step because a protonated pyridine is necessary. An alternative approach to the preparation of **5** was desired which avoided the dialkylation problem seen in the earlier work. A new approach was postulated as shown in Scheme 5, starting with the cheaper starting material 4-pyridinecarboxaldehyde **19** (ca. 50% the cost of **3**). Reaction of diethyl phosphite with 4-pyridinecarboxaldehyde gave the 2-hydroxyphosphonate **20** in >90% conversion. The idea was to hydrogenolyze the benzylic hydroxyl group of **20** and obtain **5**. Attempts to do this conversion directly were not successful under various conditions (e.g., $H_2$/Pd/C), even if the hydroxyl group was acetylated. It was found instead that **20** rearranged spontaneously to give the phosphate **21**. This reaction has been previously reported to occur on similar systems.[9]

Scheme 5: One-Pot Preparation of **5** from 4-pyridinecarboxaldehyde

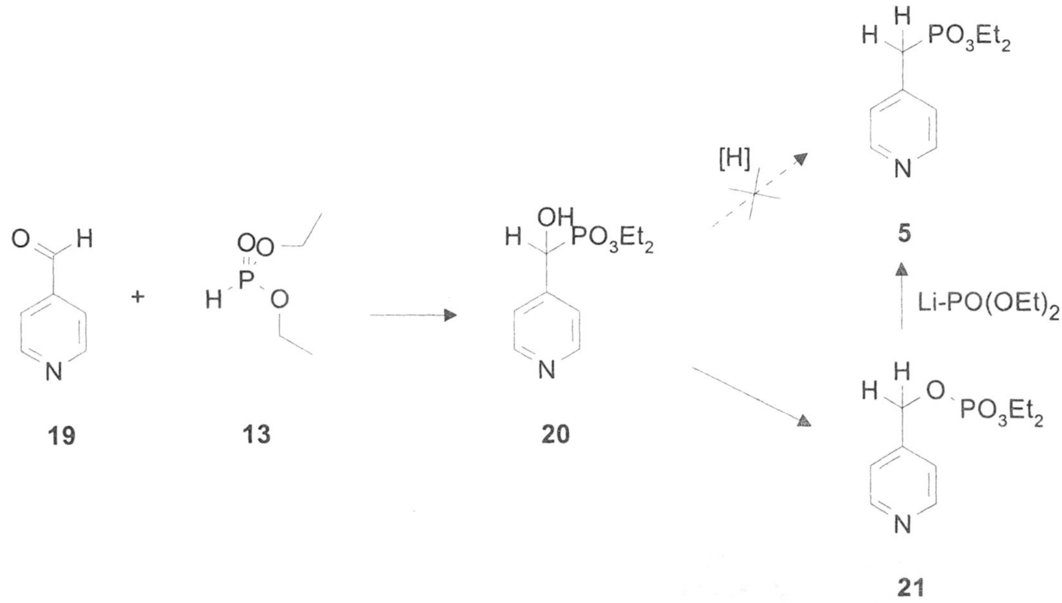

At first glance, this rearrangement product appears undesirable. We thought, however, that it may be possible to use the diethyl phosphate moiety as a leaving group as chloride was used in the previous syntheses. Indeed it proved possible then to react intermediate **21** with another mole of diethyl phosphite and obtain **5**. After initial optimization, the conversion was done in a one-pot reaction as shown in Scheme 5. Addition of the aldehyde **19** and diethyl phosphite to base in toluene initially gave the hydroxyphosphonate **20**, which rapidly rearranged to phosphate **21**. The toluene solution was dried azeotropically, and diethyl phosphite and lithium amide were added.

After the displacement reaction was complete, the product was isolated as the phosphate salt in 70% overall yield (Lab). The same reaction conditions were used on 2- and 3-pyridinecarboxaldehyde to prepare the corresponding phosphonates.[10] The reaction conditions were modified slightly in the pilot plant so that diethyl phosphite and the aldehyde **19** were added to lithium amide, as shown in Scheme 6. Addition of ethyl acetate precipitated most of the salts, which were removed by filtration. Solvent exchange to ethanol and addition of phosphoric acid gave **5.Phos** in 64% yield.

Scheme 6. Second Pilot Plant Synthesis of **1** (Batch 4)

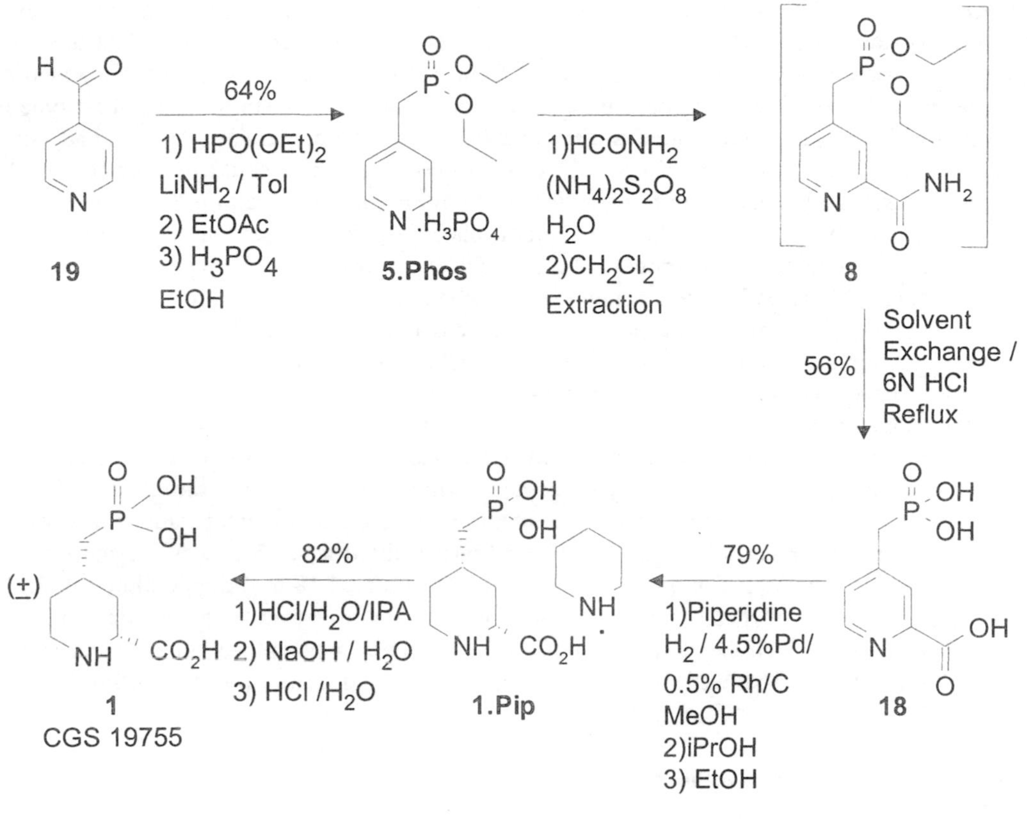

Because a large part of the yield loss in the radical carbonylation step was due to incomplete reaction and to partial hydrolysis of the SM or product in the reaction mixture (of one of the phosphorus ethyl esters), initially attempts were made to use as dry conditions as possible. It was found that potassium persulfate could be used instead of benzoyl peroxide in the presence of a solid-liquid phase transfer catalyst (TDA-1) to deal with both problems. It is interesting that the reaction did not proceed beyond ca. 40% without PTC. Although PTC has been reported in radical polymerization reactions,[11] to our knowledge this is the first case of PTC being used in a nucleophilic radical reaction. Because persulfate is reduced to sulfate, the reaction mixture becomes increasingly acidic. A feed controlled process was developed which included addition of potassium carbonate to maintain the pH between 2.5-3. The possibility of avoiding isolation of the amide **8** was found to be workable by carrying the crude material forward into the hydrolysis step and isolating the triacid **18** directly. In this way any partly hydrolyzed material would not be lost in the purification of **8**. It was found that using a saturated aqueous solution of ammonium persulfate in place of the potassium persulfate/TDA-1 system worked better in the reaction workup. Thus after the radical reaction and workup, the crude material was extracted into dichloromethane using a countercurrent centrifugal extractor, then hydrolyzed with aq. HCl, giving triacid in 74% overall yield (Lab) from **5.Phos**. When the process was scaled up to ca. 20 kg, however, problems with entrainment of formamide in the extraction step and high content of ammonium salts in crude **18** reduced the overall yield to about 56%.

Numerous catalysts and reaction conditions were investigated in the hydrogenation reaction, with the goal of improving the 90:10 *cis:trans*-isomer ratio and thus making the purification of CGS 19755 easier. The thermodynamic ratio of the isomers is estimated at 85:15 (*cis:trans*). It was found that the ratio does not change after the hydrogenation, but is fairly sensitive to catalyst type. The *cis*-isomer is formed when all 6 hydrogens are added on the same face of the pyridine moiety. The *trans*-isomer is formed when at least 2 hydrogens are added on the opposite face, possibly due to desorption/resorption on the catalyst surface. The best type of catalysts for this reaction were found to the bimetallic[12] catalysts containing 4.5% Pd and either 0.5% Pt or 0.5% Rh on carbon. Up to a 97:3 *cis:trans*-isomer ratio could be obtained on the sodium salt. A possible explanation is that the rhodium or platinum break up the aromaticity of the pyridine, and the palladium reduces the intermediate dihydropyridine with higher selectivity. To improve the efficiency of the process, it was decided to hydrogenate the piperidine salt of the pyridine triacid **18** instead of the sodium salt, and to then crystallize directly the resulting piperidine salt **1.Pip**. A process was developed using reduction in methanol, followed by isolation and crystallization in isopropanol and ethanol, giving **1.Pip** directly in up to 87% yield (79% upon scale-up). The conversion of the piperidine salt to the API was not changed because it was desired to maintain the API preparation step of the process relatively unchanged.

The advantages of the Scheme 6 synthesis versus the previous route are numerous and make it suitable for further scale-up with some additional optimizations: the cheaper SM 4-pyridinecarboxaldehyde is used in a novel reaction to prepare the phosphonate, and isolate it as a solid. The one-pot conversion of the pyridine **5.Phos** to the triacid **18** avoids isolation of the amide diester intermediate. The one-pot conversion of the triacid **18** to the piperidine salt **1.Pip** takes advantage of the better catalysts. Some areas for improvement include: elimination of $CH_2Cl_2$ from the workup of the radical reaction and addition of a recrystallization of the piperidine salt to increase purity. The overall yield (with only 4 isolations) from 4-pyridinecarboxaldehyde was 37% (Lab scale) and 23% (Pilot Plant scale).

**Batch 5 Synthesis:  Final Synthesis**

Despite the excellent results obtained following the process in Scheme 6, some slight differences in the impurity profile (new impurities below the 0.1% level, and the presence of about 0.6% of the *trans*-isomer) of the API were detected when compared with API prepared previously (0.2-0.5% *trans*-isomer). These differences have regulatory implications, so it was agreed that the early part of the Scheme 6 synthesis (up to preparation of carboxamide **8**) could be merged with the later part of the Scheme 4 synthesis for final optimization. If the product reaches the market, then the Scheme 6 synthesis could be revisited. Because of the scale-up difficulties described above, a second firm was contracted to prepare the salt **5.Phos** using literature methods (Michaelis-Becker condensation).[13] The focus at Novartis was then optimization of all steps beyond **5.Phos**. The so-called final synthesis is outlined in Scheme 7, and was demonstrated on 3 parallel batches (total 13.8 kg) which are to be used for stability and setting of the impurity profile.

After further work on the radical carbamoylation step, it was found that addition of aqueous ammonium persulfate to a solution of **5.Phos** in refluxing ethyl acetate allowed for easy temperature control and higher conversion. In addition, the use of $CH_2Cl_2$ in the workup was eliminated by isolating the crude **8** by precipitation instead of extraction (simply by addition of water to the reaction mix after removal of ethyl acetate). Solvent recycle for *n*-butyl acetate, the recrystallization solvent, was shown to be feasible. Use of the better catalyst provided a better *cis:trans*-isomer ratio than in Scheme 4. In the last steps, a considerable effort was focused on the determination of the critical variables for each step, obtaining good mass balance information, toxicological data on waste streams and isolated intermediates, all in preparation for transferring the technology to another site (Basel). The yields for the last two steps in the pilot plant were lower than expected due to the relatively small scales involved (<5 kg). A second crop of **1** (6%) was also obtained, and could be used in a future process.

Scheme 7.  Final Pilot Plant Synthesis of **1** (Batches 5 and 6)

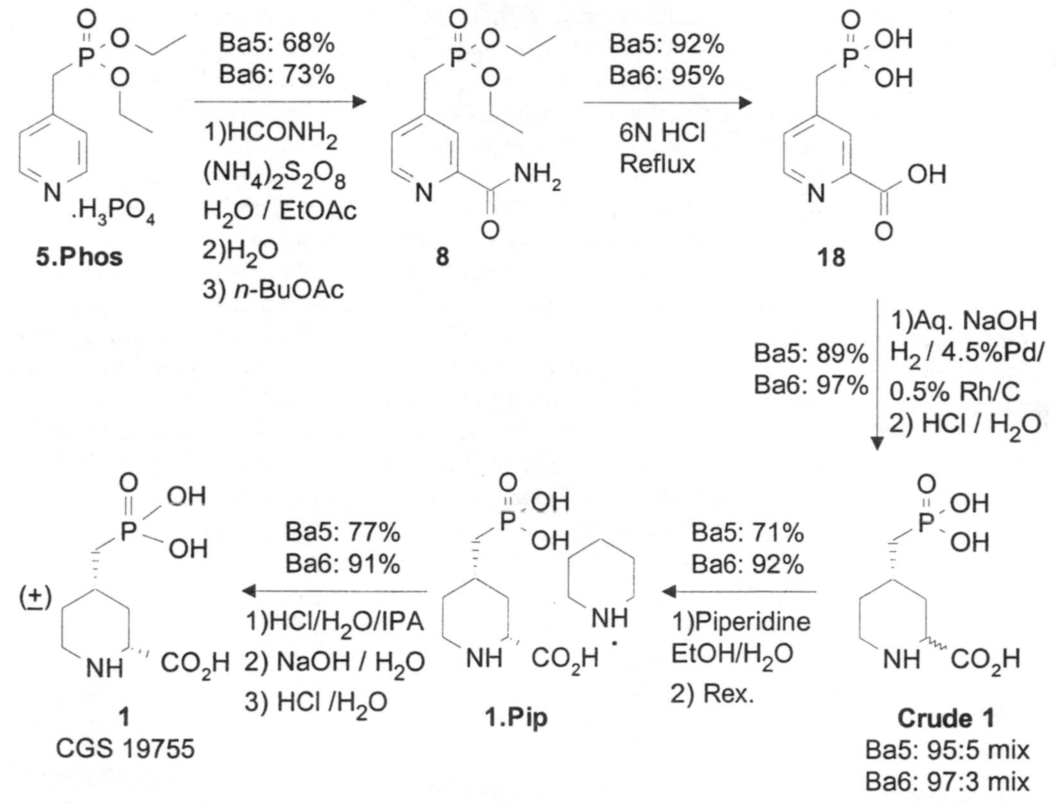

Samples of some impurities were also prepared for use in setting specifications and developing reliable analytical methods. The dicarbamoylated pyridine **22**, a small impurity occurring in **8**, was prepared by reacting isolated **8** with the ammonium persulfate/formamide while maintaining a very low pH. Hydrolysis of **22** in 6N HCl provided the tetraacid **23**. Hydrogenation of **23** under the same conditions as triacid **18**, afforded **24**, the all-*cis*-dicarboxylic acid analog of **1**.

**Batch 6 Synthesis: Final Process**

Attempts were made to solve the problems seen in the scale-up of the one-pot conversion of 4-pyridinecarboxaldehyde to **5.Phos** (Scheme 6). A fairly large variation in yields (57-70%) were experienced upon in the synthesis of Batch 4 (12.3 kg), and they were attributed to incomplete removal of lithium salts prior to acidification with phosphoric acid. Several of the components in this complex reaction were also shown to give undesired side reactions. Lithium diethyl phosphite is not stable at temperatures above 25 °C, giving diethyl ethylphosphonate. The hydroxyphosphonate intermediate **20** decomposes in the presence of base at elevated temperatures. In the presence of traces of moisture, the triester **21** forms 4-pyridinemethanol, and **5** forms the corresponding P-monoester. On the other hand, the displacement of the diethyl phosphate leaving group is very slow below 25 °C. The reaction was therefore changed so that the triester was formed under mild base conditions (cat. NEt₃), and added to preformed lithium diethylphosphite below 10 °C. To allow for easier removal of the lithium salts, a filter aid was added, and the reaction was stirred at 25 °C. The salts were then filtered, and after workup and solvent exchange to ethanol, the product was isolated in 75% yield after adding anhydrous phosphoric acid. Upon scale-up (to 50 kg of **19**), however, it was necessary to recrystallize the product **5.Phos** to remove elevated levels of hydrolysis impurities.

The last phase in the development of a process for an API before transfer to Chemical Production is definition of the final process. The final optimization is performed in the Lab via statistical experimental design (systematically changing certain process variables) for each step. This required many experiments on each step (ca. 60) and resulted in incremental improvements in yield, but also in minimizing undesired reactions and byproducts. As a result, there was a substantial improvement in the overall yield on Pilot Plant scale. A total of 440 kg of CGS 19755 were prepared for Batch 6, with an overall yield of 56% from salt **5.Phos** (compared to 30% overall yield for Batch 5). The next phase of the project would be transfer of the final process to Chemical Production and process validation in preparation for launch. In addition, the improved process for the preparation of salt **5.Phos** from 4-pyridinecarboxaldehyde would be transferred after final optimization.

**Summary**

The role of process research and development in the pharmaceutical industry is to improve and simplify the preparation of Active Pharmaceutical Ingredients (APIs) so that eventually they can be prepared on a scale large enough to meet market needs. A simple way to show this is by comparing the overall yield of an API from a common intermediate if possible. In the case of CGS 19755, we can choose diethyl 4-phosphonomethylpyridine (**5**). Table 1 summarizes the scale and overall yields obtained from discovery until the final process. It is likely that if the process of Scheme 6 is optimized further, it may offer a better overall yield and easier processing than the optimized process of Scheme 7.

**Table 1**. Evolution of Processes to Prepare **1**

| Stage of Process | Scheme No. | Kg of <u>1</u> Prepared | Overall Yield from $\underline{5}^a$ to <u>1</u> |
|---|---|---|---|
| Research | 1 | 0.02 | 46.2%$^b$ |
| Kilo Lab | 3 | 0.3 + 0.3 | 7.5% |
| Process Research | 4 | 8.1 + 51 | 24.1% |
| Process Research | 6 | 12.3 | 36.3% |
| Final Synthesis | 7 | 4.5 + 4.8 + 4.8 | 30.4% |
| Final Process | 7 | 440 (6 lots) | 56.3% |

$^a$ As free base or phosphate salt
$^b$ Purity of **1** was below 95%, therefore overall yield would be reduced if desired purity was >99% as in the other processes

The overall strategy of the various synthetic routes used to prepare **1** is the same, but they differ in the use of protecting groups, which greatly affect the properties and purification of intermediates. The conversions of intermediates have been optimized, and the use of novel but well-defined chemistry in the early steps has allowed rapid assembly of the API skeleton. Purification strategies for all isolated intermediates and the bulk drug substance have been defined. Scale-up to hundreds of kilograms has been demonstrated, and the final process is adequate for registration with regulatory authorities. The final process to prepare CGS 19755 has addressed efficiency, cost, safety, and environmental concerns while consistently delivering a high quality product.

**Acknowledgments**
The authors acknowledge that the work summarized in this chapter would not be possible without the dedicated efforts and ideas of many individuals throughout the development of the project. The key individuals are:
**Scheme 3**: H. Rodriguez, D. Malone, M. Loo
**Scheme 4**: S. Condon, F. Pita
**Scheme 6**: G. Zaunius, J. Cancelarich, A. Bach, H. Steiner, J. Skiles
**Scheme 7**: *Summit*: G. Zaunius, J. Cancelarich, J. Calienni, J. Xu, H. Meckler, A. Tedesco, G.-P. Chen, R. Winchurch. *Basel*: M. Leimer, W. Hammerschmidt, Y. Schmitt, D. Jevtic, A. Pische-Jacques, S. Tardino, M. Cireddu, H. Steiner, A. Knell, S. Gubler, M. Drechsel, P. Blumer.

**References**

[1*] Correspondence should be addressed to PG; currently at Novartis Pharmaceuticals, East Hanover, NJ 07936. Email: *peter.giannousis@pharma.novartis.com*
[⊥] Currently at Bachem Inc., Torrance, CA 90505.
[∇] Novartis was formed by the merger of Ciba-Geigy and Sandoz in 1996. The work described in this chapter was performed in the former Ciba-Geigy Pharmaceuticals.

[2](a) Lehmann, J.; Hutchinson, A.J.; McPherson, S.E.; Mondadori, C.; Schmutz, M.; Sinton, C.M.; Tsai, C.; Murphy, D.E.; Steel, D.J.; Williams, M.; Cheney, D.L.; Wood, P.L. *J. Pharmacol. Exp. Ther.* **1988**, *246*, 65. (b) Danysz, W.; Parsons, C.G.; Bresink, I.; Quack, G. *Drug News & Persp.* **1995**, *8*, 261.

[3]Hutchison, A.J.; Williams, M.; Angst, C.; De Jesus, R.; Blanchard, L.; Jackson, R.H.; Wilusz, E.J.; Murphy, D.E.; Bernard, P.S.; Schneider, J.; Campbell, T.; Guida, W.; Sills, M.A. *J. Med. Chem.* **1989**, *32*, 2171.

[4]Evans, R.H.; Watkins, J.C. *Life Sci.* **1981**, *28*, 1303.

[5] Definitions are from "*Manufacture, Processing or Holding of Active Pharmaceutical Ingredients*", FDA Draft Guidance for Industry, **August 1996**; the entire document can be downloaded from the FDA web site at "http://www.fda.gov/cder/guidance/index.htm"

[6](a) Minisci, F; Vismara, E.; Fontana, F. *Heterocycles* **1989**, *28*, 489. (b) Minisci, F; Citterio, A.; Vismara, E. *Tetrahedron* **1985** *41*, 4157. (c) Minisci, F. *Top. Curr. Chem.* **1976**, *62*, 1.

[7] Bernardi, R.; Caronna, T.; Galli, R.; Minisci, F.; Rerchinunno, M. *Tet. Lett.* **1973**, *9*, 645.

[8]Karl Gunderson, Ciba-Geigy NMR Lab, unpublished results.

[9]Timmler, H.; Kurz, Jurgen, K. *Chem. Ber.* **1971**, *104*, 3740.

[10]G. Zaunius, unpublished results.

[11]Gupta, G. N.; Mandal, B. M. *J. Indian Chem. Soc.* **1985**, *62*, 949.

[12] Cosyns, J.; Martino, G.; Le Page, J.-F. US Patent 3,954,601, **1976**.

[13]Maruszewska-Wieczorkowska, E.; Michalski, J. *Rocz. Chem.* **1964**, *38*, 625.

# Route Evaluation for the Production of *(R)*-Levoprotiline

**H.U. Blaser[1], R. Gamboni[2], B. Pugin[1], G. Rihs[1], G. Sedelmeier[2], B. Schaub[2], E. Schmidt[2], B. Schmitz[2], F. Spindler[1] and Hj. Wetter[2]**
*[1]Scientific Services and [2]Chemical and Analytical Development Pharma, Novartis AG, CH-4002 Basel, Switzerland*

Keywords:  Chiral building block, enantioselective hydrogenation, *(R)*-epichlorohydrin, *(R)*-levoprotiline, ligand synthesis, process evaluation.

## Introduction

Oxaprotiline [1-(2-hydroxy-3-methylamino-propyl)-dibenzo-[b,e]-bicylo-[2.2.2]-octadiene-hydrochloride], is a powerful and highly selective inhibitor of noradrenaline uptake in vitro and in vivo and is an effective antidepressant[1]. It is a close analog of the commercial drug maprotiline[2] and is a racemic mixture of the optical isomers *(S)*-(+)- and *(R)*-(-)-oxaprotiline. Because the *(R)*-enantiomer, *(R)*-(-)-levoprotiline showed certain therapeutical advantages[3], a technical enantioselective process had to be developed.

The present contribution describes the two most promising routes developed in parallel during the process evaluation. In both cases, the creation of the stereogenic center was considered to be the key step because the chemistry of the dibenzo-bicyclo-octadiene backbone was already well established. In *variant 1*, the chiral side chain was introduced by addition of *(R)*-epichlorohydrin to 9-Li-anthracene; in *variant 2*, the stereogenic center was created via catalytic oxidation of the racemic alcohol followed by the enantioselective hydrogenation of the α-amino ketone. In this contribution, we briefly describe the chemistry of the individual steps of the two variants, comment on the major difficulties and the optimization of the key steps and compare the over all results of the two variants. The scale up of the ligand synthesis for the enantioselective hydrogenation will be described in some detail.

## Structure Determination

In order to determine the absolute configuration, a crystal structure analysis of the hydro-bromide of levoprotiline was performed. The structure was solved by direct methods; all hydrogen atom positions were located from difference Fourier maps or calculated assuming normal geometry. The absolute configuration was determined computationally by an R-factor test on the basis of the anomalous scattering of bromine. Scheme 1 shows a perspective view of the salt. The dihedral angle between the two aromatic rings is 120.1 °; the nitrogen atom is protonated and there is a linear hydrogen bond to the bromine anion.

Scheme 1. ORTEP plot and absolute configuration of (-)-levoprotiline hydrobromide

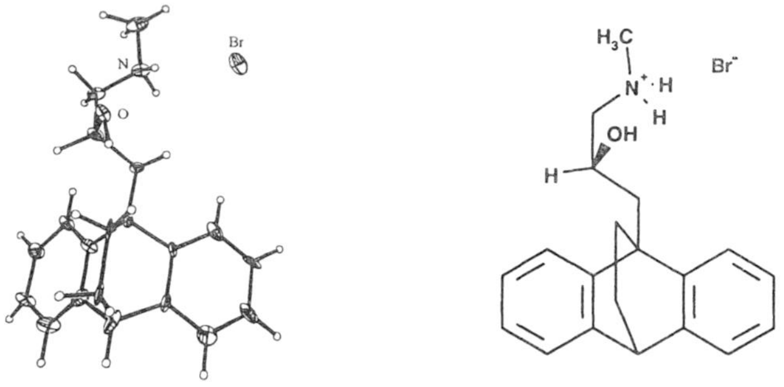

## Variant 1: The *(R)*-Epichlorohydrin Route

**Synthetic Strategy**

   This short, five step synthesis starts from inexpensive anthracene that is brominated selectively in the 9-position, followed by bromo-lithium exchange. The side chain is attached by reaction of 9-lithio-anthracene with the chiral building block *(R)*-epichlorohydrin that is available in high optical purity in commercial quantities from Daiso Corp. The ethane bridge is introduced after epoxide formation by a Diels-Alder reaction with ethylene under high pressure. Due to its shortness, this promised to be an efficient and economical route for the production of levoprotiline on a large scale without high capital investment in specialized equipment.

Scheme 2. Schematic representation of the epichlorohydrin route

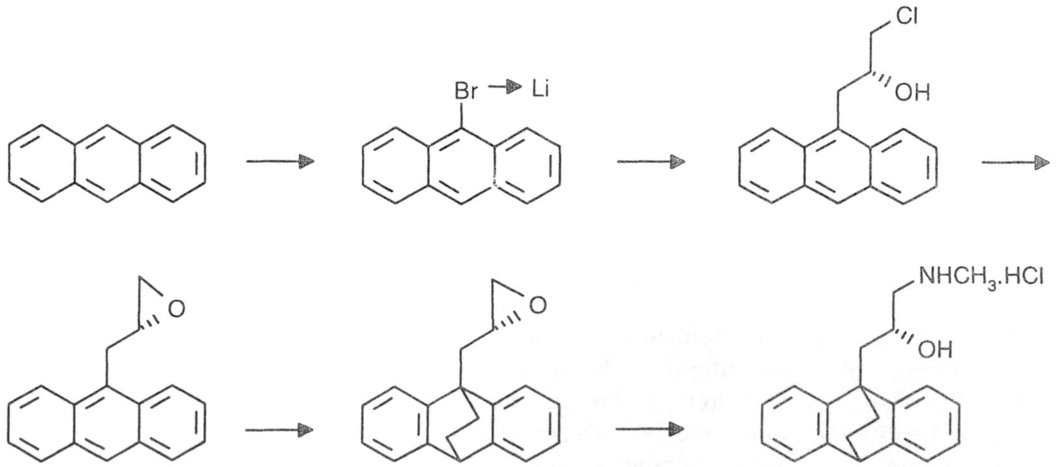

## Bromination

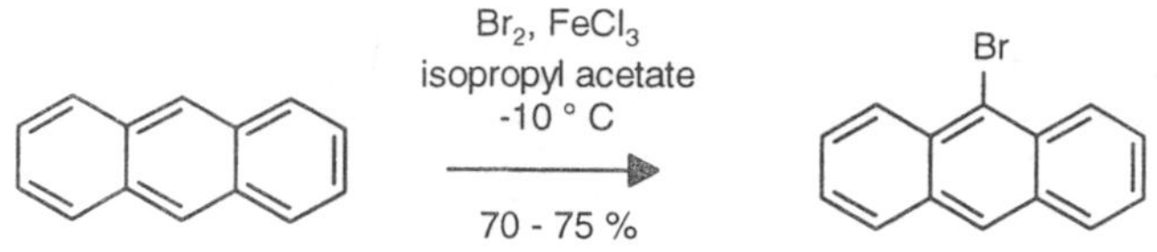

The selective bromination of anthracene **1** can be carried out either using N-bromosuccinimide[4] or with bromine and CuCl or $FeCl_3$ as catalyst[5]. For toxicological and environmental reasons, the iron catalyst was preferred and under optimized reaction conditions crude 9-bromoanthracene **2** was obtained in 75 % yield that contained up to 3-4 % of 9,10-dibromoanthracene **3** besides 2-3 % of unconverted anthracene. The purity of the 9-bromoanthracene **2** is of great importance for the quality of the final active ingredient. It was necessary to recrystallize the crude product to remove the 9,10-dibromoanthracene **3** before the next step because this would also be transformed during the following sequence of reactions to give the very insoluble compound **4**.

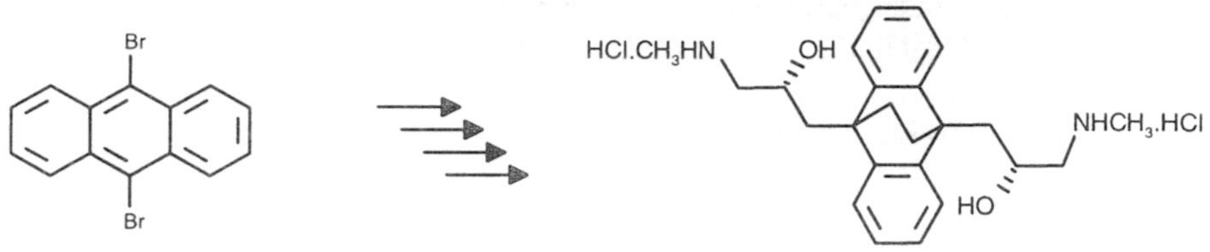

## Metalation and Epoxide Formation via Chlorohydrin Intermediate

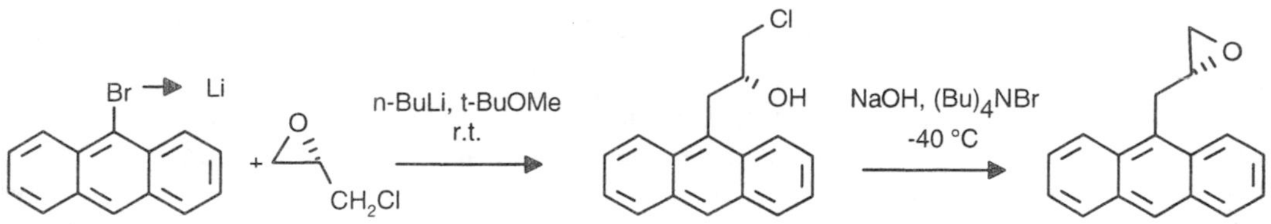

The metalation reaction, the alkylation of the 9-lithio-anthracene with (R)-epichlorohydrin (ex Daiso) to the chlorohydrin intermediate **5** and the epoxide formation was carried out as a one pot reaction. The ratio of 9-bromoanthracene **2** to the more expensive (R)-epichlorohydrin was chosen to be 1 : 0.9. In order to achieve good chemical and optical yields (tendency to racemization), the alkylation reaction had to be performed at -40 to -45 °C, whereas the bromide/lithium exchange as well as the epoxide formation were run at room temperature in *t*-butyl methyl ether. The crude epoxide was crystallized from *t*-butyl methyl ether and recrystallized from ethanol to give 75 % of the epoxide **6**. The optical purity of 96 +/- 1 %. was acceptable because the (S)-enantiomer could be removed by recrystallization of the final active ingredient from ethanol. Scale up to a batch size of 20 kg of the epoxide **6** was not problematic.

**Diels-Alder Reaction**

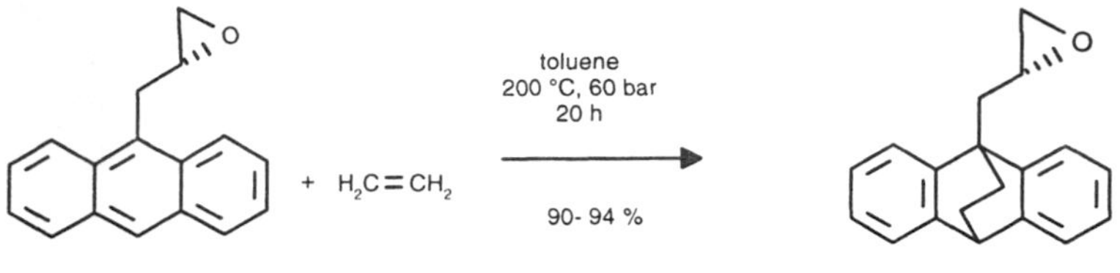

From a safety point of view this high temperature reaction was studied very carefully because the risk of decomposition of the epoxide is rather high. Calculations that take the heat of reaction into account (125 kJ/mol, i.e. 630 kJ/kg epoxide) indicated that the adiabatic temperature increase would be 140 °C. Several scenarios for running the reaction at the necessary temperature of 200 °C were investigated and it was found that the risk is highest during the first 4 hours of the reaction when the starting material is most concentrated. Therefore the reaction was carried out in toluene solution (ca. 1.7 l toluene/kg epoxide) and, in addition, only carefully purified epoxide **6** was used because impurities enhance the potential for a runaway reaction.

In the pilot plant and under optimized conditions (ethylene, 60 bar, 200 °C, 20 h, toluene) we obtained purified Diels-Alder adduct **7** with yields of 90 - 94 % after crystallization from ethanol.

**Epoxide Opening with Methylamine**

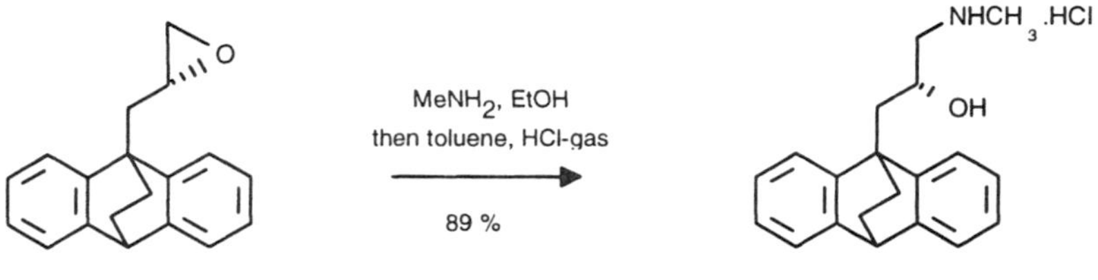

The final step to the crude active ingredient was done under the classical conditions of epoxide opening. The epoxide **7** and gaseous methylamine (10 equivalents related to **7**) was dissolved in ethanol at room temperature and the reaction solution was then heated to 50 °C. After hydrochloride formation in toluene the crude active ingredient **8** was obtained in 89 % yield. The required purity specifications, i.e., removal of all side products below the 0.1 % level and an ee of > 98 %, were reached by recrystallization from ethanol.

## Variant 2: The Amino Ketone Hydrogenation Route

**Synthetic Strategy**

Here, the stereogenic center is created via the enantioselective hydrogenation of an α-amino ketone. The amino ketone is prepared by catalytic oxidation of the racemic oxaprotiline, that can be prepared in a straightforward manner starting from a bicyclic intermediate in the synthesis of maprotiline[2] that is available in ton quantities. The key problems were the development of efficient catalytic methods, first, for the metal catalyzed oxidation of the racemic alcohol and, secondly, for the enantioselective hydrogenation of the amino ketone.

Scheme 3. Schematic representation of the amino ketone hydrogenation route

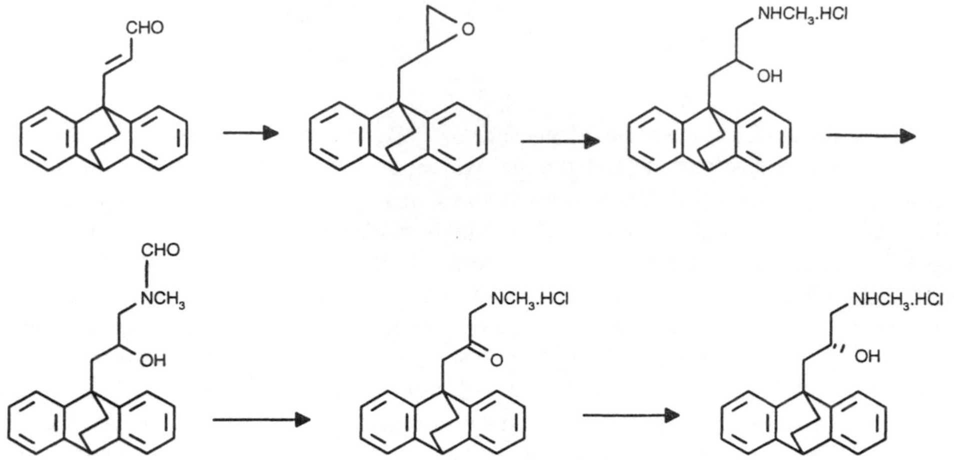

**Synthesis of the Formyl Protected Racemic Amino Alcohol**

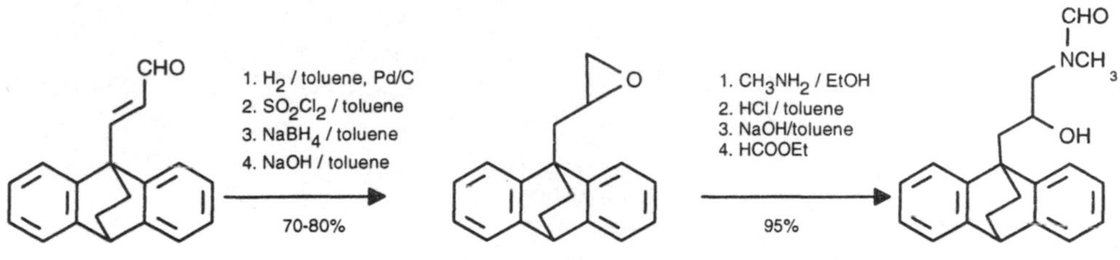

The racemic epoxide **10** was synthesized using the following sequence: Hydrogenation of the substituted acrolein **9** in presence of a palladium catalyst at 3-6 bar, 60 °C, followed by chlorination of the saturated aldehyde with sulfuryl chloride, reduction of the chlorinated aldehyde with sodium borohydride and base treatment. All transformations were performed in one pot using toluene as solvent. For the reduction and epoxide formation steps, catalytic amounts of tetrabutylammonium bromide has to be added. The racemic amino alcohol was obtained by reaction of methylamine with the epoxide under the conditions described above for *variant 1*. Finally the aminoalcohol was formylated by boiling in formic acid ethylester without solvent. The overall yield of **11** was 69-74 %.

## Oxidation and Deprotection

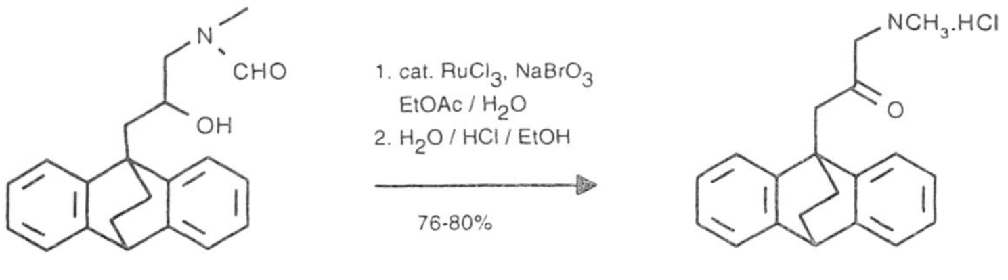

In order to avoid stoichiometric oxidants such as $CrO_3$, which were considered to be too toxic and too expensive, a catalytic method for the oxidation of the racemic alcohol was developed. Several oxidation catalysts (Ru, Ce, Mo) as well as transfer dehydrogenation systems were tested. Only the ruthenium catalyzed oxidation with sodium bromate showed promise[6] and only the formyl protected derivative was a suitable starting material. Both $RuCl_3$ and 10 % Ru/C gave isolated yields of the amino ketone in over 70 % yield. The supported catalyst showed no advantage because about 50 % of the Ru leached from the carrier. Under optimized reaction conditions (carefully purified amino alcohol, 0.3-0.7 mol % of $RuCl_3$, 1 equivalent of $NaBrO_3$ in ethyl acetate/water at 30 °C), 76 % of the formylated amino ketone was obtained. Deprotection in refluxing aqueous hydrochloric acid followed by crystallization from the concentrated reaction mixture yielded the hydrochloride of the amino ketone **12** in sufficient purity for the following hydrogenation step. The Ru content of the crystalline material was below 5 ppm. The recovery and recycling of the spent Ru catalyst was not worked out.

## Enantioselective hydrogenation

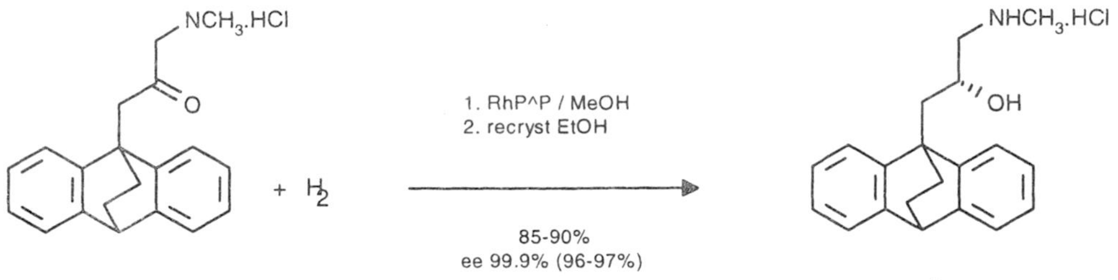

The enantioselective hydrogenation of α-amino ketones has been described with enantioselectivities >95 %[7]. For a technically feasible hydrogenation process the targets were defined as follows: ee >96 % (after recrystallization: ≥98 %); turnover number >1'000; conversion: ≥98 %.

Table 1: Enantioselective hydrogenation using various Rh and Ru diphosphine complexes.

| Entry | Catalyst | s/c | *ton* | $tof_{av}$ (h$^{-1}$) | ee (%) | abs. config. |
|-------|----------|-----|-------|------------------------|--------|--------------|
| 1 | Rh$^0$/*(R)*-(*S*)-bppfoh 13a | 100 | 92 | 4 | 98 | R |
| 2 | Rh$^0$/(*2R,4R*)-bppm 14a | 50 | 50 | 3 | 84 | R |
| 3 | Rh$^0$/*(R)*-binap 15 | 50 | 25 | 1 | 59 | S |
| 4 | Rh$^0$/(*1S,2S*)-C6-diop 16 | 50 | 47 | 2 | 45 | S |
| 5 | Rh$^0$/(*4R,5R*)-diop 17 | 50 | 42 | 2 | 7 | R |
| 6 | Rh$^0$/(*2S,3S*))-norphos 18 | 50 | 25 | 2 | 32 | S |
| 7 | Rh$^0$/(*2S,3S*))-deguphos 19 | 50 | 7 | 5 | 31 | R |
| 8 | RuCl$_2$(*S*)-binap 15 | 400 | 184 | 56 | 85 | R |

Reaction conditions: p(H$_2$): 20-100 bar; r.t.; MeOH, N(Et)$_3$

In a screening phase, 11 different commercially available catalysts were tested for the enantioselective hydrogenation of **12** (for selected results see Table 1, ligand structures see Scheme 3). The highest ee of 98 % was obtained with [Rh(nbd)Cl]$_2$/*(R)*-(*S*)-bppfoh at 0°C. Also reasonably selective were Rh/bppm (ee 84 %), Rh/binap and Ru/binap (ee's 59 % and 85 %, respectively). Rh catalysts generated from [Rh(NBD)Cl]$_2$ and chiral 1,2-diphosphines that form smaller chelates showed poor enantioselectivities (maximum ee 32 %).

For the catalyst optimization, several derivatives of the two best ligand types **13**, based on the ferrocenyl backbone and **14**, derived from hydroxyproline, respectively, were synthesized and tested. The optical yields with the diphosphines of the type **14** varied between 69 % and 89 % (achieved with ligand **14b**). It was rather surprising that in the ferrocenyl series only bppfoh **13a** and bppfsh **13b** (with 99.8 % the highest optical yield ever achieved) showed good enantioselectivities. As soon as an amine (**13d**) or an acetate (**13c**) group was introduced in the side chain or in absence of the hydroxy group (**13e**), the ee's as well as the activity dropped significantly. A hydrogen bridge seems to orient the amino ketone, thereby leading to high enantioselection and acceleration in the case of bppfoh **13a**. Unfortunately, the very selective Rh-bppfsh catalyst had a very low activity, maybe due to complexation with the metal center by the low-valent sulfur.

The influence of various reaction parameters was studied in some detail and the findings can be summarized as follows: The hydrogen pressure had no influence on the enantioselectivity, but affected the rate significantly. At 50 °C both selectivity and activity were satisfactory, at temperatures above 55 °C, ee's dropped noticeably. The solvent of choice was methanol. An excess of base (either an amine or aqueous NaOH or Na$_2$CO$_3$) was necessary for good catalyst activity. The presence of up to 4 % of water in MeOH did not inhibit the catalyst and had a positive effect on the solubility of the hydrochloride substrate **12**. In order to obtain a reasonable space-time yield, a slurry of the amino ketone **12** was used for the hydrogenation without problems. Surprisingly, traces of oxygen had a small beneficial effect on the catalyst activity. Under optimized reaction conditions at 50 °C and 80 bars H$_2$, turnover numbers in the range of up to 2'000 and 97 % ee were achieved, the average turnover frequency was in the range of 100 h$^{-1}$. Scale up of the hydrogenation to a batch size of 20 kg presented no problems; both the time for complete conversion as well as the enantioselectivity varied only little. After one recrystallization from ethanol, the optical purity of levoprotiline was >99 %, and the content of heavy metals was below 10 ppm.

Scheme 4. Structure of ligands

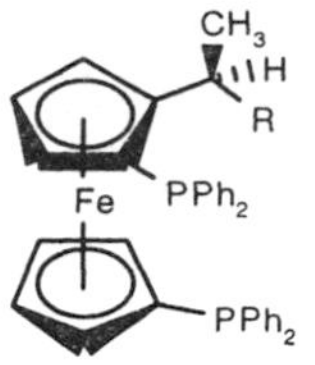

R = OH   *(R)-(S)*-bppfoh **13a**
R = SH   *(R)-(S)*-bppfsh **13b**
R = OAc  *(R)-(S)*-bppfOAc **13c**
R = NR$_2$ *(R)-(S)*-bppfa **13d**

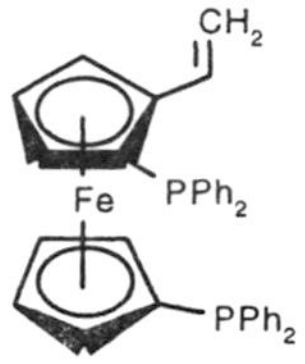

(S)-bppf-CH=CH$_2$ **13e**

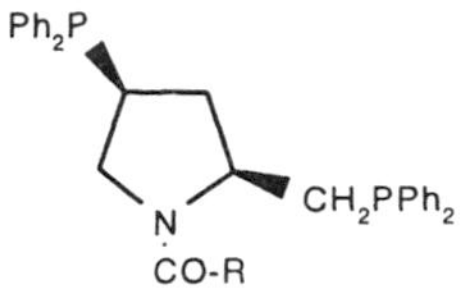

R = O-tBu: *(2S,4S)*-bppm **14a**
R = NH-iPr **14b**

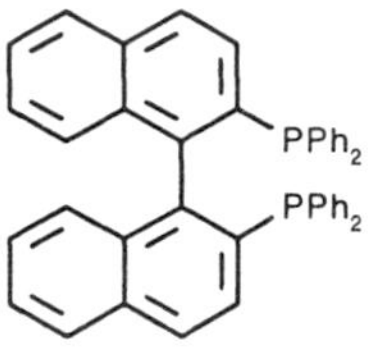

binap **15**

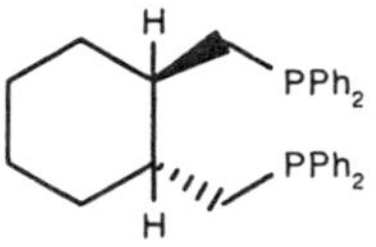

C6-diop **16**

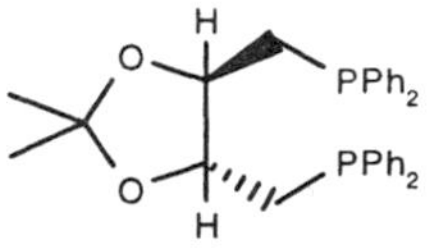

diop **17**

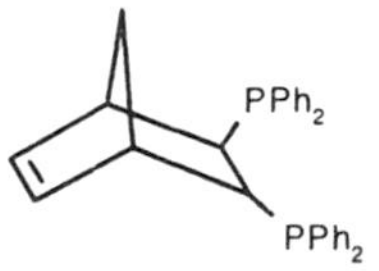

norphos **18**

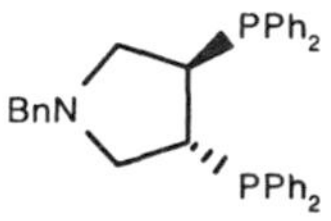

deguphos **19**

## Synthesis of the *(R)-(S)*-bppfoh 13a Ligand

When it became clear after the ligand optimization that *(R)-(S)*-bppfoh **13a** would be the ligand of choice, its preparation on a kg scale had to be developed. With the exception of the resolution step, the synthesis as described by Hayashi et al.[8] was chosen. We will describe the development of the enzyme-catalyzed kinetic resolution of 1-ferrocenylethanol **21** in more detail, however, the other steps were optimized to some degree as well. With this protocol, we prepared several kg of the optically pure *(R)-(S)*-bppfoh ligand **13a** and the process should be suitable for the production of 50-100 kg amounts of the ligand.

Scheme 5. Synthesis of the *(R)-(S)*-bppfoh ligand 13a

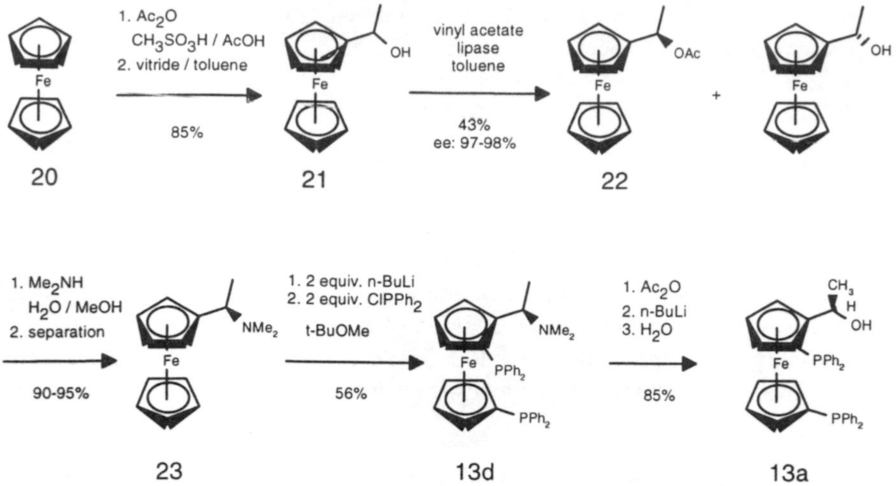

## Acylation and Reduction

Ferrocene **20** was acetylated using methane sulfonic acid in acetic acid in analogy to a procedure reported by Hauser[9]. With 3 equivalents of methanesulfonic acid at room temperature, the reaction was complete after 1-2 h. The reduction step required a stronger reducing agent than NaBH$_4$ and best results were finally obtained with NaAlH$_2$(OMe)$_2$ (Vitride R) in toluene. The resulting Al-complex of **21** was hydrolyzed with water without any special precaution and the inorganic salts were filtered off. Overall yield of the alcohol **21** was 85 %.

## Lipase Catalyzed Enantioselective Acetylation (Kinetic Resolution)

For the kinetic resolution of racemic **21**, the lipase catalyzed acylation procedure using vinyl esters as described by Wang et al.[10] and by Boaz[11] was used as starting point. The most important parameters were investigated and optimized and some engineering aspects were also studied[12].

*Screening of Various Lipases.* Of the 21 tested lipases of different origins, 13 were active for the acetylation of **21** with vinyl acetate but only three gave ee's >70 %. The lipase from *pseudomonas fluorescence* was chosen for further development.

*Effect of Solvent and Water Content.* Most organic solvents give good enantioselectivities, but the enzyme stability and activity were highest in hydrophobic solvents such as toluene and hexane. The water content had to be less than 100 ppm, otherwise both rate and ee decreased.

*Other Reaction Variables.* Up to 30 °C, ee's >98 % were obtained but a gradual decrease of the enantioselectivity was observed at higher temperatures. Re-use of the enzyme was possible at comparable conversion rates but some decline in optical yield was observed (from 97.4 % to 95.4 % ee after 3 cycles).

*Kinetic Modeling.* The following kinetic parameters were determined at 25 °C: $K_m$ (**21**) 1.4 M, $K_m$ (vinyl acetate) 0.22 M. The high $K_m$ value indicates that **21** is a "bad" substrate for the enzyme but on the other hand inhibition by **21** is not a problem ($K_i$ = 0.11 M).

*Immobilization of the Enzyme.* In order to operate a continuous fixed bed reactor, the lipase was immobilized. Best results were obtained by adsorption on a hydrophobic carrier; ee of 98->99

% and productivities about 4 times higher than the native enzyme were obtained. This is probably due to a better distribution of the enzyme on the large surface area of the carrier.

*Production of kg Amounts.* Three times 5 kg of ferrocenylethanol **21** were acetylated in a 50 l stirred tank reactor using vinyl acetate at 25-30 °C in toluene. Reaction times were 2-3 days, resulting in conversions of 38-43 % and ee's of 95-97 %. Furthermore, a small fixed bed reactor was run at 30 °C for over 200 h with a conversion of 30 % and an optical yield of 96 %.

Substitution of Acetate by a Dimethylamino Group and Introduction of the PPh$_2$ Groups

The exchange of the acetate group of **22** was carried out in methanol / aqueous dimethylamine with retention of configuration and the formed **23** was extracted with 10 % citric acid in 90-95 % yield. The following steps were carried out according to the procedure described by Hayashi[8]: The dilithio derivative was formed stepwise by addition of 1 equiv. of n-BuLi, followed by 1 equiv. n-BuLi-TMEDA at room temperature. The 1,1'-dilithio-ferrocenyl compound was then reacted at -30 °C with 2 equivalents of chloro-diphenylphosphine and *(R)-(S)*-bppfa **13d** was isolated in 56 % yield.

Transformation of the Dimethylamino Group to the Hydroxy Group

Reaction of the dimethylamino derivative **13d** with acetic anhydride followed by treatment with an excess of n-BuLi and careful hydrolysis with water gave the crystalline *(R)-(S)*-bppfoh **13a** in 85 % yield. The optical purity was determined by optical rotation and was >98 %.

### Comparison of the two Variants

A decision analysis was carried out to determine the most efficient and economical route. 13 criteria were evaluated grouped into four categories with varying degree of importance:
- i)   chemicals/starting materials (easy access, number of suppliers, toxicity); weight 20 %,
- ii)  process (robustness, number of steps, price, ecology, safety, potential for further improvement); weight 40 %,
- iii) equipment (standard multi-purpose and specialized apparatus, capital costs), weight 20 %,
- iv) schedule for introduction into production, weight 10 %.

*Variant 1* had advantages in criteria ii) and iii), whereas *variant 2* showed better results for criterion i). Surprisingly, in the end both variants had about the same number of points, i.e., both routes had overall more or less the same efficiency. In the end, the epichlorohydrin route was preferred because of the shortness of the synthesis and especially because of the low capital investment necessary for additional reactors. But the hard decision had not to be made because the project was abandoned for other reasons.

### Acknowledgements

We would like to acknowledge the skillful experimental work carried out by P. Beney, A. Eckert, M. Graf, Y. Henriquez, A. Hügel, A. Müller, U. Pittelkow, H. Schneider, K. Schreiner, N. Vostenka and Ch. Wild. We would like to thank our colleagues P. Herold, C-P. Mak and M. Studer for critical comments during the preparation of the manuscript.

## Literature

1    Schmauss, M.; Laakmann, G.; Dieterle, D.; Schmitz, R.; Wittmann, M.
     *Pharmacopsychiatry* **1985**, *18*, 86.
2    *Merck Index*, *12.* edition, **1996**, compound no. 5792.
3    Wolfersdorf, M.; Wendt, G.; Binz, U.; Steiner, B.; Hole, G. *Pharmacopsychiatry*
     **1988**, *21*, 203.
4    Weinshenker, N. M. *Organic Preparations & Procedures* **1969**, *1*, 33-34.
     Mitchell, R. H.; Lai, Y.-H.; Williams, R. V. *J. Org. Chem.* **1979**, *44*, 4733.
5    Nonhebel, D. C. *Org. Synthesis, Coll. Vol.* 5, 1973; p. 206.
6    Yamamoto, Y.; Suzuki, H.; Moro-oka, Y. *Tetrahedron Letters* **1985**, 26, 2107.
7    Hayashi, T.; Katsumura, A.; Konishi, M.; Kumada, M. *Tetrahedron Letters* **1979**,
     425.
     For a recent review see R. Noyori, *Asymmetric Catalysis in Organic Synthesis* John
     Wiley & Sons, Inc.: Chichester, 1994; p. 56.
8    Hayashi, T.; Mise, T.; Fukushima, M.; Kagotani, M.; Nagashima, N.; Hamada, Y.;
     Matsumoto, A.; Kawakami, S.; Konishi, M.; Yamamoto, K.; Kumada, M. *Bull. Chem.
     Soc. Jpn* **1980**, *53*, 1138.
9    Hauser, C. R.; Lindsay, J. K. *J. Org. Chem.* **1957**, *22*, 906.
10   Wang, Y. F.; Lalonde, J. J.; Momongan, M.; Bergbreiter, D. E.; Wong, C. H. *J. Am.
     Chem. Soc.* **1988**, *110*, 7200.
11   Boaz, N. W. *Tetrahedron Letters* **1989**, *30*, 2061.
12   Wickli, A.; Schmidt, E.; Bourne, J. R. in *Biocatalysis in Non-Conventional Media;*
     Tramper, J.; Vermüe, M. H.; Beeftink, H. H., von Stockar, U, Eds., Elsevier:
     Amsterdam, 1992; p. 577.

# The Process Research and Development of DuPont Merck's Cyclic Urea Diols, A New Class of HIV Protease Inhibitors

**Pat N. Confalone and Robert E. Waltermire**
*Chemical Process Research & Development, The DuPont Merck Pharmaceutical Company, Chambers Works, Deepwater, NJ 08023*

Key words: cyclic urea, HIV protease inhibitor, DMP 323, DMP 450.

One of the key viral enzymes that has been targeted for the discovery of HIV therapeutics is protease. Consisting of a homodimer of 99 amino acids, it is an aspartic acid protease incorporating two catalytic aspartate groups. Directed mutation of these catalytic aspartate residues to aspargine, for example, abolishes the infectivity of the HIV virus.

Potent inhibition of renin, another important aspartic acid protease in the antihypertensive area, has been accomplished in a variety of ways, employing the concept of designing stable mimics of the tetrahedral intermediate along the hydrolysis pathway. Starting from the 1,2-amino alcohols related to statine and taking advantage of the homodimeric structure of the HIV protease, DuPont Merck first synthesized C2 symmetric diols as inhibitors of the HIV protease. A structure activity relationship readily emerged and the inhibitor potency progressed from an initial $K_i$ of 300 nM for the diol **1** down to sub-nanomolar $K_i$ of 0.4 nM for **2** in the course of thirteen months (Figure 1).

**Figure 1**

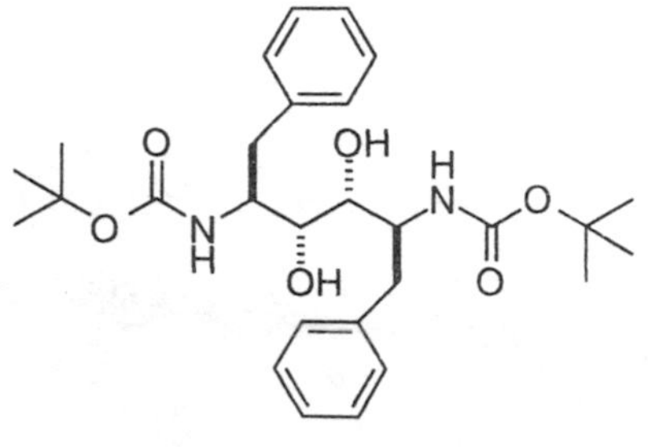

| | |
|---|---|
| **1** | **2** |
| $K_i$ = 300 nm | $K_i$ = 0.4 nm |

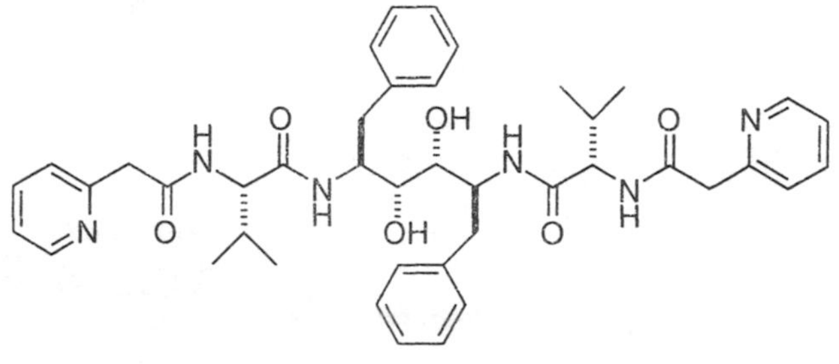

The continuing medicinal chemistry effort that eventually converted these simple linear diols to the subject cyclic urea diols as potent inhibitors of HIV protease is documented in a recent article.[1] Direct conversion of a linear diol **3** to a cyclic urea diol **4** affords initially only a 3X enhancement in potency (Figure 2). A similar result was obtained in the RSSR diastereomeric series e.g. **5**. However, substitution of both nitrogens of the cyclic urea affords a dramatic increase in potency to a $K_i$ of 150 nM for the highly lipophilic derivative **6**. In fact, the simple bis-allyl cyclic urea **7** exhibits a very potent $K_i$ of 4.7 nM.

**Figure 2**

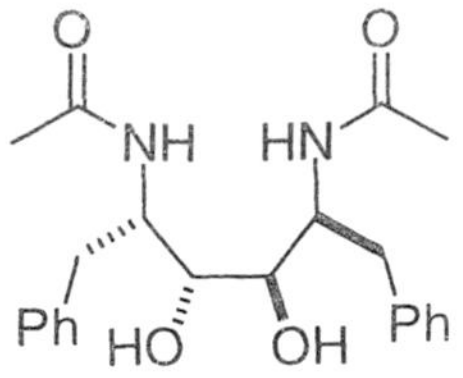

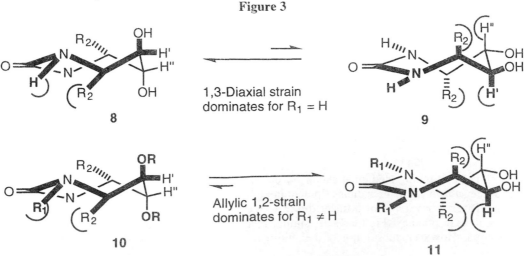

**3**
$K_i$ = 10,700 nm

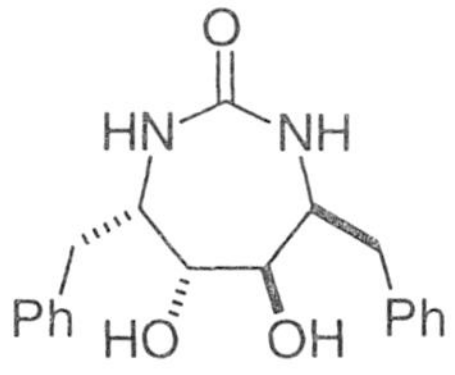

**4**
$K_i$ = 3000 nm

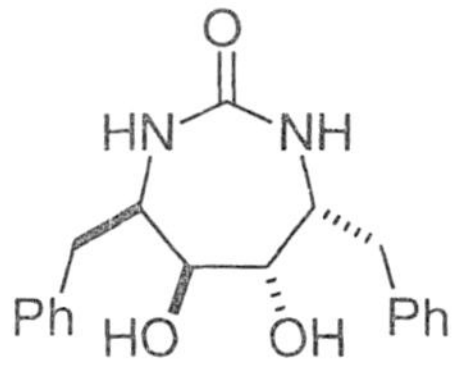

**5**
$K_i$ = 4500 nm

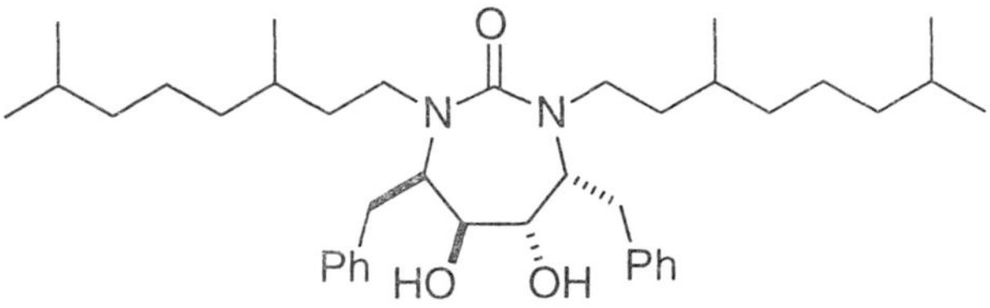

**6**
$K_i$ = 150 nm

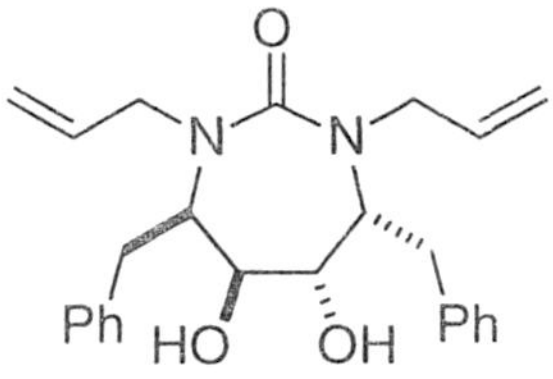

**7**
$K_i$ = 4.7 nm

Conformational analysis of these cyclic ureas indicates that a 1,3 diaxial interaction is a dominant feature in driving the equilibrium toward the diaxial hydroxyl groups in the case of the unsubstituted cyclic ureas **8** and **9**. On the other hand, when both nitrogens are substituted as **10** and **11**, an $A_{1,2}$ allylic strain becomes the dominant factor, resulting in equatorial hydroxyls. The $R_2$ groups of the cyclic urea are then oriented in the axial position. This latter conformation presents the cyclic urea diol array in a very favorable preorganized state to the enzyme active site. Since this results in a much more potent inhibitor, this precise orientation of the $R_2$ groups and the dihydroxyls are presumably set up to bind to the enzyme active site without substantial conformational adjustments.

**Figure 3**

**8**

1,3-Diaxial strain
dominates for $R_1$ = H

**9**

**10**

Allylic 1,2-strain
dominates for $R_1 \neq H$

**11**

The importance of this preorganization is illustrated by the following study (Figure 4). Simply deleting the sigma bond between the hydroxyls yields a "seco" dibenzyl cyclic urea **12** with a $K_i$ of 6,700 nM, whereas the corresponding cyclic urea analog **13** exhibits a $K_i$ of 2.5 nM. Thus, preorganization of the $P_1$, $P_{1'}$, $P_2$ and $P_{2'}$ receptor binding substituents by incorporation into a semi-rigid seven-membered ring system affords an enhancement of 268X in potency, reflecting a gain of approximately 4.8 kcal/Mol in $\Delta\Delta G$.

**Figure 4**

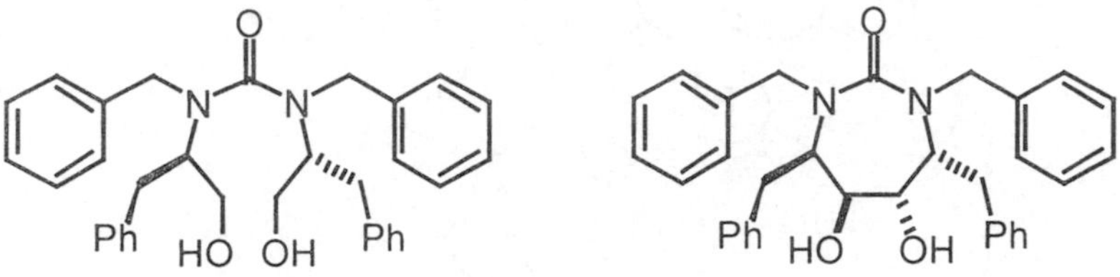

**12**
$K_i$ = 6700 nm

**13**
$K_i$ = 2.5 nm

As might be expected, the absolute stereochemistry of this class of inhibitors is also extremely important for maximal binding to the active site of HIV protease (Figure 5). For example, the corresponding SRRS diastereoisomer **14** has a $K_i$ of 4500 nM versus a $K_i$ of 4.7 nM for the preferred RSSR diastereoisomer **7**. There are only ten unique diastereomers rather than the expected sixteen, a consequence of the C2 axis of symmetry which bisects these cyclic ureas.

**Figure 5**

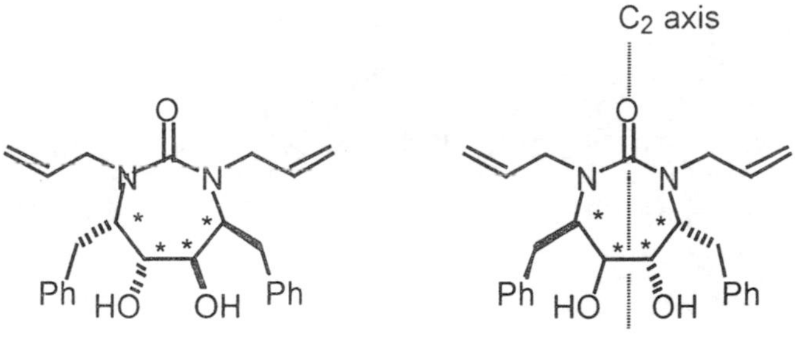

SRRS
**14**
$K_i$ = 4500 nm

RSSR
**7**
$K_i$ = 4.7 nm

In binding to the active site of HIV protease, as determined by protein x-ray crystallographic analysis, the cyclic urea carbonyl displaces an ordered water and subsequently binds to Ile 50 and Ile 150. The bis-naphthyl groups bind to the $S_2/S_{2'}$ pockets of the enzyme, whereas the bis-benzyl substituents at $P_1/P_{1'}$ bind to the corresponding $S_1$ and $S_{1'}$ sites (Figure 6). As expected, the diol interacts with the "floor" of the active site, binding to Asp 25 and Asp 125.

Figure 6

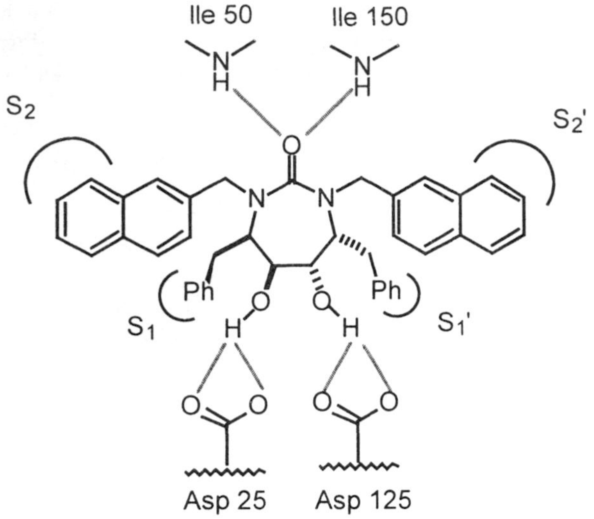

This novel structural class of HIV protease inhibitors has yielded two development candidates: DMP 323[2] (15) and DMP 450[3,4] (16) (Figure 7). Incorporating four contiguous asymmetric centers, these compounds represent challenging targets for process chemistry. The results of our synthetic efforts to discover and develop a commercial process for the manufacture of symmetrical cyclic urea diols are presented in this chapter.

Figure 7

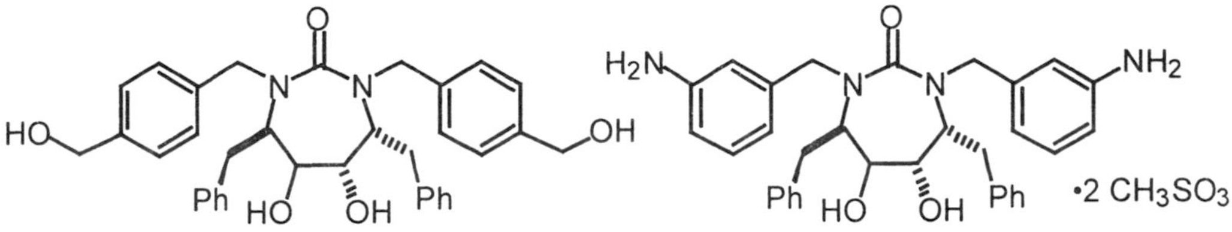

Retrosynthetically, the first approach to the $C_2$-symmetric cyclic urea diols is based on the route discovered by the medicinal chemistry effort (Scheme 1).[2] The disconnection I breaks the bonds between the urea nitrogens and the R2 substituents of 17. In the synthetic direction conditions for selective N-alkylation vs O-alkylation were known. The disconnection II on 18 provides a diol-protected diamino-diol still incorporating the required four contiguous asymmetric centers. The final disconnection III on 19 yields two moles of an N-protected D-phenylalanal from which the absolute chirality of the molecule is derived.

**Scheme 1**

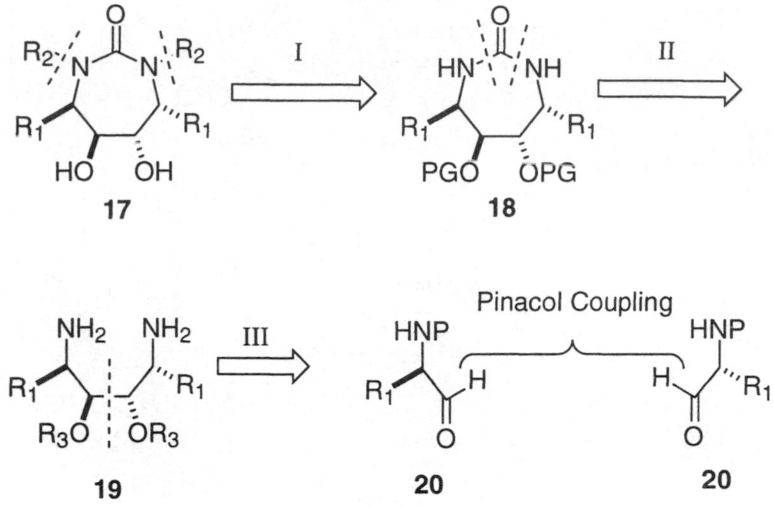

17 → (I) → 18 → (II) →

19 → (III) → Pinacol Coupling → 20 + 20

The synthesis of DMP 323 (**15**),[2] DuPont Merck's first clinical candidate, illustrates this initial cyclic urea diol synthesis (Scheme 2), and used methodology developed for the synthesis of the enantiomeric series of linear diols of the following generic structure **21**.[5]

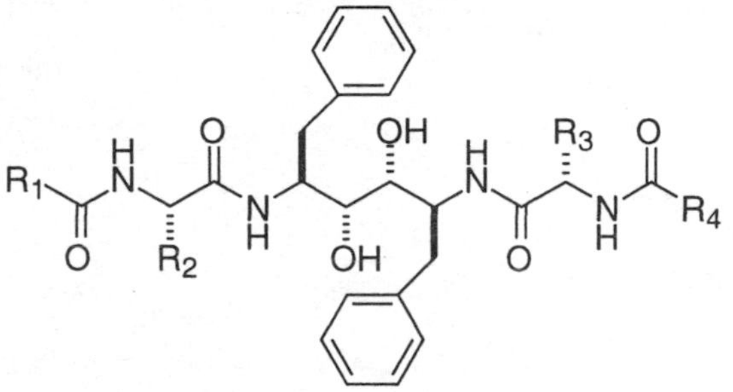

**21**

Various syntheses of the enantiomeric diol series start with D-threitol,[6] D-mannitol,[7] and L-phenylalaninol.[8] We selected the latter approach since D-phenylalaninol **22** is the only readily available enantiomeric starting material. D-Phenylalaninol **22** is converted to the N-Cbz-derivative **23** which undergoes a Swern oxidation to afford the aldehyde **24**. A highly diastereoselective pinacol coupling reaction[8] is utilized to prepare the desired diol **25** which incorporates the required asymmetric centers in the correct absolute configuration. This sequence yields **25** with a 96%de after recrystallization from THF /hexane in an overall chemical yield of 44% based on D-phenylalaninol **22**.

**Scheme 2**

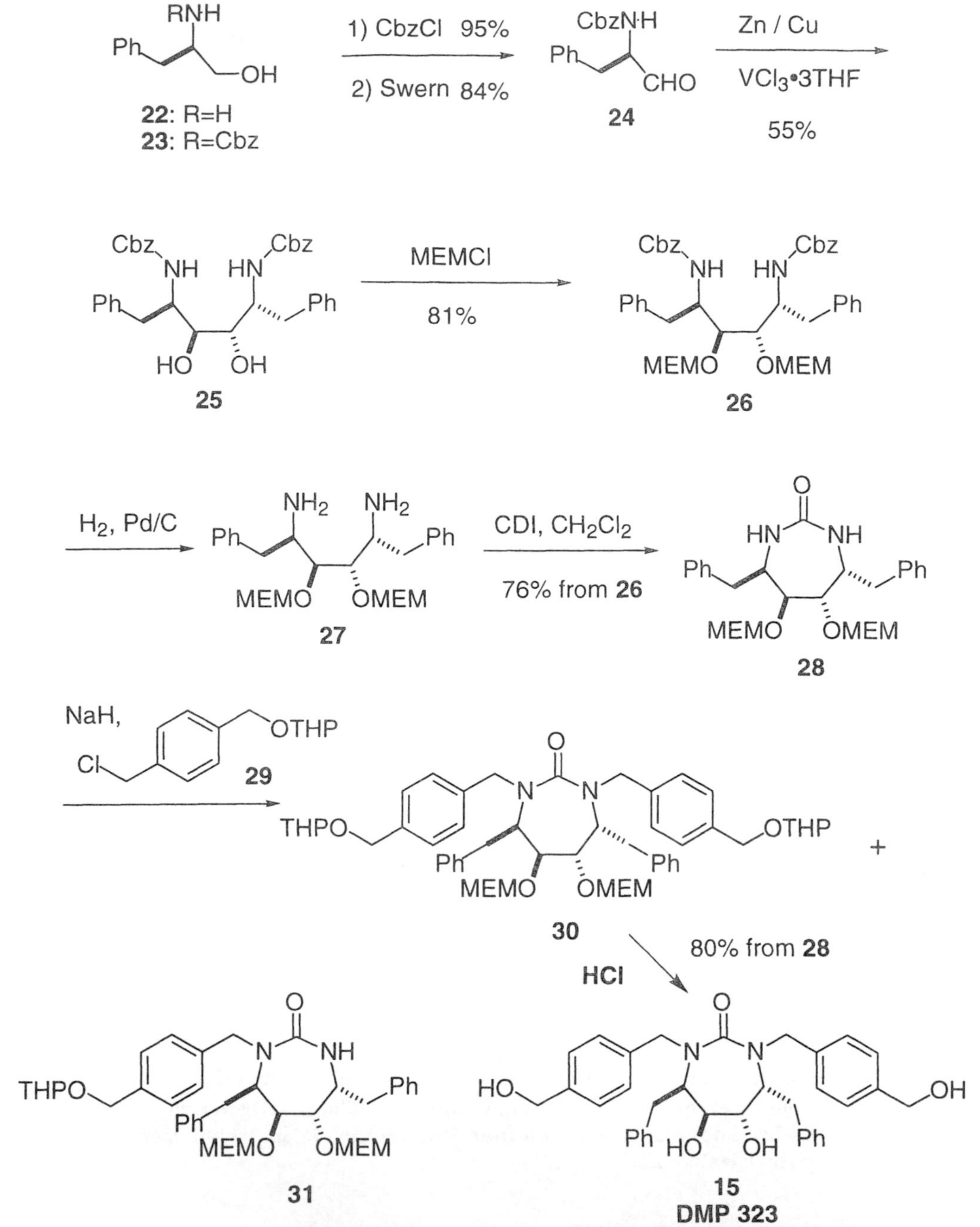

Protection of the diol **25** with methoxyethoxymethylchloride (MEMCl) provides the bis-MEM ether **26** with a >99% de. Unfortunately, **26** is the last crystalline intermediate in this synthesis of DMP 323. Hydrogenolysis of the bis-Cbz moieties affords the diamine **27** which is immediately cyclized at room temperature with carbonyldiimidazole (CDI) to yield the cyclic urea **28**. Alkylation of **28** with the benzyl chloride **29** generates the penultimate intermediate **30**. Complete deprotection of **30** with HCl in methanol, followed by chromatography, provides pure

DMP 323 in 80% overall yield from the cyclic urea **28**. This synthesis is eight steps from D-phenylalaninol (**22**) and proceeds in 23% overall yield with a final de of >99.5%. Using this methodology, five kilograms of DMP 323 was prepared by our discovery kilolab to support the rapid development of DMP 323.

**Scheme 3**

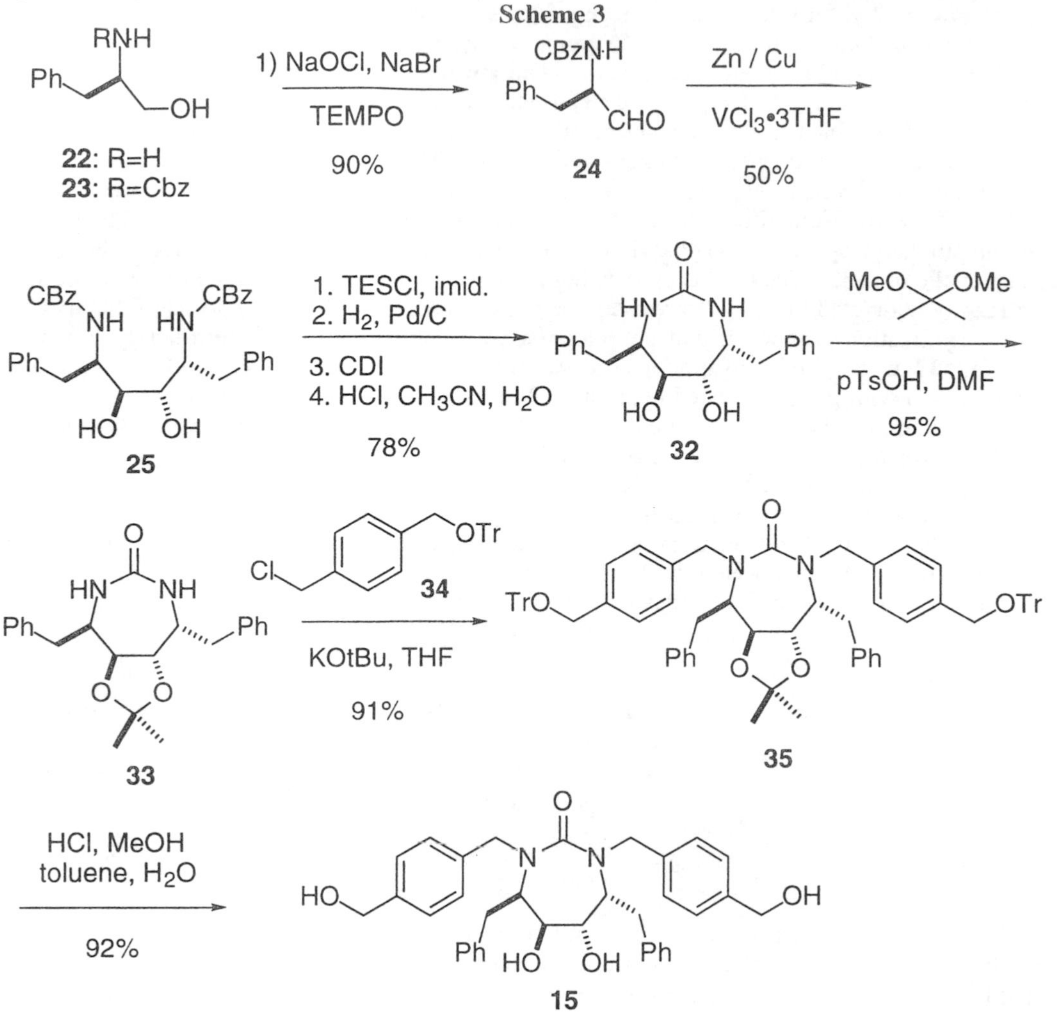

Several issues were identified with this synthesis that required attention prior to scale-up for the production of hundreds of kilograms of DMP 323. The major problem is the requirement for chromatography at the end of the synthesis. This is required to produce an acceptable purity profile, a consequence of the absence of crystalline intermediates after the bis-MEM ether **26**. Therefore, the first goal was to change the diol and/or amino protecting groups to yield a set of crystalline intermediates that would allow the preparation of DMP 323 without chromatography. Additional issues included: 1) the use of MEMCl which contains a known carcinogen, bis-chloromethylether, as an impurity at 1 to 2%, 2) the instability of the alkylating agent **29**, which led to additional impurities in the intermediate **30** and 3) the bis-alkylation of **28** was complicated by the production of >2% of the mono-alkylated derivative **31**, which was very difficult to remove without a significant yield penalty. Longer term, we still needed to address the cost of using two moles of an unnatural amino acid derivative **22** as the ultimate source of chirality, disposing of large quantities of the vanadium/zinc mixed waste generated in the pinacol coupling step and the tendency of the intermediate aldehyde **24** to epimerize and/or polymerize.

An alternative synthesis that addresses a number of these issues is shown in Scheme 3.[9] Previously, the Swern oxidation used to prepare **24** was performed at -78 °C and generated malodorous methyl sulfide. A preferred method is to carry out a TEMPO catalyzed sodium hypochlorite oxidation.[10] Initially, this reaction was quite problematic on scale and very high agitation was required to avoid overoxidation. Optimization of concentrations, especially the use of high strength bleach (12.5%), provided a consistent 90% yield of **24** at 300 gallon scale. The de of the diol **25** obtained in the subsequent pinacol coupling reaction was improved to 99% by recrystallization from methylethylketone/water instead of THF/hexane.

Our next goal was to identify a better protecting group for diol **25** and cyclic urea **32**. A number of hydroxy/dihydroxy protecting groups such as TMS, triethyl silyl (TES), *tert*-butyldimethylsilyl, acetonide, *tert*-butyl, trityl, benzoyl and THP were examined for both the diol **25** and cyclic urea **32**. The bis-TES protecting group was selected and installed on **25** which was then readily converted to **32** by a three step sequence consisting of hydrogenolysis of the bis-Cbz groups, cyclization with CDI, and deprotection with aqueous HCl. Unfortunately, bis-TES protected **32** was difficult to obtain pure *via* crystallization and alkylation led to product contamination with mono-alkylated byproduct. From our study, we concluded that the preferred protecting group for the diol functionality of **32** is clearly the acetonide. Treatment of **32** with 2,2-dimethoxypropane affords acetonide protected cyclic urea **33** as a crystalline, readily purified intermediate. To shorten the sequence, diamine acetonide **36** is prepared from diol **25** by protection with 2,2-dimethoxypropane followed by subsequent hydrogenolysis of the bis-Cbz moiety (Scheme 4). A major, unexpected problem now arose in an attempt to synthesize **33** directly from the diamine acetonide **36** as only trace quantities of **33** is obtained by attempted cyclization of **36** upon treatment with CDI (Scheme 4).

**Scheme 4**

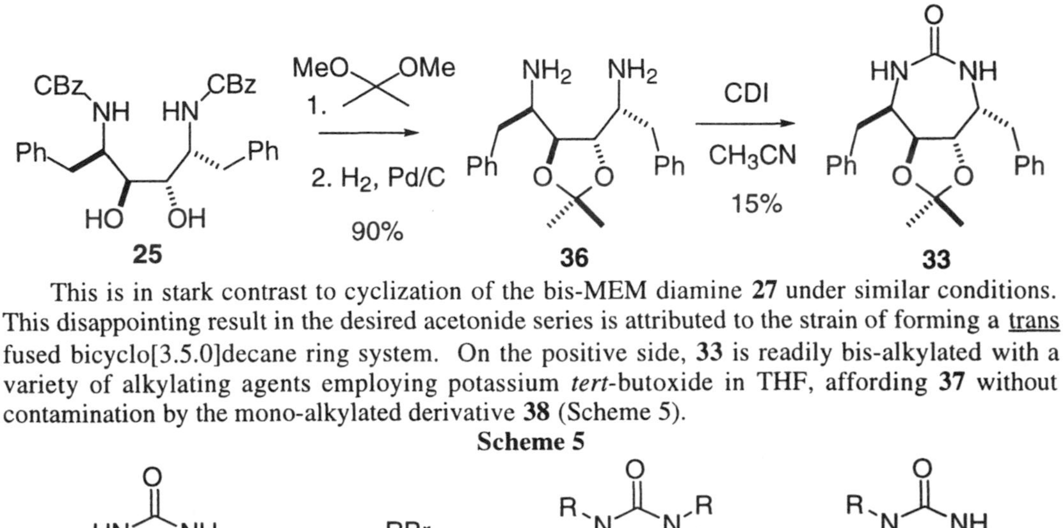

This is in stark contrast to cyclization of the bis-MEM diamine **27** under similar conditions. This disappointing result in the desired acetonide series is attributed to the strain of forming a <u>trans</u> fused bicyclo[3.5.0]decane ring system. On the positive side, **33** is readily bis-alkylated with a variety of alkylating agents employing potassium *tert*-butoxide in THF, affording **37** without contamination by the mono-alkylated derivative **38** (Scheme 5).

**Scheme 5**

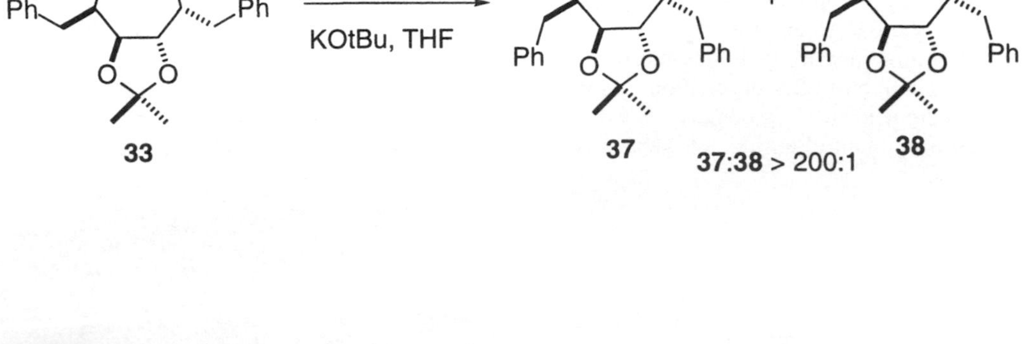

37:38 > 200:1

This result is rationalized by a conformational analysis of the two intermediates (Figure 8). X-ray crystallography and molecular modelling of these cyclic urea diols indicates that the preferred conformation of a disubstituted cyclic urea diol is one in which the diol groups are placed in an equatorial orientation and the four benzylic substituents are all axially/pseudo-axially disposed. Therefore, the acetonide derivative **33** possesses a product -ike conformation since the protected diol substituents are equatorially oriented throughout the alkylation. The bis-MEM cogener **28**, on the other hand, requires a ring flip during the alkylation process in order to adopt the conformation of the product. As discussed earlier, related considerations played a role in the design of highly potent analogs in the medicinal chemistry effort.

**Figure 8**

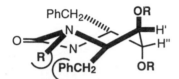

Preferred Confromation for **28**

vs.

1,3-Diaxial strain dominates for acyclic Protected Cyclic Urea

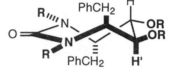

Preferred Confromation for **33**

Allylic 1,2-strain dominates

Preferred Conformation for Alkylated **28** and **33**

The instability of the alkylating agent **29** was addressed next by examining a number of hydroxy protecting groups. The tritylated benzyl chloride **34** (Scheme 3) was identified as a crystalline, stable alkylating agent.[9] Alkylation of **33** with **34** provides a 91% yield of very pure crystalline final intermediate **35**. Simultaneous deprotection of the trityl and acetonide protecting groups under acidic conditions affords a 92% yield of high quality DMP 323 (**15**) prepared in eight steps from D-phenylalaninol (**22**) in 27% overall yield without chromatography.

We next turned our attention to the undesirable protecting group switch (TES to acetonide) forced by the low yield cyclic urea formation when the diol is protected as an acetonide. Conversion of **36** to the cyclic urea **33** was analyzed as a two step process of mono-acylation followed by cyclization so possible problems with each step could be separately studied in some detail (Scheme 6). The desired product is derived from the intermediate mono-substituted diamine **39**, whereas formation of the 2:1 adduct **40** is clearly undesirable. The formation of **39** and derivatives made with phosgene and other phosgene surrogates was studied in a wide variety of solvents, temperatures, dilutions and protonated/non-protonated forms of **36** with various stoichiometries in an attempt to favor mono-acylation pathways.[11] The best conditions identified for the selective formation of **39** in 70-80% yield consist of CDI in a chlorinated solvent such as methylene chloride. The remaining material consisted of **36** and **40** in equal amounts. Cyclization to **33** requires high dilution (200 mL/g) in refluxing tetrachloroethane (TCE) bp=147 °C. Clearly, this high temperature populates the higher energy conformations which favor intramolecular ring closure. Under optimum conditions, a 67% yield of **33** is obtained from **36**. Unfortunately, the volume inefficiency of this cyclization as well as the requirement of high boiling, chlorinated solvents rendered this conversion unacceptable for scale-up.

**Scheme 6**

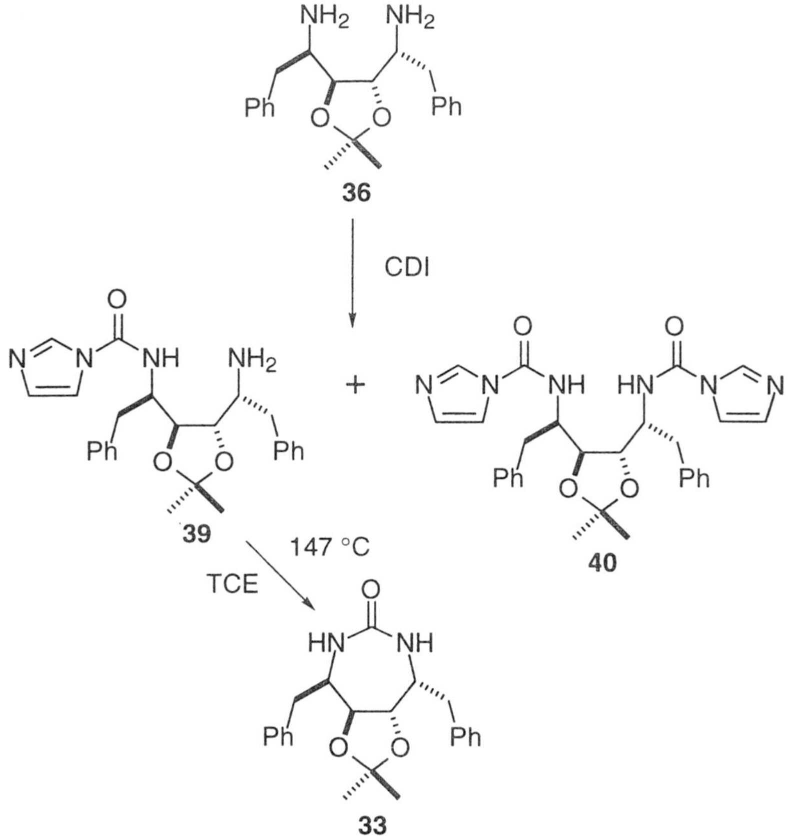

Simultaneous with our work on the D-phenylalanine route, we examined other starting materials which might afford a more efficient and economical synthesis of the cyclic urea diols. One very attractive alternative is L-tartaric acid based on the retrosynthetic analysis outlined in Scheme 7 which targets diketone **45** or dialdehyde **43** intermediates.

**Scheme 7**

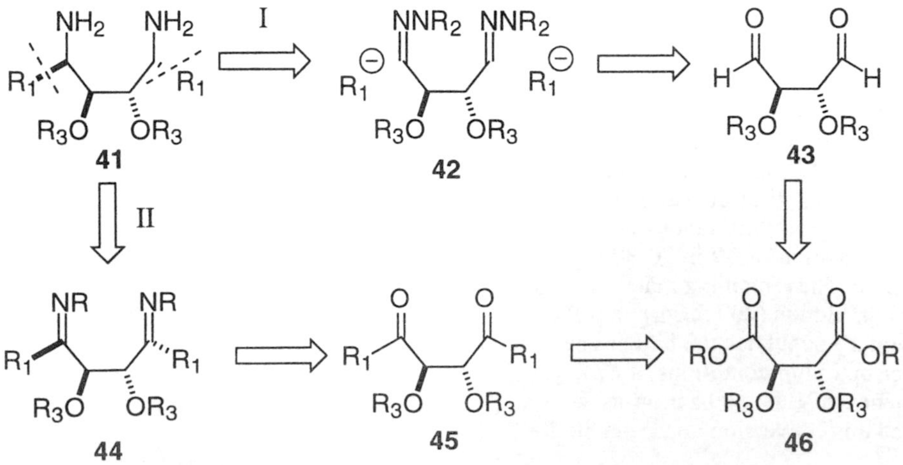

Disconnection I of the diamine **41** leads to the di-imine **42** which is likely derived from a tartrate-based dialdehyde such as **43**. In the synthetic direction, a diastereoselective addition of the $R_1$ groups to di-imine **42**, possibly chelation controlled, is required. The alternative transformation II leads to the diketone **45** via the di-imine **44**, which may originate from a tartrate ester equivalent **46**. This latter approach, which we've termed the diketone route, is presented in Scheme 8.

The naturally occuring isomer of tartaric acid, L-tartaric acid (**47**) is converted to its dimethyl ester and the diol protected as an acetonide **48** for the initial route scouting. The Weinreb amide[12] **49** is prepared via aluminum amidation[13] of **48** with N,O-dimethylhydroxylamine in 83% yield. Conversion to the diketone **50** is effected in 95% yield by reaction of **49** with benzylmagnesium bromide. The bis-oxime **51** is obtained in 82% yield by reaction of **50** with hydroxylamine hydrochloride. Reduction of the oxime **51** with twelve equivalents of diisobutylaluminum hydride (DIBAL-H) proceeds in a highly diastereoselective fashion to provide **36**, possessing the desired RSSR stereochemistry, in low yield 32%.[14] Unfortunately, other reduction conditions could not be identified that suppress the major by-product **52**, obtained via a reductive Beckmann rearrangement.[15]

**Scheme 8**

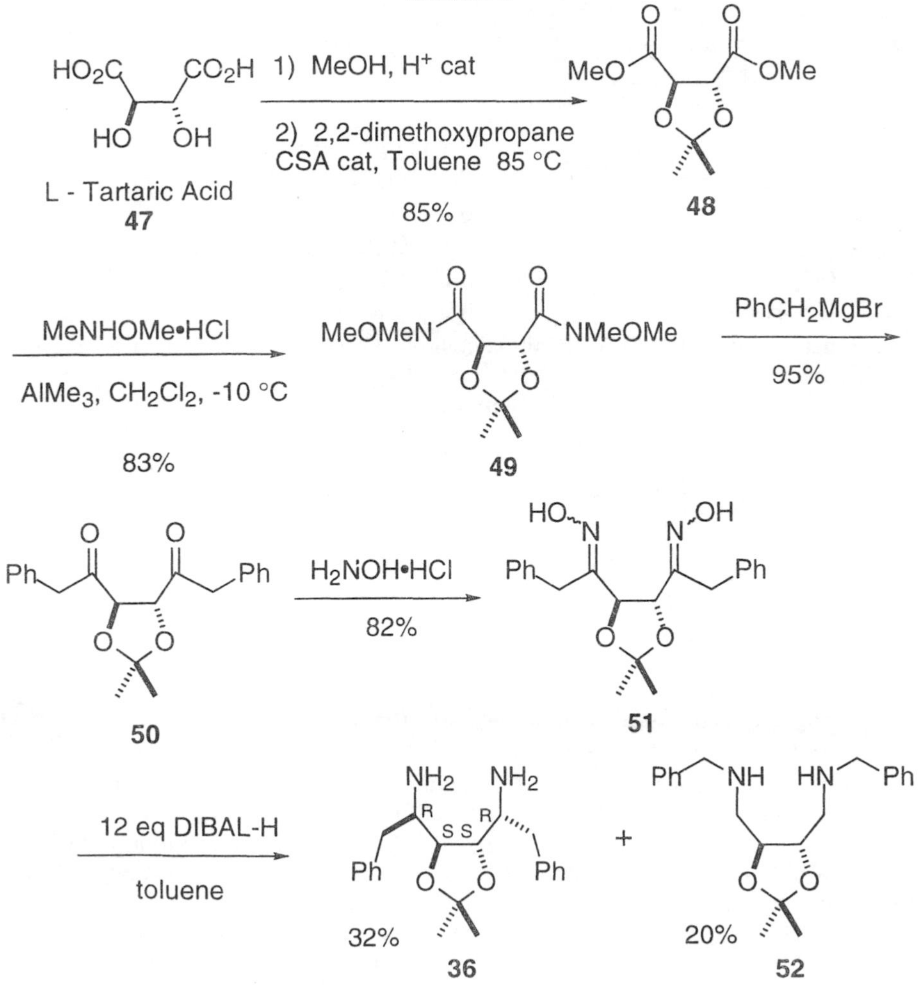

Fortunately, the dialdehyde[16,17] based routes proved more successful (Schemes 9 and 12). Reduction of the diester **48** to the dialdehyde **53** with DIBAL-H, followed by addition of a benzylamine affords di-imines **54-57**. Addition of benzyllithium to the bis-benzyl imine **54** provides diamine **58** in 70 % de, favoring the desired RSSR chelation controlled product. Upon crystallization, a 24% yield of **58** of 95% de is obtained.

**Scheme 9**

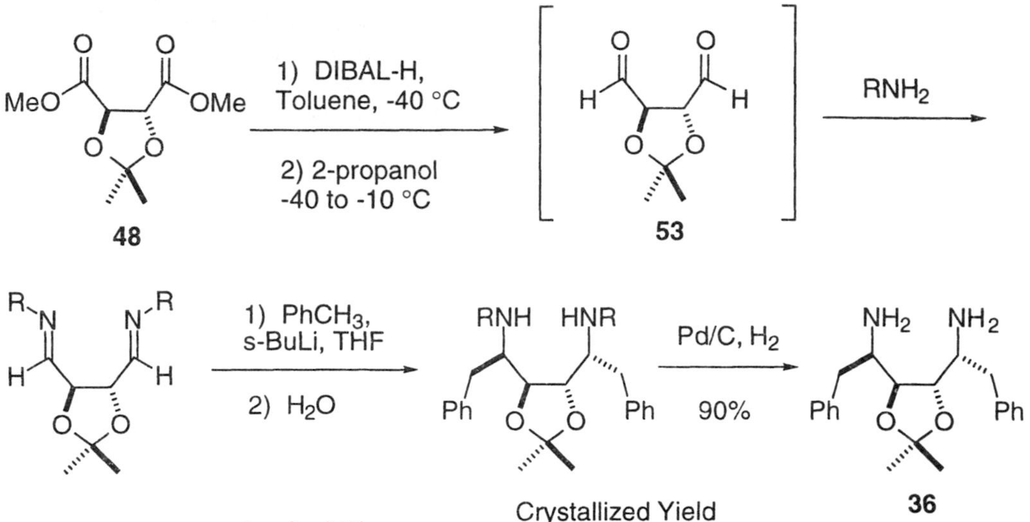

|    |              | de of addition |    | Crystallized Yield of RSSR (de) |
|----|--------------|----------------|----|-------------------------------|
| **54** | R=Bn         | 70             | **58** | 24% (95)                  |
| **55** | R=Ph$_2$CH   | 78             | **59** | 48% (96)                  |
| **56** | R=(S)-PhCHMe | 92             | **60** | 55% (100)                 |
| **57** | R=(R)-PhCHMe | 34             | **61** | 0%                        |

We reasoned that the poor yield of the addition could be attributed to the formation of an aza-allyl anion **62** (Scheme 10).

**Scheme 10**

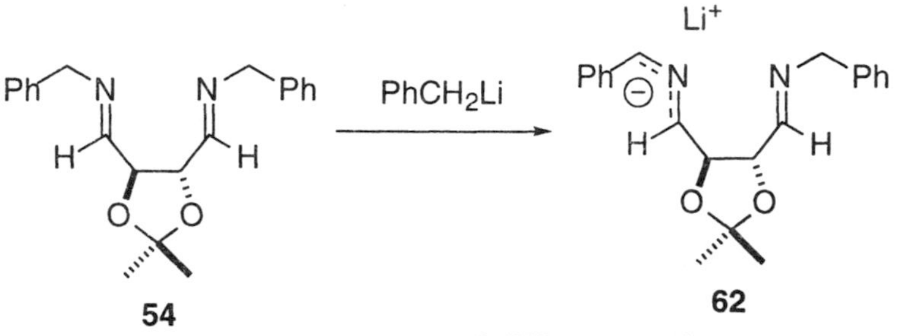

Initially, we sought to suppress this by the use of different metal counterions such as copper, cerium, magnesium and various additives such as BF$_3$-OEt$_2$, LiBr, LiClO$_4$ and TMSCl. None of these measures led to any success. A different solution to this problem was examined by adding an <u>alpha</u> -substituent to the benzylamine to reduce kinetic deprotonation to the aza-allyl anion. Formation of di-imine **55** from benzhydrylamine and addition of benzyllithium provides diamine **59** in 78% de in solution, 48% yield and 96% de upon crystallization, supporting our hypothesis. Additionally, we prepared the R and S alpha-methyl benzylimines **56** and **57** to determine if double diastereoselectivity could improve our results. As predicted, addition of benzyl lithium to S imine **56** proceeds with matched stereocontrol to provide diamine **60** in 92% de in solution, 55%

yield and 100% de upon crystallization. Addition of benzyl lithium to the mismatched R imine **57** provides a poor 34% de and no product (**61**) could be obtained <u>via</u> crystallization. We hypothesized that <u>alpha</u>-disubstituted benzylimines should be ideal since no acidic protons are available for aza-allyl anion formation. Unfortunately, a facile preparation of <u>alpha</u>-disubstituted benzylamines proved elusive. To complete the synthesis of diamine **36**, hydrogenolysis of **58-60** in the presence of a palladium catalyst affords an 85% yield of **36**. Overall, the use of S-α-methylbenzylamine provides a 50% yield of diastereomerically pure **36** from **48**. This approach is quite adequate and would have been selected as our synthesis of cyclic urea diols, however, an alternative route based on a *bis*-hydrazone route proved even better.

Simultaneous with these efforts, an Abbott group[6] published the synthesis of **66**, the enantiomer of **36**, starting from (-)-2,3-O-isopropylidene-D-threitol (**63**) (Scheme 11). Swern oxidation of **63**, followed by reaction with 1,1-dimethylhydrazine provides the *bis*-hydrazone **64**. Hydrazones derived from 1,1-dimethylhydrazine had previously been shown to afford very high diastereoselectivity for chelation controlled addition.[18]

**Scheme 11**

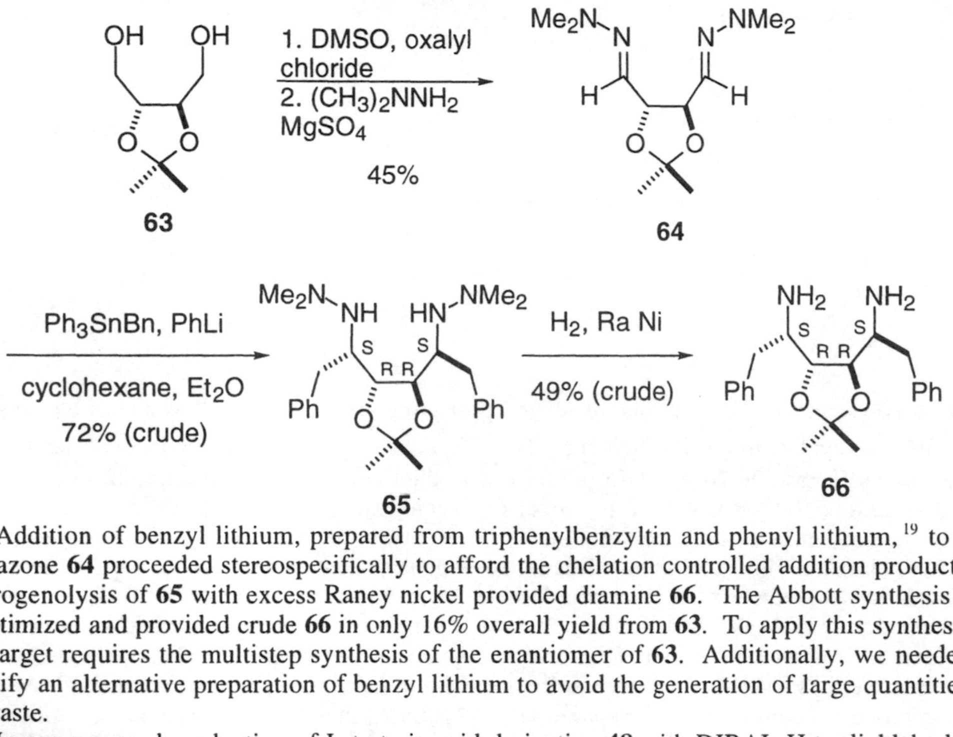

Addition of benzyl lithium, prepared from triphenylbenzyltin and phenyl lithium,[19] to *bis*-hydrazone **64** proceeded stereospecifically to afford the chelation controlled addition product **65**. Hydrogenolysis of **65** with excess Raney nickel provided diamine **66**. The Abbott synthesis was unoptimized and provided crude **66** in only 16% overall yield from **63**. To apply this synthesis to our target requires the multistep synthesis of the enantiomer of **63**. Additionally, we needed to identify an alternative preparation of benzyl lithium to avoid the generation of large quantities of tin waste.

In our approach, reduction of L-tartaric acid derivative **48** with DIBAL-H to dialdehyde **67** followed by addition of 1,1-dimethylhydrazine provides *bis*-hydrazone **68** in 85% yield (Scheme 12). Chelation controlled addition of benzyl lithium, prepared from toluene and *sec*-butyllithium in THF,[20] proceeds stereospecifically[21] to afford dihydrazine **69** as its *bis*-fumarate salt in 90% yield. Diamine **36** is obtained as its *bis* p-toluenesulfonate salt in 85% yield after hydrogenolysis of **69** with Raney nickel. Overall, this sequence affords diastereomerically and enantiomerically pure **36** in 4 steps from **48** in 65% overall yield. To consider this synthesis for large scale implementation still required a solution to our difficult cyclic urea formation from acetonide diamine **36**, or discovery of an alternative diol protecting group that would enable a more facile cyclization.

**Scheme 12**

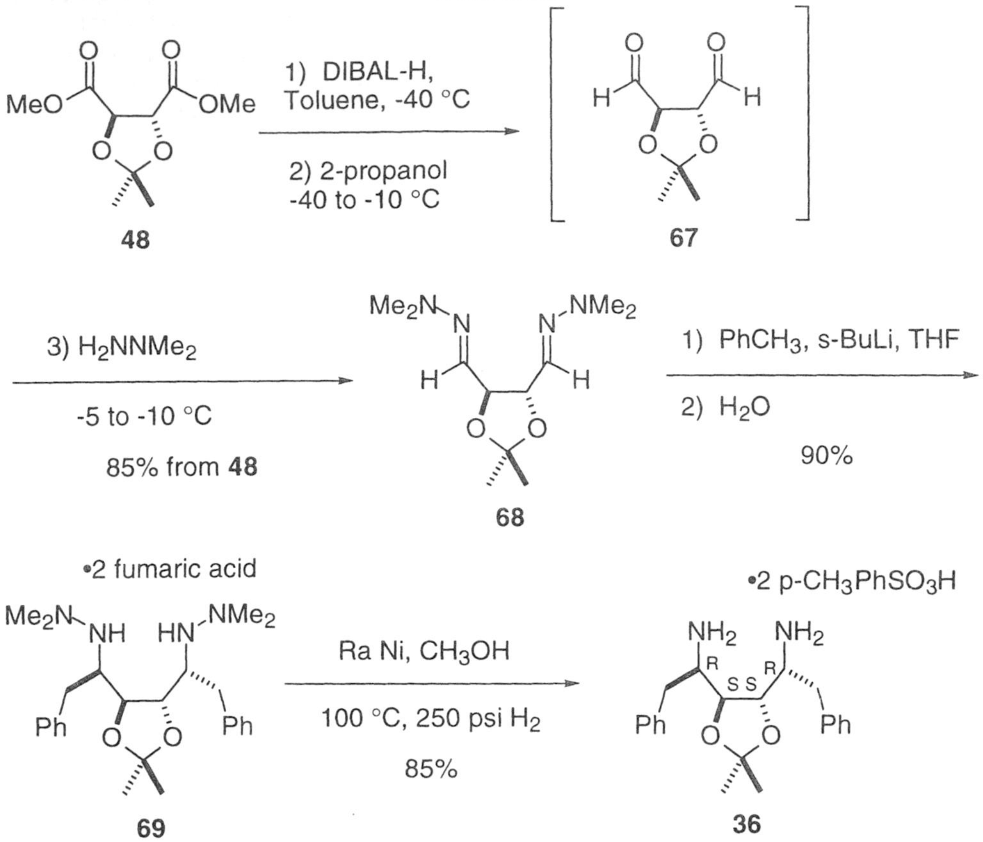

The trioxepane nucleus was one of several protecting groups examined as a possible solution to the difficult cyclization of **36** (Scheme 13).[17] Trioxepane protected derivative **71**,[22] is obtained from dimethyl L-tartrate **70** as a 94:6 mixture with the thermodynamic product **72** in 60% yield after distillation. Unfortunately, after about 65% conversion of **70** to **71**, the formation of the thermodynamic product **72** predominates. Execution of the DIBAL-H reduction, 1,1-dimethylhydrazone formation, benzyl lithium addition sequence again proceeds stereospecifically to afford the trioxepane protected dihydrazine **74** in 72% overall yield from **71** as its *bis*-methanesulfonic acid salt. Isolation of salt **74** also removes the analogous product derived from **72**. Next, **74** is converted to its free base **75** and subjected to hydrogenolysis to provide diamine **76** in 86% yield, isolated as its benzenesulfonic acid salt. Cyclization of **76** was studied and a high yielding process identified for the preparation of **77** from **76** in 92% yield. The best conditions are the sequential addition of one equivalent of triethylamine, followed by one equivalent of CDI to a solution of **76** in acetonitrile at 0 °C, followed by warming to room temperature. This protocol selectively "protects" the diamine as its mono-salt and affords the highest ratio of product to undesired oligomeric by-products. Application of this specific protocol to the difficult cyclization of the acetonide analogue **36** provides only a 25% yield of **33**. To demonstrate the utility of this approach, **77** was alkylated with **34** followed by simultaneous deprotection of both the trityl and dioxepane functionalities to provide DMP 323 (**15**) in 70% yield. Overall, this route seemed an excellent choice for scale-up. At this point, DuPont Merck's Chemical Process R&D Physical Chemistry group routinely performed thermodynamical analysis of the process. Unfortunately, thermochemical analysis revealed that the trioxepane protecting group added a severe thermochemical liability. The trioxepane group decomposes and releases two moles of formaldehyde upon heating. The combination of the trioxepane decomposition and energetics of the dihydrazone moiety of **73** combine to make it a particularly hazardous intermediate.

**Scheme 13**

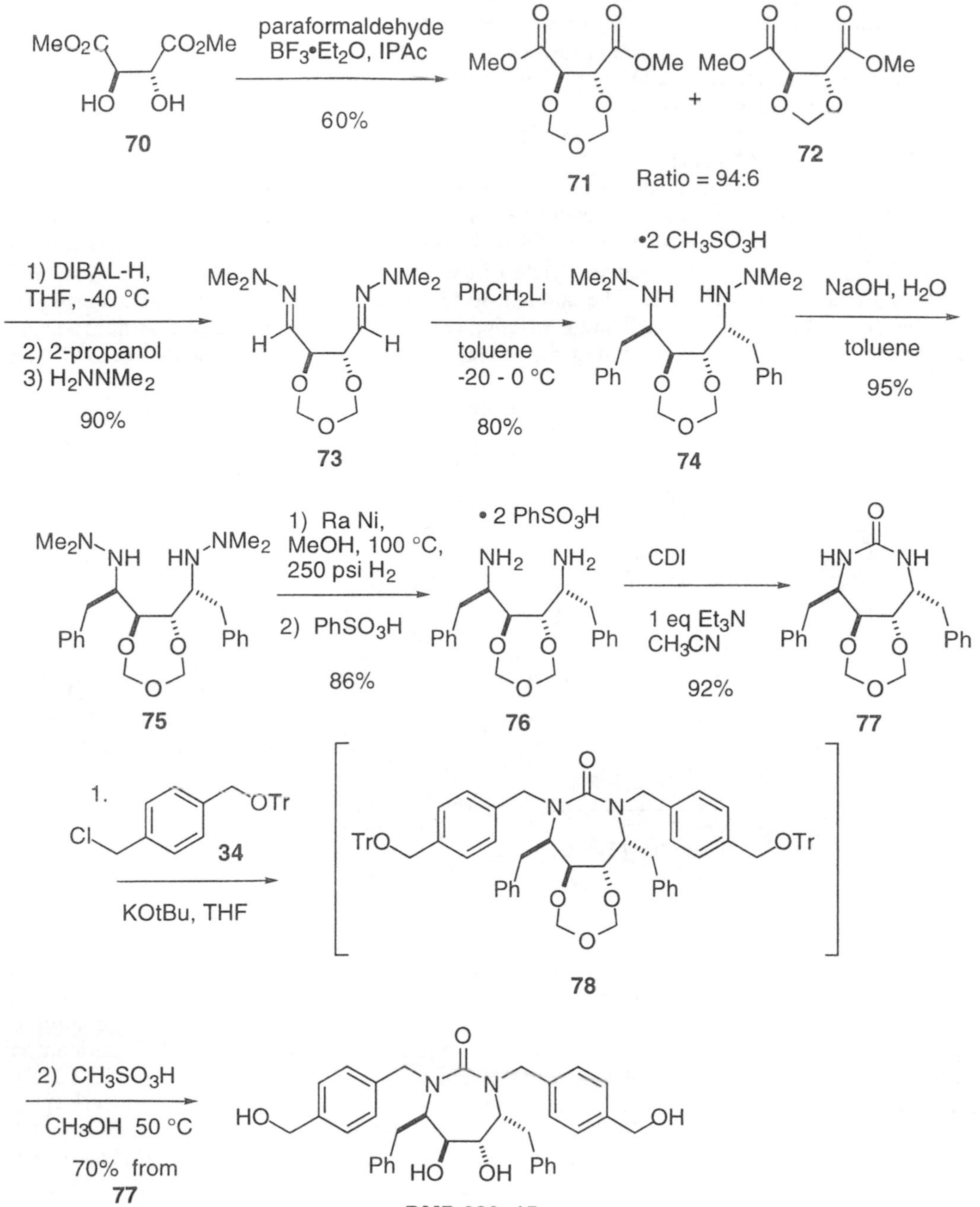

Accelerated Rate Calorimetry (ARC) of **73** exhibits vigorously energetic exothermic decomposition which is accompanied by a large increase in pressure. Dihydrazone **73** is of particular concern because its Exotherm Initiating Temperature (EIT) of 165 °C decreases to 145 °C in air and to 100 °C during an isothermal aging experiment. By comparison, the acetonide dihydrazone **68**, although having an EIT of 165 °C, is an intrinsically safer intermediate because it does not exhibit instability at lower temperatures under similar conditions and it does not release two moles of formaldehyde gas during its decomposition. At this point, DuPont Merck's second protease inhibitor, DMP 450[3] (**16**) was selected for development and large scale material requirements for this drug candidate needed to be met. We now required a rapid solution to the problem of either identifying an alternative diol protecting group or solving the challenges of the cyclization of acetonide diamine **36**.

Fortunately, a timely solution[23] was identified for the cyclization of acetonide protected secondary diamines (Scheme 14). Previously only acetonide protected diamines had been examined in the cyclization. Bis-reductive amination of **36** with benzaldehyde and sodium triacetoxyborohydride provides the crystalline *bis*-secondary diamine **58** in 90% yield, proceeding via the *bis*-Schiff base **79**. Cyclization of **58** with phosgene in the presence of triethylamine at 110°C in toluene yields a 75% isolated yield of the desired cyclic urea **80** and 10% of the *bis*-carbamoyl chloride impurity **81**. The amount of **81** could be greatly reduced by lowering the effective concentration of phosgene in the reaction medium. This facile cyclization now became the cornerstone on which a potentially commercial scale process for the synthesis of DMP 450 (**16**) could be based.

**Scheme 14**

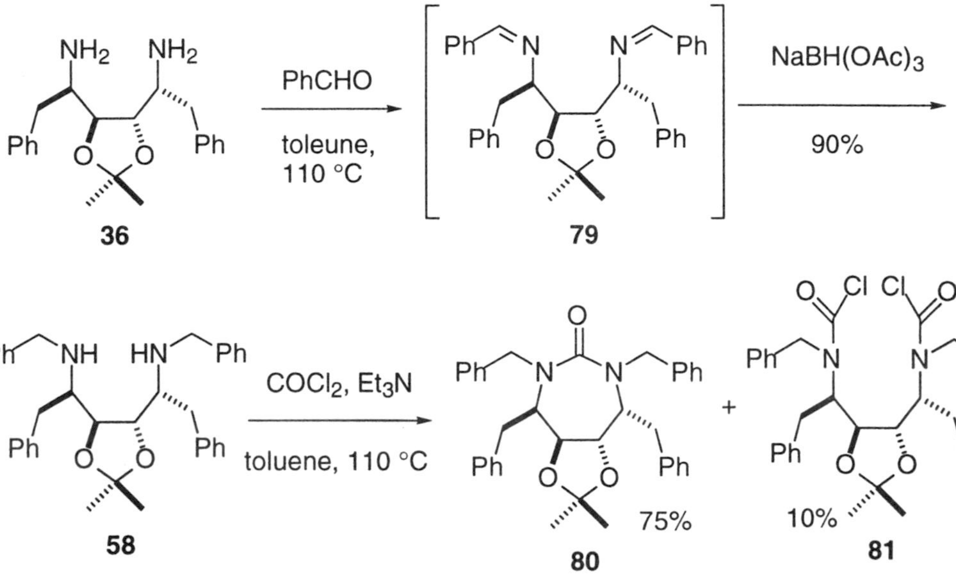

Twenty kilograms of DMP 450 (**2**) were prepared in our pilot plant using this cyclization strategy[24] (Scheme 15). Bis-reductive amination of **36** with 3-nitrobenzaldehyde and sodium triacetoxyborohydride provides crystalline **82** in 96% yield. The optimum cyclization conditions with good volume efficiency for the conversion of **82** to **83** in 94% solution yield were identified and consisted of slow addition of a solution of phosgene in chlorobenzene to a 123-127 °C solution of **82** and triethylamine in the same solvent. After cyclization, the crude product (**83**) was deprotected with methanol and sulfuric acid to provide crystalline diol **84** in 88% yield from **82**. Hydrogenation of **84** in the presence of methanesulfonic acid provided DMP 450 (**16**) in 90% yield.

Scheme 15

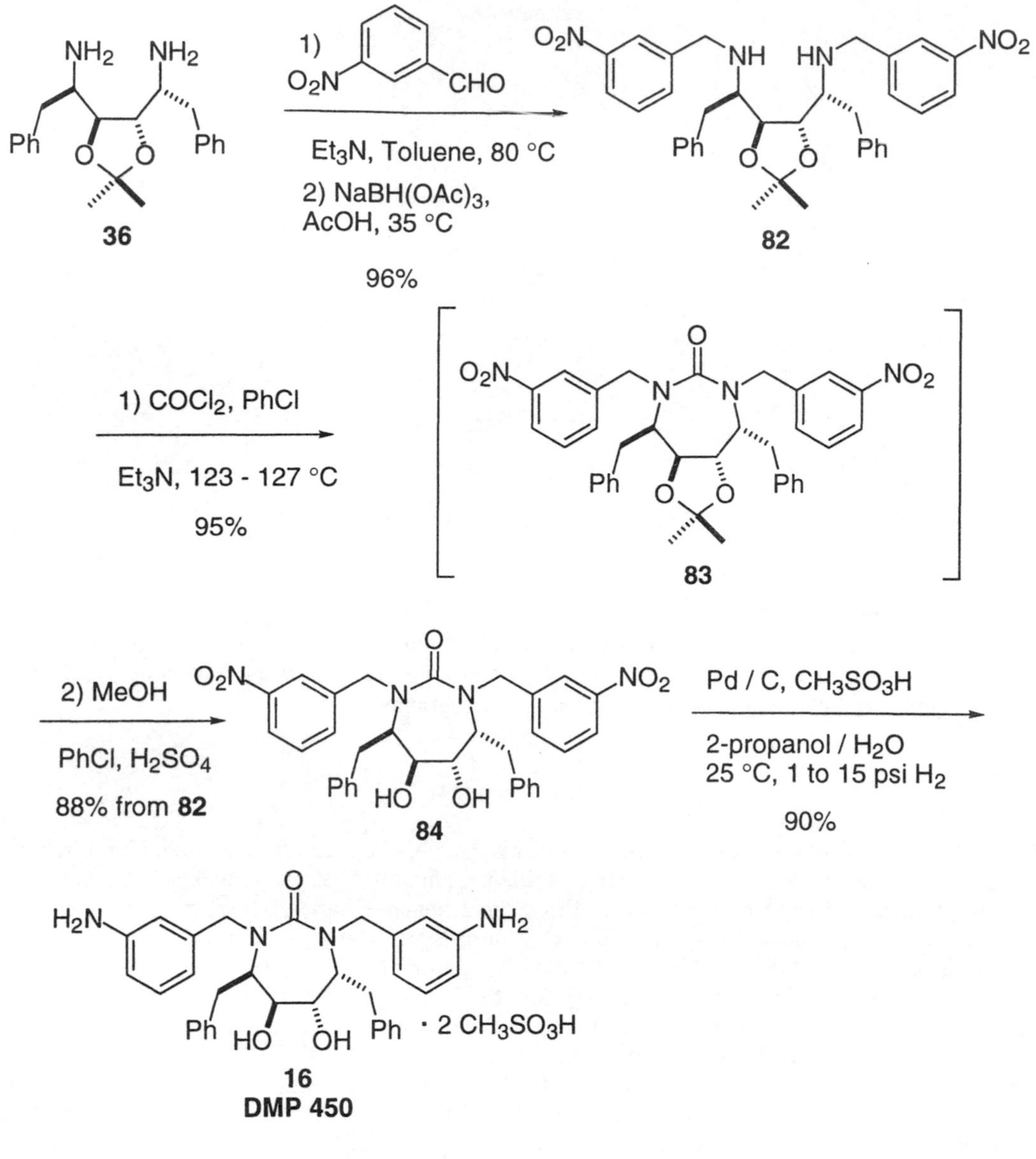
NH2 NH2
Ph
O O
Ph
36
1)
O2N
CHO
Et3N, Toluene, 80 °C
2) NaBH(OAc)3,
AcOH, 35 °C
96%
O2N
NH HN
NO2
Ph
Ph
82
1) COCl2, PhCl
Et3N, 123 - 127 °C
95%
O
O2N
N N
NO2
Ph
Ph
83
2) MeOH
PhCl, H2SO4
88% from 82
O2N
NO2
Ph HO OH Ph
84
Pd / C, CH3SO3H
2-propanol / H2O
25 °C, 1 to 15 psi H2
90%
H2N
NH2
Ph HO OH Ph · 2 CH3SO3H
16
DMP 450

Overall, this synthesis provided diastereomerically pure DMP 450 (**16**) in 12 chemical steps, with five isolated intermediates and 36% overall yield from L-tartaric acid (**47**) (Scheme 16).

**Scheme 16**

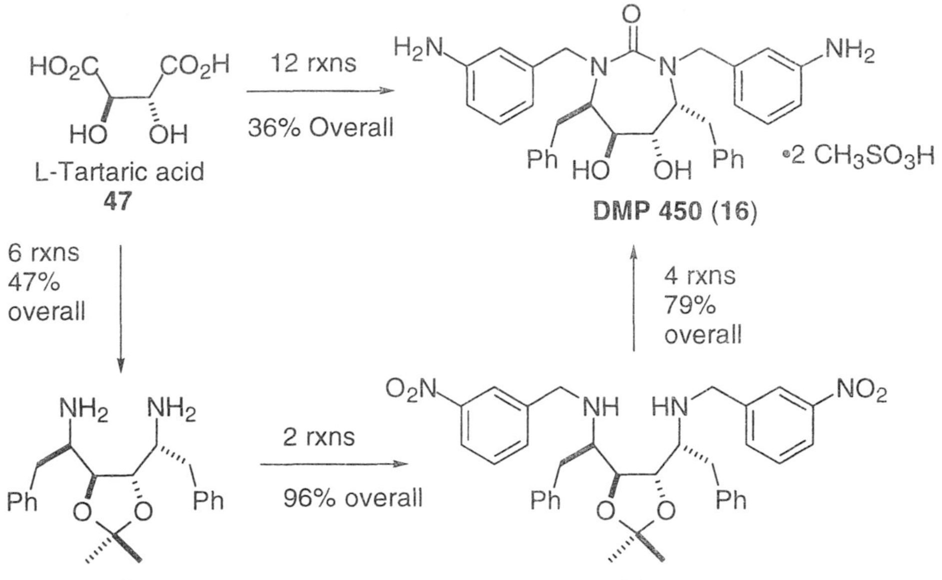

In summary, we have developed an efficient, high yield synthesis of chiral cyclic urea diols employing inexpensive starting materials, affording pure final drug substance <u>via</u> a safe, scalable synthesis without any requirement for chromatography.

**Acknowledgements**

The authors would like to acknowledge and thank Paul Aldrich, Luigi Anzalone, Chong-Hwan Chang, C. Gerald Chetkowski, J. C. Chung, William Cummings, Wayne Daneker, Wayne Davis, George DeLucca, George Emmett, Susan Erickson-Viitanen, Charles Eyermann, Joe Fortunak, Timothy Gale, C. Nicholas Hodge, Ed Holler, Qamrul Islam, David A. Jackson, Henry Jackson, Prabhakar K. Jadhav, Robert Kaltenbach III, Soo Ko, Patrick Lam, Barbara Lord, Young Lo, Phil Ma, Jay Markwalder, Dave Meloni, James Moore, T. Dzuy Nguyen, Anna Nowocin, David Nugiel, Michael Pierce, Lilian Radesca, James Rodgers, Lucius Rossano, Steven Seitz, Thomas Smyser, Louis Storace, Rodger Stringham, Ionnis Valvis, Ed Wat, Gary Koolpe, Gregory McDermott, Marianne Asaro, Robert Wilson Jr., Ralph Harris III, and all other members of the HIV protease inhibitor biology and medicinal chemistry working groups and CPR&D project teams, all of whom's efforts contributed to this manuscript.

**References**

[1] Lam, P. Y. S.; Jadhav, P. K.; Eyermann, C. J.; Hodge, C. N.; Ru, Y.; Bacheler, L. T.; Meek, J. L.; Otto, M. J.; Rayner, M. M.; Wong, Y. N.; Chang, C.-H.; Weber, P. C.; Jackson, D. A.; Sharpe, T. R.; Erickson-Viitanen, S. *Science*, **1994**, *263*, 380.
[2] Lam, P. Y. S.; Ru, Y.; Jadhav, P. K.; Aldrich, P. E.; DeLucca, G. V.; Eyermann, C. J.; Chang, C.-H.; Emmett, G.; Holler, E. R.; Daneker, W. F.; Li, L.; Confalone, P. N.; McHugh, R. J.; Han, Q.; Li, R.; Markwalder, J. A.; Seitz, S. P.; Sharpe, T. R.; Bacheler, L. T.; Rayner, M. M.; Klabe,

R. M.; Shum, L.; Winslow, D. L.; Kornhauser, D. M.; Jackson, D. A.; Erickson-Viitanen, S.; Hodge, C. N. *J. Med. Chem.*, **1996**, *39*, 3514.

[3] Hodge, C.N.; Aldrich, P.E.; Bacheler, L.T.; Chang, C.-H.; Eyermann, C.J.; Garber, S.; Grubb, M.F.; Jackson, D.A.; Jadhav, P.K.; Korant, B.; Lam, P.Y.-S.; Maurin, M.B.; Meek, J. L.; Otto, M.J.; Rayner, M.M.; Reid, C.; Sharpe, T.R.; Shum, L.; Winslow, D.L.; Erickson-Viitanen, S. *Chemistry and Biology* **1996**, *3*, 301.

[4] DMP 450 is currently under development by Avid Therapeutics, Inc., Philadelphia, PA.

[5] Kempf, D. J.; Sowin, T. J.; Doherty, E. M.; Hannick, S. M.; Codavoci, L. M.; Henry, R. F.; Green, B. E.; Spanton, S. G.; Norbeck, D. W. *J. Org. Chem.*, **1992**, *57*, 5692.

[6] Baker, W. R.; Condon, S. L. *J. Org. Chem.*, **1993**, *58*, 3277.

[7] Jadhav, P. K.; Woerner, F. K. *Bioorg. Med. Chem Lett.*, **1992**, *2*, 353.

[8] Freudenberger, J. H.; Konradi, A. W.; Pedersen, S. F. *J. Am. Chem. Soc.*, **1989**, *111*, 8014.

[9] Pierce, M. E.; Harris, G. D.; Islam, Q.; Radesca, L. A.; Storace, L.; Waltermire, R. E.; Wat, E.; Jadhav, P. K.; Emmett, G. C., *J. Org. Chem.* **1996**, *61*, 444.

[10] Leanna, M. R.; Sowin, T. J.; Morton, H. E. *Tetrahedron Lett.*, **1992**, *33*, 5029.

[11] Confalone, P. N., Smyser, T. S., Rossano, L. T., unpublished results.

[12] Nahm, S., Weinreb, S. M. *Tetrahedron Lett.* **1981**, *22*, 3815.

[13] Levin, J. I., Turos, E., Weinreb, S. M. *Synth. Commun.* **1982**, *12*, 989.

[14] No other diastereomer was observable by [1]HNMR. R. Kaltenbach US Patent 5,559,252 1996.

[15] Sasatani, S., Miyazaki, T., Maruoka, K., Yamamoto, H. *Tetrahedron Lett.* **1983**, *24*, 4711.

[16] Research initiated under contract with SRI International, Menlo Park, CA, Final Report June 21, 1993.

[17] Rossano, L. T.; Lo, Y. S.; Anzalone, L.; Lee, Y-C.; Meloni, D. J.; Moore, J. R.; Gale, T. M.; Arnett, J. F. *Tetrahedron Lett*, **1995**, *36*, 4967 references cited therein and unpublished results.

[18] Claremon, D. A.; Lumma, P. K.; Phillips, B. T. *J. Am. Chem. Soc.*, **1986**, *108*, 8265.

[19] Gilman, H.; Rosenberg, S. D. *J. Org. Chem.* **1959**, *24*, 2063.

[20] Optimized procedure based on: Screttas, C. G.; Estham, J. F.; Kamienski, C. W. *Chimia* **1970**, *24*, 109.

[21] No other diastereomers could be detected by [1]H NMR or HPLC (mixtures of diastereomers were prepared to demonstrate the utility of the HPLC method) analysis.

[22] Beck, A. K.; Bastani, B.; Plattner, D. A.; Petter, W.; Seebach, D.; Braunschweiger, H.; Gysi, P.; La Vecchia, L. *Chimia* **1991**, *45*, 238.

[23] P. N. Confalone, T. E. Smyser manuscript in preparation.

[24] Waltermire, R. E.; Anzalone, L.; Confalone, P. N.; Chung, J. C.; Davis, W. P.; Harris, G. D.; Kauffman, G. S.; Jackson, H.; Lord, B. S.; Nguyen, T. D.; Sheeran, P. J.; Smyser, T. E.; Rossano, L. T.; Stringham, R. W.; Valvis, I. I.; Zhang, L. H., Manuscript in preparation.

# Practical Enantio- and Diastereo-selective Processes for Azetidinones

T. K. Thiruvengadam, Anantha R. Sudhakar and Guangzhong Wu

*Chemical Process Research and Development, Schering Plough Research Institute, Union, NJ 07083*

Keywords: azetidinone; chiral oxazolidinone; N-acyloxazolidinone enolate; aldol reaction; imine addition; N-β-aminoacyl oxazolidinone; cyclization; valine sulfonamide; Grignard reaction; selective hydrogenolysis.

## Introduction

Azetidinone structure constitutes an important pharmacophore of potent cholesterol lowering agents operating via an entirely novel mechanism.[1] Among the many azetidinones examined, Sch 48461 (**1**) and Sch 58053 (**2**) were actively being pursued as development candidates for our cholesterol absorption inhibitor program. Known azetidinone syntheses can be divided into four major categories depending on the particular bond ($N_1$-$C_2$, $N_1$-$C_4$, $C_2$-$C_3$ or $C_3$-$C_4$) that is used to construct the ring.[2] We have developed a number of useful stereoselective methodologies for the large scale production of azetidinones in high optical purity based on $N_1$-$C_2$ and $N_1$-$C_4$ bond formation in the cyclization step. This review summarizes some of our synthesis work in this area, which utilizes a chiral auxiliary based approach exemplified by the synthesis of **1** and a catalytic asymmetric approach as illustrated by the synthesis of **2**.[3]

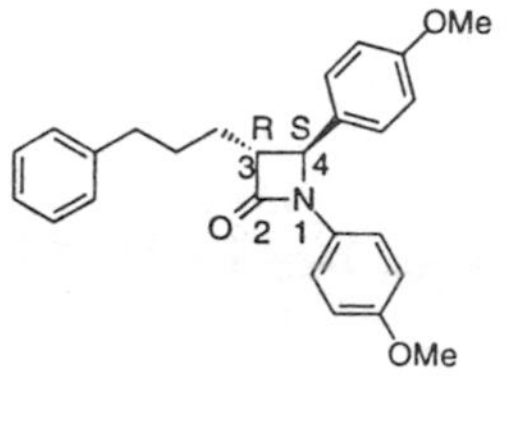

1, Sch 48461

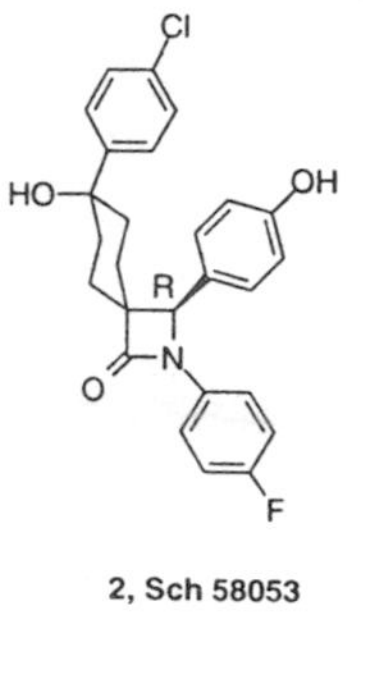

2, Sch 58053

Synthesis of Sch 48461 (1)

Sch 48461 is a *trans* β-lactam with (3R, 4S) configuration.[1e,f] It is a low melting solid (mp 42 °C) which proved very difficult to crystallize in the presence of small amounts of impurities.

As a result, it is also difficult to purge diastereomers from **1** by crystallization. Lack of a useful functional group prevented the use of chemical resolution as a viable option for its synthesis from racemic **1**. Because of these reasons, an enantioselective methodology (>99% *ee*) was necessary for a useful synthesis of Sch 48461.

Our colleagues in the Drug Discovery Group prepared **1** by condensation of imine **3** with the enolate derived from the chiral ester **4**.[1f] This method produced mainly the *cis* isomer in 93% enantiomeric excess (*ee*), which was isomerized to the desired **1** with a ratio of 4:1 (trans:cis) upon base treatment. This procedure was not very practical for large scale production of **1**, because preparative chiral HPLC is needed to obtain **1** in >98% *ee*, as required for the clinical program.

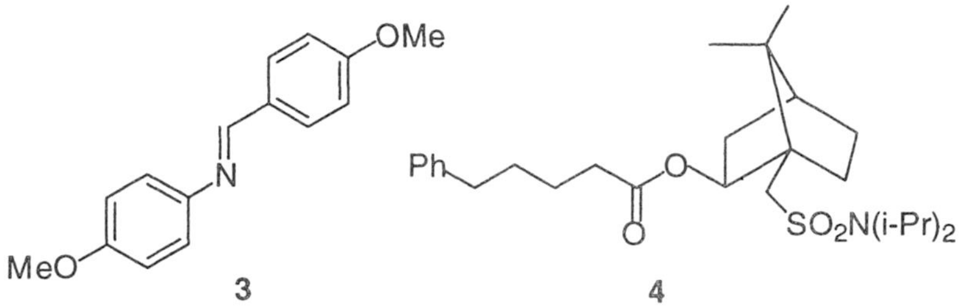

Our strategy is to pre-fix the absolute configuration of C-3 and C-4 (so that cyclization will lead directly to *trans* Sch 48461) at an early stage of the synthesis, where the other isomers can be purged by crystallisation. The later steps should maintain optical integrity leading to >99% *ee* of Sch 48461 even in the crude reaction mixture. This strategy was successfully applied in the following two approaches in developing practical enantioselective syntheses of Sch 48461.

**Aldol Approach**

Our first approach, called the aldol approach, is based on the retrosynthetic analysis shown in Figure 1. The key steps are the aldol reaction of the acid derivative **C** (Xc = chiral auxiliary) with

**Figure 1. Aldol Approach**

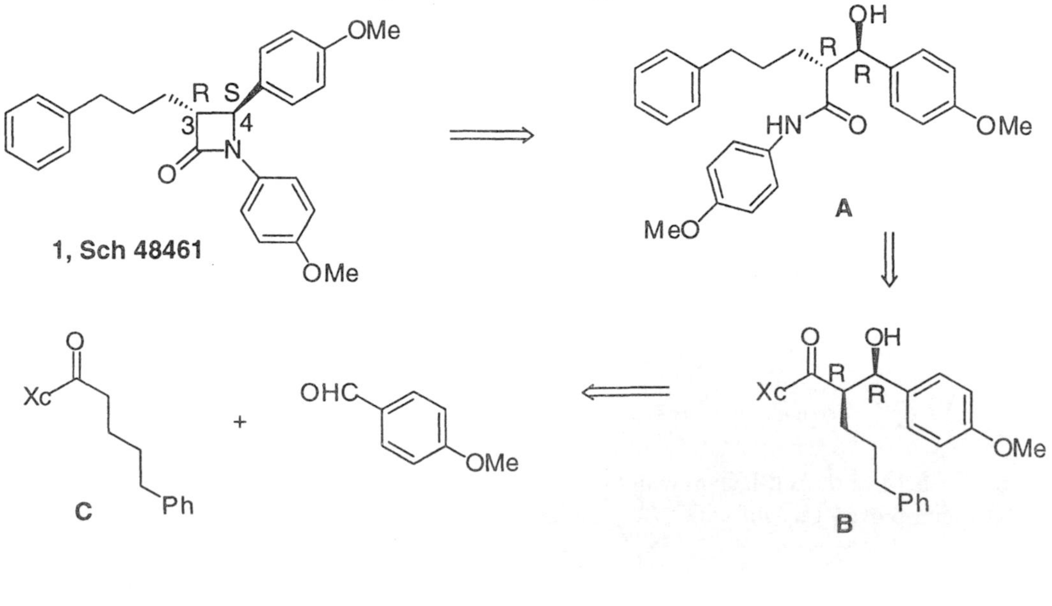

*p*-anisaldehyde to give the (R,R)-aldol (*syn* aldol) product **B** and the cyclization of the (R,R) β-hydroxyamide **A** to the *trans* β -lactam, Sch 48461, (**1**).

A number of useful methodologies were available for the preparation of enantiomerically pure *syn* aldols.[4] Among them, aldol reactions of chiral N-acyloxazolidinones, developed by Evans *et. al.*,[5] and that of chiral sultams, developed by Oppolzer *et. al.*,[6] are very highly stereoselective. To the best of our knowledge, none of these methodologies have been used on a large scale until our work. We decided to use the chiral oxazolidinone as the chiral auxiliary because both enantiomers were readily available.

The approach used to prepare the initial requirement of 500 g for 2-week toxicology studies used (R)-benzyloxazolidinone **5** as the starting material (Scheme 1). Treatment of **5** with n-BuLi followed by 5-phenylvalerylchloride (prepared from 5-phenylvaleric acid) gave the N-acyloxazolidinone **6** as a low melting solid. This compound was treated with freshly prepared di-n-butylboron triflate and Hunig's base (DIPEA) at -78 °C in $CH_2Cl_2$ followed by the addition of *p*-anisaldehyde according to Evans's procedure [5a] gave a mixture (>99:1) of two *syn* aldol products from which the pure RRR isomer **7** was isolated by crystallization in 85-87% yield from **5**. Hydrolysis of **7** (LiOH, $H_2O_2$)[7] followed by coupling with *p*-anisidine (DCC, HOBT) gave the β–hydroxyamide **9** in 80% yield.[8]

**Scheme 1**

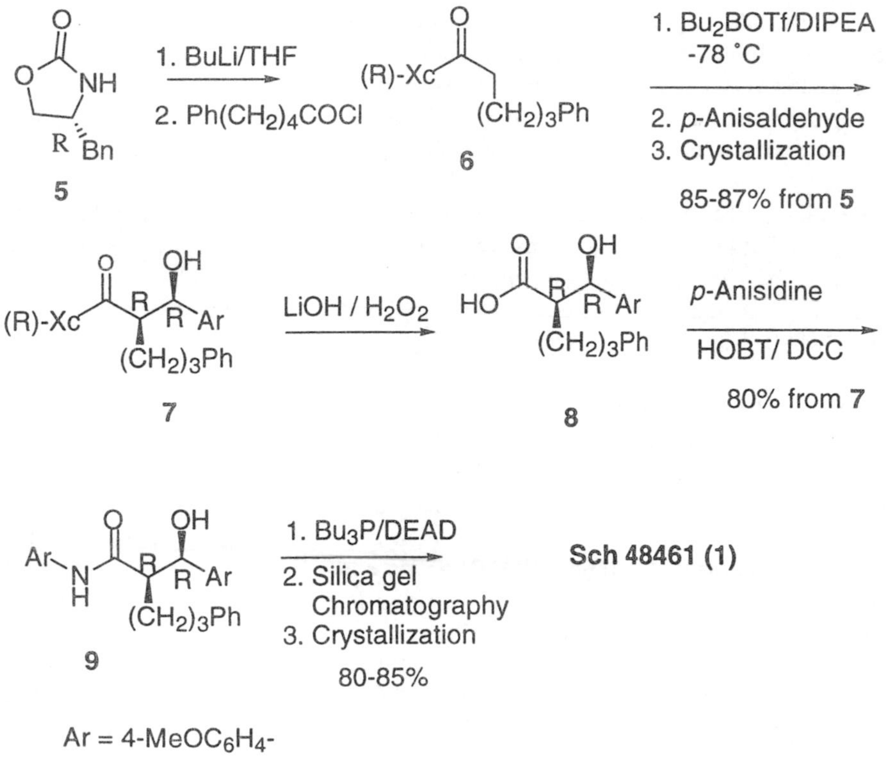

Ar = 4-$MeOC_6H_4$-

For the initial delivery, we evaluated two methods for azetidinone ring formation: (1) mesylation of the hydroxy group followed by cyclization,[10a] and (2) cyclization under Mitsunobu reaction conditions.[10b] Although the mesylation methodology was shown to be useful by Evans and coworkers for azetidinone ring formation, mesylation of **9** resulted in loss of optical integrity due to the lability of the *p*-methoxybenzylic center. However, treatment of the amide **9** with triphenylphosphine and diethylazodicarboxylate (DEAD) in THF under Mitsunobu reaction conditions gave Sch 48461 (**1**) in about 33% yield, but in high optical purity. Two elimination products were observed (structures **D** and **E** below) as impurities. Use of the more nucleophilic tri-n-butylphosphine minimized the formation of these impurities and resulted in 80-85% yield of Sch 48461, which was purified by silica gel chromatography (twice) and crystallized from ether-hexane. The overall yield of Sch 48461 was 55-60% from **5** and the enantiomeric purity was very high (>99.9% *ee*).

When the development program advanced further, we investigated alternate synthetic procedures for the following reasons: (1) n-BuLi was somewhat undesirable in the acylation step; (2) dibutylboron triflate is pyrophoric and expensive and introduces safety issues; (3) it was desirable to eliminate the use of $Bu_3P$ and (toxic) DEAD in the cyclization, and (4) multiple chromatographies were needed to obtain pure **1**. Also the low temperatures (-78 °C) required for aldol reactions were not readily available in our pilot plant.

The literature procedures for the acylation of oxazolidinones involve deprotonation with n-butyllithium or sodium hydride followed by treatment with an acid halide or mixed anhydride.[5b] We discovered that the acylation can be effected in high (>90%) yield simply by treating oxazolidinones with acid chlorides or mixed anhydrides in the presence of triethylamine (TEA) (1.5 eq.) and cat-DMAP (5 mol%).[11]

As an alternative to the boron promoted aldol reaction, we investigated the $TiCl_4$ mediated reaction. Evans and coworkers reported[13] that treatment of N-acyloxazolidinones with aldehydes in the presence of $TiCl_4$ (1.1 eq.) and a base (DIPEA or tetramethylethylenediamine (TMEDA), 2 eq.) gave high yields of *syn* aldol products in high diastereomeric excess (*de*). Treatment of **6** under these conditions ($TiCl_4$, 2 eq DIPEA) with *p*-anisaldehyde at -25 °C or at -78 °C gave, however, a mixture of two *syn* aldols in 1:2 ratio (RRR/RSS). In contrast to the reported work with similar substrates, the major aldol product was the *undesired* RSS diastereomer. However, when 2.2 equivalent of TMEDA was used in place of DIPEA, the reaction at -25 °C gave the desired RRR isomer **9** in excellent yield (>90%) and selectivity (>99:1). This reaction worked very well in the laboratory on a "small" scale (100-500 g), but upon scale-up to about 20-40 kg in the pilot plant, incomplete reactions (10-15% starting material) were often experienced. Although this did not impact the quality of the product, because both starting materials were removed completely during the crystallization step, it was important that we understood the role of TMEDA before further scale-up.

Two equivalents of TMEDA were necessary to obtain the desired selectivity. We speculate that the first equivalent of TMEDA simply acts as a base in deprotonating the complex of **6** with TiCl$_4$ to form the 'chelated' enolate **F** whereas the second equivalent of TMEDA causes the disruption of chelation and results in the formation of the 'open' enolate **G** as shown in Figure 2. In the absence of this disruption of the chelation, the aldol reaction proceeds substantially via pathway **A** to give the undesired RSS diastereomer predominantly. *Clearly the second equivalent of TMEDA was critical for the high selectivity of our aldol reaction.*[15]

**Figure 2**

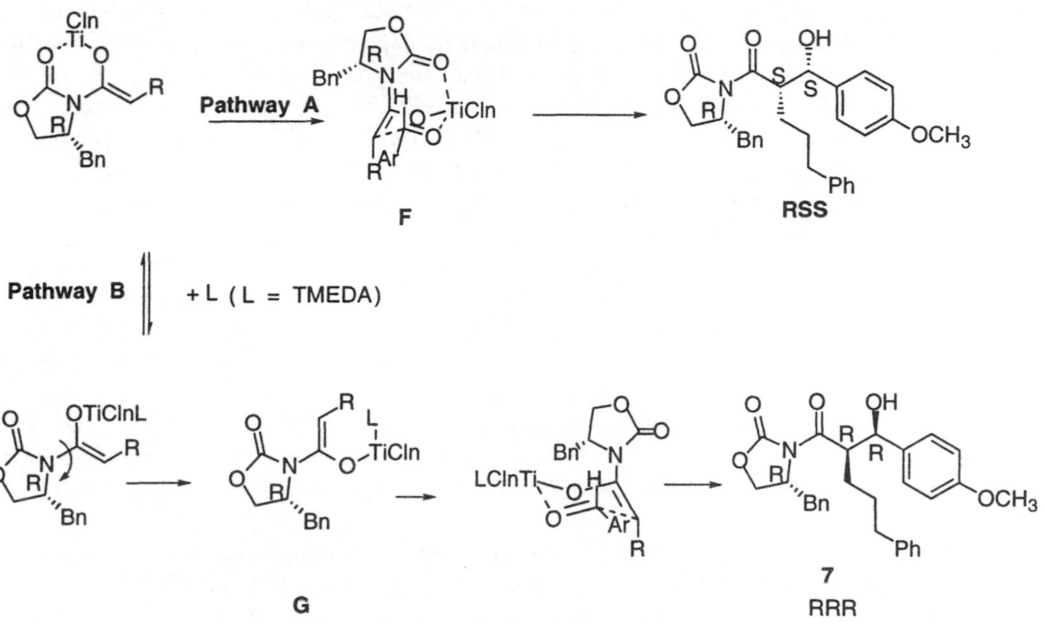

In accordance with this hypothesis, when the initial deprotonation was carried out with triethylamine (1.1 eq.) and then TMEDA (1.1 eq.) was added prior to the aldehyde addition, the same high selectivity for the RRR isomer was obtained. Using this protocol for the reaction, the Lewis acidity of TiCl$_4$ is not reduced to the same extent as that using 2.2 eq. TMEDA and the reactions went almost to completion (>97%) with >97:3 selectivity on large scale (40 kg) (Scheme 2).[15]

Hydrolysis of the aldol product **7** to the acid **8** in THF-water was straightforward on a large scale. A solution of the hydroxy acid **8** was carried directly onto the coupling step with *p*-anisidine using essentially the laboratory procedure to give the amide **9**.

Scheme 2

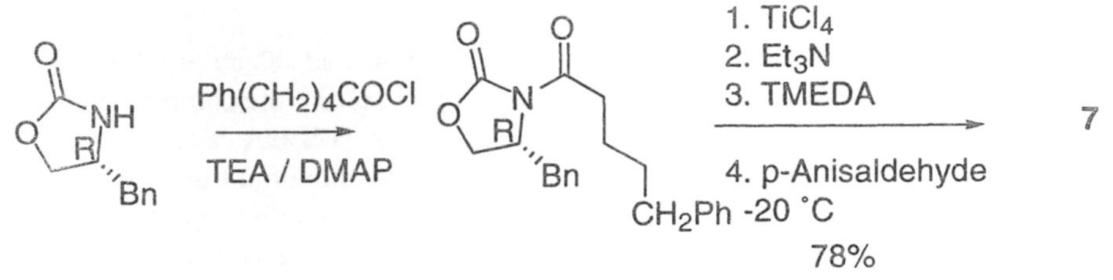

We have investigated in great detail replacement of the Mitsunobu reaction for the critical cyclization to the azetidinone. Due to the problems encountered earlier with the mesylate leaving group, it was felt that a *poorer* leaving group might prevent opportunities for epimerization of the benzylic center. Hindered ester groups have been used as effective leaving groups in macrocyclization to form macrolides.[16] To test this approach, the hydroxyamide **9** was treated with 2,4,6-trichlorobenzoyl chloride in the presence of sodium hydride in $CH_2Cl_2$/DMF. Depending on the amount of NaH used, either the ester **10a** or the azetidinone **1** was obtained in good yield and enantiomeric purity. Having established the feasibility of this reaction, efforts were made to develop this into a practical process. We chose the 2,6-dichlorobenzoate as the leaving group because it was more readily available. The NaH/Ar'COCl procedure was not attractive for large scale preparation of the ester **10b**. The hydroxyamide **9** did not react readily with 2,6-dichlorobenzoyl chloride under normal acylating conditions such as using TEA and catalytic DMAP. Finally, the esterification was achieved by treatment of **9** with 2,6-dichlorobenzoyl chloride in $CH_2Cl_2$ - 50% NaOH under phase transfer catalysis (PTC)[17] conditions using catalytic amount (5 mol %) of tetrabutylammonium hydrogen sulfate (TBAH), to give the ester **10b** containing about 2% of the azetidinone **1** (Scheme 3). Crystallization purged all impurities from the crude product and gave **10b** in 85-87% yield from **9**.

The formation of the azetidinone in the acylation under PTC hinted that even the cyclization might be carried out under PTC.[18] However, treatment of the ester **10b** in $CH_2Cl_2$ with 50% NaOH in the presence of catalytic amount (5 mol%) of TBAH, gave only a low yield (<5%) of Sch 48461 **1**; mainly starting material was recovered unchanged. Changing the phase transfer catalyst did not alter the outcome significantly. It was discovered that the low conversion was due to the high solubility of the $Bu_4N^+Ar'CO_2^-$ in the organic media. In other words, $Bu_4N^+Ar'CO_2^-$ and (not $Bu_4N^+OH^-$) was transported to the organic media. In support of this hypothesis, it was found that use of stoichiometric amounts of PTC resulted in high conversion and excellent yield (90-95%) of the product (Scheme 3). After screening a number of PTC, benzyltriethylammonium chloride (BTAC) was chosen for scale-up work because it gave the cleanest reaction. Main impurities in this reaction were starting material (~2%) and the "hydrocarbon" (~2%) impurity **E** noted above. Less than 0.2% loss of stereochemical integrity was observed under this strongly alkaline condition.

## Scheme 3

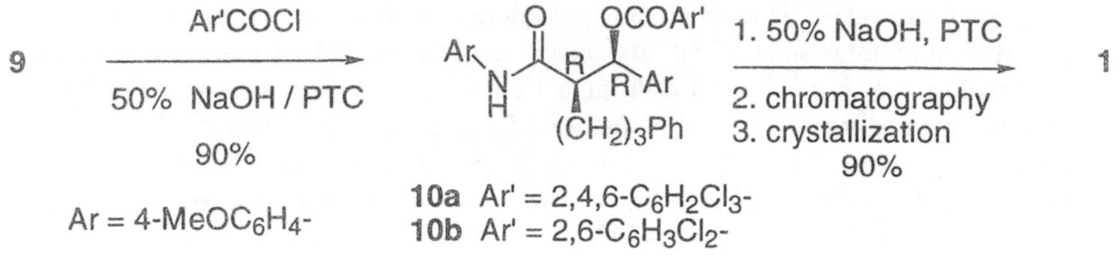

9 → ArCOCl / 50% NaOH/PTC / 90% → (intermediate 10a/10b) → 1. 50% NaOH, PTC / 2. chromatography / 3. crystallization / 90% → 1

Ar = 4-MeOC₆H₄-

10a  Ar' = 2,4,6-C₆H₂Cl₃-
10b  Ar' = 2,6-C₆H₃Cl₂-

All our attempts to purge the remaining starting material **10b** (2%) by crystallization were not successful. A maximum of 0.5% **10b** was tolerated in the crystallization. As a result, we were forced to use silica gel chromatography to purify the crude product. Indeed, this procedure was implemented on a large scale and 10-20 kg lots of pure Sch 48461 were produced by this process.

As described above, use of the hydrophobic benzoate leaving group necessitated the use of a stoichiometric amount of the PTC in the cyclization step. It was clear that a *hydrophilic*, yet poor, leaving group might result in a cyclization reaction that is truly *catalytic* in PTC. Because acylation is catalytic in PTC, this might result in a single step conversion of the hydroxyamide **9** to the azetidinone **1** using catalytic amounts of the phase transfer catalyst. Towards this goal, the hydroxyamide **9** was treated with diethyl chlorophosphate and 50% NaOH in CH₂Cl₂ at room temperature in the presence of catalytic amount of TBAH.[19] Under these conditions, the azetidinone **1** was obtained as the major product (Scheme 4).[20] In this case also, use of BTAC

## Scheme 4

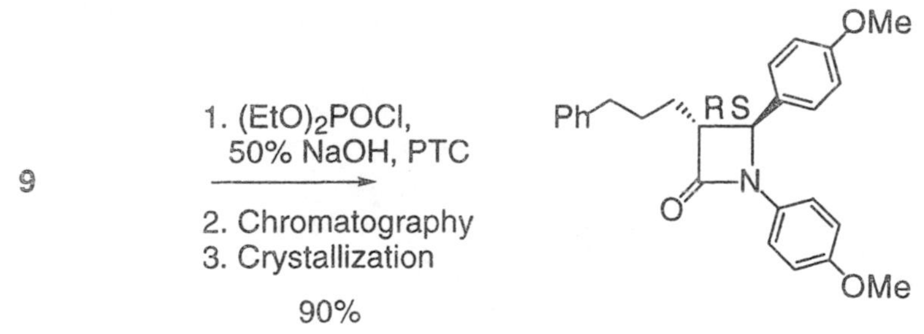

9 → 1. (EtO)₂POCl, 50% NaOH, PTC / 2. Chromatography / 3. Crystallization / 90% → **1, Sch 48461**

resulted in cleaner reactions. The main impurities in this reaction were the elimination product **D** (0.2% in laboratory experiments) and the hydrocarbon impurity **E** (~1-2%). On a large scale (10-20 kg), however, the elimination product increased to about 2% and was not purged during the crystallization. Unlike the ester **10b**, this amide impurity was completely removed during filtration of a toluene solution of the crude product through a short silica gel column. Concentration of the filtrate and crystallization (see below) from a mixture of methyl t-butyl ether (MTBE) and isopropanol (IPA) gave pure Sch 48461 in about 80-90% yield from **9**.

One of the most challenging issues that we faced during the development of Sch 48461 was its crystallization. Because of its low melting point, a number of crystallization conditions were not successful even when pure **1** was used. The initial procedure involved addition of a solution of

**1** in MTBE to heptane. The product oiled out and the mixture was cooled and stirred (1-4 days) until crystallization was complete. On a laboratory scale, this was successful but this method was irreproducible on a large scale. In addition, this procedure often resulted in formation of hard lumps which were difficult to break and difficult to transfer out of the reactor. Effect of temperature, rate of addition, seeding, etc. had no significant impact on the oiling phenomenon. Finally, it was discovered that slow addition of an MTBE solution of **1** to a pre-cooled (0 to 5 °C) suspension of 5% seeds in isopropanol resulted in excellent reproducible crystallization. This procedure not only scaled up well, but resulted in purging of the hydrocarbon impurity from 2% to <0.2% levels.

The synthesis of Sch 48461 using the aldol approach can be summarized in Scheme 5. The overall process involves five chemical steps, two isolated intermediates (**7** and **9**), one silica gel plug filtration and gave an overall yield of about 50-55% based on the chiral auxiliary.

**Scheme 5**

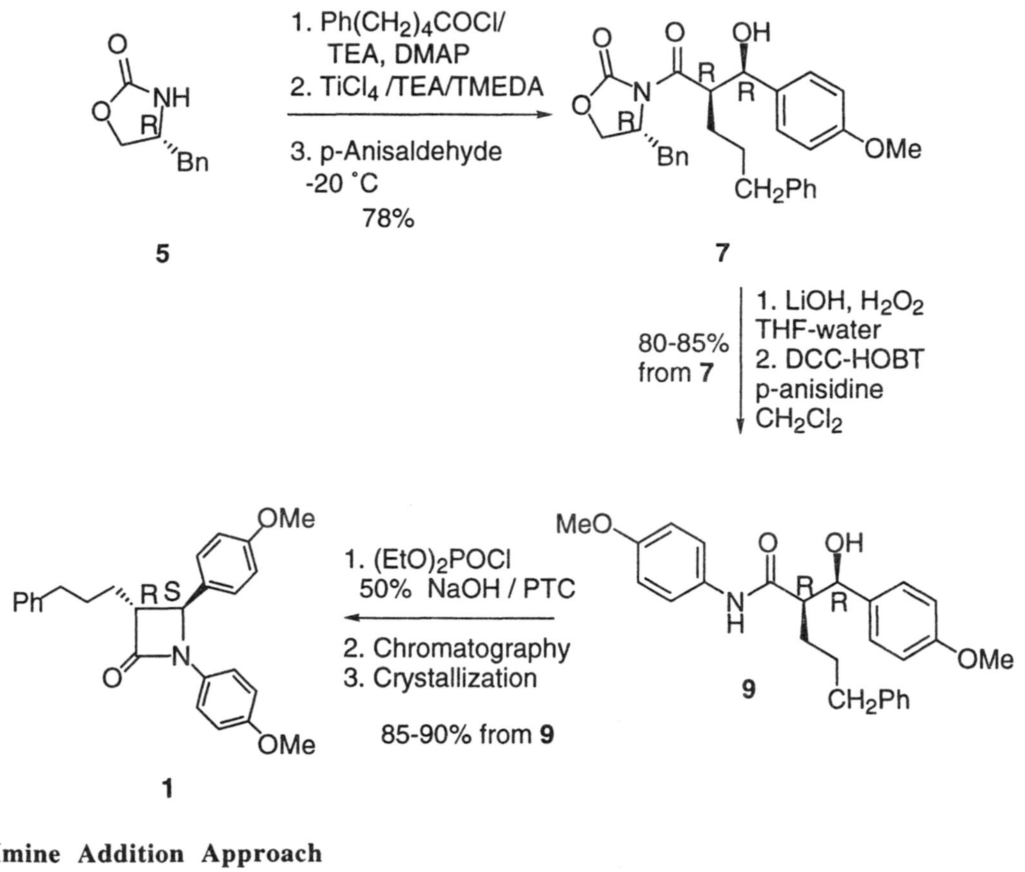

## Imine Addition Approach

All of the above processes although simple, high yielding and practical to scale up, still did not meet the criteria for a commercial process. These processes are longer (5 steps), require chromatographic purification and use the expensive chiral oxazolidinone of unnatural configuration. Therefore a much simpler, shorter and cost efficient enantioselective commercial process was invented and developed. This approach, called the imine addition approach, is based on the retrosynthetic analysis shown in Figure 3. The key reactions are the imine addition to a metal enolate of N-acyloxazolidinone followed by an intramolecular cyclization leading to a β-lactam.

While aldehyde addition to metal enolates (lithium, boron, titanium) of N-acyloxazolidinones is widely reported, imine addition to these imide enolates has not been fully studied.[21] As noted above, the first synthesis of Sch 48461 involved an imine addition, although this led primarily to the *cis* isomer, most likely as a result of the intermediacy of the (E)-ester enolate.[3b] Since enolization of N-acyloxazolidinones results in (Z)-enolate,[5a] imine addition should give the desired *trans* azetidinone. However, when the lithium enolate of **6** was treated with imine **3** in THF at -78 °C, no product was obtained.

### Figure 3: Imine Addition Approach

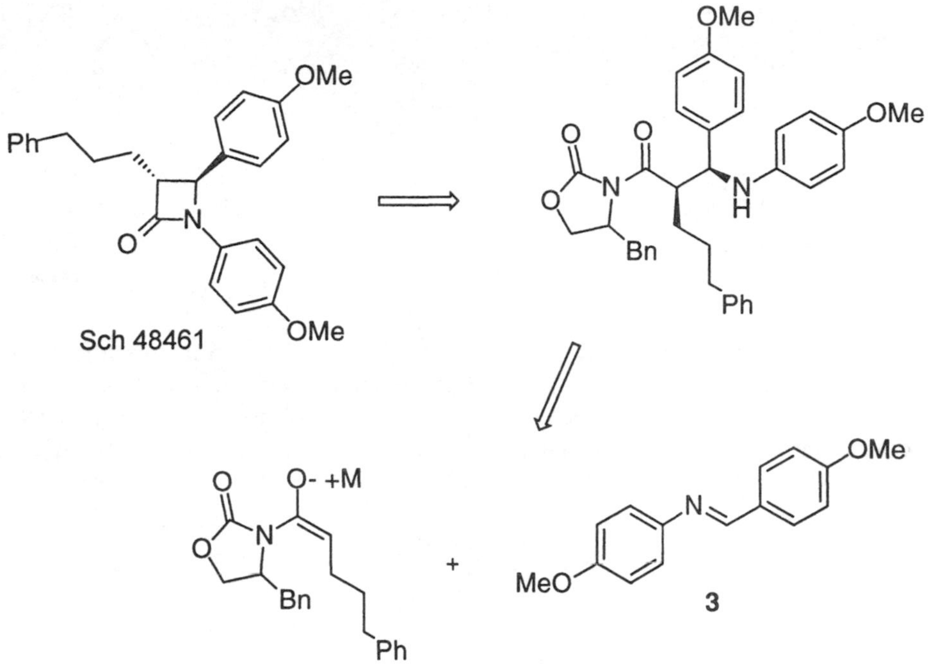

Sch 48461

Despite this discouraging result, the imine addition reaction was tried on the titanium enolate of **6**, generated by the treatment of **6** with TiCl$_4$ (1 eq.) with TMEDA (2 eq.) in methylene chloride at -20 °C as in the aldol reaction. The reaction gave two major imine addition products in the ratio of 95:5 (${}^1$HNMR and HPLC). The major isomer was assigned the *anti* configuration based on a coupling constant of J$_{2,3}$ = 7.8 Hz. The minor component (J$_{2,3}$ = 4.9 Hz) was assigned the *syn* structure.[22] Of all the four possible isomers (**12**, **13**, **14** & **15**), compounds **14** and **15**, previously prepared[23] from (RRR)-aldol product **7**, were formed in only trace amounts (Scheme 6). This clearly suggested that the configuration at C-2 in the two imine addition products is opposite to that in the aldol product **7** and is therefore (S). This C-2 configuration is opposite to that required for Sch 48461. Based on these facts, the imine addition products were assigned the structures **12** and **13**.

Scheme 6

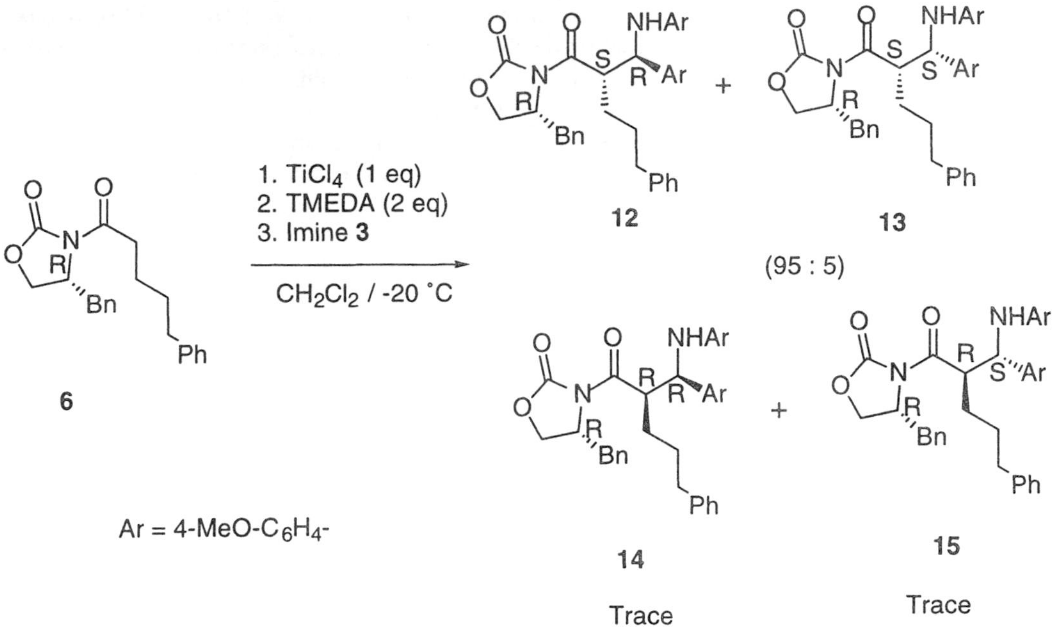

Ar = 4-MeO-C$_6$H$_4$-

6

1. TiCl$_4$ (1 eq)
2. TMEDA (2 eq)
3. Imine 3

CH$_2$Cl$_2$ / -20 °C

12                    13

(95 : 5)

14                    15

Trace                Trace

The observed stereoselectivity of this reaction can only be explained if the reaction takes place via a chelated transition state **H** (Figure 4). Although the enolate was prepared in exactly the same way as in the aldol reaction, the stereoselectivity, unlike the aldol reaction, is independent of the nature of the tertiary amine used in the enolate formation. Thus use of TMEDA, triethylamine or DIPEA for enolate formation gave mainly **12** and **13**.

Figure 4

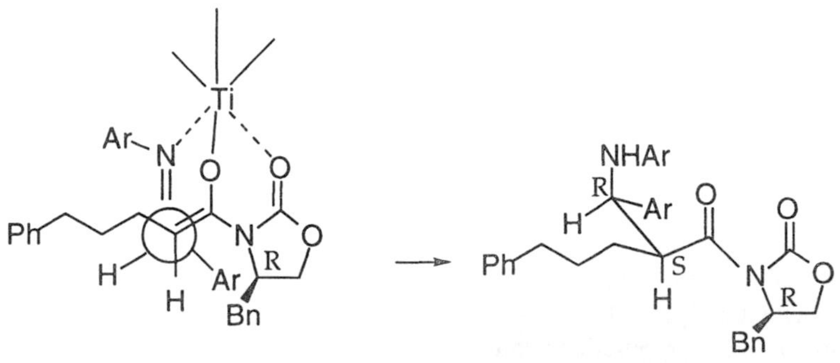

H                    12

To prepare Sch 48461, the above reaction was repeated with the (S)- benzyloxazolidinone to give the corresponding *anti* and *syn* imine addition products (enantiomers of **12** and **13**) in 95:5 ratio. However, all attempts to purge the minor *syn* isomer by crystallisation were not successful. In order to change crystallisation properties, the reaction with other chiral oxazolidinones were investigated.

Thus, commercially available (S)-4-phenyloxazolidinone **16** was acylated with 5-phenylvaleryl chloride (according to the procedure described above for **7**) to give **17** in quantitative yield. The titanium enolate of **17** was prepared using 1 mol eq of $TiCl_4$ and 2 mol eq of DIPEA in $CH_2Cl_2$ at below -20 °C . The addition of imine **3** to the titanium enolate at -20 °C gave, after quenching with 10% aqueous tartaric acid solution at room temperature, a 95:5 ratio of *anti* and *syn* products **18** and **19**. To our delight, the undesired *syn* isomer **19** was purged by a single crystallisation from ethyl acetate.

**Scheme 7**

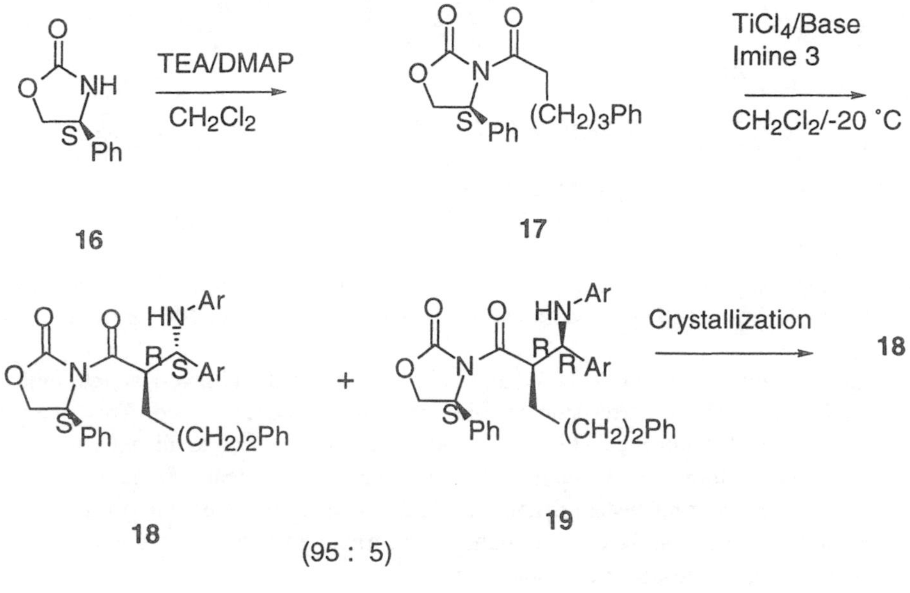

Ar = 4-MeO-$C_6H_4$-

In the early phase of the development work, we noticed that the yield of the imine addition reaction varied from 30 to 80%. Reactions quenched at room temperatures gave the lower yields and lower conversions. When the reactions were quenched with aqueous tartaric acid solution at 0 °C, consistently 79 to 83% solution yield of products **18** and **19** were obtained. Apparently, the initial adduct **20A** undergoes "retro-aldol" reaction at higher temperatures (Scheme 8) to give the starting materials **17** and **3**. According to this hypothesis, protonation of the Ti-N bond of the complex at lower temperature would prevent the retro-aldol reaction *via* **20B** and might result in higher yield. Indeed, when the cold reaction mixture was quenched with glacial acetic acid at -20 °C, and then worked up as usual, the reactions gave 90-93% yield of the products in 87:13 ratio. We believe the higher selectivity obtained under room temperature quench to be the result of selective decomposition (*via* retro-aldol pathway) of the minor isomer. The minor isomer **19** was

**Scheme 8**

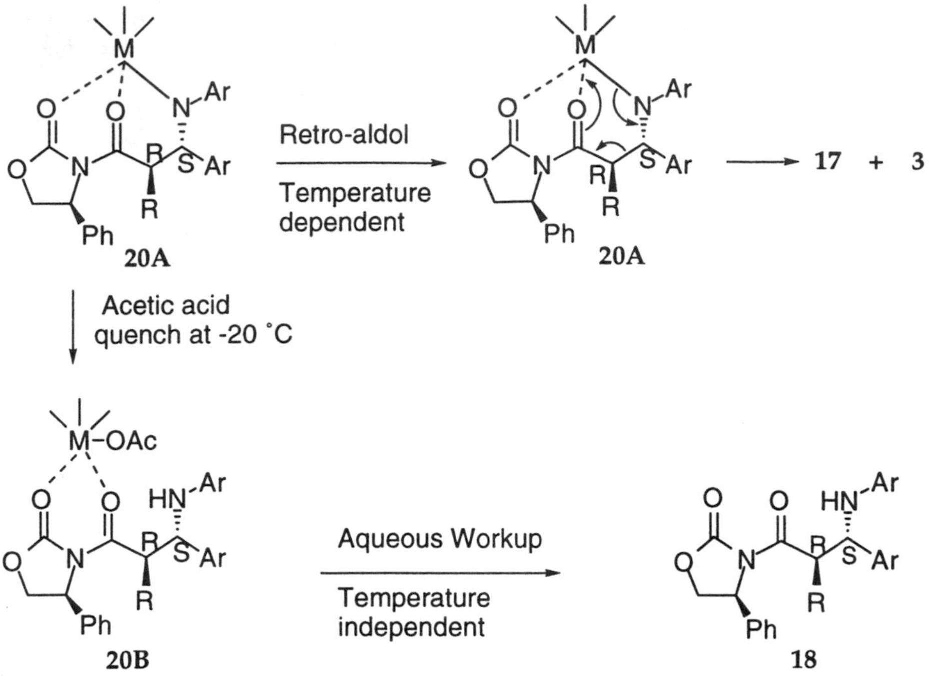

purged by crystallisation of the crude product to give enantiomerically and diastereomerically pure **18** in 73 to 75% yield from **16**. This process was successfully scaled up to 100 kg in the plant.

For this imine addition reaction to be of any value, an efficient and practical intramolecular cyclization of **18** leading to the corresponding β-lactam **1** had to be developed. This cyclization requires the generation of a nitrogen anion followed by an intramolecular displacement of the oxazolidinone moiety to form the β-lactam. Intramolecular displacement of the oxazolidinone moiety leading to a β-lactam ring system is unknown and the outcome is not obvious, since the nitrogen anion can attack either at the exo- or endo-cyclic carbonyl group to give the β-lactam **1** or 9-membered ring product **21**, respectively (Scheme 9).[24]

**Scheme 9**

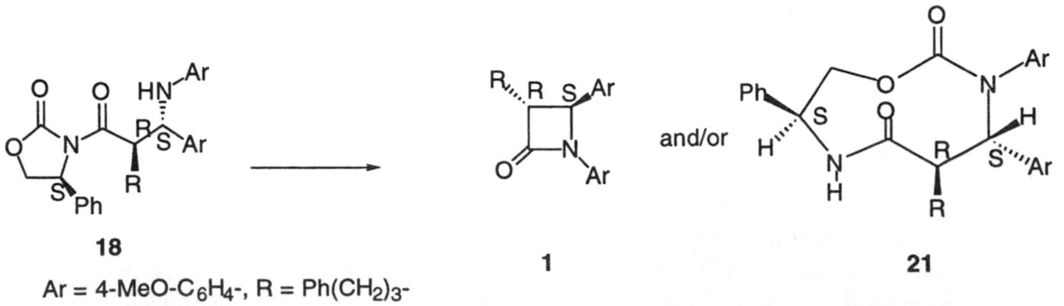

Ar = 4-MeO-C$_6$H$_4$-, R = Ph(CH$_2$)$_3$-

After considerable experimentation with various bases and solvents, it was discovered that the cyclization can be achieved in 87% yield using sodium or lithium bistrimethylsilylamide in methylene chloride at -78 °C. This reaction gave enantiomerically pure Sch 48461, which met the criteria of maintaining the stereochemical integrity during the cyclization, and chiral auxiliary as the by-product, along with small amounts of impurities such as **21** (Scheme 10). Purification of the product, however, still required silica gel chromatography. Interestingly, changing the cyclization solvent to THF at 0 °C gave a lower yield (48%) of **1** along with **21** (20%).

**Scheme 10**

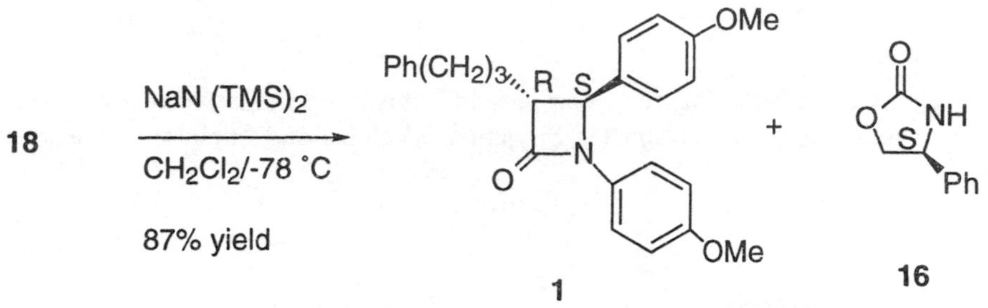

To further minimize the formation of impurities and to eliminate chromatography, efforts were focused on generating a nitrogen anion of different nucleophilicity. One such approach involved silylation of the amino group of **18**, followed by treatment with a fluoride ion to generate the nitrogen anion (Scheme 11). Initial experiments with silylation of **18** with N,O-bistrimethylacetamide (BSA) in refluxing THF followed by treatment with 1 mol eq of tetra-n-butylammonium fluoride trihydrate (TBAF•3H$_2$O), gave **1** in only 5% yield; the major product was the hydrolysis product **22** formed by the hydrolysis of the endocyclic carbonyl of **18** with water present in the reagent.

**Scheme 11**

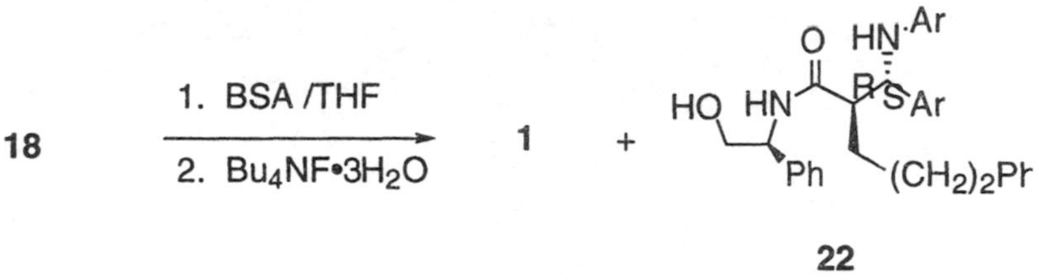

Although the yield was only 5%, it was encouraging that the hypothesis worked. Therefore, the above reaction was repeated with a catalytic amount (10 mole %) of TBAF•3H$_2$O to minimize the amount of water present. Indeed, the yield of product **1** increased to about 90% with <1% of compound **22**. After experimentation with different solvents, we found toluene and CH$_2$Cl$_2$ to be the solvents of choice for minimizing(<0.5%) epimerization. Under the optimum condition shown in Scheme 12, Sch 48461 was obtained in almost quantitative yield with the chiral auxiliary as the by-product. In contrast to the cyclization using strong bases, this reaction is conveniently carried out at ambient conditions and is complete within an hour.

## Scheme 12

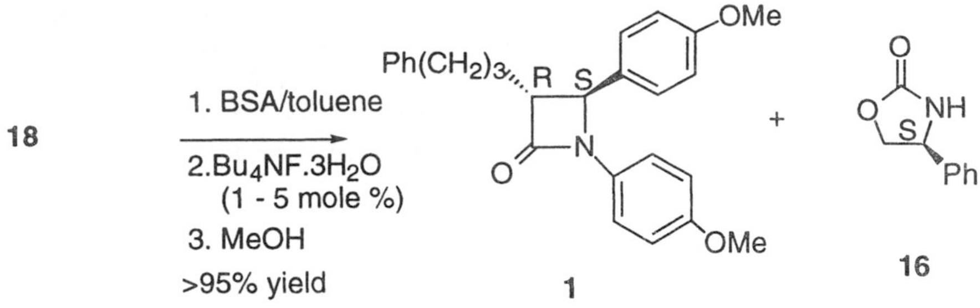

The mechanism shown in Figure 5 is proposed for this novel intramolecular cyclization. Two observations support this mechanism. (1) Silylation is critical and important - treatment of 18

## Figure 5

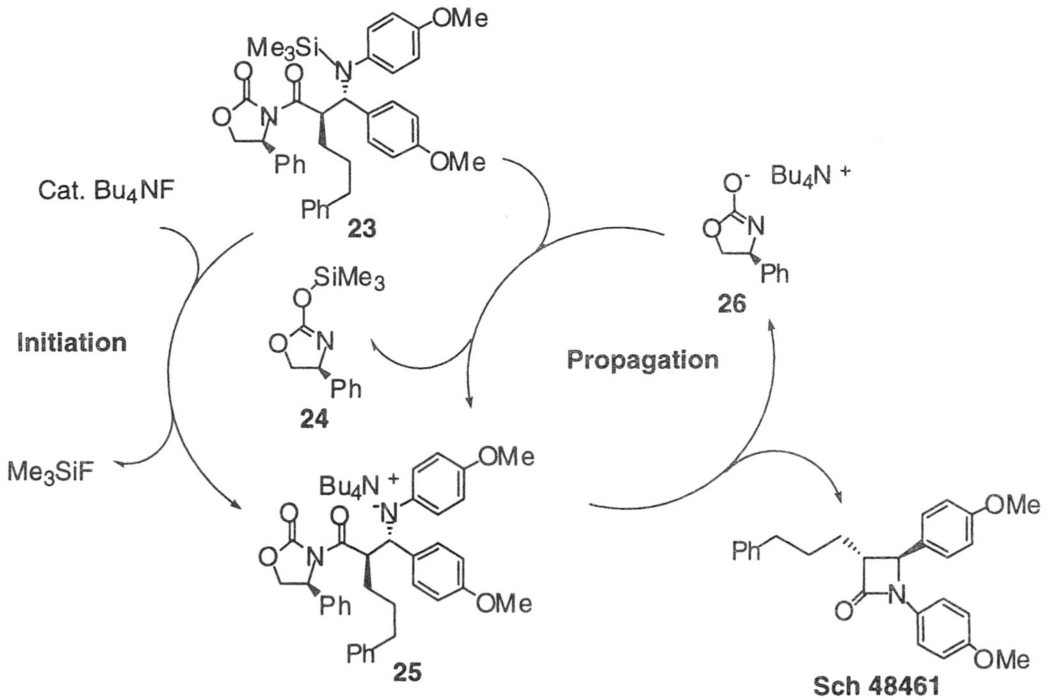

with TBAF.3H$_2$O alone gives very little β-lactam product, and (2) the cyclization can take place in the absence of fluoride ion as long as a catalytic amount of tetrabutylammonium salt 26 of the chiral auxiliary was added to the silylated 18.

As a result of this clean cyclization reaction, it became possible to isolate pure Sch 48461 by simple crystallisation of the crude product without using any column chromatography. The reaction mixture containing Sch 48461 and silylated chiral auxiliary 24 was treated with methanol to effect the desilylation of 24. Then it was concentrated to crystallize 16 along with acetamide and filtered. Pure chiral auxiliary **16** was recovered in 80 - 85 % yield from the crude solid mixture and reused. The filtrate containing Sch 48461 and <10 mol % of **16**, after aqueous work-up, was concentrated to an oil which was then crystallized from *t*-butyl methyl ether and isopropanol at -5 to -8 °C to give Sch 48461 in 87% isolated yield from **18**. This intramolecular cyclization was also scaled up successfully on 100 kg scale.

The overall manufacturing process for Sch 48461, summarized in Scheme 13, involves just one isolated intermediate (**18**) and three steps from the chiral auxiliary **16**. In addition, no chromatography was necessary to obtain very high purity (>99.5%) product. The overall yield of **1** was 55-60% based on the chiral auxiliary **16**.

Scheme 13

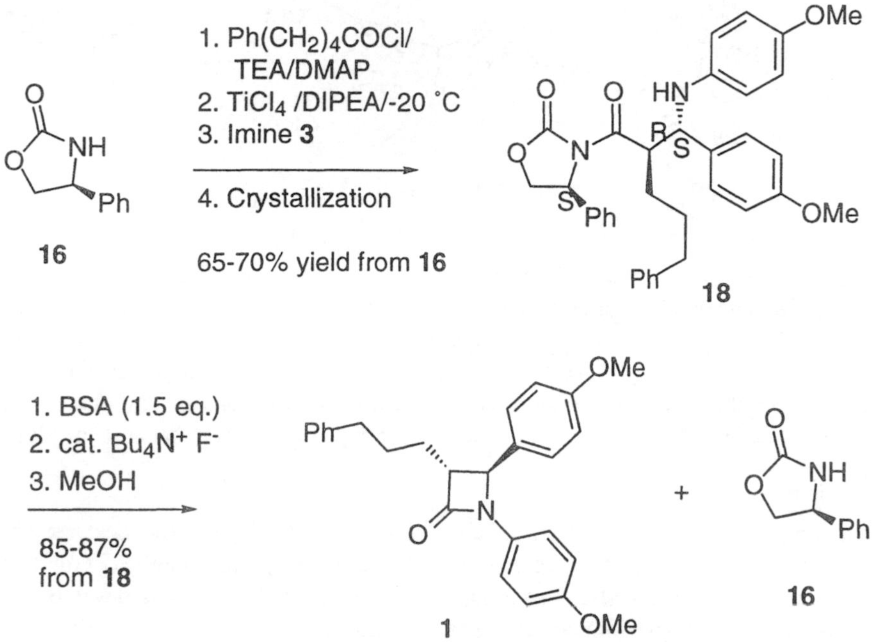

## Synthesis of Sch 58053 (2)

This section describes a practical synthesis of the spiro-fused azetidinone **2** based on catalytic asymmetric aldol condensation. Using the imine addition approach discussed above, our colleagues in research reported the first synthesis of **2**.[25] An alternate *catalytic* approach to **2** was developed based on the retrosynthetic analysis shown in Figure 6. We sought to establish the chiral center on the cyclohexyl ring *via* a Grignard addition to ketone **30** while the β-lactam ring would be constructed by an intramolecular cyclization of hydroxyamide **29** as described above for Sch 48461. The key intermediate **28** would be obtained *via* a *catalytic* enantioselective aldol condensation of silyl enol ether **27** and an aromatic aldehyde.

**Figure 6**

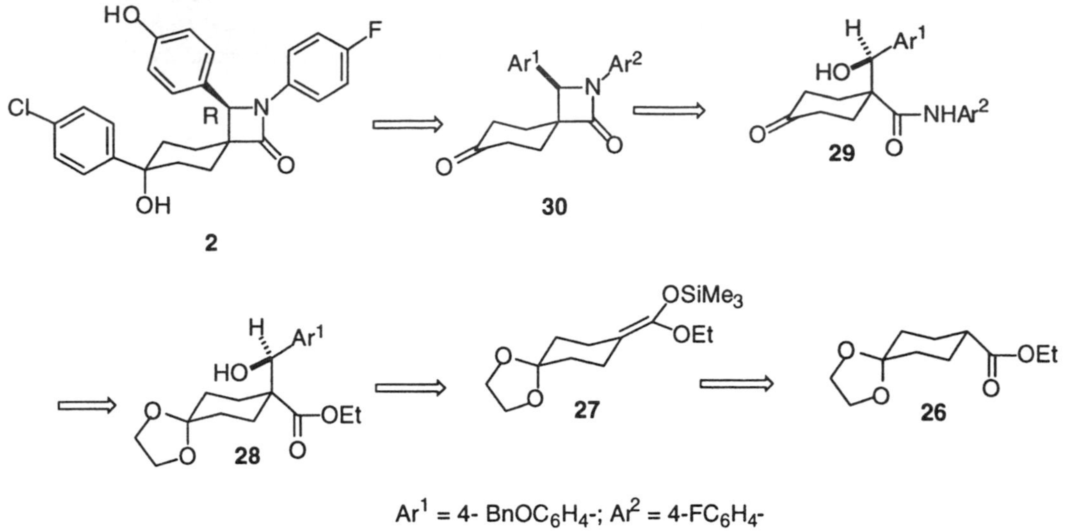

$$Ar^1 = 4\text{-} BnOC_6H_4\text{-}; \ Ar^2 = 4\text{-}FC_6H_4\text{-}$$

We first studied the catalytic Mukaiyama-type aldol condensation[26] to establish the hydroxy chiral center in **28**. The catalytic asymmetric aldol reaction shown in Figure 6 emerged only in the last few years and only a limited number of examples has been reported.[27] Catalysts employed in the literature can be divided into two major categories, the amino acid derivatives[27] and the artificially designed chiral ligands exemplified by the binaphthol-titanium complexes.[28] From an industrial process point of view, we chose the former type of catalyst because of their ease of preparation.

Among the various catalysts screened, Kiyooka catalyst **31**[29] gave the best result and was selected for further studies. For the preparation of **31**, a one-step procedure (Scheme 14) starting from commercially available D-valine and 2-naphthalenesulfonyl chloride was developed in lieu of the three-step method.[29] After work-up, **31** was isolated in 90% yield and in 99% purity. The active catalyst **31a** is formed *in situ* by reaction with BH₃•THF.

**Scheme 14**

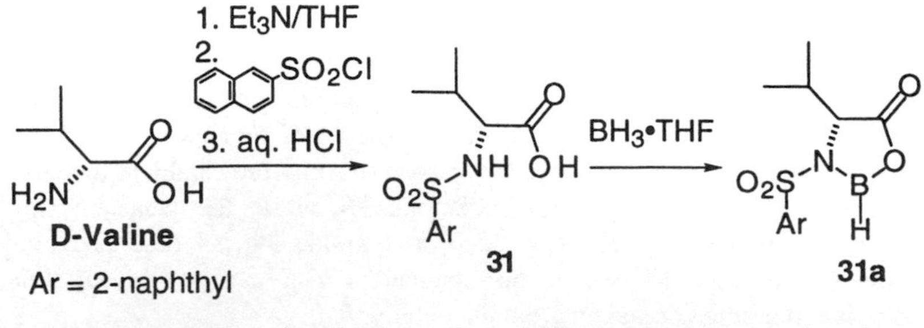

Initial attempts to carry out the aldol reaction of **27** (prepared by treating the ester **26** with LDA and trimethylsilyl chloride) with aldehyde (Ar¹CHO) in the presence of catalyst **31a** resulted in a very poor selectivity of S:R on the hydroxy chiral center. During process development, we identified three critical factors that control the enantioselectivity: 1) the ratio of $BH_3 \cdot THF$ to **31**, 2) addition rate of the aldehyde and 3) the amount of catalyst. The optimum result was obtained using 40 mol% catalyst (generated by adding 0.9 eq. of $BH_3 \cdot THF$ to **31**) while the aldehyde was being added very slowly. The increase in the size of ester from methyl to ethyl improves the ratio of S:R from 93:7 to 96:4. Under these conditions, the catalytic chiral aldol reaction was successfully scaled up to multikilogram batches. Although 0.4 eq. of catalyst is used (90% of which is easily recovered and reused) the net consumption in each run is only 5 mol%. Crystallization of crude **28** further enhanced the S:R ratio to 97:3.

## Scheme 15

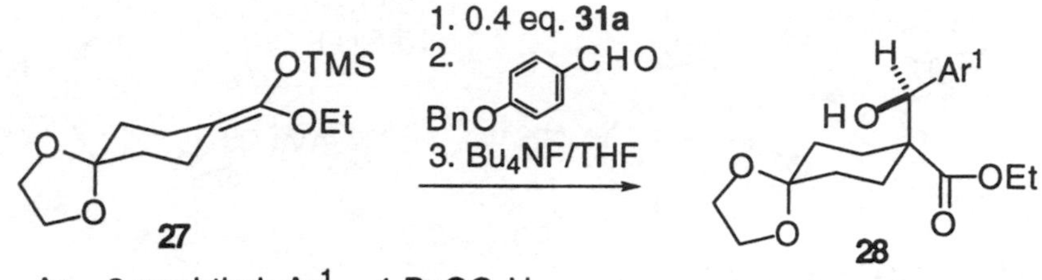

Ar = 2-naphthyl; $Ar^1$ = 4-$BnOC_6H_4$-

The following mechanism for the catalytic enantioselective aldol reaction is proposed as shown in Figure 7. The $sp^3$ hybridized boron in **31a** has an empty orbital which shares the lone pair electron from the carbonyl group in *p*-benzyloxybenzaldehyde, forming the coordination bond shown in intermediate I.     As a result of the asymmetric environment created by the preexisting

## Figure 7

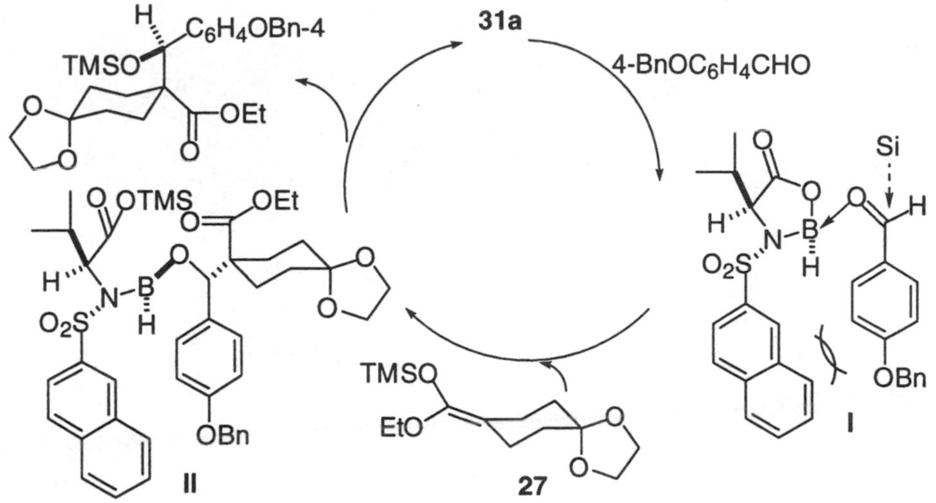

isopropyl group, the pro-R face of the carbonyl is blocked by the bulky naphthyl moiety while the pro-S face is open to attack by enol silyl ether **27**. Two observations are indicative of this mechanism: 1) the initial aldol product exists primarily in the $Me_3Si$ protected form; and 2) the Me-B or Bu-B version of the active catalyst can be isolated.[27]

Amination of **28** to **29** was achieved in high yield by reaction with $p$-$FC_6H_4NH_2$ under Weinreb conditions[9c] using $Me_3Al$/toluene. Formation of the β-lactam ring was accomplished by treating **29** with $(EtO)_2P(O)Cl$ and NaOH in the presence of BTAC, as described above for the synthesis of **1**. A simple crystallization not only produced the spiro-fused β-lactam **3** in 59% yield over 2 steps but also enriched the *ee* to 99.5%.

### Scheme 16

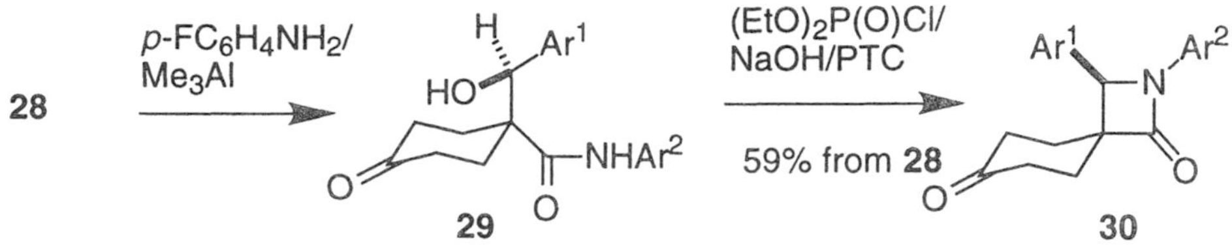

$Ar^1$ = 4-$BnOC_6H_4$-; $Ar^2$ = 4-$FC_6H_4$-

Next, we studied the diastereoselective addition of the Grignard reagent. The addition of $p$-$ClC_6H_4MgBr$ takes place from both faces of the carbonyl group in ketone **30** producing a diastereomeric mixture of **35** and **35a**. The ratio of **35:35a** was 83:17 when $p$-$ClC_6H_4MgBr$ was added to a solution of compound **30** at -20 °C in THF. Worse yet was the low yield and the incomplete reaction.

### Scheme 17

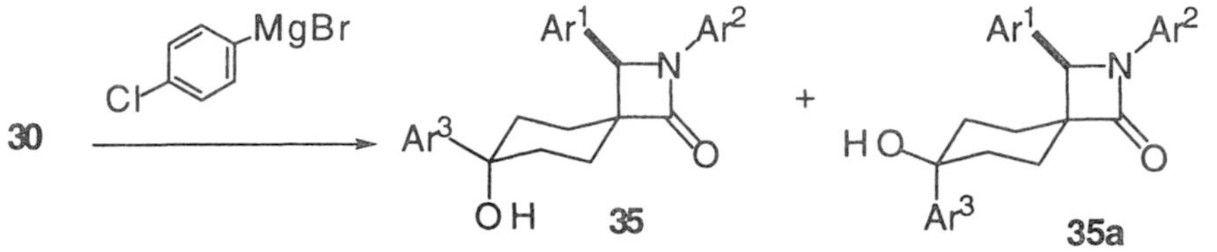

$Ar^1$ = 4-$BnOC_6H_4$-; $Ar^2$ = 4-$FC_6H_4$-; $Ar^3$ = 4-$ClC_6H_4$-

We speculated that the incomplete reaction was due to enolization, which may be minimized by a reverse addition. Indeed, the reaction proceeded to >98% completion when **30** was added dropwise into 1.5 equivalent of the Grignard reagent. Also, reactions run at higher temperatures generally gave better selectivity. Thus when compound **30** was added to the Grignard reagent at 80 °C in MTBE-toluene, the reaction gave a 94:6 mixture of **35** and **35a**. In this case, the product was isolated by a simple precipitation in 90% yield.

Debenzylation of compound **35** to produce **2** under normal hydrogenolysis conditions using Pd/C as a catalyst gave up to 7% of deschloro impurity. Crystallisation did not purge this impurity to an acceptable level. We thought that the dechlorination could have been taking place *via* a palladium(0) oxidative insertion into the C-Cl bond followed by a hydrogenative reduction of the newly-formed C-Pd-Cl. Addition of a Lewis acid may be able to slow down the oxidative insertion. To our delight, the dechlorination process was completely blocked when 0.6 eq. of $ZnBr_2$ was introduced. In addition, $ZnBr_2$ did not increase the des-OH impurity. To the best of our knowledge, this is an unprecedented selective debenzylation reaction.[30] A simple recrystallization produced the final drug in 99.5% chemical purity and 99.5% *ee*.

**Scheme 18**

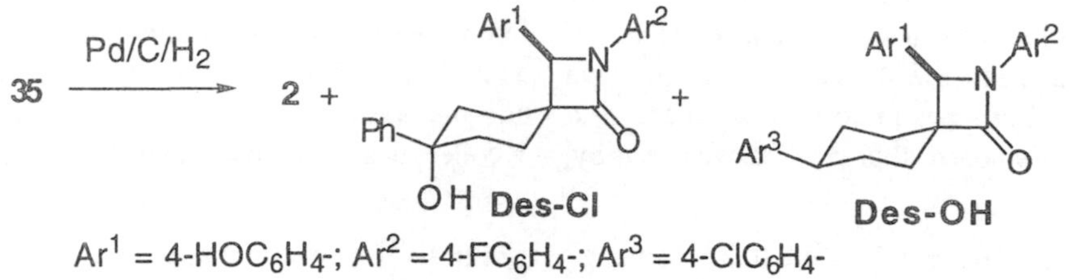

$Ar^1$ = 4-$HOC_6H_4$-; $Ar^2$ = 4-$FC_6H_4$-; $Ar^3$ = 4-$ClC_6H_4$-

The synthesis of Sch 58053, **2**, using this catalytic approach (Scheme 19) involves six steps with four isolated intermediates and no chromatography. The overall yield of **2** from **26** was 32% and the process was successfully scaled up to produce 1.5 kg of **2** needed for toxicology studies.

**Scheme 19**

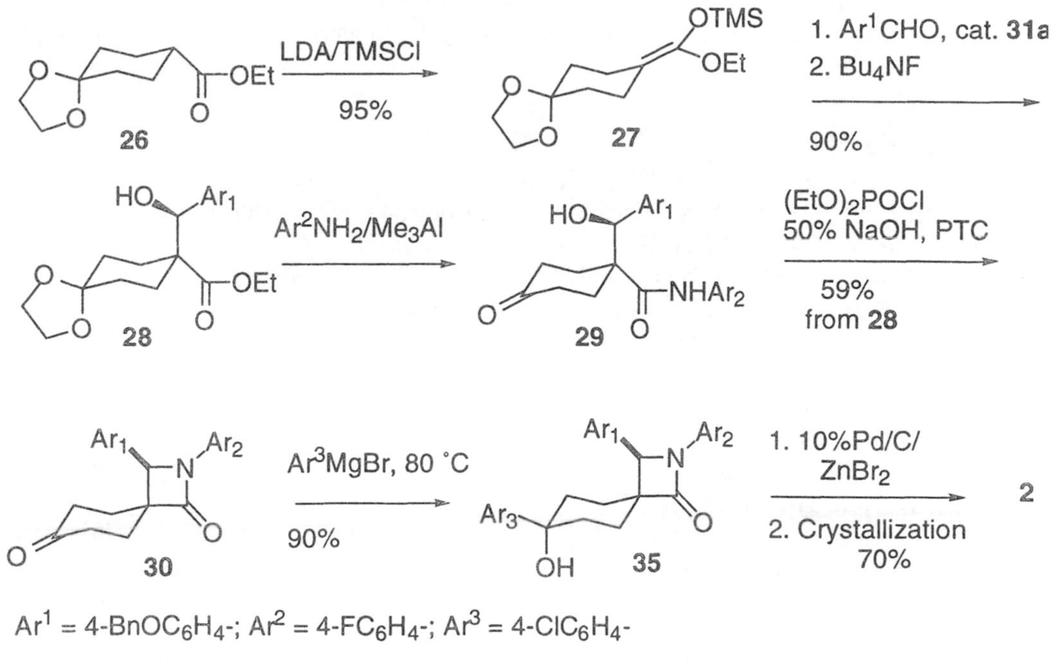

$Ar^1$ = 4-$BnOC_6H_4$-; $Ar^2$ = 4-$FC_6H_4$-; $Ar^3$ = 4-$ClC_6H_4$-

## Conclusions

We have discovered several practical enantioselective routes to azetidinones. In the aldol approach to Sch 48461, Evans' aldol reaction was used to set the stereochemistry. For the synthesis of **2**, the required aldol **28** was prepared by a catalytic asymmetric approach. Three cyclization methods were developed for the β–lactam ring formation from the β-hydroxyamides. We have also developed the first example of an imine addition reaction to titanium enolates of chiral N-acyloxazolidinones of type **17** to form β-aminocarbonyl derivatives. A novel fluoride catalyzed cyclization of N-silyl protected β-aminocarboxylic acid derivatives such as **18** provided azetidinones in excellent yield and enantiomeric purity. The shortness and simplicity of the imine addition process has led us to choose the imine addition approach as the commercial process. This process was successfully scaled up in the plant to produce ~ 620 kg. The ready availability of chiral oxazolidinones in both enantiomers makes this methodology quite powerful to make chiral azetidinones. Indeed, this process was used by our colleagues to prepare analogs of **1** for SAR studies.

It should be emphasized that changing a synthesis (or even changing a crystallization procedure) during the clinical studies itself poses an enormous challenge unfamiliar to many in that the impurity profile of the product of the new process must be as good or better to within <0.1% levels of the old process. In spite of this stringent specifications, all of the above processes clearly met the challenge and produced quality material for clinical studies.

## Acknowledgments

We thank the many skillful chemists who worked and contributed to this project as well as the plant personnel in the scale up efforts. Special thanks go to Drs. M. Mitchell, D. Walker and Mr. B. Shutts for their valuable suggestions in proof-reading this manuscript and Ms. Kathy Torpey for typing.

## References and Notes

(1) (a) Clader, J. W.; Burnett, D. A.; Caplen, M. A.; Domalski, M. S. Dugar, S.; Vaccaro, W.; Sher, R.; Browne, M. E.; Zhao, H.; Burrier, R. E.; Salisbury, B.; Davis, H. R. Jr. *J. Med. Chem.* **1996**, *39*, 3684. (b) Salisbury, B. G. Davis, H. R. et al. *Atherosclerosis*, **1995**, *115*, 45. (c) McKittrick, B. A.; Dugar, S.; Clader, J. W.; Davis,     H. Jr.; Czarniecki, M. *Bioorg. Med. Chem. Lett.* **1996**, *6*, 1947. (d) Dugar, S.; Clader, J. W.; Chan, T. M.; Davis, H. Jr. *J. Med. Chem.* **1995**, *38*, 4875. (e) Burnett, D. A.; Caplen, M. A.; Davis, H. R.; Burrier, R. E.; Clader, J. W. *J. Med. Chem.* **1994**, *37*, 1733. (f) Burnett, D. A. *Tetrahedron Lett.* **1994**, *35*, 7339.

(2) (a) Ternansky, R. J.; Morin, Jr., J. M. in *The Organic Chemistry of β-lactams*, VCH Publishers, New York, 1993, pp.257-293.

(3) (a) Thiruvengadam, T. K.; Tann, C. H.; Lee, J. McAllister, T.; Sudhakar, A. US patent. **1994**, 5306817. (b) Thiruvengadam, T. K.; Tann, C. H.; Lee, J. McAllister, T. US patent. **1996**, 5561227. (c) Thiruvengadam, T. K.; Tann, C. H.; McAllister, T. PCT Int. Appl. **1995**, WO 95 01,961. (d) Wu, G.; Tormos, W. *J. Org. Chem. Submitted*

(4) (a) Heathcock, C. H. in *Comprehensive Organic Synthesis*, Volume 2, Pergamon Press, Oxford, 1991,    pp.133-238; (b) Kim, B. M.; Williams, S. F.; Masamune, S. in *Comprehensive Organic Synthesis*, Volume 2, Pergamon Press, Oxford, 1991, pp.239-275;

(c) Paterson, I. *Comprehensive Organic Synthesis*, Volume 2, Pergamon Press, Oxford, 1991, pp.301-319.

(5) (a) Evans, D. A.; Bartroli, J.; Shih, T. L. *J. Am. Chem. Soc.,* **1981**, *103*, 2127. (b) For a review of the use of oxazolidinones in asymmetric synthesis, see Ager, D. J.; Prakash, I.; Schaad, D. R. *Aldrichimica Acta*, **1997**, *30*, 3.

(6) Oppolzer, W.; Blagg, J.; Rodriguez, I.; Walther, E. *J. Am. Chem. Soc.,* **1990**, *112*, 2767.

(7) (a) Evans, D. A., Britton, T. C., Ellman, J. A. *Tetrahedron Lett*, **1987**, *28*, 6141. (b) Evans, D. A., Chapman, K. T., Bisaha, J. *J. Am. Chem. Soc.,* **1988**, *110*, 1238. (C) Gage, J. R.; Evans, D. A. *Org. Synth.,* **1989**, *68*, 83.

(8) Reaction of **7** with *p*-anisidine in the presence of trimethylaluminum[9] gave low yield of **9**.

(9) (a) Nahm, S.; Weinreb, S. M. *Tetrahedron Lett.,* **1981**, *22*, 3815. (b) Evans, D. A.; Bender, S. L. *Tetrahedron Lett.,* **1986**, *27*, 799. (c) Basha, A.; Lipton, M.; Weinreb, S. M. *Tetrahedron Lett.,* **1977**, 4171.

(10) (a) Evans, D. A.; Sjogren, E. B. *Tetrahedron Lett.,* **1986**, *27*, 3119. (b) Miller, M.J.; Mattingly, P. G.; Morrison, M. A.; Kerwin, J. F. *J. Am. Chem. Soc.,* **1980**, *102*, 7026.

(11) After this procedure was disclosed in our patent[2a], other research groups reported similar procedures [12].

(12) (a) Ager, D. J.; Allen, D.R.; Froen, D. E.; Schaad, D. R. *Synthesis*, **1996**, 1283. (b) Ho, G.-J.; Mathre, D. J. *J. Org. Chem.* **1995**, *60*, 2271. (c) Lee, J. Y.; Chung, Y. J.; Kim, B. H. *Synlett* **1994**, 197.

(13) The same *syn* aldol was obtained from boron enolates and from titanium enolates: Evans, D. A.; Rieger, D. L.; Bilodeau, M. T.; Urpi, F. *J. Am. Chem. Soc.,* **1991**, *113*, 1047.

(14) Evans, D. A.; Ennis, M. D.; Mathre, D. J. *J. Am. Chem. Soc.,* **1982**, *104*, 1737.

(15) Similar but less dramatic effects of amine bases on the stereochemistry of aldol reactions have been reported. For example, see Baker, R.; Castro, J. L.; Swain, C. J. *Tetrahedron Lett.,* **1988**, *29*, 2247; Shirodkar, S.; Nerz-Stormes, M.; Thornton, E. R. *Tetrahedron Lett.,* **1990**, *31*, 4699; Nerz-Stormes, M.; Thornton, E. R. *J. Org. Chem.,* **1991**, *56*, 2489; Pridgen, L. N.; Abdel-Magid, A. F.; Lantos, I.; Shilcrat, S.; Eggleston, D. S. *J. Org. Chem.,* **1993**, *58*, 5107.

(16) For the use of 2,4,6-trichlorobenzoyl chloride in macrocyclizations, see Inanaga, J.; Hirata, K.; Sacki, H.; Katsuki, T.; Yamaguchi, M. *Bull. Chem. Soc. Japan*, **1979**, *52*, 1989; Inanaga, J.; Katsuki, T.; Takimoto, S.; Ouchida, S.; Inone, K.; Nakano, A.; Okukuda, N.; Yamaguchi, M. *Chem. Lett.,* **1979**, 1021; Hikota, M.; Tone, H.; Horita, K.; Yonemitsu, O. *J. Org. Chem.,* **1990**, *55*, 7.

(17) Szeja, W. *Synthesis*, **1980**, 402.

(18) For the phase transfer catalyzed cyclization of 3-bromoalkanoic amides to non-chiral azetidinones, see Keller, W. E. in *Phase-Transfer Reactions*, Fluka-Compendium Volume 2, Georg Thieme Verlag Stuttgart, 1987, pp. 699-704.

(19) For the phase transfer catalyzed phosphorylation of alcohols, see Keller, W. E. in *Phase-Transfer Reactions*, Fluka-Compendium Volume 2, Georg Thieme Verlag Stuttgart, 1987, pp. 827.

(20) Macrolactonization using diphenylphosphate as leaving group has been reported: Kaiho, T.; Masamune, S.; Toyoda, T. *J. Org. Chem.,* **1982**, *47*, 1612.

(21) (a) After this work was disclosed[3b], reactions of chiral oxazolidinone enolates with *activated* imines has been reported: Abrahams, I.; Motevalli, M.; Robinson, A.J.; Wyatt, P. B. *Tetrahedron*, **1994**, *50*, 12755. (b) For a review of enolate additions to imines, see Hart, D. J.; Ha, D-C. *Chem. Rev.* **1989**, *89*, 1447. (c) For addition of boron enolates of thiolesters to imines see Corey, E. J.; Decicco, C. P.; Newbold, R. C. *Tetrahedron Lett.,* **1991**, *32*, 5287.

(22) Heathcock, C. H. in *Asymmetric Synthesis*, J. D. Morrison Ed., Volume 3, Academic Press, 1984, pp. 111-212.

(23) A mixture of **14** and **15** was prepared according to the following scheme:

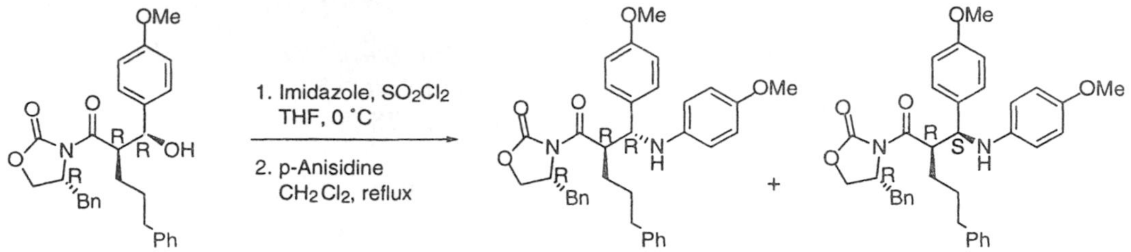

1. Imidazole, SO₂Cl₂
   THF, 0 °C

2. p-Anisidine
   CH₂Cl₂, reflux

**7**     **14**     **15**

(15 : 85)

The structures of **14** and **15** were further confirmed by conversion to the known azetidinones.

(24) (a) Intramolecular displacement of oxazolidinone to form γ-lactam has recently been reported: Mulzer, J.; Zuhse, R.; Schmiechen, R. *Ang. Chem. Int. Ed. Engl.*, **1992**, *31*, 870. (b) β-Lactam formation by intramolecular cyclization of chiral β-aminothiolesters by treatment with t-BuMgCl has been reported.[21c]

(25) Chen, L.; Zaks, A.; Chackalamannil, S.; Dugar, S. *J. Org. Chem. Soc.* **1996**, *61*, 8341.

(26) (a) Mukaiyama, T.; Kobayashi, S.; Sano, T. *Tetrahedron*, **1990**, *46*, 4653. (b) Saigo, K.; Osaki, M.; Mukaiyama, T. *Chem. Lett.* **1975**. 989.

(27) (a) Corey, E. J.; Cywin, C. L.; Roper, T. D. Tetrahedron Lett. 1992, 33, 6907. (b) Corey, E. J.; Loh, T.-P. *J. Am. Chem. Soc.* **1991**, *113*, 8966. (c) Parmee, E. R.; Tempkin, O.; Masamune, S. *J. Am. Chem. Soc.* **1991**, *113*, 9365. (d) Kiyooka, S. I.; Kaneko, Y.; Kume, K.I. *Tetrahedron Lett.*, **1992**, *33*, 4927. Kobayashi, S.; Horibe, M. *J. Am. Chem. Soc.* **1994**, *116*, 9805.

(28) (a) Sodeoka, M.; Ohrai, K.; Shibasaki, M. *J. Org. Chem.* **1995**, *60*, 2648. (b) Carreira, E. M.; Singer, R. A.; Lee, W. *J. Am. Chem. Soc.* **1994**, *116*, 8837. (c) Mikami, K.; Terada, M.; Nakai, T. *J. Am. Chem. Soc.* **1989**, *111*, 1940.

(29) Kiyooka, S.; Kaneko, Y.; Komura, M.; Matsuo, H.; Nakano, M. *J. Org. Chem.* **1991**, *56*, 2276.

(30) For leading references see: *Catalytic Hydrogenation in Organic Syntheses*, Rylander, P. N. Academic Press, New York, 1979. *Catalytic Hydrogenation*, Augustine, R. L. Marcel Dekker, New York, 1965.

# Synthesis of the Common Lactone Moiety of HMG-CoA Reductase Inhibitors

Ian M. McFarlane and Christopher G. Newton

*Dagenham Research Centre, Rhône-Poulenc Rorer Central Research, Rainham Road South, Dagenham, Essex, Essex RM10 7XS, UK*

Philippe Pitchen

*Cawthorn Centre, Rhône-Poulenc Rorer Central Research, 500 Arcola Road, Collegeville, PA 19426, USA*

Key words: HMG Co A Reductase, asymmetric synthesis, carbohydrates, glucose, galactose, mannose, process chemistry.

## Introduction

Hypercholesterolemia is an important risk factor in the development of coronary heart disease, the major cause of death in the western world. The enzyme which governs the rate limiting step in the biochemical assembly of cholesterol is 3-hydroxy-3-methylglutaryl-Coenzyme A reductase (HMG CoA reductase, EC 1.1.1.34).

The discovery by Endo[1] working in conjunction with the Sankyo company of Japan, of the potent inhibitor of HMG CoA reductase called compactin (**1**) from the metabolites of *Penicillium citrinium* was a breakthrough in the search for inhibitors of this enzyme, highlighting the key structural features required for enzyme inhibition. Sankyo did not develop compactin (generic name: mevastatin) to market, but subsequently, and in conjunction with Bristol-Myers Squibb, developed and launched in 1989 the microbially transformed mevastatin analogue, pravastatin (**2**).

Pravastatin (**2**) was not however the first HMG-CoA reductase compound to be marketed. The discovery of compactin/mevastatin had initiated the search for other potent HMG-CoA reductase inhibitors from microbes. A second, potent inhibitor, initially called mevinolin (**3**) was isolated independently by two groups from *Monascus ruber*[2] and *Aspergillus terreus*.[3] Mevinolin (**3**) (generic name lovastatin), was launched in 1987. A semi-synthetic derivative of mevinolin/lovastatin (**3**) was also developed by Merck, and this compound, simvastatin (**4**) is also marketed. The success of these drugs can be measured by their sales figures; 1995:[4] lovastatin $1,260M, simvastatin $1,955M; 1996:[5] combined sales of lovastatin and simvastatin $4,056M.

These three marketed compounds (**2-4**) are all derived from fermentation processes. Structure-activity conclusions drawn from chemical explorations in the 1980s[6] identified the core features of these molecules, for HMG CoA reductase inhibition activity. In particular was noted the absolute requirement for a 4-hydroxy tetrahydropyranone structure (or its open chain 3,5-dihydroxypentanoic acid), of specific absolute chirality, which could be attached via a 2-carbon bridge to a variety of heterocycles. The first such totally synthetic analogue, fluvastatin (**5**) (launched in 1995) contains an appropriately substituted indole attached to the above key dihydroxypentanoic acid. For a variety of reasons, perhaps partly related to product stability,[7] fluvastatin is manufactured and marketed as a racemate, so only 50% of fluvastatin is a potent inhibitor of HMG CoA reductase.

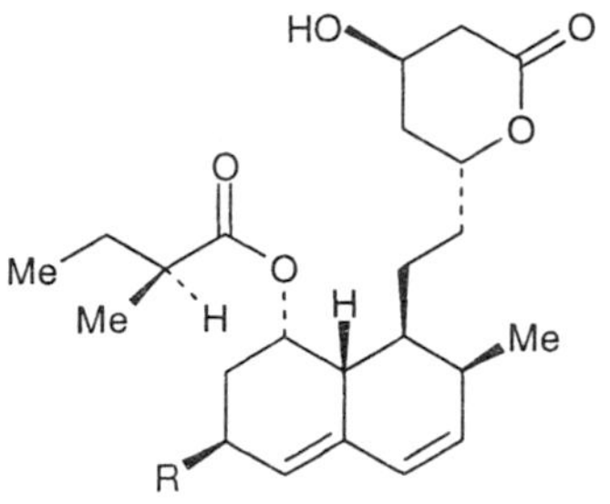

(**1**), R = H, compactin/mevastatin
(**2**), R = OH, pravastatin (marketed as the
       sodium salt of the 3,5-dihydroxy acid).

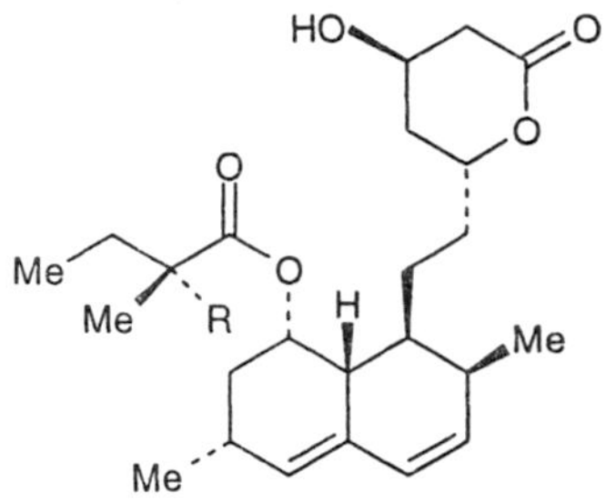

(**3**), R = H, mevinolin/lovastatin.
(**4**), R = Me, simvastatin.

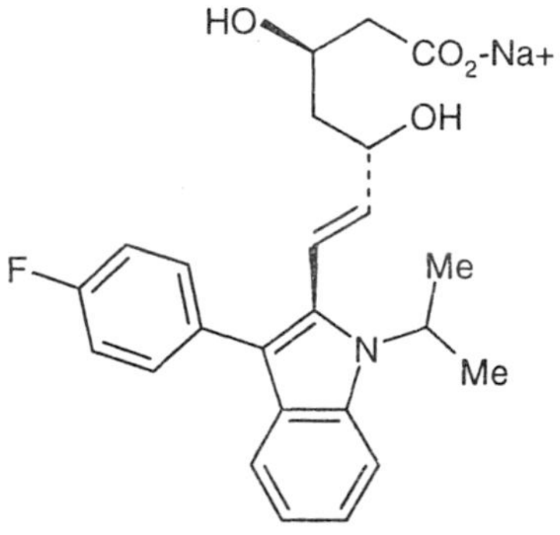

(**5**), fluvastatin.

The structural feature which is necessary (but by no means sufficient) for HMG CoA reductase inhibitory activity in all the above compounds (**1-5**) is the lactone (**6**), or its open chain dihydroxy acid (**7**). It is the total synthesis of this core unit, and its subsequent attachment to a heterocyclic or carbocyclic southern substituent which is the subject of this review. More specifically, this review will focus upon our attempts to manufacture the lactone in homochiral form, with the key target synthon a protected version of the tetrahydropyran (**8**).

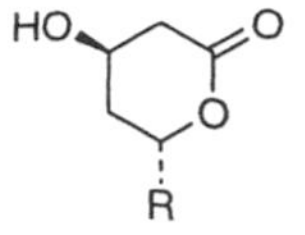

(6) 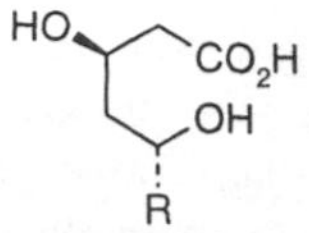          (7) 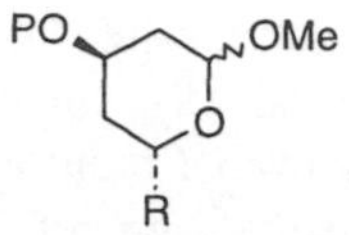          (8)

Within Rhône-Poulenc Rorer (RPR), a research programme designed to discover new inhibitors of HMG CoA reductase identified two candidate compounds: RP 61969 (**9**) and RG 12561 (**10**). As can be seen from inspection of the structures (**9**) and (**10**), both contain the key warhead component (**6**).

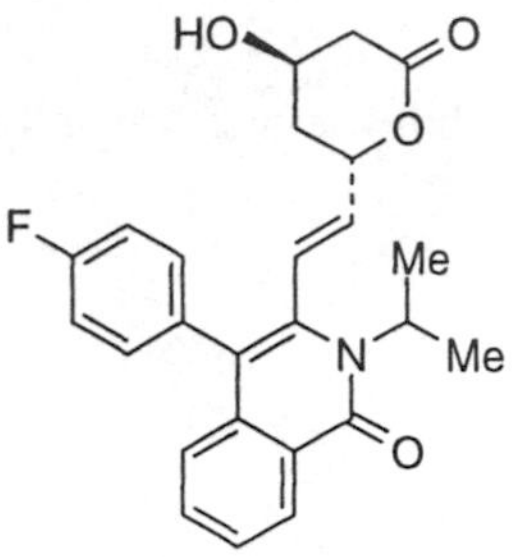

(**9**) RP 61969 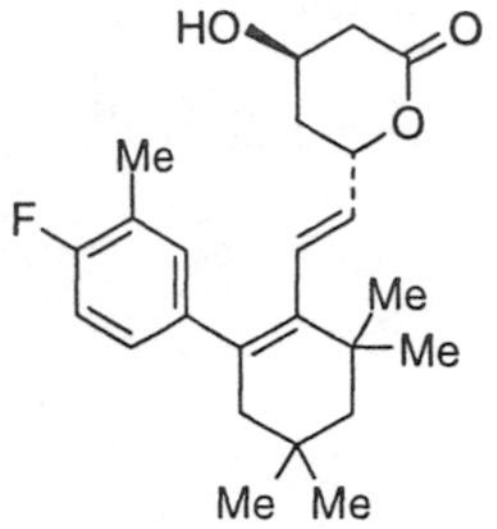          (**10**) RG 12561

Many approaches to the synthesis of statins are viable, and this publication summarizes a part of our work, which focused upon carbohydrates as a chiral pool approach to the synthesis of **8** and thus **9** and **10**.

## Carbohydrate strategy

The essential strategy was a disconnection of the type shown below in Scheme 1. Thus, a synthon of six carbon atoms was required to be connected to the southern halves of each molecule, implying a hexose sugar, as carbohydrate precursor of (**8**).

Scheme 1: Disconnection approach to RP 61969 (**9**) and RG 12561 (**10**)

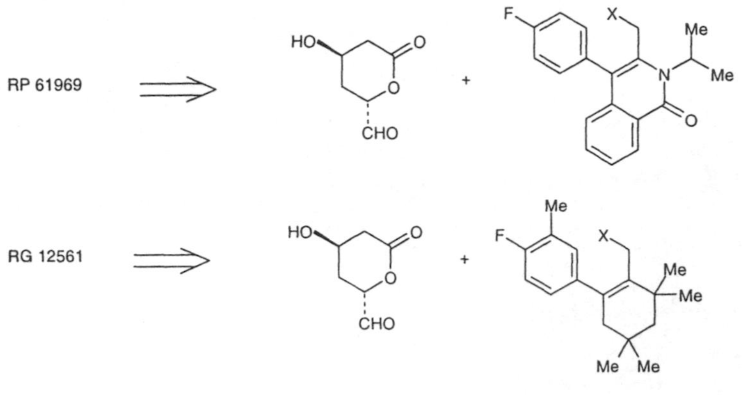

Of the sixteen possible aldohexoses that may be considered as a precursor of the lactone (**6**)/tetrahydropyran (**8**), eight belong to the unnatural class of sugars with the opposite stereochemistry to that required at C-6 of the lactone (**6**). This, together with the fact that these eight unnatural sugars are not available in commercially attractive prices and quantities, rules them out as viable industrial precursors. Of the eight natural reducing hexoses, five (talose, allose, gulose, altrose and idose) can also be rejected on the grounds of cost and availability. This leaves three natural six-carbon sugars as very attractive, viable members of the chiral pool as precursors of our target, which is simply depicted as structure (**11**) (Scheme 2). These three are: galactose (**12**), mannose (**13**) and especially glucose (**14**). The relationship between the stereochemical centres of these three viable carbohydrate precursors and of the HMG-CoA reductase inhibitor is explicitly indicated in Scheme 2.

Scheme 2: Stereochemical relationships between three carbohydrate precursors and the target synthon (**11**)

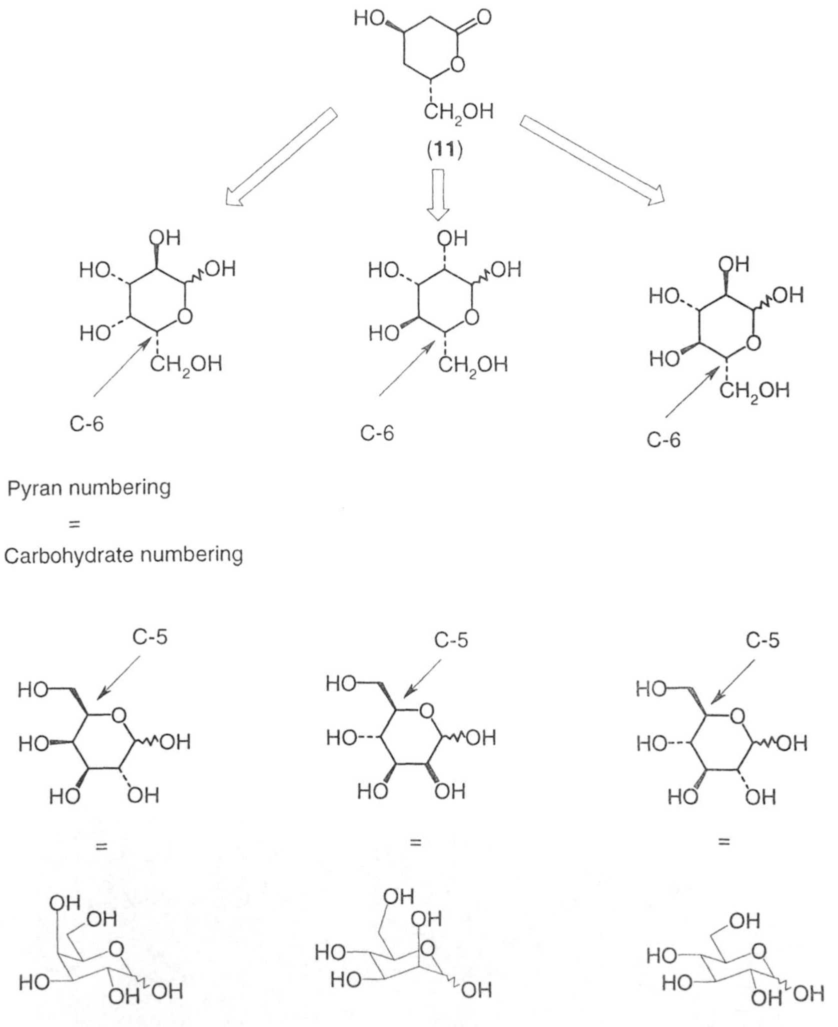

Pyran numbering

=

Carbohydrate numbering

(**12**) galactose          (**13**) mannose          (**14**) glucose

The three viable starting materials all possess the correct stereochemistry at C-5 of the carbohydrate ring, to become C-6 of the lactone ring. All three, however, require deoxygenation at sugar positions C-2 and C-4 (whatever the original orientation of these hydroxy groups), and all require the hydroxy group at C-3 of the carbohydrate to be inverted, as depicted in Scheme 3.

Scheme 3: General transformations required for conversion of carbohydrate precursors to target synthon (**16**)

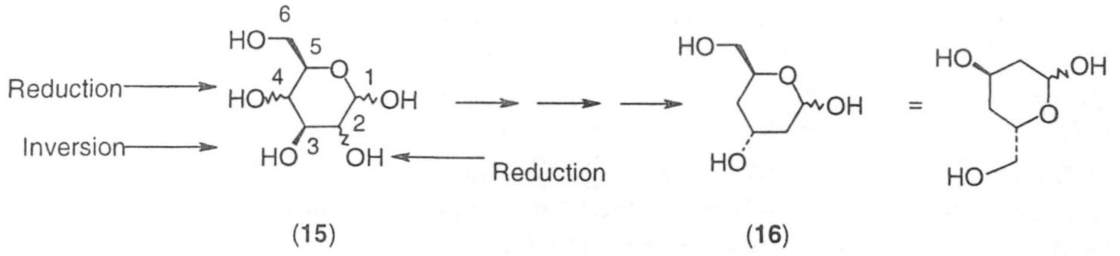

Initial Route Used in Discovery, description and problems:

The first route used in Discovery Chemistry to prepare RP 61969 (**9**) in homochiral form used a glucose derivative (**17**) as a chiral pool synthon, in an extended route first pioneered by Heathcock[8] (Scheme 4).

Scheme 4: The original discovery route to chiral RP 61969 synthon

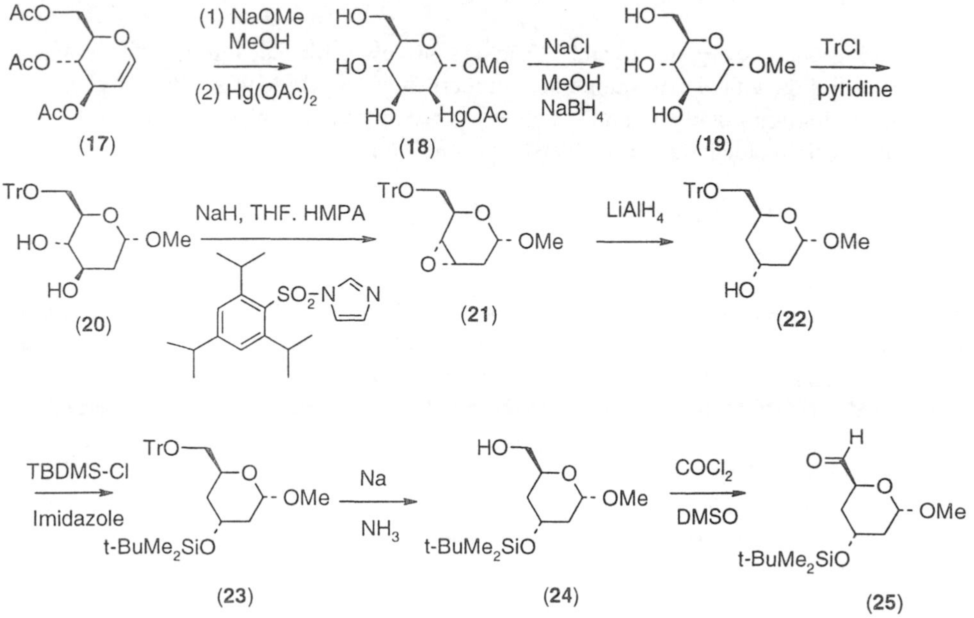

In this case, a downstream intermediate tri-*O*-acetyl-D-glucal (TADG) (**17**) was commercially available, since it is easily prepared from glucose in three steps.[9] Discovery Chemistry (and later Process Chemistry) were able to purchase this material (Pfanstehl Laboratories, USA and Genzyme Laboratories, UK). Using this route in an essentially unchanged form, gram quantities of the key six-carbon synthon (**25**) could be synthesised, as described below and depicted in Scheme 4.

The first steps of the sequence essentially complete the C-2 deoxygenation already begun with the transformation of glucose to tri-O-acetyl-D-glucal (**17**). First, **17** is deacetylated with a catalytic quantity of sodium methoxide in methanol and then regiospecifically and stereospecifically methoxymercurated with mercuric acetate in methanol to give the mercury bearing sugar, **18**. We regarded the production of a single anomer at C-1 as important, since although this chirality is lost in the final product, homogeneous diastereoisomers were going to be required at later stages of the synthesis to facilitate isolation and purification. After exchange at mercury of the acetoxy functionality with chloride, the mercurial group was reductively eliminated with sodium borohydride to give the C-2 reduced sugar (**19**). The primary hydroxyl function was then protected from reacting in the subsequent transformations, by making the trityl derivative (**20**).

Following the Heathcock precedent, the de-oxygenation at C-4 of the carbohydrate and inversion at C-3 was performed as follows: the diol (**20**) was closed to the epoxide (**21**) via a bulky sulfonating agent (tri-isopropyl benzenesulfonyl imidazole) in a reaction conducted at -23°C, with powdered sodium hydride as base to give regiospecific sulfonation at C-3. Treatment of the sulphonate with an excess of base caused epoxide formation by attack of the C-4 hydroxy group, to generate the alpha C-3 C-4 epoxide (**21**). Then, the epoxide (**21**) was opened with LiAlH$_4$ in a regioselective reaction to give **22** as a 92:8 mixture with the corresponding 4-OH regioisomer. This mixture of products was then treated with tertiary butyl dimethyl silyl chloride, and the resulting 3-TBDMS ether was then crystallised cleanly from the 92:8 mixture, yielding (**23**).

With C-2 and C-4 deoxygenation and C-3 inversion now achieved, it remained to prepare the synthon for coupling with the isoquinolone southern half of RP 61969. The trityl group was removed with sodium in liquid ammonia in ether to give the primary alcohol (**24**) which was then oxidized to the desired aldehyde (**25**) under Swern conditions.

Many features of this synthesis rendered it unsuitable for even moderate scale-up, and these were removed as follows: after isolation of the organomercurial compound (**18**), the extremely water soluble triol (**19**) was not isolated but was tritylated directly to give **20**. The epoxide closure was carried out at room temperature (as opposed to -23°C) and N,N-dimethylimidazolidinone (DMI) substituted for HMPA. Sodium hydride powder was successfully substituted by a 60% dispersion in mineral oil. We had noted that the crude epoxide (**21**) could be reduced directly by addition of 1M LiAlH$_4$·2THF in Toluene (Aldrich). However, on scale-up, commercial LiAlH$_4$ solution (Callery) as obtained was more concentrated (3.95M) and contains a greater percentage of THF as co-solvent. Used directly, as purchased, Callery material gave 28% of the undesired 4-OH isomer. Fortunately, the simple expedient of diluting with toluene gave the aforementioned, acceptable 7-8% of the 4-OH isomer. A further isolation was also avoided, and crude **22** material could be directly silylated and crystallised from MeOH to give **23** in an overall yield of 72% from **20**.

The preparation of **24** was a clean reaction, but separation of the product from the triphenylmethane by-product was problematical. Approximately 80% of the triphenylmethane could be removed by slurrying the crude product with methanol, but the remainder had to be removed by chromatography. The existence at RPR of a very large scale automated HPLC machine (K-Prime 3000 from Amicon Ltd.) solved this problem. Finally, we found that the Swern oxidation to (**25**) could be carried out successfully at -30°C and not -60°C, so making the early scale-up much simpler.

Obviously this route had many disadvantages, particularly length (eight steps as drawn), toxicity (especially of mercury reagents, intermediates and effluent) thermal safety, and environmental problems associated with the preparation of the organomercurial compound (**18**), but it did allow us to prepare 20 kg of **25** and 10 kg of **5**.

The above problems had to be resolved, and we began the examination of many other possible routes to **25**. Many of these routes were highly speculative, whilst others made use of known chemistry, in order to overcome the difficulties indicated above to allow us to prepare more material on a larger scale and more quickly.

### Improvement to the Heathcock Routes from glucose: the use of alternative C-2 deoxygenation methodologies.

Starting from TADG (**17**), another method of preparing 2-deoxy derivatives is via reduction of 2-iodo sugars,[10] according to the two possibilities outlined in Scheme 5, which differ in the order in which the transformations are performed.

Scheme 5: Reductive iodination at C-2 of glucose derived intermediates

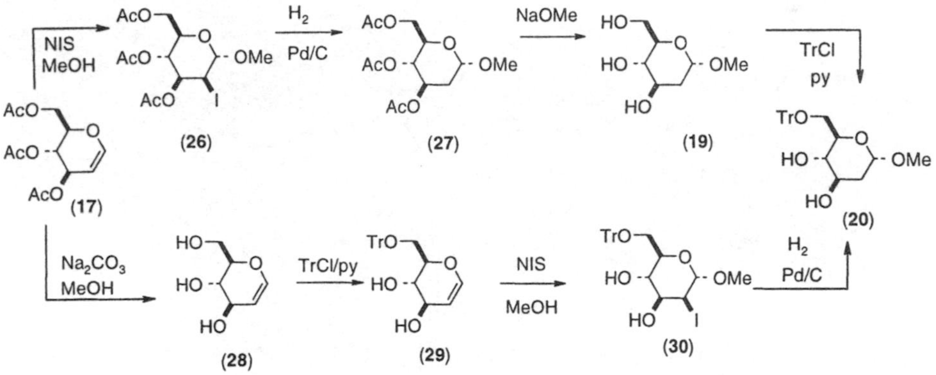

A regiospecific and stereoselective iodomethoxylation takes place when TADG (**17**) is treated with N-iodosuccinimide (NIS) in methanol to give **26**. The 2-iodo function is easily reduced using catalytic hydrogenation, giving **27**. Deacetylation and tritylation then gives access to **20**. This route to the key intermediate **20** (Scheme 5) thus avoids completely the mercury containing transformations. An alternative *modus operandi* was to run the synthesis in the second sequence, **17** → **28** → **29** → **30** → **20**, which in fact gives a better diastereoselectivity at C-1 in the iodomethoxylation step **29** → **30**. In this second case, access to **30** was achieved in almost quantitative yield, with only the 1-α anomer detected by NMR. Catalytic hydrogenation in a mixture of triethylamine/toluene/methanol then gave **20** in an overall yield of 59%.[11]

This route did have the great advantage that it allowed the 2-deoxygenation without the use of any mercury containing intermediates. All the steps up to the intermediate (**20**) were scaleable, and the good diastereoselectivity of addition allowed an easy crystallization at the end.

The drawbacks were mainly the length. Now we had four steps from **17** to **20**, whereas previously there were three. Some other minor problems were a number of solvent exchanges and the high cost of NIS. Thus although this route met some of our requirements, we were still not satisfied and continued to search for alternatives.

### Routes from mannose: via Methyl-α-D-Mannopyranoside

In an attempt to devise a shorter synthesis of our key six carbon synthon, our attention was drawn to a paper by Tatsuta *et al*[12] which describes the tritylation of methyl-α-D-mannopyranoside (**31**), regioselective dimesylation and conversion to an epoxy mesylate (**34**) (Scheme 6). We realised that by treatment with a suitable reducing agent a series of reductions and ring closures could take place which would lead to an advanced intermediate (**22**) (Scheme 6).

Scheme 6: A route from a mannose derived precursor

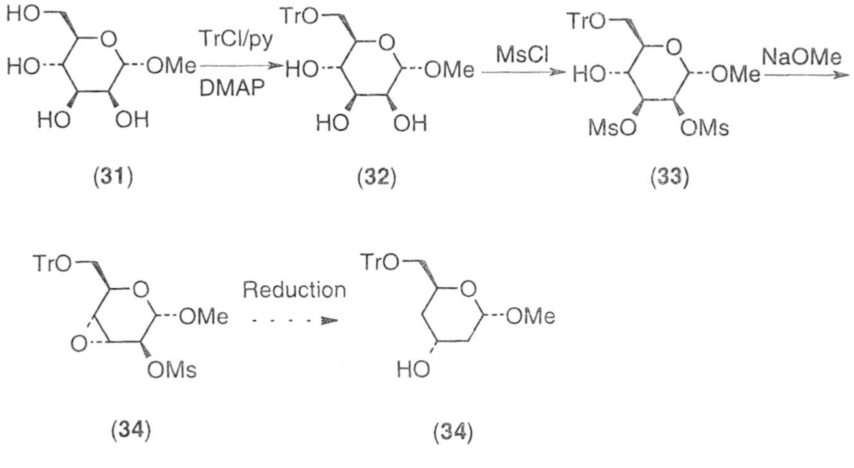

According to the literature[12] methyl α-D-mannopyranoside (**31**) can be treated with trityl chloride (TrCl) in pyridine with a catalytic quantity of 4-dimethylaminopyridine (DMAP) to give **32**. After stirring the mixture overnight, an excess of mesyl chloride is added and the mixture stirred for a further eight hours. The reaction mixture is evaporated to dryness, dissolved in ethyl acetate, washed, and evaporated to dryness to give the 6-trityl derivative (**33**) as a syrup.

This syrup was dissolved in chloroform and treated with sodium methoxide. The organic layer is separated, washed with water, dried, evaporated to dryness and the residue purified by chromatography followed by recrystallisation. This gave **34** in a claimed overall yield of 73%.[12]

However, in our hands repetition of the above literature procedure for the dimesylation did not prove as regioselective as expected and led to a mixture (Scheme 7) containing the desired product

**33**, together with the other products **35**, **36** and **37** which could all be separated by column chromatography and isolated.

Scheme 7: Mesylation Products

| | R1 | R2 | R3 |
|------|-----|-----|-----|
| (33) | Ms | Ms | H |
| (35) | H | Ms | Ms |
| (36) | Ms | Ms | Ms |
| (37) | H | Ms | H |

Treatment of the crude reaction mixture containing **33** with sodium methoxide using the reported conditions led to a complex mixture from which **34** could indeed be isolated by column chromatography, but in only 16% overall yield. We found that selectivity during the mesylation could be improved by carrying the reaction out at -20°C. The epoxidation was then carried out by adding aqueous sodium hydroxide to the crude reaction mixture in pyridine thus providing a simple direct one-pot transformation of **31** to **34**, also avoiding the use of chloroform. Using this process, pure crystalline **34** could be obtained without chromatographic purification in 35-40% overall yield. So with **34** in hand, we turned to its reduction.

When compounds such as **34** are treated with lithium triethylborohydride (Superhydride®) it is known that they can undergo ring opening of the epoxide ring to give **38** followed by ring closure to **39** and finally further reduction to give **22** (Scheme 8).[13]

Scheme 8: Mechanism of reduction of (**34**)

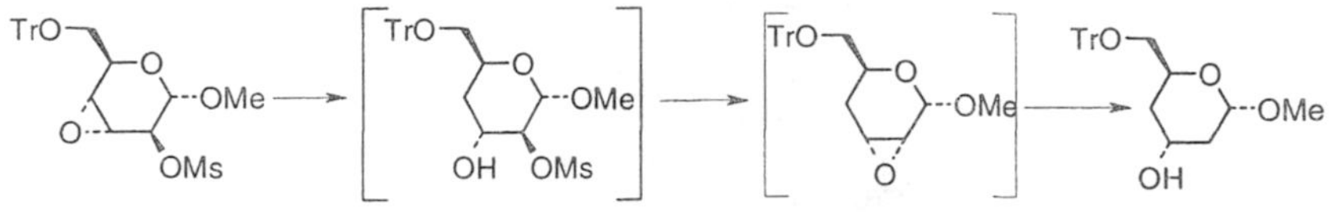

Obviously to carry out the reaction successfully and to arrive at the correct stereochemistry it is necessary to have the right epoxide stereochemistry and a *trans* relationship with the mesylate, hence the need to start from a mannose derivative.

Accordingly, when **34** was treated with 4 equivalents of lithium triethylborohydride in THF at room temperature **22** could be isolated by column chromatography in 69% yield. By silylating the crude product directly it was possible to prepare **23** from **34** without chromatography in 63% overall yield.[14]

Unfortunately, the use of lithium triethylborohydride has some very severe drawbacks: (i) it is expensive, (ii) it only contains one mole of reducing hydrogen per mole of reagent, and in practice a minimum of four equivalents are needed, (iii) it has a very limited commercial availability and (iv) the work-up requires the use of a very exothermic reaction with hydrogen peroxide to oxidatively hydrolyse the pyrophoric boranes. In an attempt to find other reducing agents we screened a large quantity of alternatives (Table 1).

Table 1. Alternatives to LiEt$_3$BH

| Conditions | Results |
|---|---|
| LiAlH$_4$ 2eq. THF RT | **39** (23%) + **40** (12%) |
| LiAlH$_4$ 4eq. THF RT | **38** + **40** (by TLC) |
| BH$_3$ NaBH$_4$ 2.4eq THF Rfx | No reaction |
| NaBH$_3$CN 3eq. BF$_3$.Et$_2$O 1eq THF Rfx | No reaction |
| REDAL® 1.8eq THF RT | Complex Mixture |
| DIBAL® 5eq. THF RT | **38** (63%) + **34** |
| CaCl$_2$ NaBH$_4$ 4eq MeOH Rfx | No reaction |
| LiBH$_4$ 4eq. RT | No reaction |
| ZnCl$_2$ NaH DME t-AmOH Rfx | detritylation |
| NaBH$_4$ 10eq. DMSO 140°C | **22** + **41** (~1:1) |
| NaEt$_3$BH 4eq. THF | Mixture of **22, 38, 40** and **41** |

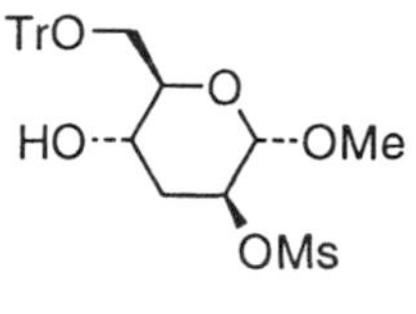

(40) 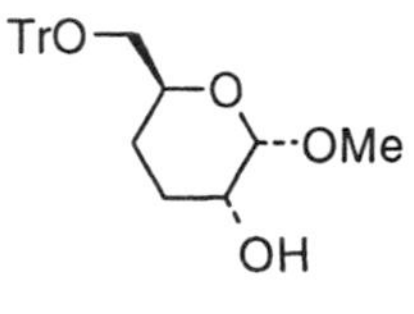                          (41)

Aluminium based reagents led to incomplete reduction (Table 1). Sodium borohydride in DMSO did reduce the first epoxide, but the regioselectivity of the reduction of the second epoxide (**39**) was poor, leading to substantial amounts of **41**.

Subsequent to this work, colleagues in the Rhône-Poulenc group discovered an alternative reductant: lithium borohydride/titanium tetraisopropoxide.[15] This reagent combination cleanly accomplishes the transformation **34** → **22**, and solves many of the above problems associated with the original lithium triethyl borohydride reductions.

**Speculative routes from glucose, mannose and galactose: via 3,6-anhydro-sugars**

Recognising the need to remove the hydroxyl groups at C-2 and C-4, whatever the configuration in the starting sugars, whilst protecting and manipulating C-3 and C-6, we examined chemistry to "tie" together C-3 and C-6, thus allowing C-2 and C-4 to be manipulated easily. The strategy behind this approach with a glucose precursor is shown in Scheme 9.

Scheme 9: General methodology for the 3,6-anhydro sugar approach

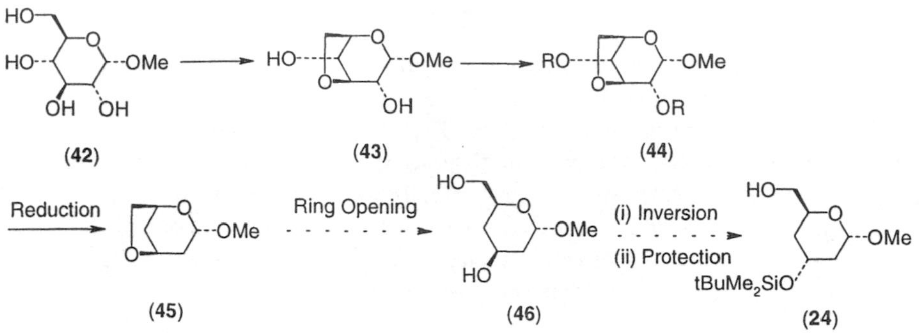

Starting from methyl-α-D-glucopyranoside (**42**), the 3,6-anhydro compound (**43**) was easily prepared[16] by converting the 6-hydroxy group into a leaving group (chloride) and treating with base. With the C-3 and C-6 hydroxy groups now removed from play, derivatisation of the naked C-2 and C-4 hydroxy groups to make them suitable for reductions was easily achieved (**43**→ **44**). However, when **44** (R=Ms or Ts) was treated with normal hydride based reducing agents, reduction was unsuccessful. Furthermore, Barton radical dehydroxylation[17] of the phenoxythiocarbonyl derivatives[18] ((**44**) R=PhOCS) with tributyl tin hydride was unsuccessful as the first formed radical at the 2 position was captured intramolecularly by the 4-phenoxythiocarbonyl group.[19] This problem was duplicated in the case of mannose as precursor. In both these two cases, with glucose and mannose as precursors, the incipient radical at C-2 is being trapped by the phenoxythiocarbonyl group at C-4, which being in the alpha configuration, is well positioned for this side reaction to become predominant.

However, when galactose was used as precursor (where C-4 has the beta configuration) the intramolecular trapping is not possible and we did manage to isolate **45**.[19] However, despite valiant attempts, opening of the ether bridge to generate **46** was unsuccessful and this route was abandoned.

**Return to the glucose routes: direct addition of methanol to TADG**

The methods we have described earlier for the preparation of 2-deoxy glycosides from TADG (**17**) involve three (in the case of mercuration) or four (in the case of iodination) steps. Unlike the addition of alcohol to dihydropyran to make tetrahydropyranyl ethers, it is well known that the direct addition of alcohol to TADG leads to 2,3-unsaturated glycosides, through the Ferrier rearrangement[20] as shown in Scheme 10. Indeed our preliminary attempts to achieve the direct methanolysis of TADG led to mixtures.

Scheme 10: The Ferrier rearrangement on attempted addition of methanol to TADG (**17**)

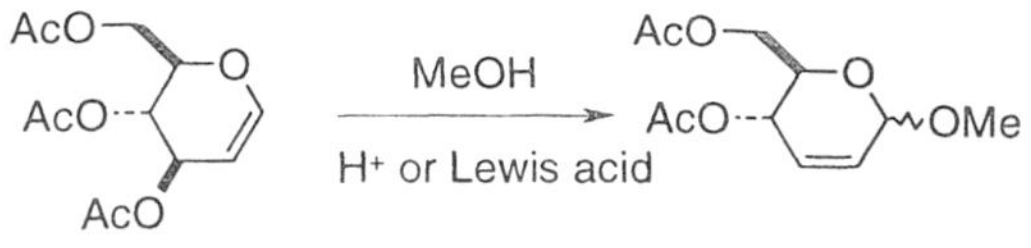

(**17**)                                                              (**47**)

However, a timely publication by Mioskowski and Falck[21] offered a method of resolving the problem of direct addition of methanol to **17**, thus solving many of the inherent problems encountered in the C-2 deoxygenation of glucose. The key feature of the work of Mioskowski and Falck in the addition of alcohols to glucals in dichloromethane was the use of triphenylphosphine hydrobromide (TPPHBr) as catalyst. This gave directly the 2-deoxy compounds, with no Ferrier product (Scheme 11). The only drawback was that when using simple alcohols, e.g. ethanol, the anomeric ratio was poor (only 78/22 α:β) although the yield was good (88%). Thus this method appeared ripe for some classical process development work.

Scheme 11. The direct addition of alcohols to TADG (**17**)

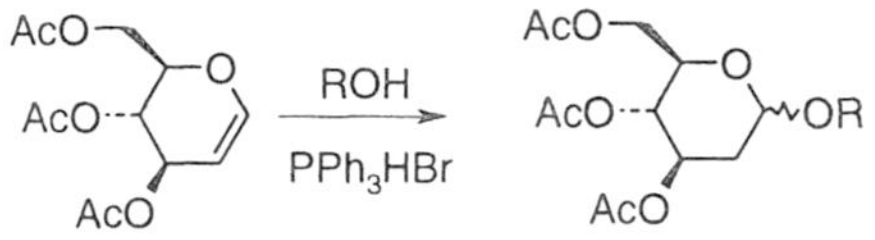

(**17**)                                                              (**48**)

Our first task was to prepare some TPPHBr, which at this time was not commercially available (it is now available from various laboratory suppliers). We rapidly developed a simple method which involved gassing a solution of triphenylphosphine in toluene with hydrogen bromide. The product is then simply isolated by filtration. This was scaled up to 45 kg of triphenylphosphine in a 500 litre vessel giving 95-99% yield of TPPHBr.

When the Mioskowski/Falck[21] conditions were applied to TADG and methanol the α/β ratio in our hands was found to be only 2:1 (see Table 2, entry 1). On further investigation, we found the reaction to be extremely solvent dependent, with the best stereoselectivities obtained in acetonitrile or dimethoxyethane (DME).[22] Surprisingly, we also found that aqueous hydrogen bromide catalyses the reaction (aqueous hydrochloric acid does not), although with poorer selectivity (entries 7-9). A solution of dry hydrogen bromide in DME resulted in a slower reaction and poorer selectivity than aqueous hydrogen bromide (entry 10), although this may be due to the expected instability of the DME solution of hydrogen bromide. A better result was obtained by use of a solution of hydrogen bromide in methanol in acetonitrile (entry 11).

The fact that hydrogen bromide catalyses the addition supports the view that TPPHBr acts as a hydrogen bromide carrier in the reaction. Thus, we believe that methanol catalyses the release of hydrogen bromide which then adds to the TADG to form the α-bromo derivative (**49**). The attack of such nucleophiles is known to give preferentially α-anomers.[23] As the major product of the reaction is also the α-anomer this suggests that the methanol does not displace the α-bromo substituent directly, but that the reaction must proceed via an oxonium intermediate (**50**) (Scheme 12).

Table 2. Direct methoxylation of TADG

| Entry | % | Catalyst | Solvent | Yield | α:β ratio |
|-------|-----|----------------|---------|---------|-----------|
| 1 | 5 | TPPHBr | CH₂Cl₂ | 75 | 2:1 |
| 2 | 5 | TPPHBr | MeCN | 86 | 4:1 |
| 3 | 10 | TPPHBr | MeCN | 81 | 8:1 |
| 4 | 20 | TPPHBr | MeCN | 81 | 9:1 |
| 5 | 50 | TPPHBr | MeCN | 82 | 10:1 |
| 6 | 20 | TPPHBr | DME | 81 | 10:1 |
| 7 | 20 | aq.HBr (48%w/w) | DME | 53 | 5:1 |
| 8 | 5 | aq.HBr (48%w/w) | DME | 38 | 2:1 |
| 9 | 5 | aq.HBr (48%w/w) | MeCN | 64 | 4:1 |
| 10 | 20 | HBr | DME | Mixture | 3:1 |
| 11 | 20 | HBr | MeCN | 93 | 5:1 |

Scheme 12: Proposed mechanism of direct methanolysis of TADG (**17**)

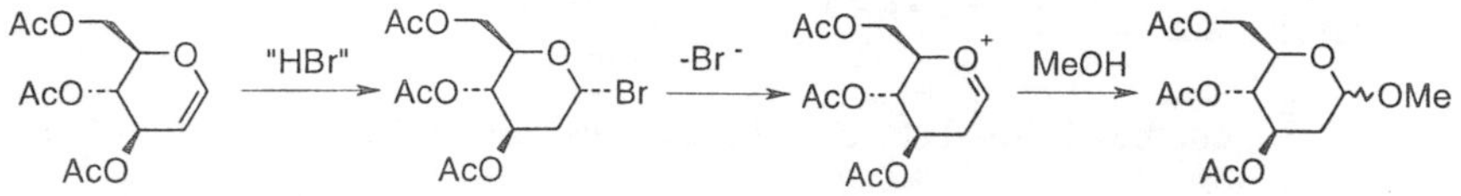

α > β

(17)          (49)          (50)          (48)

As DME is less toxic than acetonitrile this was selected as the solvent of choice using the conditions of entry 6, Table 2. As the mixture of anomers (**48**) and their hydrolysed products were non-crystalline, we developed a one pot process to enable TADG (**17**) to be transformed into **20** and for **20** to be isolated without any chromatography (Scheme 13).

Scheme 13: One-pot synthesis of key derivative (**20**) from TADG (**17**)

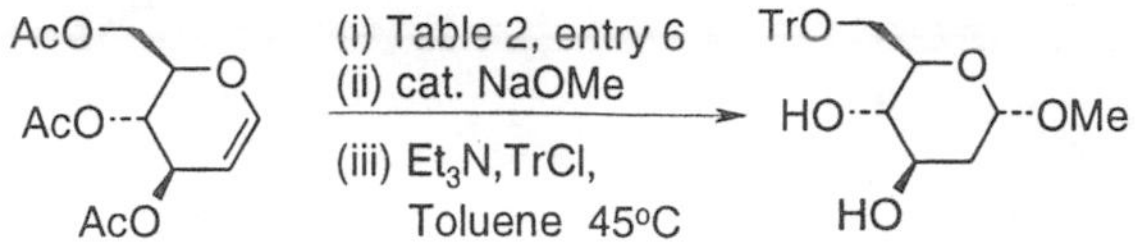

(i) Table 2, entry 6
(ii) cat. NaOMe
(iii) Et₃N,TrCl,
    Toluene  45°C

(17)                              (20)

Thus, after reaction with TPPHBr and methanol, neutralisation of excess acid together with the simultaneous methanolysis of the three acetyl groups were performed by addition of 0.3 equivalents of sodium methoxide solution. After stirring for one hour the excess sodium methoxide was quenched with 0.2 equivalents of triethylamine hydrochloride. This effected the neutralisation without the formation of water. Any residual methanol was removed by displacement of the solvents by co-distillation with toluene. Tritylation was carried out by addition of 3 equivalents of triethylamine and 1 equivalent of trityl chloride, stirring at 45°C for 12 hours. After work-up, **20** could be isolated by crystallisation in 60-65% overall yield, with α:β anomeric ratio of more than 30:1.

Thus, this procedure modification gave a 3-step one-pot synthesis of **20** which had many advantages: (i) the starting material is cheap and available (ii) the key reaction is highly diastereoselective (iii) the product is easily isolated in a purified form (iv) there are no toxic, expensive or poorly available reagents to use.

This synthesis was successfully scaled up in the Pilot Plant to 500 litre scale, using 27kg of TADG as batch input. A total of more than 200kg of **20** was produced, and this was subsequently transformed into the synthon (**25**) for coupling with the precursors for RP 61969 (**9**) and RG 12561 (**10**) respectively.

## Completion of the syntheses

The synthesis of RP 61969 (**9**) and RG 12561 (**10**) were completed in different ways, with the key coupling in the RP 61969 synthesis (Scheme 14) being a Wittig reaction[24] and for RG 12561 (Scheme 16) a Heck reaction.

Scheme 14: Final stages in the synthesis of RP 61969 (**9**)

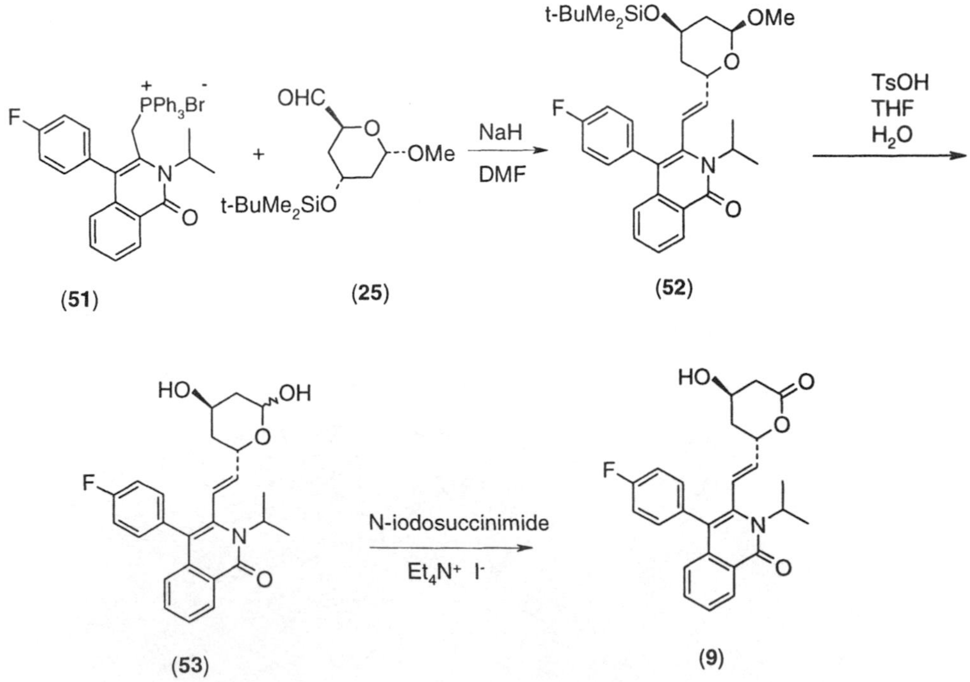

The aldehyde (**25**) was coupled with the phosphonium bromide (**51**) in a Wittig reaction to give the desired *E* double bond configuration of **52**. Acid hydrolysis of the protecting groups gave the lactol (**53**). Oxidation of the lactol (**53**) to the lactone (**9**) was carried out using tetraethylammonium iodide/N-iodosuccinimide (NIS) to complete the asymmetric synthesis of RP 61969 (**9**).

The strategy for the completion of the synthesis of RG 12561 (**10**) had to be somewhat different as the phosphonium bromide (**54**) required for a Wittig coupling proved very unstable in basic media due to its facile conversion into the diene (**55**) (Scheme 15).

Scheme 15: Decomposition of RG 12561 precursor phosphonium bromide

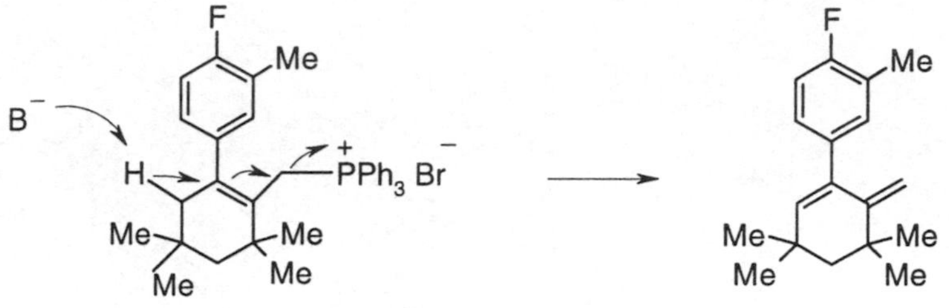

(54)               (55)

A Heck reaction proved to be the key reaction to couple the two halves of the molecule. Thus the aldehyde (**25**) was converted to the alkene (**56**) with methyl triphenylphosphonium bromide/BuLi. This alkene (**56**) could then be coupled with an alternative RG 12561 specific synthon, the vinyl iodide (**57**) using palladium bistriphenylphosphine dichloride in DMF with triethylamine as base to give **58**. Only the *E* double bond could be detected. Now the deprotection had to be undertaken in a stepwise fashion to overcome epimerisation at C-6 (tetrahydropyran numbering), and because of problems encountered due to the extreme hydrophobicity of the molecule. Thus the silyl protecting group was removed using tetrabutylammonium fluoride (TBAF) to give **59** followed by conversion to the lactol (**60**) with a catalytic amount of tosic acid in aqueous THF. The synthesis was again completed by oxidation of the lactol (**60**) with the tetraethylammonium iodide/N-iodosuccinimide (NIS) system used previously, giving finally RG 12561 in homochiral form (Scheme 16).

Scheme 16: Final stages in the synthesis of RG 12561 (**10**)

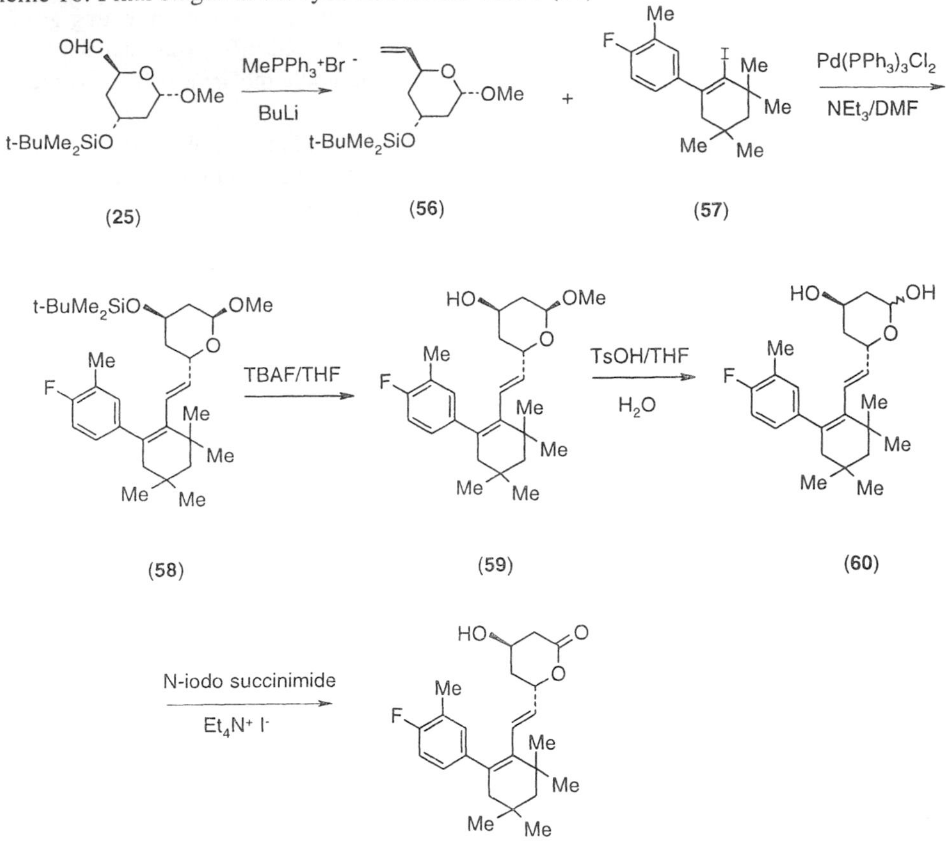

(25)                          (56)                          (57)

(58)                          (59)                          (60)

(10)

## Conclusion

There was a large synthetic effort in the 1980s and early 1990s towards the synthesis of the *statins*. Many of these routes start from precursors more expensive than glucose, or are very long. Alternative methods to homochiral compounds require the synthesis of the 3,5-dihydroxyacid chain onto the desired heterocycle (or carbocycle) and then resolving the open chain acid before ring closure,[25] but there is no easy method for racemisation of the unwanted enantiomer cleanly to the desired enantiomer, making resolution extremely expensive.

Of the routes described in this review, two are particularly noteworthy.

1. The route from glucose involving the direct addition of methanol to TADG catalysed by TPPHBr which was very rapidly scaled up from laboratory to 500 litre scale without problems to allow production of material quickly and safely.

2. The route from mannose, involving the regioselective dimesylation of methyl-D-mannopyranoside and its epoxidation and reduction. This route compares attractively as an industrial route of synthesis, to that above from glucose.

## Acknowledgements

We would like to acknowledge the many people in the Process Chemistry group at Dagenham who contributed to the work described in this review, both in the laboratory and the pilot plant. They are Jane Callow, Robert Chambers, Stan Clark, Geoff Darnborough, Chris France, Nigel Griffiths, Panicos Hadjigeorgiou, Bob Hewett, Ron Jones, John Myers, Clive Pemberton, Terry Perchard, Alan Thatcher, Colin Vale and Mike Webster.

We would also like to thank Process Chemistry colleagues at RPR (Collegeville, USA) and RP (CRIT, Lyons, France) for valuable ideas and support.

## References

1. Endo, A.; Kuroda, M.; Tsujita, I. *J. Antibiot.* **1976**, *29*, 1346.
2. Endo, A.; *J. Antibiot.* **1979**, *32*, 852-4.
3. Alberts, A. W.; Chen, J.; Kuron, G.; Hunt, V.; Huff, J.; Hoffman, C.; et al. *Proc. Nat. Acad. Sci. USA.* 77, 3957-61.
4. Merck and Co. Annual Report 1995. See http://www.Merck.com/overview.
5. Merck and Co. Annual Report 1996. See http://www.Merck.com/overview.
6. Ashton, M. J.; Fenton, G. In *Design of Enzyme Inhibitors as Drugs Vol. 2*; Sander, M.; Smith, H. J., Ed.; Oxford University Press, Oxford, 1994.
7. Stokker, G. E., Pitzenberger, S. M. *Heterocycles* **1987**, *26*, 157-162.
8. Rosen, T.; Taschner, M. J.; Heathcock, C. H. *J. Org. Chem.* **1984**, *49*, 3994-4003.
9. See for example Helferich, B.; Mulcahy, E. N.; Ziegler, H. *Chem. Ber.* **1954**, *87*, 233-7.
10. See for example Lemieux, R. U.; Levine, S. *Can. J. Chem.* **1964**, *42*, 1473.
11. WO 91 10,772. See *Chem. Abs.* **1991**, *115*, 183795y.
12. Tatsuta, K.; Koguchi, Y.; Kase, M. *Bull. Chim. Soc. Jpn.* **1988**, *61*, 2525-30.
13. For the use of LiEt$_3$BH in the reduction of $\alpha,\beta$-epoxy alcohols see: Kelly, A.G.; Roberts, J.S.; *Carbohydr. Res.* **1979**, *77*, 231. Edwards, M. P.; Ley, S. V.; Lister S. G.; Palmer, B. D. *J. Chem. Soc. Chem. Comm.* **1983**, 630. Baker, R.; Boyes, R. H. O.; Broom, D. M. P.; O'Mahony, M. J.; Swain, C. J. *J. Chem. Soc. Perkin I.* **1987**, 1613. Fox, C. M. J.; Hiner, R. N.; Warrier, U.; Whitc, J. D. *Tet. Lett.* **1988**, *29*, 2923. Paquette, L. A.; Oplinger, J. A. *J. Org. Chem.* **1988**, *53*, 2953.
14. WO 91 10,673. See *Chem. Abs.* **1991**, *115*, P183790t.
15. WO 92 08711 See *Chem. Abs.* **1992**, *117*, P69732s.
16. Castro, B.; Chapleur, Y.; Gross, B. *Bull. Soc. Chim. Fr.* **1973**, 3034.
17. Barton, D. H. R.; McCombie, S. W. *J. Chem. Soc. Perkin I* **1975**, 1574.
18. Robins, M. J.; Wilson, J. S.; Hansske, F. *J. Am. Chem. Soc.* **1983**, *105*, 4059.
19. Barton, D. H. R.; France, C. J.; McFarlane, I. M.; Newton, C. G.; Pitchen, P. *Tet.* **1991**, *47*, 6381-8.
20. Ferrier, R.J. *J. Adv. Carbohydr. Chem. Biochem.* **1969**, *24*, 199-266. Ferrier, R. J.; Prasad, N. *J. Chem. Soc. C.* **1969**, 570.
21. Bolitt, V.; Falck, J. R.; Lee, S.G.; Mioskowski, C. *J. Org. Chem.* **1990**, *55*, 5812-3.
22. France, C. J.; McFarlane, I. M.; Newton, C. G.; Pitchen, P.; Webster, M. *Tet. Lett.* **1993**, *34*, 1635.
23. Michalska, M.; Borowieka,J.; Lipka, P.; Rokita-Trygubowicz, T. *J. Chem. Soc. Perkin Trans. I* **1989**, 1619. Pelyvas, I.; Sztaricskai, F.; Szilagyi, L.; Bognar, R.; Tamas, J. *Carbohydr. Res.* **1979**, *68*, 321.
24. EP 326,386 **1989** See *Chem. Abs.* **1990**, *112*, 158072r
25. US 4,939,143 **1990** See *Chem. Abs.* **1991**, *114*, 23802z.

ENZYMATIC INTERVENTION

AND

PHASE TRANSFER CATALYSIS

# Chemistry, Biocatalysis and Engineering: An Interdisciplinary Approach to the Manufacture of the Benzodiazepine Drug Candidate LY300164

**Benjamin A. Anderson, Marvin M. Hansen, and Jeffrey T. Vicenzi**
*Lilly Research Laboratories, A Division of Eli Lilly and Company, Chemical Process Research and Development Division, Indianapolis, IN 46285*

**Milton J. Zmijewski**
*Lilly Research Laboratories, A Division of Eli Lilly and Company, Natural Products Discovery Development Research, Indianapolis, IN 46285*

Key words: AMPA receptor antagonist; yeast; reduction; resin; isochroman; autoxidation

## Introduction

Benzodiazepines serve a central role in drug therapies which target central nervous system disorders. Structure activity relationship (SAR) studies involving these agents have inspired the generation of thousands of analogs. The 1,4-benzodiazepine ring system, represented by clinically important drugs such as Diazepam (Valium®), has been the principal focus of much of this work.[1] Hungarian researchers at the Institute of Drug Research, however, have centered a drug discovery program on the structurally and pharmacologically distinct class of 2,3-benzodiazepine compounds. From this effort, LY300164 was selected for clinical development as a therapy for epilepsy and neurodegenerative disorders.

**Diazepam**                    **LY300164**

Several hurdles were presented to the chemical development of LY300164. The limited synthetic studies involving the 2,3-benzodiazepine ring system documented only a few strategic options for the preparation of the heterocyclic system.[2] These options were further limited by the need to produce the candidate in optically pure form. The account presented herein details both strategic and tactical issues which confronted our development effort and reviews how these challenges were met. The undertaking culminated in the design of a novel synthetic route for the production of the optically pure 5H-2,3-benzodiazepine ring system on multikilogram scale with outstanding control of stereochemistry. An unusually effective combination of chemistry and biocatalysis which was enabled through inventive engineering was pivotal to the successful execution of the process.[3]

## Mechanism of action

Glutamate is the principal excitatory neurotransmitter in the mammalian brain and spinal cord. The excitatory effects of glutamate in many synapses are countered from inhibitory input by γ-aminobutyric acid (GABA). The balance of input from these neurotransmitters dictates the degree of excitation at a synapse. Seizure disorders, collectively described as epilepsy, result as a consequence of excessive excitatory neurotransmission.[4]

Drugs that elicit inhibitory GABA input are associated with several anticonvulsant therapies.[5] Conversely, glutamate antagonists might also ameliorate excessive excitatory neurotransmission. From this rational, LY300164 emerged as a clinical candidate. LY300164 is a non-competitive, or allosteric, antagonist of a subtype of glutamate receptor which is characterized by its selective affinity for α-amino-3-hydroxy-5-methyl-4-isoxazolepropionic acid (AMPA).

Glutamate antagonists have clinical potential beyond the treatment of epilepsy. Prolonged excitatory neurotransmission, a process described as glutamate excitotoxicity, has also been implicated in a variety of acute and chronic neurodegenerative disorders. Selective glutamate antagonists may also be effective in the treatment of amyotrophic lateral sclerosis (ALS), cerebral ischemia and Parkinson's disease. However, such targeted glutamate antagonists remain to be clinically validated.

## Medicinal chemistry

GYKI 52466, a non-competitive AMPA antagonist discovered by researchers at the Institute of Drug Research, was the progenitor for the series of derivatives generally represented by **1**.[6] Extensive pharmacological testing of GYKI 52466 demonstrated its potency as a muscle relaxant and anticonvulsant.

The SAR from which LY300164 emerged was fueled by the discovery that the double bond between the 3 and 4 position of the benzodiazepine could be selectively reduced.[7] Acylation of the resulting ring system provided access to several new compounds (**1**) which proved to have potent *in vivo* anticonvulsant activity.[8] The activity of the *N*-acetyl derivative was notable among these new derivatives.

GYKI 52466                   1

The new stereocenter thus generated bore the potential to influence binding affinity of the compound. Indeed, upon separation of the resulting enantiomers, the (-) isomer (LY300164) was significantly more potent than the (+) isomer in blocking maximal electroshock induced convulsions in mice following oral administration. In addition to seizure disorders, preclinical models also support the candidacy of LY300164 as a therapy for a variety of acute and chronic neurodegenerative disorders.

## Process evaluation

The original preparation of LY300164 clearly reflected its evolution from GYKI 52466.[7] Furthermore, as is the case with many drug candidates, the synthesis was born from a synthetic strategy designed primarily to support the SAR development by providing facile access to many related structures.[7,8] When a candidate is selected for clinical development, the synthetic approach pursued for its production does not suffer the same constraints as a discovery route but must instead accommodate the familiar boundaries imposed by economy, safety and environmental impact. However, the original synthesis need not necessarily be abandoned in favor of an alternative for large scale production.

Indeed, optimization of the discovery synthesis of LY300164 positioned it as a reasonable candidate as a production process (Scheme 1). The eight step synthesis began with the sodium borohydride-mediated reduction of commercially available 3,4-methylene-dioxyphenyl acetone **2**. An acid-catalyzed Prins-type cyclization of the resulting alcohol **3** with *p*-nitrobenzaldehyde generated an isomeric mixture of the crystalline isochroman **4**. The heterocycle was then subjected to oxidation by Jones reagent to yield a diketone intermediate **5**.

Direct incorporation of the nitrogen constituents of the benzodiazepine nucleus could not be accomplished cleanly. Rather, the acid-catalyzed reaction of **5** with hydrazine led to competitive production of *N*-aminoisoquinoline **11**.[2h] Stepwise activation of the substrate by conversion to the oxonium salt **6** by reaction with tetrafluoroboric acid provided a suitable substrate into which hydrazine was readily incorporated to yield the desired benzodiazepine nucleus **7**.[9] This intermediate provides the differentiation point between GYKI 52466 and the newer series of compounds including LY300164.

Reduction of the nitrogen-carbon double bond between positions 3 and 4 of **7** could be accomplished stereo-randomly with sodium borohydride. More desirably, reduction employing the combination of the amino alcohol ligand **14** and borane dimethylsulfide led to generation of intermediate **8** which possessed the stereochemistry required for LY300164.[7]

Scheme 1. Modified discovery synthetic route to LY300164

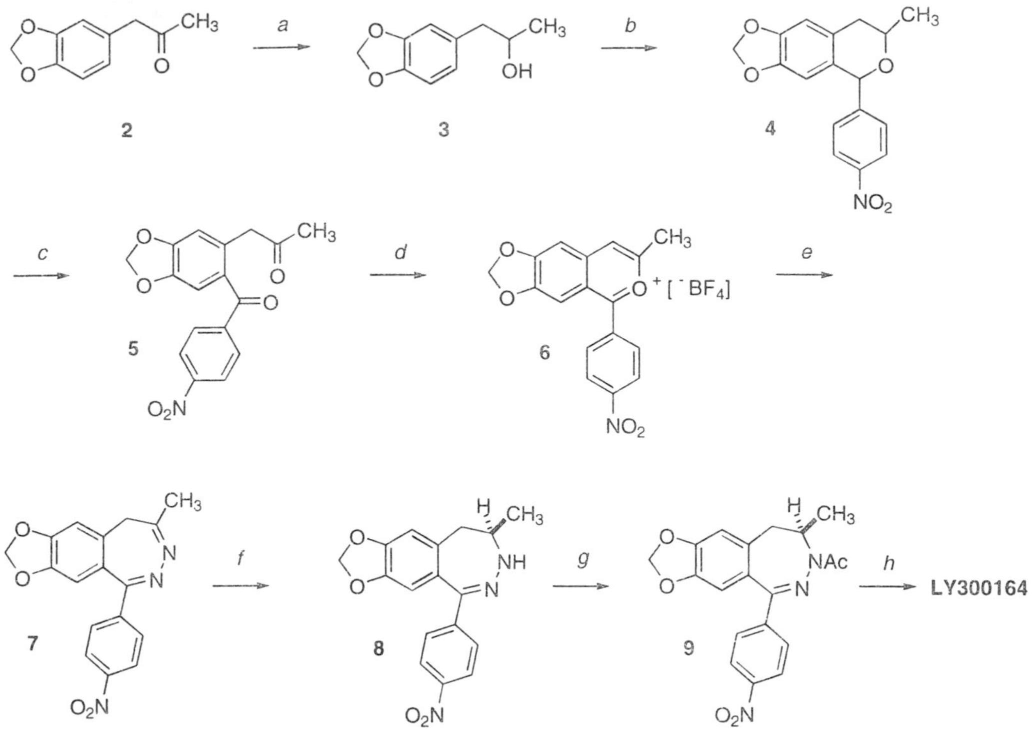

Reaction conditions: (*a*) NaBH$_4$ (*b*) *p*-NO$_2$PhCHO, HCl (*c*) CrO$_3$, H$_2$SO$_4$ (*d*) HBF$_4$·OMe$_2$ (*e*)N$_2$H$_4$ (*f*) BH$_3$·SMe$_2$, **14** (*g*) Ac$_2$O (*h*) H$_2$, Pd/C

Optimization of the challenging asymmetric reduction was the focus of significant study. The use of stoichiometric quantities of **14** and borane was required to obtain high levels of enantioselection. The nature of the amino alcohol ligand influenced the reaction with optimal stereoselectivities achieved by employing the chiral modifier derived from D-leucine (eq 1). Extended reaction times, initially as long as 3 days at room temperature, could be reduced to 3.5 hours if the reaction was conducted at 60 °C with little impact on selectivity.[10] The optical purity of the product could be enriched by recrystallization from 73% ee to 96% ee in moderate yield (56%). Acylation of **8** was effected by reaction with acetic anhydride to give the penultimate intermediate **9**. Reduction of **9** provided LY300164 in approximately 90% yield.[11]

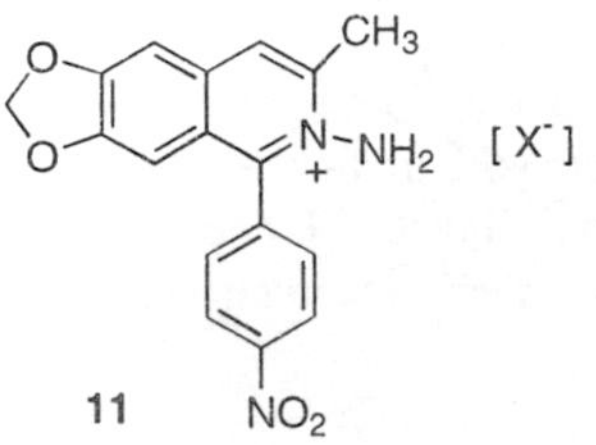

The modified discovery synthesis proved quite reproducible and reliably delivered the desired product in approximately 14% overall yield. Importantly, the route provided the early kilogram quantities necessary to fuel the initial stages of clinical development.

The synthesis could not be considered a practical manufacturing route, however, without several key process modifications. The most significant issues were symptoms of an overall strategic problem which centered on excessive manipulation of oxidation state. The carbon represented as C-4 in the final product underwent two redox changes prior to the final reduction step which set the required stereochemistry. The modest efficiency (56% yield) of the asymmetric reduction was suffered late in the synthesis and the use of stoichiometric quantities of an expensive chiral modifier was required.

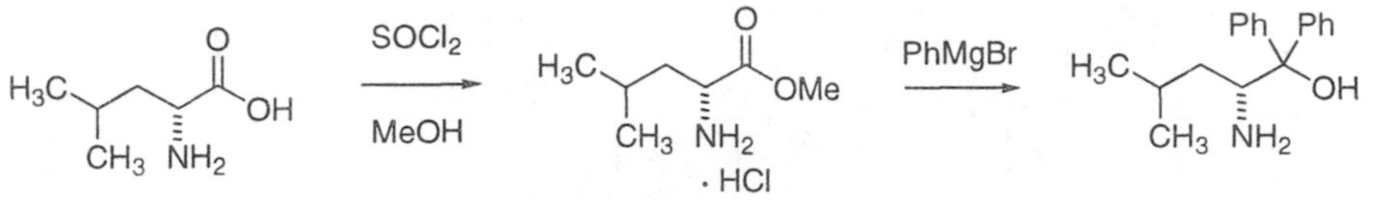

The oxidative cleavage of **4** employing Jones reagent represented an even greater tactical problem. The reaction was inefficient (47% yield) and only moderately reliable. However, the large quantity of chromium by-product was the foremost concern. For every kilogram of LY300164 that was generated, approximately three kilograms of chromium waste was produced. Modification of this step became the focus of our effort since identification of a favorable alternative to the metal-based oxidant was expected to have both short and long term benefits.

An unanticipated consequence of our focus on the oxidation was the emergence of an entirely new synthetic route to LY300164. The new synthesis successfully met all the criteria sought in a manufacturing synthesis and provided the stage for an exceptionally productive collaboration between microbiology, chemistry and engineering.

## Process research

### Isochroman oxidation

Although oxidation reactions involving isochromans have been studied in some detail,[12] chromium-based oxidants are the only reagents known to effect ring cleavage to yield the corresponding diketone.[13] Our attempts to replace chromium were unsuccessful. Reagent combinations led to one of two seemingly undesirable results. Products arising from either selective C-1 oxidation or fragmentation of the substrate were invariably encountered (eq 2). Application of halogen-based reagents led to reaction at the electron rich aromatic ring.

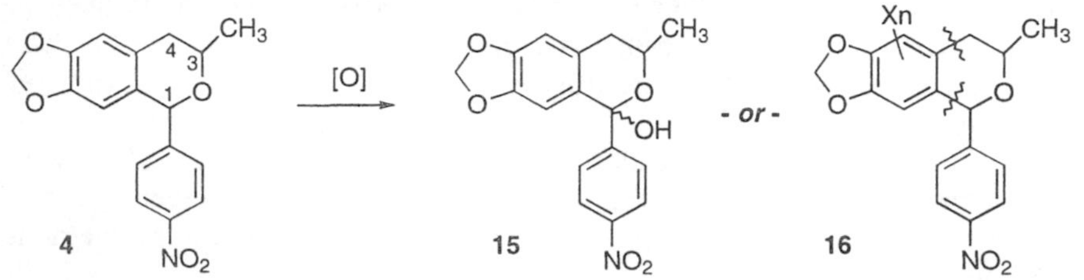

Several benefits were realized from the isolation and characterization of the hemiketal **15** which resulted from selective C-1 oxidation. Thibault and Maitte postulated such compounds were intermediates in the $CrO_3$ / acetic acid mediated oxidative cleavage reactions of isochromans to diketones.[13b]    Indeed, HPLC analysis of reactions in which the isochroman was treated with aliquots of Jones reagent at 0 °C revealed the initial formation of **15** without evidence of **5**. The hemiketal disappeared and **5** was observed upon addition of the standard excess of the chromium reagent and warming to room temperature (eq 3). Reproducibility of the process was improved by carefully monitoring for complete disappearance of **15** which ensured that the reaction endpoint was consistently achieved.

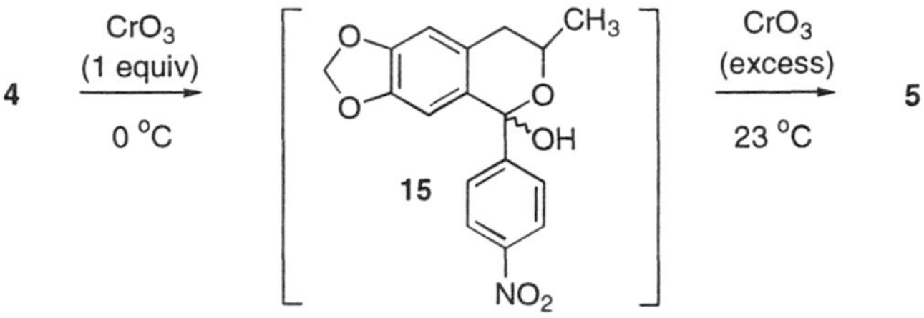

## Strategy options

Accomplishing the second oxidation at C-4 was critical to the success of the synthesis. As discussed earlier, the transformation was strategically uneconomical since a reduction of C-4 was ultimately necessary to install the required stereocenter. Careful analysis of **15**, which was at the correct oxidation state of the final product, prompted consideration of a radically different synthetic approach. Construction of the required benzodiazepine ring system from the hemiketal simply required insertion of the nitrogen constituents by means of hydrazone formation at the masked carbonyl followed by dehydrative cyclization (Scheme 2). Retrosynthetic analysis suggested that the C-4 stereocenter might be derived from a relatively simple chiral alcohol.

Scheme 2. Proposed retrosynthesis of LY300164

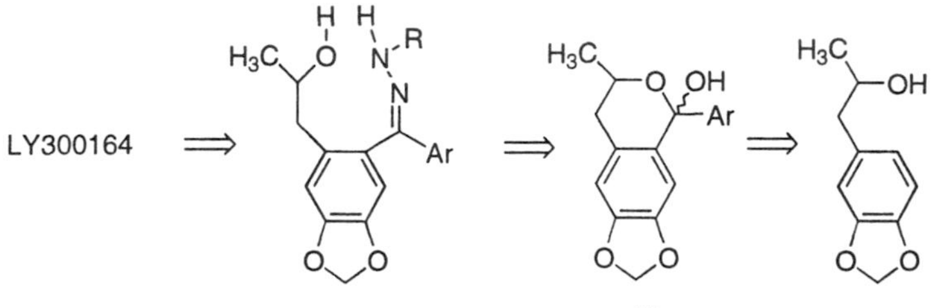

Several options for preparation of **15** were available and the viability of the strategy was first established by the preparation of racemic penultimate intermediate **20** (Scheme 3). Oxidation of **4** with an unbuffered aqueous potassium permanganate solution in acetone at 0 °C afforded **15** in 70% yield. Analysis by $^1$H and $^{13}$C NMR spectroscopy indicated that **15** existed as a ~4:1 mixture of isomers. Facile interchange between the diastereomers was presumed and no efforts were taken to separate the racemic diastereomers.[14]    Treatment of the hemiketal diastereomers with acetic hydrazide cleanly produced the corresponding hydrazone **19** as a 3:1 mixture of geometric isomers about the C=N bond.

Scheme 3. Synthesis of racemic LY300164

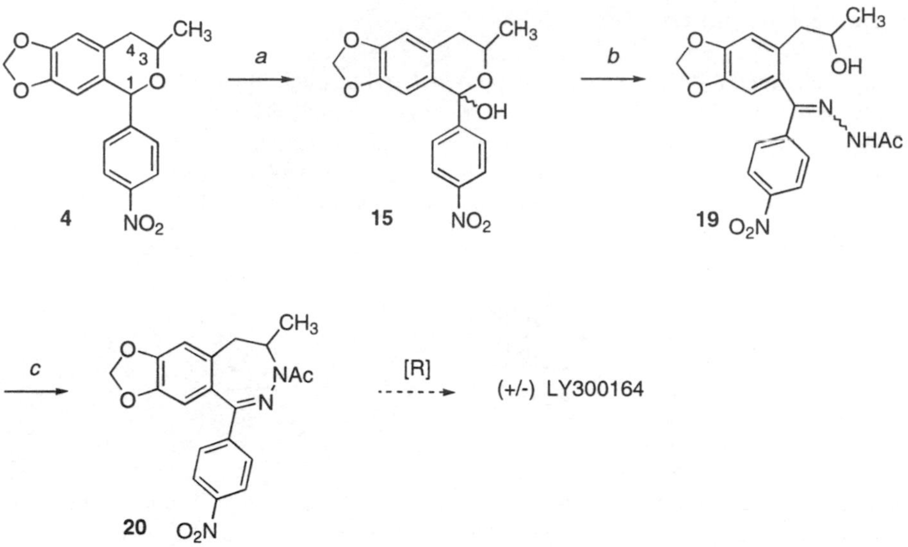

Reaction conditions: (a) KMnO$_4$, 0 °C (b) H$_2$NNHAc (c) DIAD, PPh$_3$

Ring closure to **20** was successfully accomplished employing standard Mitsunobu conditions. High conversion to **20** (>90%) indicated that the two hydrazone isomers interconverted through a common tautomeric structure which converged to a reactive geometry (Figure 1). Isolation of the product was hampered by the typical by-product issues associated with the Mitsunobu reaction and resulted in only 70% yield. Most importantly, the few unoptimized experiments provided a clear direction for development of a new synthesis of LY300164 through the hemiketal.

Figure 1. Resonance structures of hydrazone anion intermediate

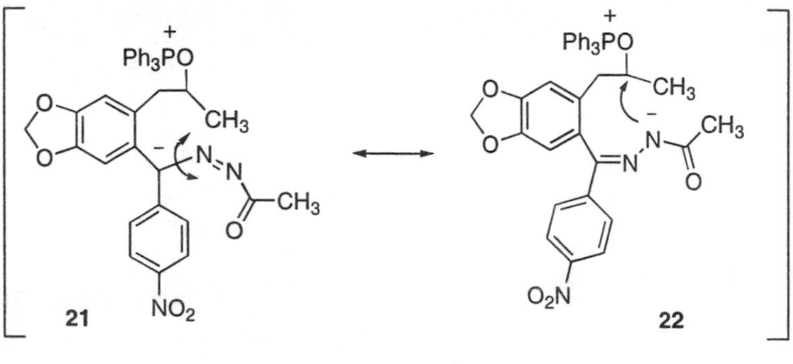

## Process development

### Asymmetric alcohol synthesis

The challenge remained to advance optically pure material through the synthesis. Hence, access to the optically active isochroman via the corresponding alcohol **24** was necessary. Targeting a specific antipode of the alcohol was not an immediate concern since the absolute configuration of LY300164 was unknown at the time. Copper iodide-catalyzed reaction of the Grignard reagent derived from 1-bromo-3,4-methylenedioxybenzene with (*S*)-propylene oxide gave the (*S*) alcohol **24** in 91% yield (>98 % ee) (eq 4) .

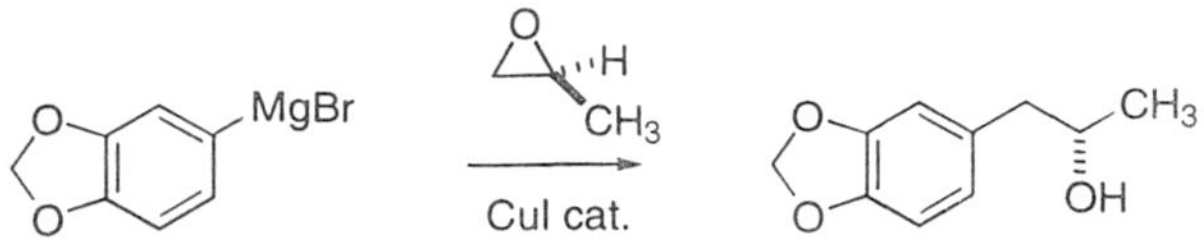

$$(4)$$

         **23**                                              **24**

Compound **24** was then processed to the penultimate intermediate **9** employing the reaction conditions summarized above (Scheme 3). High fidelity of stereochemistry was confirmed by chiral HPLC analysis of **9** which coincidentally corresponded to the required enantiomer. The absolute configuration of LY300164 could be assigned as the (*R*) enantiomer if uncomplicated inversion of the stereocenter in the Mitsunobu reaction was assumed. This proved to be the case as the absolute configuration of the product was unequivocally proven by single crystal X-ray analysis of a diastereomerically pure derivative **25** (Figure 2).[3d]

Figure 2. X-ray crystal structure of diastereomerically pure derivative **25**

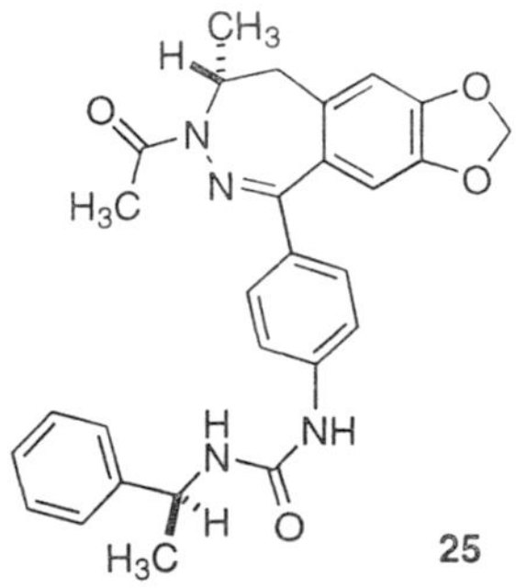

         **25**

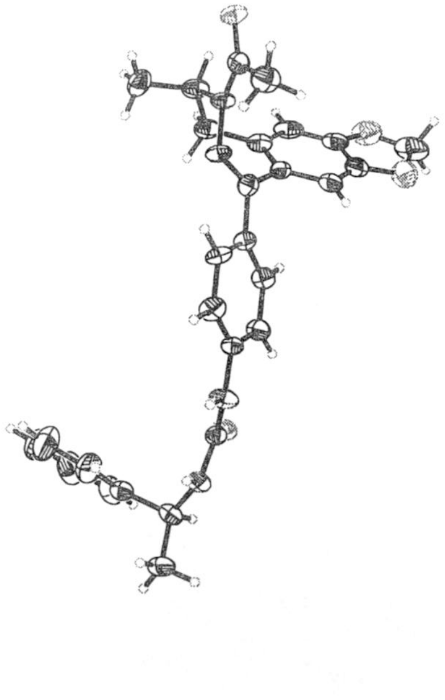

The new synthesis therefore required a method which would accommodate large scale production of the (*S*) enantiomer of the alcohol **24**. (*S*)-Propylene oxide was expensive and available only in lab scale quantities (1 g). Synthesis of the epoxide has been reported, but the multiple steps which are required for its preparation limited the appeal of this strategy.[15]

## Asymmetric ketone reduction

A more appealing approach to the optically active alcohol involves the asymmetric reduction of ketone **2**. Chemical methods for the enantioselective reduction of phenyl acetone derivatives are typically unselective or require the use of costly reagents.

The stoichiometric applications of optically pure boron (**26**)[16] or aluminum hydrides (**27**),[17] for example, are reported to effect asymmetric reduction of phenyl acetone with good stereocontrol. However, preparation of these reducing agents is labor intensive and demands the recovery and regeneration of the expensive reagents.

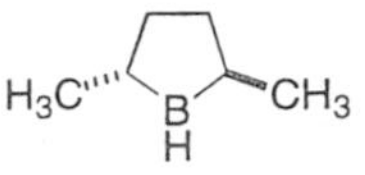

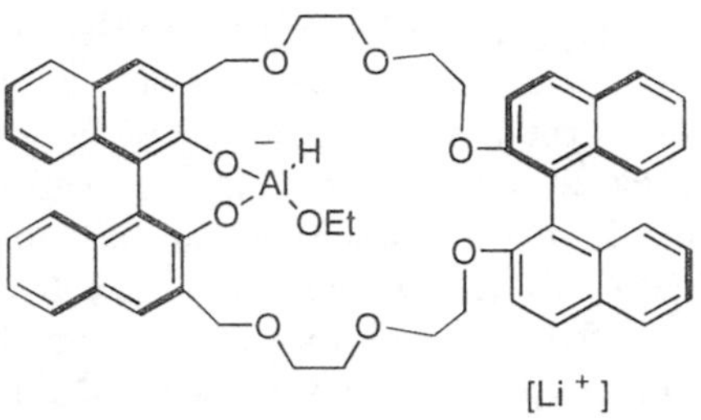

Limited success has been reported for the catalytic reduction of related ketones. Oxazaborolidine-catalyzed ketone reductions have had significant impact on the preparation of enantiomerically pure pharmaceutical agents but are generally unsuccessful for benzylic ketones.[18] Moreover, catalysts prepared from naturally occurring amino acids typically result in production of alcohols with absolute (*R*) configuration. A notable exception is the catalyst derived from (*S*)-indoline-2-carboxylic acid (**28**) which is reported to deliver the corresponding (*S*) alcohol of phenylacetone in high yield (91%) with good stereocontrol (86% ee).[19] Our attempts to employ **28** for the preparation of **24** were unsuccessful. Alternatively, rhodium-catalyzed hydrosilylation employing **29** effects reduction of phenyl acetone in 84% yield with reasonable stereocontrol (80% ee). Once again, reagent costs and waste stream concerns limited our interest in pursuing large scale development of this or related technologies. Instead, our attention was drawn to biocatalytic methods since the most successful examples of the reduction of phenyl acetone had been reported employing baker's yeast.[20]

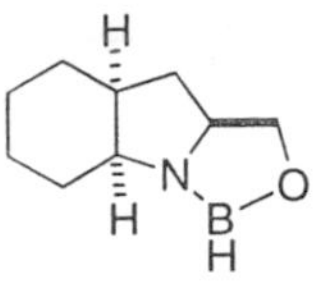

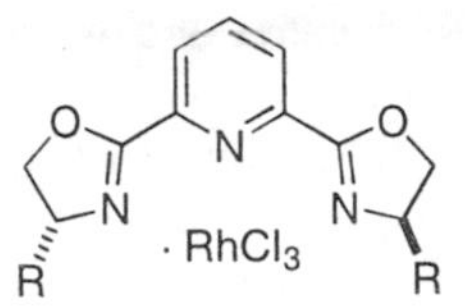

## Biocatalytic ketone reduction

The use of baker's yeast to mediate asymmetric ketone reductions is well known, but like other biocatalysts, its general application to organic synthesis is often limited by low yields and modest scope.[21] Functional issues associated with dilute conditions and complicated reaction workups often represent further practical limitations of biocatalysts. It was therefore disappointing, yet not unanticipated, that baker's yeast was an ineffective reducing agent for **2** (11% yield, 24 hours). Undeterred by the initial result, microbial libraries were screened in search of a suitable enzyme.[3c] The screen was limited to whole cell microorganisms which offer the advantage of *in vitro* recycling of the NAD(P)H cofactor required for reductase activity. Although successful large scale reductase reactions which employ purified enzymes with *in-situ* co-factor recycling have been described,[22] this approach is typically more complicated, expensive and development intensive compared to the use of living cells.

The reductase activity of *Candida famata* was distinguished from a preliminary screen of approximately fifty organisms. The yeast produced the alcohol in gram quantities with the correct absolute configuration and excellent optical purity ( >98% ee). However, additional work with this organism was disappointing. The enzyme levels were low but sufficient to support development of a dilute (2 gm/l) process. The organism was sensitive to ketone or alcohol concentrations above 2 gm/l. Cell death induced by the toxicity of both reaction components prevented recycle of the co-factor and resulted in incomplete reaction.

Evaluation of the *C. famata* mediated process on a 1000 L scale provided convincing evidence of its limited long term utility. The reduction was achieved by adding 2 g/L of **2** directly to *C. famata* whole broth. After a reaction time of 24 hours, the broth was centrifuged to remove cell solids. Extraction of the product from the centrate with organic solvents was complicated by persistent emulsions and yielded only 69 wt % of low quality material (98% ee, 54% chemical purity).

The inadequacies of the dilute process involving *C. famata* clearly pointed to the need for a more resilient and active microorganism. Continued screening of yeast and fungi libraries led to the identification of *Zygosaccharomyces rouxii* as a more suitable candidate. The yeast had several attributes which prompted a more thorough exploration. The initial screen of the organism resulted in 78% conversion to the alcohol at 10 g/L substrate concentration with excellent stereocontrol (>99.9% ee). Examination of the enzymatic conversion indicated that *Z. rouxii* had 6 fold greater enzymatic activity compared to *C. famata*. Furthermore, the yeast is classified as a *generally regarded as safe* (GRAS) organism and has been used in the food industry for centuries to prepare miso and soy sauce. Therefore, operations involving the microbe did not require special safety and disposal precautions.

## Biocatalytic process scale-up

The generally favorable characteristics of *Z. rouxii* were compromised by its limited viability in the presence of high concentrations of organic species. Albeit not as dilute as required for *C. famata*, 5-7 g/L represented the maximum concentration at which complete reaction occurred (Figure 3). The sensitivity to high product concentrations was particularly burdensome. Slow addition would circumvent the poor tolerance of high concentration of **2**. Product inhibition, however, demanded that the concentration of **24** be controlled by its removal upon generation. Biphasic reaction mediums employing immiscible organic solvents have been successfully applied to biocatalytic reactions in which the active organism suffers similar intolerances.[23] Organic solvents, as is common to many microorganisms, proved too toxic for *Z. rouxii* to be employed in a

Figure 3.  Effect of ketone concentration on *Z. rouxii* mediated reduction

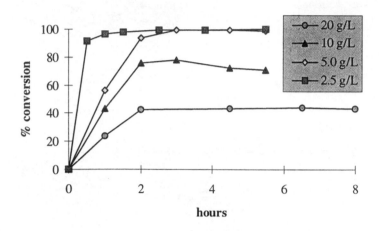

two solvent phase system.[24]

A related biphasic reaction design proved to be a successful alternative. [3b, 23e] The reductase reaction could be conducted at 7-fold greater substrate concentrations  by addition of a polymeric resin (XAD-7) to the aqueous medium (Figure 4).  Furthermore, this modification yielded a simple solution to problems associated with product isolation and purification.  The protocol involved charging **2** to an aqueous slurry containing the resin, buffer and glucose.  Approximately 80 grams of **2** was added for every liter of resin.  The ketone, which has only sparing water solubility (3.5 g/L), was primarily adsorbed on the non-polar resin surface.  *Z. rouxii* was added to the agitated mixture and allowed to react with the equilibrium concentration of approximately 2 g/L of **2** remaining in the aqueous phase.  Upon its production, **24** was adsorbed to the resin surface thereby limiting its aqueous phase concentration to sublethal levels.  At the end of the reaction , the phase equilibrium provided ~ 75-80 g/l of resin-bound **24**  with ~ 2 g/L of **2**  remaining in the aqueous phase (Figure 5).

Figure 4. Schematic of 3 phase reaction medium

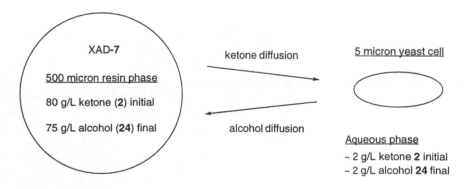

Figure 5.  Phase concentration diagram of reduction reaction

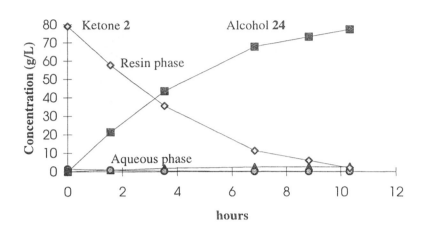

Several factors played important roles in the success of the reaction.  High product affinity for the resin was desirable, but to ensure complete reaction, it was necessary to select a resin for which the substrate association was not too great.  The use of highly polar resins such as HP20 or SP207, for example, resulted in incomplete reactions.  The activity of *Z. rouxii* was not lost when the reaction was conducted under anaerobic conditions.  This was fortuitous since the high agitation shear necessary to achieve aeration shut down enzymatic reduction.  It appeared that the highly agitated slurries of the resin mechanically destroyed the *Z. rouxii* cells.

The size differential of the resin (~ 500 mm) and the yeast cells (~ 5 mm) enabled the product containing resin to be collected on a 150 micron filter screen at the end of the reaction.  After washing with water, **24** was liberated from the resin by rinsing with acetone.  Approximately 85% of **24** was thus recovered.  An equilibrium concentration of 2-3 grams per liter, representing 10-15% of the total yield,  remained in the aqueous phase.  To recover this yield, the resin was first washed with additional water to displace the acetone.  Then, the original water washes were passed over the resin, adsorbing ~95% of **24** back onto the resin.  The partially product loaded resin was then immediately reused for subsequent reductions.  The resin could be recycled in this manner many times, bringing the overall isolated yield to 95%.

**Chemistry issues**

Pilot studies were conducted to ensure that **24** could be forward processed to deliver LY300164 in good yield with high stereochemical fidelity.  Several transformations which were relied upon in that exercise were untenable for large scale application.  The potassium permanganate oxidation of **4** presented waste issues and produced material of only marginal purity.  The key Mitsunobu reaction which effected ring closure provided even more troublesome purification issues.  A two step process involving silica gel chromatography followed by recrystallization was required to liberate **9** from phosphine and azodicarboxylate derived by-products.  Fortunately, a variety of options were available to address each issue.

### Oxidation scale-up

The oxidation of the ioschroman was revisited in an effort aimed toward rendering an optimized process for the production of optically pure hemiketal. The propensity of isochromans to undergo autoxidation to yield the corresponding peroxide and peroxide dimer provided an appealing direction for these studies (eq 5). Both the parent and 1-alkyl isochroman derivatives undergo facile autoxidation exclusively at C-1 by the action of atmospheric oxygen under UV irradiation to give mixtures of the hydroperoxide and peroxide dimer.[25] A related process would be a particularly appealing alternative to the chemical oxidation upon which our synthesis relied. Unfortunately, C-1 aryl isochromans have been reported to resist reaction under the irradiation conditions.[25b]

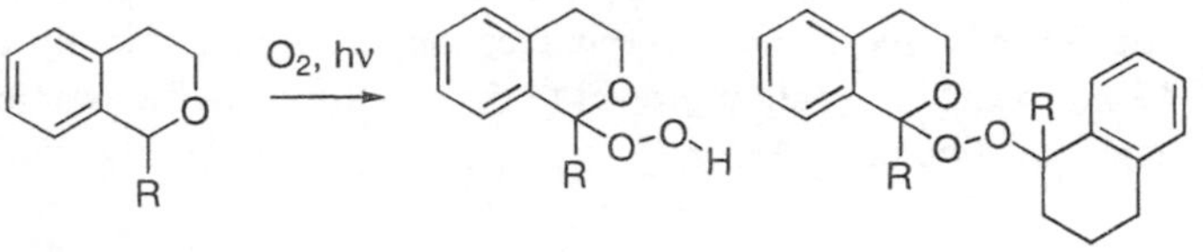

| | | |
|---|---|---|
| **30a** R = H | **31a**, 51% | **32a**, 42% |
| **30b** R = CH$_3$ | **31b**, 37 | **32b**, 55 |
| **30c** R = Ph | **31c**, 0 | **32c**, 0 |

Russell's seminal work involving related systems suggested that **33** might succumb to a base-mediated autoxidation reaction.[26] Indeed, complete conversion to **35** was accomplished within 2 h when sodium hydroxide was added to an air saturated DMSO solution of **33** (eq 6). Carbanion formation was facilitated by conducting the reaction in the polar aprotic solvent. Furthermore, the dimethylsulfoxide served a critical role by reducing the initially formed peroxide anion **34**.[27]

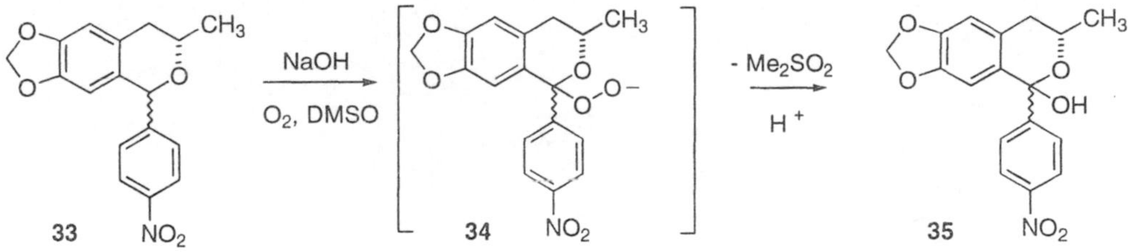

A long-lived intermediate, presumed to be the peroxide anion **34**, was observed when the reaction was carried out in dimethylformamide. The intermediate disappeared immediately when dimethylsulfoxide was added to the mixture. Since conducting the reaction at lower temperatures (0-15 °C) provided cleaner products, a mixed DMF/DMSO solvent system was adopted which not only suppressed the freezing point of the polar medium but also provided the capacity to reduce the peroxide anion.

Clean conversion of **33** to **35** required generation of the carbanion in the presence of oxygen. Poor air flow into the reaction mixture led to a competing process in which a pseudo-dimeric compound **36** was generated (eq 7). Formation of **36** could be explained by oxidation of **33** at C-1 by the nitroaromatic rather than oxygen (Scheme 4).[28] A second related electron transfer process would be required to generate the hydroxylamine **40**.

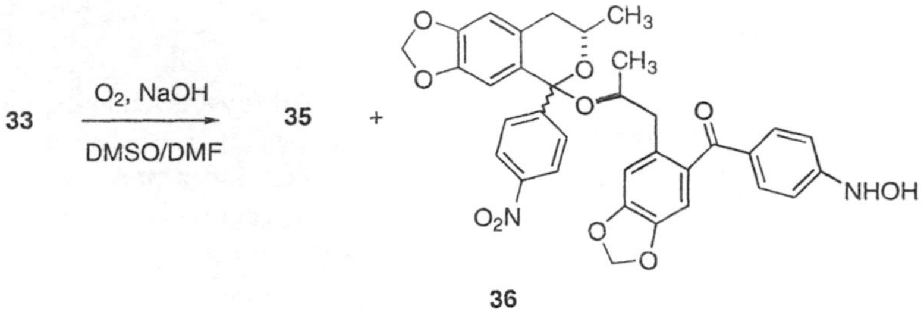

(7)

36

Reduction of nitroaromatic groups by carbanions are well known but typically result in the production of azo and azoxy dimers.[29] The connectivity of the hemiketal constituents is explained by the relative electrophilicity of the corresponding *para*-hydroxylamine and *para*-nitro carbonyls. Although production of **36** was mechanistically interesting, sufficient air flow avoided the competing process. The autoxidation reaction proved to be a significantly favorable alternative to the original permanganate oxidation process.

Scheme 4. Proposed mechanism for the production of **36**

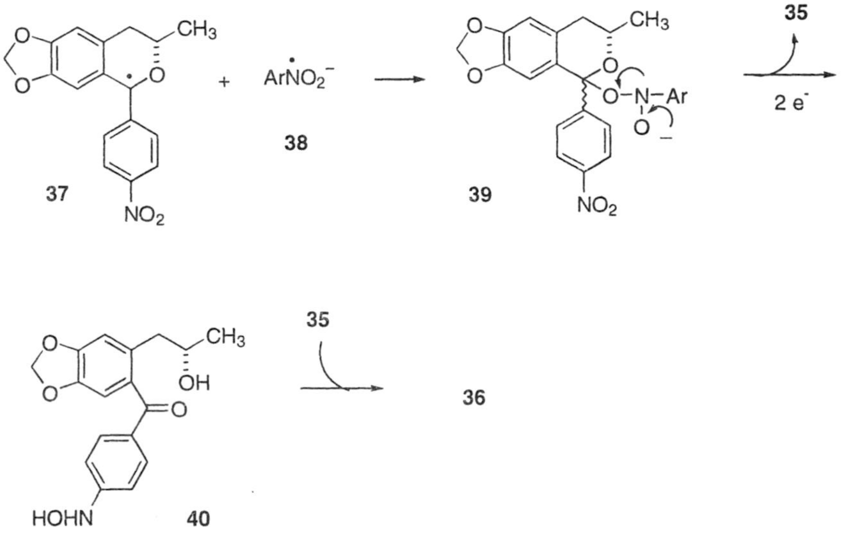

### Cyclization scale-up

Having overcome issues associated with the isochroman oxidation, our attention turned to development of a ring-forming protocol that avoided the inherent purification problems of the Mitsunobu reaction. The most elementary strategy was attempted first. Treatment of the isomeric mixture of hydroxy-hydrazone **41** with methanesulfonyl chloride in the presence of triethylamine led to the production of a mixture of the corresponding mesylate esters **42** (eq 8). Products arising from *N*-mesylation (*e.g.* **43**) were not detected. Addition of potassium *tert*-butoxide to a tetrahydrofuran solution of **42** effected rapid and clean cyclization to the desired benzodiazepine nucleus **9** (eq 9). Chiral HPLC analysis of the resulting product indicated that complete inversion of stereochemistry was accomplished.

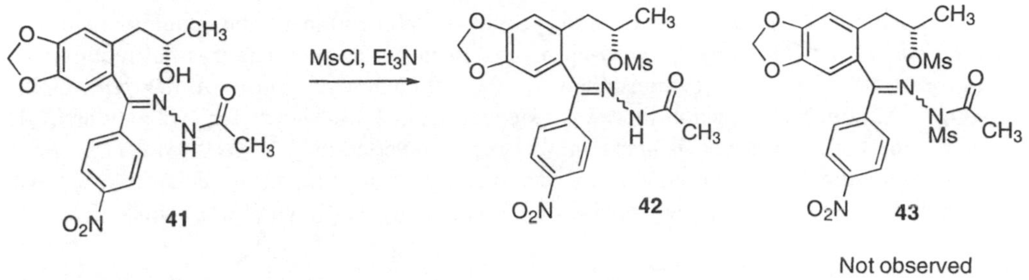

$$(8)$$

Not observed

The facility with which ring closure was realized for the stabilized hydrazone anion was surprising given that only alkylations of *N*-acetyl hydrazones by displacement of benzylic[30] and carbonyl-activated[31] leaving groups had been previously documented. Competing processes, including O or C- alkylation[32] and azoxazoline production,[33] were not observed. Most notably, only minimal quantities (<4 %) of the elimination product **44** were produced. Competitive elimination could be further minimized by the use of lithium *tert*-butoxide (<1%). Although somewhat higher levels of elimination product were encountered (2-3%), the use of sodium hydroxide in ethanol proved to be the most economically and operationally practical method for effecting cyclization (*vide infra*).

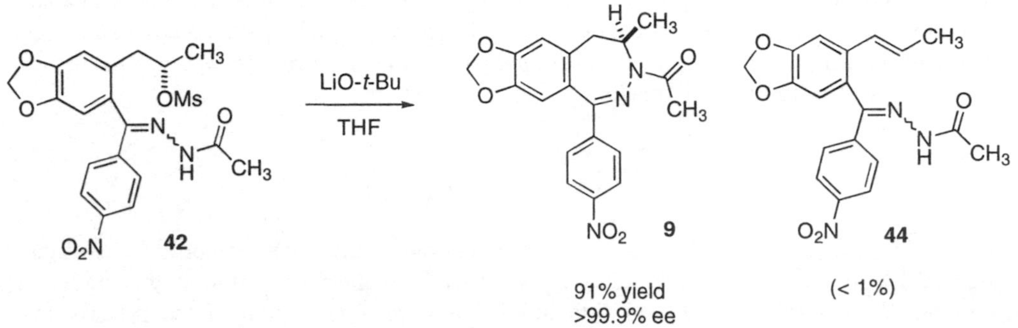

$$(9)$$

**9**
91% yield
>99.9% ee

**44**
(< 1%)

Significant levels of elimination were observed when halogens were employed as the leaving group (*e.g.* **45**). In cases involving either the bromo or iodo analog, dehydrohalogenation was the major reaction pathway upon attempts to effect ring cyclization.

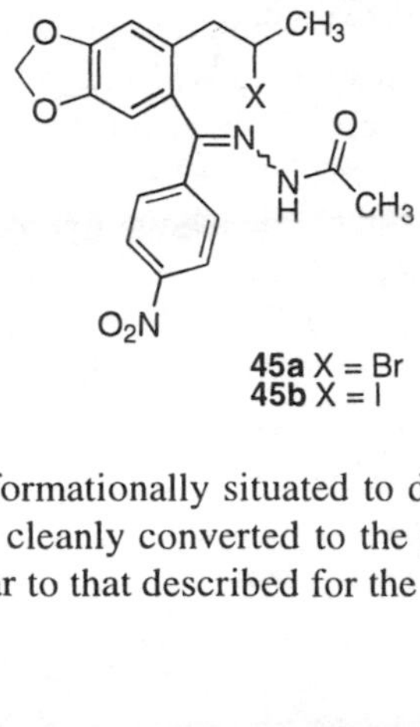

**45a** X = Br
**45b** X = I

Only the *Z* isomer of **42** is conformationally situated to displace the mesylate group. Since a mixture of the *E* and *Z* isomers was cleanly converted to the product, the two hydrazone isomers must interconvert in a manner similar to that described for the Mitsunobu reaction (Figure 1). The bis-benzylic position of the delocalized anion provided stabilization of the required tautomer. Stabilization was also afforded by the *para*-nitro substituent but its presence was not required for successful reaction. This was demonstrated by the efficient cyclization of the *para*-amino derivative 46. Although the reaction time was extended from 1 to 4 hours for cyclizations of 42 compared to 46, both hydrazone isomers of 46 were converted to LY300164 in 71% yield. Cyclization employing Mitsunobu conditions was attempted with the aniline derivative but was met with limited success. Conversion stopped after 2 h and only a 30% yield was achieved.

$$\text{46} \xrightarrow[\substack{\text{THF, 4h} \\ 71\ \%}]{\text{LiO-}t\text{-Bu}} \text{LY300164} \qquad (10)$$

Reduction of the nitroaromatic group was most favorably effected as the final step. This provided the most seamless transition between the discovery synthesis and the newer route since it employed the same penultimate intermediate. Use of the common intermediate minimized the differences in the quality of material prepared by the two routes. As was demonstrated in the original synthesis, the transformation could be accomplished by a number of methods. Palladium-catalyzed hydrogenation, employing either hydrogen gas or transfer methods, was satisfactory and provided material of uniformly high quality.

## Pilot Plant Scale Up

The entire process was scaled up to several hundred gallon scale to produce sufficient quantities of LY300164 for clinical trial studies. Development focus prior to this point had addressed the tactical issues associated with individual steps. Transition of the process from laboratory to pilot plant scale demanded careful orchestration between steps. This was necessary to ensure optimal efficiency and safety of the process and maintain the utmost control over the quality of the final product. Isolation of only selected compounds accomplished many goals of the pilot plant scale-up.

Only three of the six intermediates generated in the synthesis were isolated (Scheme 5). Alcohol 24 produced by the biocatalytic reduction was carried directly into the second step in which 33 was isolated in 87% overall yield. Solutions of the products resulting from the oxidation of 33 and the hydrazone forming step were carried into subsequent steps. Precipitation of the crystalline mesylate 42 led to only the second isolated intermediate in the process. The three steps which employed isochroman 33 as starting material were accomplished in 75% overall yield. Cyclization of 42 was effected in ethanol with sodium hydroxide. Generation of the elimination product (44) increased to approximately 4% of the reaction mixture. Selective precipitation of 9 by the addition of water as an antisolvent led to efficient product recovery (93% yield) with the exclusion of 44 (<1%). Palladium-catalyzed transfer hydrogenation employing potassium formate successfully delivered LY300164 in 91% yield. Overall, the optically pure LY300164 was prepared in 55% yield calculated from ketone 2.

Scheme 5. Pilot plant synthesis of LY300164

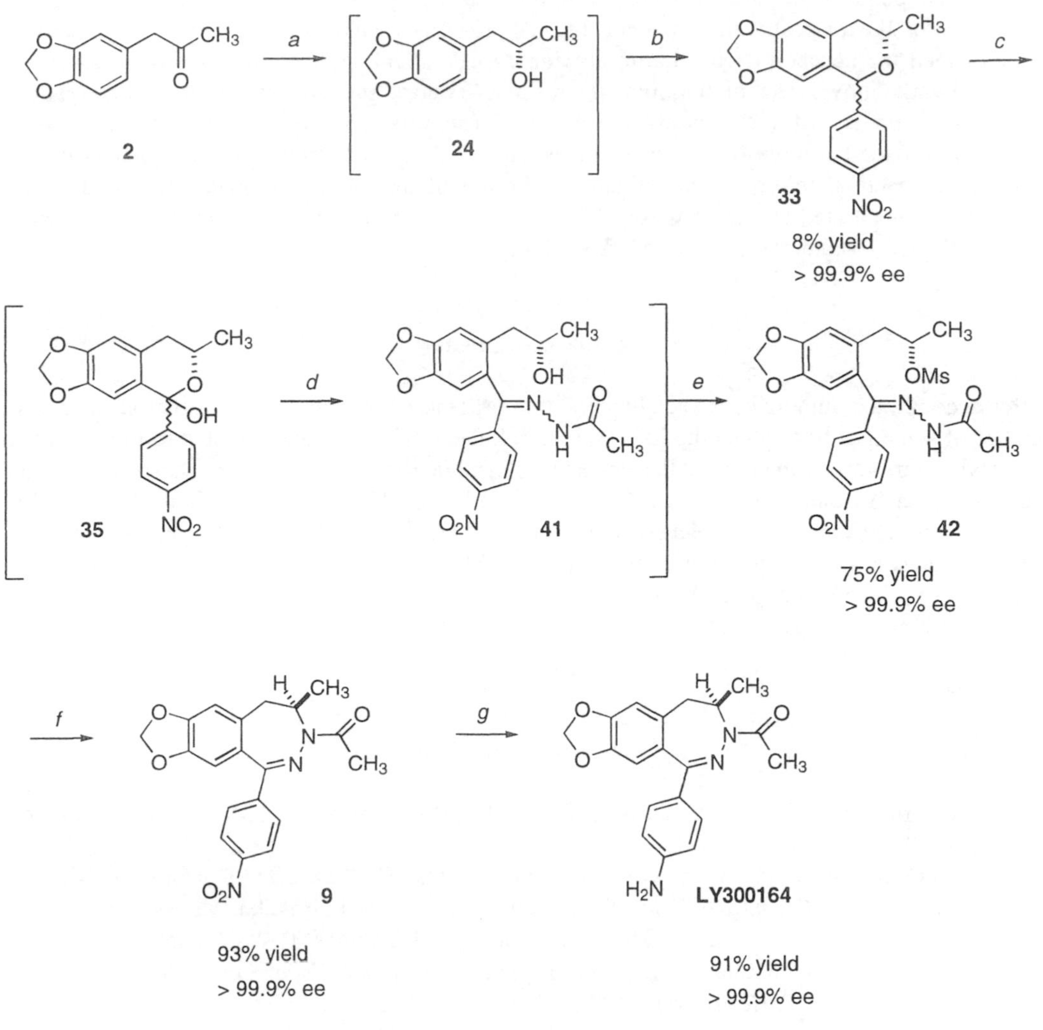

Reaction conditions: (a) *Z. rouxii*, XAD-7 (b) *p*-NO$_2$PhCHO, HCl (c) air, NaOH, DMSO (d) H$_2$NNHAc (e) MsCl, Et$_3$N (f) NaOH, EtOH (g) KO$_2$CH, Pd/C

## Summary

    The discovery synthetic route, albeit designed to support the SAR program, was reproducible and reliably delivered multiple gram quantities of high quality final product. Several factors, however, compromised the long term utility of the process. Most notably, a chromium-based oxidation reaction proved inefficient and represented significant health and environmental hazards. This and related efficiency issues prompted the development of an alternate synthetic route to the optically active benzodiazepine.

The ensuing work established the viability of a concise strategy which intersected the original route at the penultimate intermediate. The practicality of the route hinged on the efficient production of the optically pure alcohol **24**. A previously unknown ketoreductase, *Z. rouxii*, accomplished the necessary reduction to provide **24**. Practical application of the biocatalyst was enabled by the novel use of a polymeric resin to control the concentration of the reaction constituents and afford facile product isolation. The efficiency of this and the subsequent chemistry provided a three-fold increase in overall yield as compared to the original synthesis. Furthermore, tactical improvements eliminated the use of metal-based oxidants and reduced the overall waste generated by the synthesis. These factors positioned the new synthesis as a feasible route for the large scale manufacture of LY300164.

## Acknowledgments

In this account, the authors have been honored to represent the outstanding contributions of a large number of scientists throughout the Lilly Research Laboratories. We are indebted to the important technical contributions made by Allen Harkness, Cynthia Henry, Lisa Becke, Nancy Harn, Matt Reinhard, Lisa Gaebler, Jim Dunigan, Sara Gallo, Phil Poor, Ken McDowell, Craig Kemp, Steve Maples, Doug Dorman, David Marks, Brian Scherer, David Varie, Kevin Sullivan, John Grutsch, Greg Stephenson, Jim Carty, Bryan Landen, William Muth, Paul Marler, Avi Lagu, Jim Kelley and Mark Strege. Professors Paul Wender, Ted Taylor, Marvin Miller, William Roush and Leo Paquette are also acknowledged for their important intellectual contributions to the development effort.

## References and Notes

1.   (a) Sternbach, L. H. *J. Med. Chem.* **1979**, *22*, 1. (b) Williams, M. *J. Med. Chem.* **1983**, *26*, 619.

2.   For other 2,3-benzodiazepine syntheses, see: (a) Gatta, F.; Piazza, D.; Del Giudice, M. R.; Massotti, M. *Il Farmaco-Ed. Sc.* **1985**, *40*, 942. (b) Kurita, J.; Enkaku, M.; Tsuchiya, T. *Chem. Pharm. Bull.* **1982**, *30*, 3764. (c) Chenard, B. L.; Bulter, T. W.; Menniti, F. S.; Prochniak, M. A.; Richter, K. E. G. *Bioorganic & Medicinal Chem. Lett.* **1993**, *3*, 1991. (d) Lida, S.; Mukai, T. *Heterocycles* **1978**, *11*, 401. (e) Reid, A. A.; Sharp, J. T.; Sood, H. R.; Thorogood, P. B. *J. Chem. Soc. Perkin Trans. 1* **1973**, 2543. (f) Bendall, V. I. *J. Chem. Soc., Chem. Commun.* **1972**, 823. (g) van der Stelt, C.; Hofman, P.S.; Nauta, W. T. *Rec. Trav. Chim. Pays-Bas* **1964**, *84*, 2039. (h) Fryer, R. I.; Walser, A. *Bicyclic Diazepines: Diazepines with an Additional Ring*; Fryer, R. I., Ed., John Wiley and Sons Inc.: New York, 1991; pp 1-88.

3.   Previous accounts of this work have been published. See, (a) Anderson, B. A.; Hansen, M. M.; Harkness, A. R.; Henry, C. L.; Vicenzi, J. T.; Zmijewski, M. J. *J. Am. Chem. Soc.* **1995**, *117*, 12358. (b) Vicenzi, J. T.; Zmijewski, M. J.; Reinhard, M. R.; Landen, B. E.; Muth, W. L.; Marler, P. G. *Enzyme Microb. Technol.* **1997**, *20*, 0000. (c ) Zmijewski, M. J.; Vicenzi, J.; Landen B. E.; Muth, W.; Marler, P.; Anderson, B. *Applied Microb. Biotech.* **1997**, *47*, 162. (d) Anderson, B. A.; Hansen, M. M.; Harkness, A. R.; Henry, C. L.; Vicenzi, J. T.; Zmijewski, M. J, Manuscript in preparation.

4.  Goldensohn, E. S.; Glaser, G. H.; Goldberg, M. A. *Merrit's Textbook of Neurology*, 8th ed.; Rowland L. P., Ed.; Lea and Febiger Publishers: Philadelphia; pp. 780-806.

5.  Bloom, F. E. *The Pharmacological Basis of Therapeutics*, 7th ed.; Gilman, A. G.; Goodman, L. S.; Rall, T. W.; Murad, F., Ed.; MacMillan: New York, 1985; Chapter 12.

6.  Tarnawa, I.; Engberg, I.; Flatman, J.A. in *Amino Acids: Chemistry, Biology and Medicine,* ed. Lubec, G.; Rosenthal, G. A., Escom Science Publishers B. V.: Leiden, 1990, 538-546.

7.  (a) Ling, I.; Podanyi, B.; Hámori, T.; Solyom S. *J. Chem. Soc. Perkin Trans. 1*, **1995**, 1423 and references cited therein.  (b) Andrási, F.; Berzsenyi, P.; Botka, P.; Farkas, S.; Goldschmidt, K.; Hámori, T.; Kórösi, J.; Moravcsik, I.; Tarnawa, I. US Patent 5,536,832  (c) Ling, I.; Hámori, T.; Botka, P.; Sólyom, S.; Simay. A.; Moravcsik, I. WO 95/01357.

8.  Tarnawa, I.; Berzsenyi, P.; Andrási, F.; Botka, P.; Hámori, T.; Ling, I.; Kórösi, J. *Bioorg. Med. Chem. Lett.* **1993**, *3*, 99 and references cited therein.

9.  The original discovery synthetic route (Hungarian IDR) used perchloric acid which generated a potentially explosive oxonium salt.

10. The optimal conditions reported by the Hungarian IDR utilized **14** and borane tetrahydrofuran complex which afforded complete conversion in 24 hours.  By changing to borane dimethylsulfide, the reaction time was reduced to 3.5 hours.

11. The conditions reported by the Hungarian IDR (Raney nickel, $H_4N_2$) were modified to use hydrogen and catalytic Pd/C.

12. Markaryan, E. A.; Samodurova, A. G. *Russ. Chem. Rev.* **1989**,*58*, 479.

13. (a) Gatta, F.; Piazza, D.; Delqiudice, M. R.; Massotti, M. *Farm. Ed. Sci.* **1985**, *40,* 942.  (b) Thibault, J.; Maitte, P. *Bull. Chim. Soc. Fr.* **1969**, 915.

14. Valters, R. E.; Flitsch, W. *Ring-Chain Tautomerism*; Katrizky, A. R., Ed.; Plenum Press: New York, 1985; pp 108-114.

15. (a) Rossen, K.; Simpson, P. M.; Wells, K. M. *Synth. Commun.* **1993**, *23*, 1071.  (b) Nicolaou, K. C.; Randall, J. L.; Furst, G. T. *J. Am. Chem. Soc.* **1985**, *107*, 5556.

16. Imai, T.; Yamamuro, A.; Sato, T.; Wollman, T. A.; Kennedy, R. M.; Masamune, S. *J. Am. Chem. Soc.* **1986**, *108*, 7402.

17. Yamamoto, K.; Ueno, K.; Naemura, K. *J. Chem. Soc. Perkin Trans. 1*, **1991**, 2607.

18. Deloux, L.; Srebnik, M. *Chem. Rev.* **1993**, *93*, 763.

19. Kim, Y. H.; Park, D. H.; Byun, I. S. *J. Org. Chem.* **1993**, *58*, 4511.

20. (a) Fronza, G.; Grasselli, P.; Mele, A.; Fuganti, C. *J. Org. Chem.* **1991**, *56,* 6019.  (b) Cai, Z. Y.; Ni, Y.; Sun, J.-K.; Yu, X.-D.; Wang, Y. -Q. *J. Chem. Soc.* **1985**, 1277.

21. (a) Csuk, R.; Glanzer, B. I. *Chem. Rev.* **1991**, *91*, 49.  (b) Servi, S. *Synthesis,* **1990**, 1.

22. (a) Ohsima, T.; Wandrey, C.; Conrad, D. *Biotechnol. Bioeng.* **1989**, *34*, 394.  (b) Wichman, R.; Wandrey, C. *Biotechnol. Bioeng.* **1981**, *23*, 2789. (c) Wong, C.; Whitesides, G. *J. Org. Chem.* **1982**, *47*, 2816. (d) Vasic-Racki, D.; Jonas, M.; Wandrey, C.; Hummel, W.; Kula R. *Appl. Microbiol. Biotechnol.* **1989**, *31*, 215.

23. (a) Holst, O.; Mattiasson, B. *Extractive Bioconversions;* Marcel Dekker, Inc.: New York, 1991.  (b) Van Sonsbeek, H. M.; Beeftink, H. H.; Tramper, J. *Enzyme Microb. Technol.* **1993**, *15,* 722.  (c) Lilly, M. D. *J. Chem. Tech. Biotechnol.* **1982**, *32*, 162.  (d) Carrea, G. *Trends in Biotechnol.* **1984**, *2*, 102. (e) Aldercreutz, P.; Mattiason, B. *Biocatalysis* **1987**, *1*, 99.  (f) Jayasinghe, L.; Kodituwakku, D.; Smallridge, A.; Trewhella, M. *Bull. Chem. Soc. Jpn.* **1994**, *67*, 1.  (g) Tramper, J.; Vermue, M. H. *Chimia* **1993**, *47*, 110.

24.  (a) Vermue, M.; Sikkema, J.; Verheul, A.; Tramper, J. *Biotechnol. Bioeng.* **1993**, *42*, 747.
     (b) Laane, C.; Boeren, S.; Vos, K. *Trends Biotechnol.* **1985**, *3*, 251. (c) Bruce, L.;
     Daugulis, A. *Biotechnol. Prog.* **1991**, *7*, 116.

25.  (a) Schmitz, E.; Rieche, A. *Chem. Ber.,* **1956**, *10*, 2807. (b) Reiche, A.; Schmitz, E.
     *Chem. Ber.* **1957**, *90*, 1082.

26.  Russell, G. A.; Moye, A. J.; Janzen, E. G.; Mak, S.; Talaty, E. R. *J. Org. Chem.* **1967**, *32*,
     137.

27.  Russell, G. A.; Bemis, A. G. *J. Am. Chem. Soc.* **1966**, *88*, 5491.

28.  Russell, G. A.; Janzen, E. G.; Strom, E. T. *J. Am. Chem. Soc.* **1964**, *86*, 1807.

29.  Guthrie, R. G. *Comprehensive Carbanion Chemistry*, Vol. 5A; Vuncel, E.; Durst, T., Ed.;
     Elsevier: New York.; 1980; Chapter 5.

30.  Kallay, F.; Janzsco, G.; Koczor, I. *Tetrahedron*, **1965**, *21*, 3037.

31.  Matsumura, N.; Kunugihara, A.; Yoneda, S. *Tetrahedron Lett.* **1984**, *40*, 4529.

32.  Sartym I. *Tetrahedron*, **1972**, *28*, 2307.

33.  Hassan, E.; Al-Ashmawi, M. I.; Abdel-Fattah, B. *Pharmazie,* **1983**, *12,* 833.

# Benefits and Challenges of Applying Phase-Transfer Catalysis Technology in the Pharmaceutical Industry

**Marc E. Halpern**

*PTC Technology, Suite 627, 1040 N Kings Hwy, Cherry Hill, New Jersey 08034*

Key words: phase-transfer catalysis

Phase-Transfer Catalysis, "PTC," technology is currently used in many commercial manufacturing processes for pharmaceuticals and pharmaceutical intermediates. The pharmaceutical industry was one of the first chemical segments to take advantage of the unique process performance attributes offered by PTC. A review article entitled "Phase-transfer catalysis in the production of pharmaceuticals" was already published in 1980.[1] At that time the authors cited advantages of PTC in the following order: "solvent economy", replacing sodium metal, NaH or $NaNH_2$ with NaOH and simplified workup. The authors also cited as challenges catalyst degradation and catalyst separation/recycle. Since then, pharmaceutical process chemists have expanded the advantages of applying PTC to commercial processes and have found innovative methods for overcoming some of the challenges of commercializing PTC processes.

## Criteria for Selecting PTC for Commercial Pharmaceutical Process Applications

In order for PTC to be considered for any commercial organic process, the candidate process must meet, at a minimum, all of the following three criteria: (1) the reaction is within the scope of PTC applications, (2) PTC is capable of providing the desired process performance for that reaction and (3) the advantages which PTC provides are sufficient to justify a change in the process, which in turn is highly dependent on how early in the overall reaction sequence the change is proposed and how early in the development cycle the change is proposed. In addition to meeting these criteria, commercialization of PTC processes may encounter barriers which need to be overcome. Challenging barriers to commercialization of PTC processes usually relate to technical[2] and organizational issues.[3] These criteria, benefits and challenges of commercializing PTC processes will be discussed below and illustrated with examples from the pharmaceutical patent literature.

## Criterion 1: Reactions

The reactions of pharmaceutical interest which are most commonly successful using PTC include: N-alkylation and C-alkylation of organic N-H and C-H acids up to a pKa of 23 (e.g., many heterocyclic N-alkylations and many C-alkylations of diactivated methylene groups substituted by nitriles, ketones, sulfones and imines); esterification and transesterification; etherification and thioetherification; nucleophilic aliphatic and aromatic substitutions involving "inorganic" anions such as fluoride, cyanide, iodide, azide and others; dehydrohalogenation; base-promoted condensations such as Michael addition, aldol condensation, Wittig, Darzens and carbene additions; selected oxidations, mostly with hypochlorite and $H_2O_2$; borohydride reduction; and even some transition metal co-catalyzed reactions such as CO insertion. Examples of the application of PTC to these reactions of interest to pharmaceutical processes will be described in detail below.

## Criterion 2: Desired Process Performance - Benefits

The process performance results which are most often achieved using PTC are: high yield, short(er) reaction time, flexibility in choosing solvent (including no solvent), high(er) selectivity, easier workup (less unit operations), control of exotherms and use of less hazardous, less quantity, less expensive and/or easier to handle raw materials. These and other advantageous process benefits will be highlighted in the examples below.

These criteria and benefits of PTC may be categorized as enhancing productivity, environmental performance, quality, safety and plant operability. Productivity gains are realized using PTC by several means. PTC is often associated with high yield reactions, but in fact, most of the productivity gains achieved by commercial PTC processes result from reduced cycle time. The reduced cycle time is achieved by exploiting several of PTC's unique attributes. Lower solvation of nucleophiles, bases and redox agents relative to non-PTC systems results in enhanced activation during the rate determining step leading to shorter reaction times. The great flexibility PTC provides in choosing solvent leads to productivity gains as well as plant operability and environmental improvements. Appropriate PTC solvents can be chosen from the full range of polarity from cyclohexane through DMSO. Therefore, solvents for PTC can be chosen based on ease of recovery, cost, boiling point, toxicity, polarity, flammability and other factors. Very often, one can employ "solvent-free PTC" conditions in which one of the liquid reactants (e.g., alkylating agents, which are usually very good PTC solvents) or product serves as the "solvent." In such cases, productivity is improved by filling the reactor with more reactant/product instead of occupying valuable reactor volume with added solvent. Choosing an easily recoverable solvent such as toluene or MIBK instead of DMSO, DMF, NMP or HMPA reduces cycle time and improves plant operability by reducing distillation times or eliminating extraction, recovery and/or solvent drying unit operations (see the streamlined Parke-Davis cyanation below). Increased productivity can save both operational costs as well as initial capital investment (higher plant capacity in smaller equipment), if the PTC process option is identified prior to the building of the plant. When the PTC process option is discovered after the plant is operational (e.g., for debottlenecking, reducing plant variability, compliance with new environmental constraints, etc.), often the capital investment for retrofitting with PTC is very low or non-existent. Again, an on-stream plant will change the process only if the driving force is *very* strong. So if process changes are to be considered at all for a certain plant process, PTC options should be considered among them if the desired process improvement meets any of the criteria cited here.

The benefits of PTC relating to quality usually are for improving selectivity. For example, the selectivity of C-alkylation vs O-alkylation can be modified by many more handles using PTC than non-PTC systems, such as varying catalyst structure or being able to use a wider range of solvents or hydration levels. PTC provides the opportunity to perform conversions of functional groups in the presence of other functional groups which are hydrolytically sensitive and are on the same molecule. PTC plants also tend to be more controllable than non-PTC plants and thus tend to be more consistent. Consistency is an important production quality issue in all industries and for pharmaceuticals in particular.

The benefits of PTC relating to safety usually are better control of exotherms and the ability to use less of and less hazardous raw materials. Due to the two phase nature of PTC systems, it is sometimes possible to limit the rate of reaction by limiting the quantity and rate of transfer of the anionic reactant into the organic reaction phase by choosing a low catalyst level and low agitator speed. Reducing hazardous raw materials is often achieved since PTC often requires less excess of reagents such as cyanide (see Parke Davis cyanation below), dimethyl sulfate and phosgene (e.g., phosgene usage was reduced from 30 mole% excess to 2 mole% excess[4]). PTC sometimes provides the opportunity to replace hazardous or flammable bases (e.g., $NaNH_2$, NaH or methanolic NaOMe made using Na metal with $H_2$ by-product) with aqueous NaOH (hazardous but more easily handled, thus safer) or replacing transition metal catalyzed oxidations with NaOCl. Environmental performance is enhanced by PTC primarily by generating less waste due to higher yield, reducing solvent emissions by eliminating solvent or using a more easily recoverable solvent and by using less or alternate raw materials (e.g., less cyanide or eliminating transition metal usage).

### Criterion 3: Stage of Development Cycle and Reaction Sequence

In the pharmaceutical industry, in order to choose PTC for a candidate reaction, it is not enough for the reaction to be one of those listed in Criterion 1 with the desired performance listed in Criterion 2. The decision to use PTC further depends on the stage of development and the placement of the reaction in the sequence. In both cases, the earlier PTC is considered, the better.

Before the IND stage, there is considerable freedom in choosing process parameters (including the introduction of PTC technology), although usually at this stage, the chemists are less concerned with optimizing process performance and are more concerned with making enough material for studies. Even during pre-IND stages, PTC can help by providing high yields, especially of the most common highly successful PTC reactions. When PTC reaction application is relatively trivial (i.e., it works the first time), higher yield may translate into less lab batches needed to prepare samples. After the IND, the impurity profile begins to become more rigid or even finally set. During the IND stages it is easier to consider PTC for reactions which are earlier in the reaction sequence. For example, if catalyst decomposition is a concern (quaternary ammonium salts may produce amines), if there are three distillations and two recrystallizations between the candidate PTC step and the final product, then PTC may be easily considered. One must be careful of pitfalls individual to each case. For example, if one is attempting to perform a C-butylation using benzyl triethyl ammonium chloride as the catalyst, one may produce small quantities of the C-benzylated product, which may not be easily separated from the desired product. In most cases, appropriate phase-transfer catalysts can be chosen to meet the requirements of the specific process. Generalized guidelines cannot be given since each process is different. Often a dedicated industrial PTC expert consultant can help in choosing the appropriate catalyst and other conditions.

In the NDA stage, all changes become much more difficult to justify, including incorporating PTC where it was not previously used. Nevertheless, sometimes it is possible to prove to the FDA that a better impurity profile and/or a safer process can be achieved. Such factors (and/or very strong cost advantages) can provide incentive to change the process at almost any stage of its development (except the last step or two). The driving forces for change usually relate to reducing impurities, controlling production quality consistency, solvent ramifications, difficult unit operations (especially separations and work up) or low yield of an intermediate which has a lot of cost tied up in it. This author has witnessed serious consideration of retrofitting processes with PTC at the IND and NDA stages as well as after years of commercial production. Companies do not approach a PTC retrofit with PTC in mind. Rather, the driving force for change comes from the desire to significantly improve a given process or from the decision not to live with a nearly unbearable production situation. At such time, PTC options should be considered along with other potential solutions. Another practice is to source an intermediate from a custom manufacturer as a raw material, then have the custom manufacturer "innovate" the PTC process.

## Challenges in Developing PTC Processes

Given the powerful benefits of PTC cited above, there must be significant barriers to commercializing PTC processes, or else almost every commercial reaction involving an anion would be performed by PTC. In addition to the process recertification issues discussed above, the challenges in developing PTC processes are primarily technical, relating mostly to the catalyst. Secondary challenges are organizational.

Once a PTC reaction is determined to "work,"[5] a major concern for pharmaceutical process development is purity. The catalyst can have direct impact on the impurity profile of the product or intermediate in two major ways, separation from product and catalyst decomposition. Since many commercial phase-transfer catalysts are at least somewhat lipophilic as are most pharmaceutical intermediates, the catalysts and the intermediates may be soluble in many of the same solvents. The catalyst may be separated by the product by washing/extraction into water, recrystallization of the product, distillation of the product or adsorption of the catalyst onto a sorbent. If the catalyst is not properly chosen, extensive water washings may be necessary to adequately remove the catalyst from the product. In many of the examples below, relatively hydrophilic catalysts, such as benzyl triethyl ammonium chloride and others, were chosen just for the ability to effectively wash out the catalyst. In some cases, distillation, recrystallization or sorption are feasible methods for catalyst separation (examples will be shown below). In rare cases in which none of these separation procedures are feasible, special procedures may be implemented (such as precipitation of the catalyst from a specially devised solvent system), but usually at this point the PTC option is no longer considered.

Catalyst decomposition can be a serious challenge in commercializing a PTC process. For example, quaternary ammonium ("quat") salts decompose, especially in the presence of strong base, to liberate tertiary amines and other compounds. PTC nucleophilic substitutions which require high temperature must be carefully considered when using quats. For example, if PTC esterifications are performed at 100°C and are left to react for extended periods in the presence of $Bu_4NBr$, then it is not uncommon to detect the butyl ester in the product matrix. If such byproducts, resulting from the reaction of the substrate with the catalyst, cannot be separated from the desired product, one must consider alternate more thermally stable catalysts, such as crown ethers or polyethylene glycols. It may be appropriate to note here that chemists often overlook considering crown ethers for commercial processes due to their cost and toxicity. Regarding toxicity, before rejecting crown ether for commercial processes, this author recommends obtaining the latest toxicology literature relating to 18-crown-6 and its derivatives and asking for review and opinion from a qualified industrial hygienist (a non-comprehensive list of crown ether toxicology literature may be found in ref[6]). Regarding cost, it is recommended that the cost of the catalyst be calculated on the basis of *cost of catalyst per kg of product*, not cost of catalyst per kg of catalyst.

Scale up of two phase systems are usually complicated by agitation, complex kinetics, phase separation and heat transfer issues. When a PTC reaction is "Transfer Rate Limited"[7] it is important to ensure high agitation efficiency to maintain productivity (unless an uncontrollable exotherm would result). If the reaction is not transfer rate limited, agitation is not important beyond a threshold to provide adequate mass transfer. The concept of only moderate agitation for a two phase system is sometimes anti-intuitive for engineers, but in fact, it is the catalyst that is performing the phase-transfer and elaborate capital intensive agitation systems are rarely needed for commercial PTC reactions. Depending on the identity of the specific application, heat transfer can be enhanced by PTC or can be the source of significant hazard. Heat transfer issues can be a barrier to commercializing a PTC process if for example, an exothermic reaction must be performed under solid-liquid conditions without an aqueous phase serving as a heat sink. Rag layer and perfect phase cuts may pose additional challenges. Each application must be considered on a case-by-case basis.

Sometimes initial screening studies do not produce the desired results and the project is dropped. One must be careful in reaching negative conclusions too early since different applications may require radically different optimal catalysts, hydration levels, solvents, leaving groups, etc. Even the catalyst counteranion can render an otherwise good PTC application not viable. For example, the bromide of $Bu_4NBr$ can react with the substrate to form small quantities of a brominated substrate which can react further to form impurities. In such cases, the catalyst counteranion is generally not suspected. Another common oversight is disregarding (or lack of awareness) the deactivating ability of high levels of hydration of the reacting anion. Progressive companies effectively consult with industrial phase-transfer catalysis experts and are able to better commercialize PTC opportunities and minimize the real challenges described.

Organizational factors which can hinder commercialization of a PTC process include lack of allocated development time to solve PTC process challenges, lack of awareness of the scope of PTC reactions and advantages, corporate culture or individual bias unfavorable toward PTC (usually resulting from an initial failed attempt to apply PTC), lack of expertise in developing PTC processes and even simple resistance to change. Fortunately, in many pharmaceutical companies and/or specific departments, most of these organizational barriers are not too widespread although they occur often enough to result in lost opportunity for some companies. In most cases, these lost opportunities are never recognized, especially when they are overlooked due to lack of expertise or an inflated perception of expertise.

## Examples of Pharmaceutical Reactions: Benefits and How Challenges Were Overcome

Many commercial processes for the manufacture of pharmaceutical intermediates benefit from using PTC. The author is not at liberty to disclose these elegant proprietary processes, however some of these advantageous processes have been published or patented. Following are examples highlighting the advantages and challenges of using PTC for selected applications.

**Table 1**: Parke-Davis Cyanation - Excellent Example of Process Improvement by PTC

|                                          | DMSO Process | PTC Process |
| ---------------------------------------- | :----------: | :---------: |
| **Materials Usage per kg Product**       |              |             |
|                                          |              |             |
| 2,4,6-triisopropyl-1-chloromethylbenzene (kg) | 1.29         | 1.04        |
| *excess* NaCN (kg)                       | 0.17 (70 mole%) | 0.009 (5 mole%) |
| $Bu_4NBr$ (kg)                           |              | 0.016       |
| DMSO (L)                                 | 1.86         |             |
| toluene (L)                              |              | 1.04        |
| hexane (L)                               | 3.42         |             |
| water (L)                                | 11.03        | 1.56        |
| **Other Process Parameters**             |              |             |
|                                          |              |             |
| temperature/reaction time                | 90-100°C/1h (exotherms to 135°C with each of 4 additions) | 80-85°C/7h (exotherms effectively controlled by stirring) |
| workup                                   | 3h water wash; filtration; water wash; dissolution & precipitation from hexane; large aq DMSO/cyanide waste stream with much cyanide needs to be treated | simple phase separation; water wash; aq NaOH destroys aq cyanide |
| solvent recycle                          | DMSO recycle not feasible; hexane recycle easier | single solvent toluene passed through to next step during which it is easily recovered |

Scientists at Parke-Davis provided a rare peek into the thought process for taking a developing process from a published procedure acceptable for preparing an initial sample to achieve outstanding results using a PTC process on a 40 kg scale prior to the IND submission.[8] Table 1 illustrates many of the advantages of applying PTC to the **cyanation** of the hindered benzyl chloride, **1**. Relative to the original process (the "DMSO process"), the PTC process provided nearly 20% yield increase, a 20-fold reduction in cyanide waste per kg product produced, an 85% reduction in aqueous waste generated per kg of product, replacement of non-recyclable DMSO (in this application) with easily recovered toluene, much simpler workup (many less unit operations and elimination of an additional workup solvent), enhanced safety through more control of the exothermic reaction and easier handling of aqueous cyanide waste (vs skin absorption hazard of DMSO/cyanide) and, finally, obtaining a product with greatly reduced color.

The only tradeoffs appear to be the use of 1.5 wt% (per wt of starting material) Bu$_4$NBr, which is easily separated in the water wash and a longer reaction time. If important, the reaction time may be shortened by increasing the temperature to that of the DMSO process, using more catalyst, adjusting the water content of the system or by other means. When considering overall cycle time, one must consider the workup time, which includes a long (3h) water wash and many unit operations for the DMSO process as compared to a direct transfer of the organic phase of the PTC process to the next step after a rapid water wash.

Scheme 1: PTC Cyanation

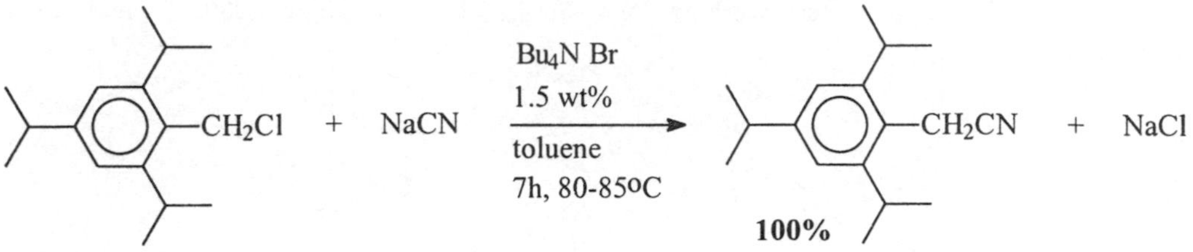

Fluorination by nucleophilic aromatic substitution of activated haloaromatics is an important reaction for the production of pharmaceuticals. It is interesting that most PTC aromatic fluorination patents and publications report the use of polar aprotic solvents, such as sulfolane[9] and DMSO[10] together with the phase-transfer catalyst. Often overlooked is a patent about to expire (at the time writing) which describes the high yield fluorination of 2-chloronitrobenzene with NO solvent and with only a 20 mole% excess of fluoride (Scheme 2).[11] Since the phase transfer of fluoride reduces the energy of activation, the reaction can be performed at temperatures up to 100°C less than the corresponding non-PTC fluorination. As a result, the inventor notes "negligible attack on stainless steel or glass enamel test pieces and very low corrosion rate on mild steel test pieces." Separation of the catalyst and inorganic salts from the product and product purification in this case were achieved by filtration (aided by toluene) and distillation of the product. Solid-liquid PTC conditions were used without special agitation. High reactivity was achieved using KF with less than 0.2 wt% water.

Scheme 2: PTC Fluorination

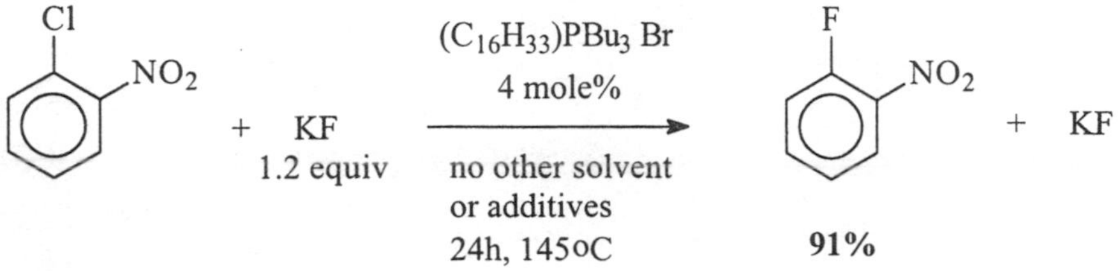

Esterification of penicillin derivatives may be the first commercial PTC reaction used in the pharmaceutical industry identified as such.[12] In this large scale esterification, benzyl penicillin was reacted with a hydrolytically sensitive alkylating agent, α-chlorodiethylcarbonate. The use of PTC

(Bu$_4$NHSO$_4$) provided 86% yield. The ability of PTC to react and protect hydrolytically sensitive compounds has since been well documented (in non-pharmaceutical applications) for such reactants as phosgene[13] and benzoyl chloride.[14] Carbamylation of estradiol (at the phenolic OH) with a carbamoyl chloride was performed on a 9 kg scale in 90% yield of **5** in the early years of PTC.[15] The non-PTC process used pyridine gave only 50% yield and required a large excess of the nonrecoverable carbamoyl chloride.

PTC was applied to the high yield esterification (97%) of cephalosporin derivatives.[16] The esterification of the cephalosporin **6** (R=H) with pivaloyloxymethyl iodide was complete within 0.5h in CH$_2$Cl$_2$ using Bu$_4$NOH as base and phase-transfer agent.[17] The catalyst was removed and the product purified by water washing and flash chromatography (89% yield after chromatography).

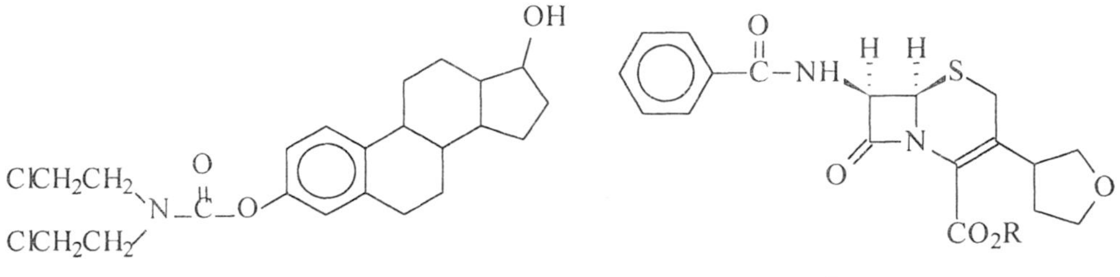

**N-alkylation** is one of the reactions most amenable to PTC and useful in the synthesis and manufacture of pharmaceuticals. Many N-alkylations are performed on activated N-H groups, often with pKa in the range of 16-23. N-H groups in this pKa range are usually easily deprotonated by aq NaOH under PTC conditions and the resulting N-anions are usually stable enough to participate in nucleophilic substitutions (less acidic N-H "acids" with pKa > 23 can be deprotonated under PTC/OH conditions but not easily alkylated). PTC N-alkylations are often high yield in short reaction time and are often performed without additional solvent. For example, indole is N-ethylated in 98% yield after 10 min at room temperature in the presence of 1 mole% Bu$_4$NBr, KOH, diethyl sulfate and no added solvent (Scheme 3).[18]

Scheme 3: PTC N-Alkylation

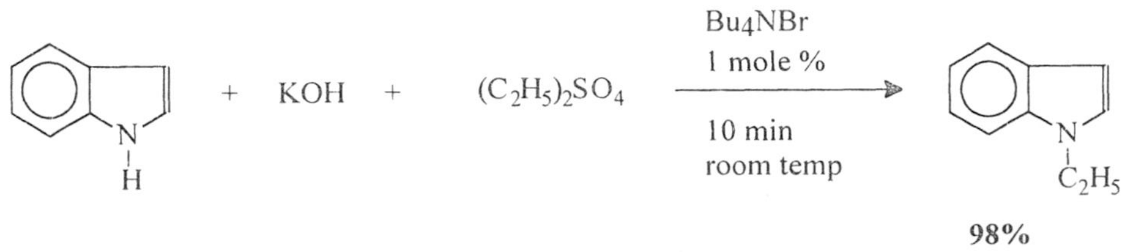

An additional attractive feature of N-alkylations which is particularly suitable for PTC processes is flexibility in separating the catalyst from the product. Many heterocyclic compounds and other N-containing compounds can be recrystallized or extracted into acidic water or be retained in an organic phase, thereby providing a variety of effective means for separation from the catalyst. As with most PTC processes, a common method for separating the catalyst from the product of N-alkylations is extraction of the catalyst into water. For example, compound **10** is an analgesic prepared by an intramolecular N-alkylation of a diactivated N-H group.[19] The catalyst chosen is effective at room temperature and is very soluble in both water (it is actually quite hygroscopic) and in methylene chloride. Product purification was performed by aqueous wash and recrystallization to give 90% isolated yield. The same hydrophilic catalyst was chosen for an N-alkylation/cyclization to a lactam (hypocholesterolemic agent) and was separated primarily by extraction into water.[20] The separation of a tetrabutylammonium salt from an N-alkylation of non-activated amine could also be performed by extraction of the catalyst into water.[21]

Scheme 4: PTC N-Alkylation

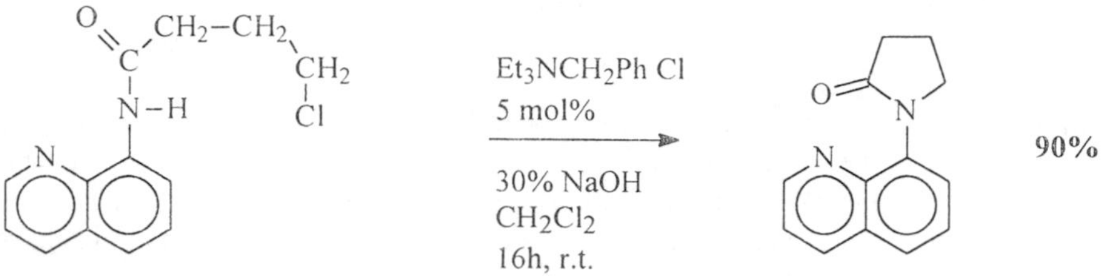

Many N-heterocycles have been alkylated under PTC conditions including pyrroles,[22] pyrazoles,[23] azepines,[24] phenothiazines,[25] carbazoles,[26] purines and pyrimidines,[27] and many other heterocycles. PTC N-alkylations quite often enjoy commercialization due to the ability of PTC to deliver the desired process performance parameters (such as high yield with desirable reaction times and solvent system), without excessive process challenges (such as isolation procedures) cited at the beginning of this article.

**Chiral C-alkylation** is the subject of some of the most fascinating PTC work ever published. Chiral PTC is based upon the assumption that a chiral phase-transfer catalyst can bind stronger to one side of an organic anion (e.g., carbanion generated under PTC/OH⁻ conditions) than the other, thereby inducing the subsequent chiral alkylation or chiral addition. The most comprehensive study of the interactions between chiral quats and a substrate concluded that the key binding regions between the relatively rigid planar deprotonated indanone substrate **12** and the chiral cinchona alkaloid quat occurred at three sites on each molecule,[28] none of which were the optically active nitrogen of the quat or the prochiral carbanion. This chiral alkylation is further complicated by the discovery that the active catalyst system is actually a dimer/zwiterion of the catalyst. The elegant and complex study will not be described here and should be read in its

Scheme 5: PTC Chiral Alkylation

**Chiral Phase-Transfer Catalyst:**

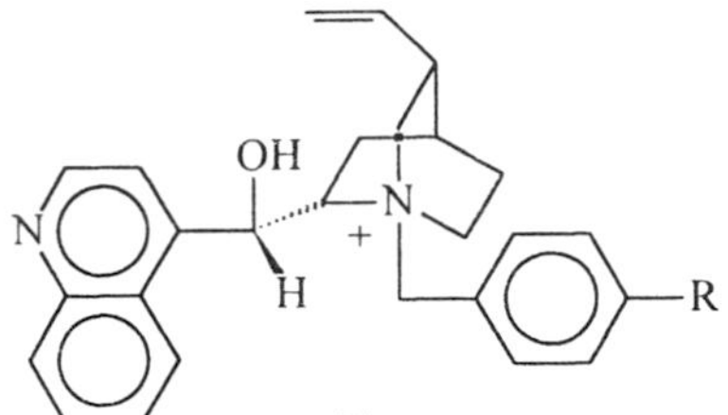

**11**

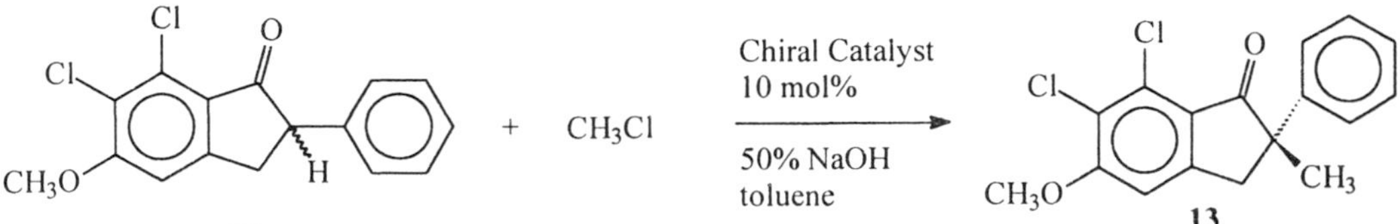

**12**          +     CH$_3$Cl

Chiral Catalyst
10 mol%

50% NaOH
toluene

**13**
**95% yield**
**92% ee**

entirety. Impressive results were obtained as shown in Scheme 5, however it must be noted that very significant resource was invested to achieve these results and the underlying understanding of the system. When embarking upon a development program for chiral PTC, corporate commitment must be high at the outset of the project and remain high and results are not guaranteed.

Non-rigid planar methylene groups can also be alkylated by chiral PTC, but the enantiomeric excess is lower. O'Donnell et al have been able to achieve 60-70% ee in the chiral alkylations, followed by enantiomeric enrichment by recrystallization (e.g., Scheme 6).[29] In contrast to conventional non-PTC alkylations (e.g., using LDA as base), the base must be added *after* the alkylating agent is added to avoid side reactions of the substrate in these PTC chiral alkylations.[30]

Scheme 6: PTC Chiral Alkylation

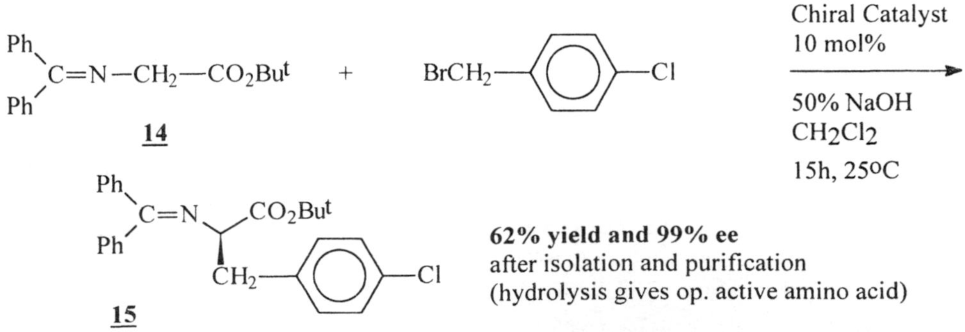

Ph
  \
   C=N—CH$_2$—CO$_2$Bu$^t$     +     BrCH$_2$—⬡—Cl
  /
Ph
        **14**

Chiral Catalyst
10 mol%

50% NaOH
CH$_2$Cl$_2$
15h, 25°C

Ph
  \
   C=N         CO$_2$Bu$^t$
  /        \  /
Ph          \
         CH$_2$—⬡—Cl
   **15**

**62% yield and 99% ee**
after isolation and purification
(hydrolysis gives op. active amino acid)

Even though the chemists in the radiolabelling department of a pharmaceutical firm have very different objectives than the process chemists and chemical engineers, some of their reaction constraints can be very similar. For example, in the **C-alkylation** shown in Scheme 7,[31] using a very short half-life isotope ($^{11}$C) , the methylation needed to be *high yield, short cycle time and easy workup*. Phase-transfer offered these advantages and was the method of choice. Large amounts of quat were used in this case. Economical C-methylations are routinely performed under PTC conditions using dimethyl sulfate and low catalytic quantities of quat.

Scheme 7: PTC C-Alkylation

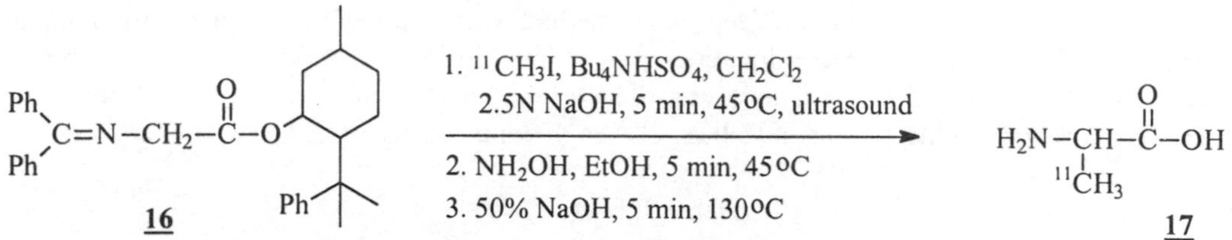

1. $^{11}$CH$_3$I, Bu$_4$NHSO$_4$, CH$_2$Cl$_2$
   2.5N NaOH, 5 min, 45°C, ultrasound

2. NH$_2$OH, EtOH, 5 min, 45°C
3. 50% NaOH, 5 min, 130°C

$$H_2N-CH-\overset{\overset{\displaystyle O}{\|}}{C}-OH$$
$$^{11}CH_3$$

**16**          **Ph**                                              **17**

total synthesis time = 35-55 min, 40% radiochemical yield, 98% radiochemical purity, 52% ee

**C-condensation** (Aldol) using strong base under PTC conditions was performed commercially in high yield in the manufacture of chloramphenicol (Scheme 8).[32] The patent notes that the PTC process offers the advantages of increasing the yield from 50-78% (earlier non-PTC patents) to over 90%, avoiding the use of flammable alcoholic solvents and simpler, less time consuming workup. This patent describes a special case in which the product is extracted into water (acidic) and precipitated from water. This product separation is a good example of designing the workup of the PTC system around the specific attributes (i.e., solubility) of the system components. The best catalyst found for this process was methyl tributyl ammonium chloride. This catalyst is quite soluble in water (commercially available as a 75% aqueous solution) and in methylene chloride. This catalyst and the nitrobenzaldehyde starting material were recycled.

Scheme 8: PTC Condensation

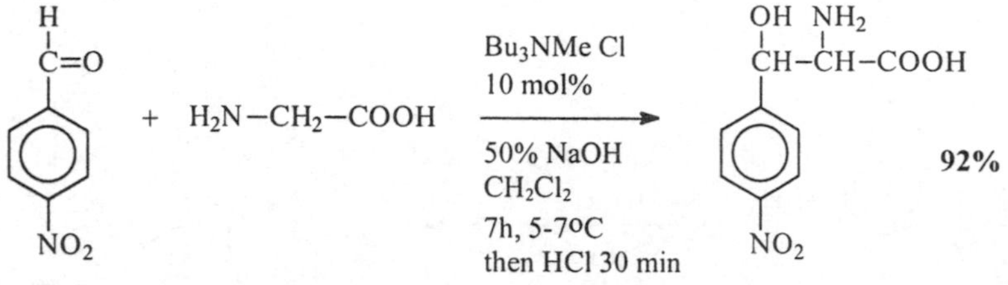

+ H$_2$N$-$CH$_2$$-$COOH

Bu$_3$NMe Cl
10 mol%

50% NaOH
CH$_2$Cl$_2$
7h, 5-7°C
then HCl 30 min

OH  NH$_2$
CH$-$CH$-$COOH

**92%**

**18**                                                              **19**

**Hydrolysis** using PTC was originally reported by Starks in his classic first paper.[33] Ester hydrolysis sometimes seems like a trivial reaction, but it is usually an interfacial reaction between aqueous hydroxide and a lipophilic ester. Reactions such as hydrolysis, which are between water soluble inorganic anions and lipophilic organic substrates are usually good candidates for PTC application. The reaction temperature of a hydrolysis can often be reduced by 30°C using PTC or maintain the same temperature and reduce the cycle time using PTC. Complete hydrolysis of a thiazolidine ester was obtained within 30 min with no heat using 14% NaOH and 5 mole%

Bu$_4$NHSO$_4$.[34] Two isomeric N-alkylated imidazole carboxylic esters were prepared in a non-selective N-alkylation then the desired intermediate (for an angiotensin II antagonist) was separated after selective hydrolysis of only one of the isomers using Bu$_4$NBr/NaOH.[35]

Hydrolysis of alkyl halides to the corresponding alcohols with NaOH cannot be performed directly with PTC due to low reactivity and reaction of the resulting alcohols with the alkyl halides to form ethers. However, Sasson et al[36] developed a very good method for converting alkyl halides to alcohols utilizing the facile PTC esterification of formates and the even more facile PTC hydrolysis of formate esters in one step. This method was applied in the preparation of an intermediate for an HIV protease inhibitor shown in Scheme 9.[37] Aliquat 336® (2 mole%; Trademark of the Henkel Corporation) was used as the catalyst and was separated from the product by adsorption onto carbon and silica gel then diatomaceous earth.

Scheme 9: Hydrolysis

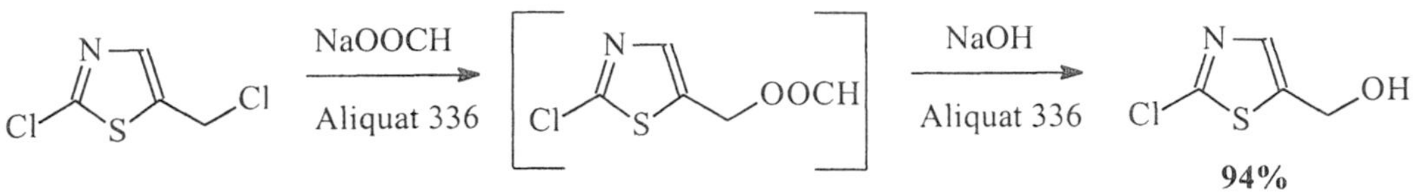

**20**                                    **21**                                    **22**

A series of interesting nucleophilic attacks of acetate, thioacetate, azide, cyanide and bromide on steroidal bis-epoxides was reported (Scheme 10).[38] PTC provided the opportunity to perform these reactions in high yield under mild reaction conditions leaving a carbonate and other functionalities intact. The catalyst was separated from the product by extraction into water and further purification was performed by chromatography to achieve 86% isolated yield for the reaction with acetate. Nucleophilic substitutions using inorganic nucleophiles such as azide and cyanide built upon the strengths of PTC to transfer and react these anions with organic substrates in high yield in the patented preparations of taxane derivatives[39] and 1-H tetrazole acetic acid.[40] The application of PTC to many nucleophilic reactions involving glycosides has been reviewed.[41]

Scheme 10: PTC Nucleophilic Attack

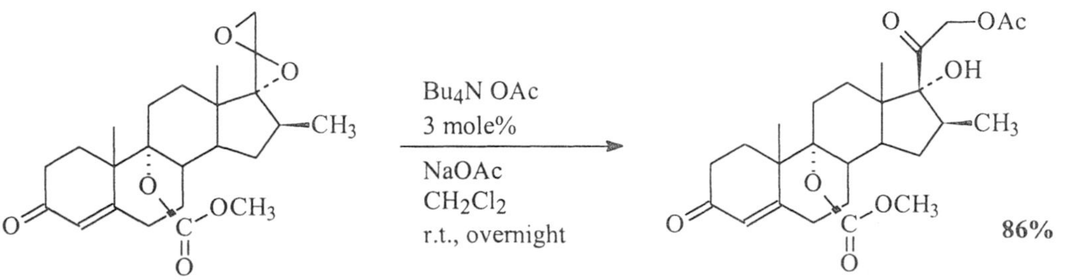

**23**                                                                    **24**

**Carbonylation** is not yet widely recognized as a candidate for commercial PTC application, though it should, and it is recommended that the readers familiarize themselves with the many unique advantages of using PTC for carbonylation.[42] The phase-transfer catalyst will aid in solubilizing hydroxide and metal complexes in the organic phase, in which CO is usually an order of magnitude more soluble than in water. An application of PTC carbonylation to a steroid is shown in Scheme 11.[43] The catalyst was separated from the product by extraction into water and the product was further purified by recrystallization.

Scheme 11: PTC Carbonylation

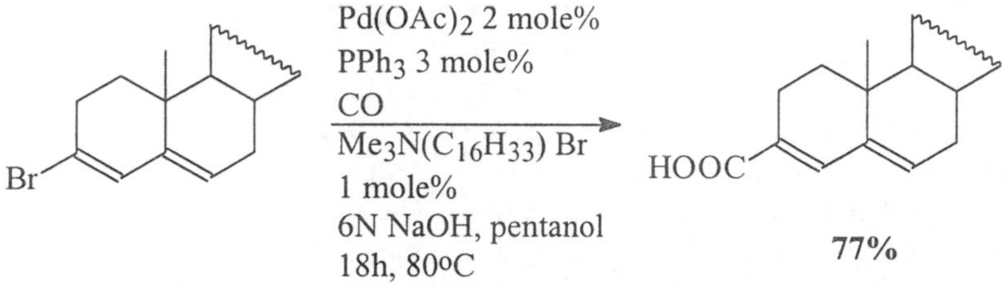

$$
\begin{array}{c}
\text{Pd(OAc)}_2 \text{ 2 mole\%} \\
\text{PPh}_3 \text{ 3 mole\%} \\
\text{CO} \\
\overline{\text{Me}_3\text{N(C}_{16}\text{H}_{33}) \text{ Br}} \\
\text{1 mole\%} \\
\text{6N NaOH, pentanol} \\
\text{18h, 80}^{\circ}\text{C}
\end{array}
$$

**25**

**77%**

**26**

Borohydride **reduction** can usually be performed without PTC, but sometimes the nature of the two phase system provides special processing advantages. For example, in the conversion of terfenadone to terfenadine in the presence of alcohol, one must use a large excess of NaBH$_4$ and even then "the known processes for the preparation of terfenadine give products with variable polymorphic compositions with melting points which are neither constant nor controllable." The use of PTC allows flexible choice of solvent which "permits obtaining terfenadine in each of the two substantially pure polymorphic forms directly."[44] In addition, "the alkaline borohydride is much more stable in the aqueous alkaline solutions than in the organic ones (e.g., the lower boiling alkanol solutions); this permits utilizing the reducing agent in a complete way; while with the use of alcoholic solvents, a part of the reducing agent cannot be effectively utilized because of its instability and this implies a large consumption of reducing agent for obtaining comparable yields of the final product." The hydrophilic catalyst is separated from the product by extraction into water. The product is purified by filtration through carbon and precipitation by reducing the volume of the xylene solvent.

Scheme 12: PTC Reduction

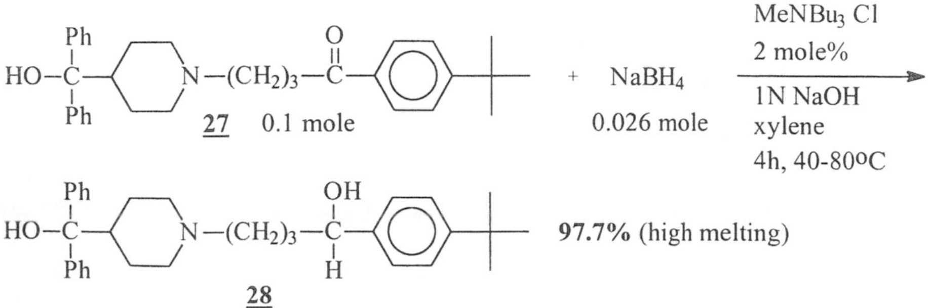

**27**  0.1 mole

+ NaBH$_4$

0.026 mole

MeNBu$_3$ Cl
2 mole%

1N NaOH
xylene
4h, 40-80$^{\circ}$C

**28**

**97.7%** (high melting)

**Carbene** reactions have been classical PTC applications since the 1960's. PTC carbene reactions have been applied to convert the 3-OH of a steroid to 3-Cl with retention of configuration[45] and to the preparation of spirocycloproyl derivatives of glycosides.[46] The catalyst is usually the hydrophilic benzyl triethyl ammonium chloride and is separated from the product by extraction into water.

## Summary

In summary, phase-transfer catalysis continues to enjoy growing commercial application in the pharmaceutical industry as may be inferred from the list of companies to which PTC process are assigned. Many of the fundamental driving forces for obtaining desired process performance match the strengths of PTC well. Barriers always exist in commercializing any new reaction and PTC systems provide additional challenges such as separating the catalyst from the product. These barriers as well as other barriers are routinely solved. The rule of thumb for retrofitting existing pharmaceutical processes with PTC is the same as for non-PTC processes, i.e., the earlier in the reaction sequence and the earlier in the product development cycle, the better. Industrial organic process chemists are integrating their knowledge of PTC with their creativity and experience to achieve the commercial goals of their pharmaceutical employers.

Acknowledgement: The author would like to thank Dr. Neal Anderson for his input as a pharmaceutical process development expert.

References

[1] Lindblom, L.; Elander, M. *Pharmaceutical Technology*, October **1980**

[2] Halpern, M. *Phase Trans. Catal. Comm.*, **1996**, *2*, 1

[3] Halpern, M. *Phase Trans. Catal. Comm.*, **1996**, *2*, 33

[4] Boden, E.; Phelps, P. (General Electric) **1994** US Patent 5,300,624

[5] see "Choosing a Phase-Transfer Catalyst for the First Experiment" Halpern, M. *Phase Trans. Catal. Comm.*, **1997**, *3*, 1

[6] Halpern, M. *Phase Trans. Catal. Comm.*, **1995**, *1*, 17

[7] Starks, C.; Liotta, C.; Halpern, M. *Phase-Transfer Catalysis: Fundamentals, Applications and Industrial Perspectives*; Chapman and Hall: New York, **1994**; Chapter 1-3, 6

[8] Dozeman, G.; Fiore, P.; Puls, T.; Walker, J. *Org. Proc. Res. Dev.*, **1997**, *1*, 137

[9] Gonzalez-Trueba, G.; Paradisi, C.; Scorrano, G. *Gazz. Chim. Ital.*, **1996**, *126*, 457

[10] White, C. (Mallinckrodt) **1987** US Patent 4,642,399

[11] North, R. (Boots) **1981** US Patent 4,287,374

[12] (Astra) **1977** Swed Patent 397, 981

[13] Boden, E.; Phelps, P.; Ramsey, D.; Sybert, P.; Flowers, L.; Odle, R. (General Electric) **1995** US Patent 5,391,692

[14] Krishnakumar, V.; Sharma, M. *Ind. Eng. Chem. Proc. Des. Dev.*, **1984**, *23*, 410

[15] Fex, H.; Kristensson, S.; Stamvik, R. (Leo) **1978** US Patent 4,081,461

[16] Ganboa, I.; Palomo, C. *Synthesis*, **1986**, 52

[17] Burton, G.; Naylor, A. (Pfizer) **1996** WO 96/17847

[18] Barry, J.; Bram, G.; Decodts, G.; Loupy, A.; Pigeon, P.; Sansoulet, J. *Tetrahedron Lett.*, **1982**, 5407

[19] Hodgson Jr., C. (Burroughs Wellcome) **1982** US Patent 4,366,158

[20] Thiruvengadam, T.; Tann, C.; Lee, J.; McAllister, T.; Sudhakar, A. (Schering) **1994** US Patent 5,306,817

[21] Wilkerson, W.; Rodgers, J. (DuPont Merck Pharmaceutical) **1996** US Patent 5,508,400

[22] Diez-Barra, E.; de la Hoz, A.; Loupy, A.; Sanchez-Migallon, A. *Heterocycles*, **1994**, *38*, 1367

[23] Diez-Barra, E.; de la Hoz, A.; Sanchez-Migallon, A.; Tejeda, J. *Synth. Comm*, **1990**, *20*, 2849

[24] Gozlan, I.; Halpern, M.; Rabinovitz, M.; Avnir, D.; Ladkani, D. *J. Heterocyclic Chem.*, **1982**, *19*, 1569

[25] Gozlan, I.; Ladkani, D.; Halpern, M.; Rabinovitz, M.; Avnir, D. *J. Heterocyclic Chem.*, **1984**, *21*, 613

[26] Fleming, M. (Syntex) **1982** US Patent 4,332,723

[27] Phillipposian, G. (Nestle) **1984** US Patent 4,450,163

[28] Hughes, D.; Dolling, U.; Ryan, K.; Schoenwaldt, E.; Grabowski, E. *J. Org. Chem.*, **1987**, *52*, 4745; Dolling, U.; Davis, P.; Grabowski, E. *J. Amer. Chem. Soc.*, **1984**, *106*, 446; Dolling, U.; Grabowski, E.; Pines, S. (Merck, Sharpe and Dohme) **1986** US Patent 4,605,761

[29] O'Donnell, M.; Bennett, W.; Wu, S. *J. Amer. Chem. Soc.*, **1989**, *111*, 2353

[30] O'Donnell, M.; Esikova, I.; Mi, A.; Shullenberger, D.; Wu, S. In *Phase-Transfer Catalysis: Mechanism and Syntheses*; Halpern, M. Ed., American Chemical Society: Washington DC, **1997**; Chapter 10

[31] Fasth, K.; Antoni, G.; Langstrom, B. *J. Chem. Soc. Perkin I*, **1988**, 3081

[32] Koch, M.; Magni, A. (Gruppo Lepetit) **1985** US Patent 4,501,919

[33] Starks, C. *J. Amer. Chem. Soc.*, **1971**, *93*, 195

[34] Poli, S.; Magni, A.; Bocchiola, G. (Poli) **1996** WO 96/10036

[35] Harris, G. (DuPont Merck Pharmaceutical) **1996**US Patent 5,510,495

[36] Zahalka, H.; Sasson, Y. *J. Chem. Soc. Chem. Comm.*, **1984**, 1652

[37] Leanna, R.; Morton, H. (Abbott) **1996** WO 96/16050

[38] Andrews, D.; Sudhakar, A. (Schering) **1996** US Patent 5,502,183

[39] Menichincheri, M.; Ceccarelli, W.; Ciomei, M.; Fusar Bassini, D.; Mongelli, N.; Vanotti, E. (Pharmacia) **1996** WO 96/14309

[40] Wright, I.; (Eli Lilly) **1985** US Patent 4,539,422

[41] Roy, R.; Tropper, F.; Cao, S.; Kim, J. In *Phase-Transfer Catalysis: Mechanism and Syntheses*; Halpern, M. Ed., American Chemical Society: Washington DC, **1997**; Chapter 13

[42] Starks, C.; Liotta, C.; Halpern, M. *Phase-Transfer Catalysis: Fundamentals, Applications and Industrial Perspectives*; Chapman and Hall: New York, **1994**; Chapter 13

[43] Baine, N.; McGuire, M.; Yu, M. (SmithKline Beecham) **1996** WO 96/11206

[44] Magni, A. (Gruppo Lepetit) **1989** Eur. Pat. EP 0 346 765 A2

[45] Ikan, R.; Markus, A.; Goldschmidt, Z. *Israel J. Chem.*, **1973**, *11*, 591

[46] Lakhrisi, M.; Chaouch, A.; Chapleur, Y. *Bull. Soc. Chim. Fr.*, **1996**, *133*, 531

# ASYMMETRIC SYNTHESIS

# AND

# ENANTIOSELECTIVITY

# Enantioselective Synthesis: The Optimum Process

**John J. Partridge and Brian L. Bray**
*Chemical Development Department, Glaxo Wellcome Inc, Research Triangle Park, North Carolina 27709*

## I. Introduction

In the field of synthetic organic chemistry, chiral chemical entities are being launched as breakthrough pharmaceutical specialties (1), agrochemicals (2) and flavor and fragrance ingredients (3). A review of the 1997 edition of the Physician's Desk Reference (4) reveals that well over half of the drugs used in clinical medicine are chiral compounds. Yet the majority of these chiral compounds are still prescribed as racemates.

As the next century and millennium approach, the emphasis on single chiral compounds will increase. In the United States, the Food and Drug Administration (FDA) has enunciated a specific policy on stereoisomers since 1992. This policy continues to evolve with periodic Internet updates (5). Authorities and health providers are now questioning the need for racemic drugs when one enantiomer does not contribute to efficacy, but may contribute to toxicity. As a result of this shift in thinking, a significant niche in the chemical industry has been created for chiral substances.

Reactions of high enantioselectivity are defined here as having enantiomeric ratios of greater than 95:5 (i.e. >90% enantiomeric excess or >90% ee). Such reactions were sparse in the 60's and 70's, much more abundant in the 80's and commonplace in the 90's. In parallel fashion, the number of scientific papers on asymmetric synthesis or enantioselective synthesis has spiraled upward in the last 40 years.

In addition to teaching several generations of chemists about chirality and stereochemistry (6), Eliel set out some guidelines for good enantioselective synthesis in 1974 (7). To these three guidelines is added a fourth, that the process be economically viable (Scheme 1). By adding this fourth guideline, the term good enantioselective synthesis can be broadened to include classical resolutions with recycling of the unwanted enantiomers (8, 9), dynamic kinetic resolutions (10) and the like.

Scheme 1. Guidelines for Good Enantioselective Synthesis

I.   The desired compound should be obtained in high enantiomeric and chemical yields.

II.  The chiral product should be readily separable from any chiral auxiliary, chiral catalyst or chiral reagent.

III. The chiral auxiliary, catalyst or reagent should be recovered in good yield and undiminished chemical and chiral purity.

IV.  The process should be economically viable.

The field of highly enantioselective synthesis has moved from a theoretical curiosity in the first half of this century to an integral component of organic synthesis. The wealth of recent monographs on this subject (11), as well as more focused texts on industrial applications (12), testify to the coming of age of this important area. Review articles are regularly found in specialty journals as diverse as *Advances in Asymmetric Synthesis, Aldrichimica Acta, Applied Catalysis, Chemical Technology & Biotechnology Reviews, Organic Process Research & Development, Pure and Applied Chemistry* and *Tetrahedron Asymmetry*. Less noticed in Western countries is the large amount of creative methodology being disclosed in local language journals such as *Farumashia (Pharmaceutical Chemistry, Japan), Hecheng Huaxue (Synthetic Chemistry, China), Huaxue Shiji (Chemistry Reagents, China), Kagaku to Kogyo, (Chemistry and Industry, Japan), Youji Huaxue (Organic Chemistry, China)* and particularly *Yuki Gosei Kagaku Kyokaishi (Journal of Synthetic Organic Chemistry, Japan)*.

## II. Enantioselective Methods

### a. Stoichiometric Chiral Reagents

Many of the earliest successes in highly enantioselective synthesis occurred with chiral stoichiometric reagents. Brown and Zweifel pointed the way with the discovery of (+)- and (-)-pin$_2$BH (13). Their conversion of cis-2-butene into 2R-butanol with 87-98% ee started the movement toward economical highly enantioselective methods (Scheme 2).

Scheme 2. Preparation and Use of (pin)$_2$BH

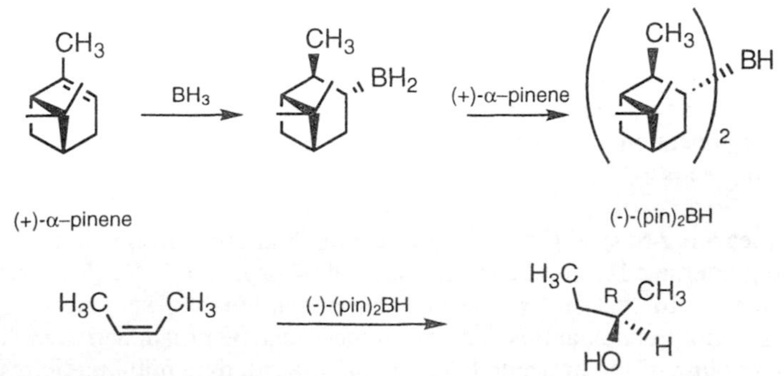

The reagent (pin)$_2$BH was employed to good effect in the hydroboration of 5-substituted cyclopentadienes (14). The corresponding R, R-cyclopenten-3-ols were converted by a Roche group into some prostaglandins (15). The S, S-cyclopenten-3-ols were transformed to carbocyclic nucleosides by Glaxo Wellcome colleagues (16) (Scheme 3).

Scheme 3. Enantioselective Hydroboration of 5-Substituted Cyclopentadienes

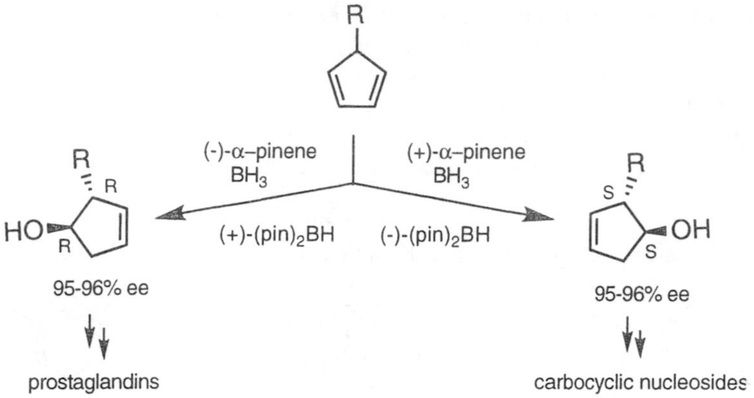

Merck workers (17) have used (pin)$_2$BCl to efficiently reduce a prochiral ketone to the desired alcohol precursor of L-699,392, an anti-asthma drug candidate. The robust (+)- and (-)-(pin)$_2$BCl reagents were disclosed in recent patents (18). The related R- and S-alpine boranes have been made and employed as chiral reducing agents. These reagents were particularly valuable in reducing α-ketoacetylenes to the corresponding propargylic alcohols (19) (Scheme 4).

Despite the non-catalytic nature of the pinene-based reagents, they remain commercially viable for several important reasons. (+)-α-Pinene is a major component of terpentine produced by the Naval Stores industry from Florida pine trees. Similarly, (-)-α-pinene is a major terpentine ingredient from Eastern European pine trees. In addition to low cost and availability, a self selection process takes place during the formation and equilibration of BH$_3$ + α-pinene ----> (pin)$_2$BH, leading to a crystalline chiral reagent. Relatively low % ee α-pinene can produce very high % ee (pin)$_2$BH when a slight deficit of such α-pinene is used. The chiral crystalline reagent can be harvested by filtration, then stirred with more α-pinene of the same quality to give a further enantiomerically enriched reagent (14, 17, 20). Serendipitously, the (+, +) and (-, -) reagents are more reactive than the (+, -) *meso* reagent (see Scheme 2).

Scheme 4. Enantioselective Reductions with (pin)$_2$BCl and Alpine Borane

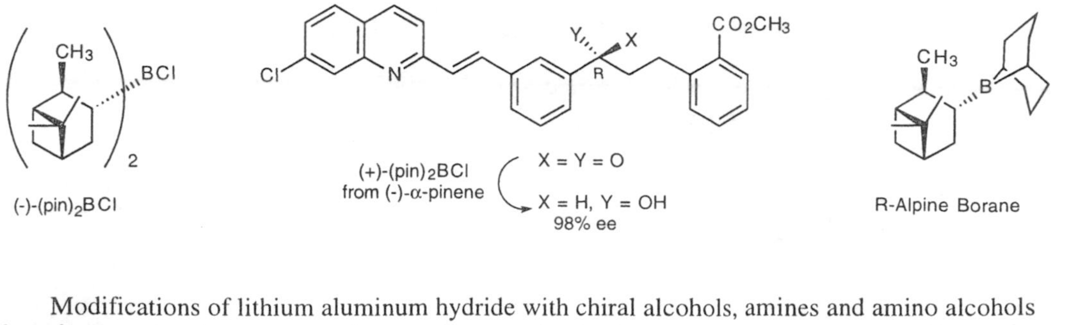

Modifications of lithium aluminum hydride with chiral alcohols, amines and amino alcohols have been explored as a means of generating enantioselective reducing agents (Scheme 5). To date, most of these methods have failed as widely used reagents for lack of generality and predictable enantioselectivity. Specific examples are adaptable to large scale synthesis. Using "Darvon alcohol" (21) as the chiral modifier, Mosher (22a) produced a reagent which reduced acetophenone to give the (R)-methylphenyl carbinol in up to 75% ee. Aging this reducing agent for 24 hr. prior to the reduction, yielded (S)-methylphenyl carbinol in up to 75% ee. This reagent was successfully used by a Roche team to reduce an α-ketoacetylene to a 90% ee alcohol intermediate for vitamin E (22b). Lithium aluminum hydride-tartaric acid combinations have been examined by a number of groups (23) and one of the more scalable classes of reagents was named TADDOL by Seebach (24). Another popular reagent, BINAL-H, was created by Noyori (25). Related $C_2$ symmetry complexes have also been studied. In addition to reducing prochiral ketones to alcohols with high % ee values, R- and S-BINAL-H have been used to produce 95+% ee α-hydroxy stannanes (26).

Scheme 5. Stoichiometric Enantioselective Reducing Agents

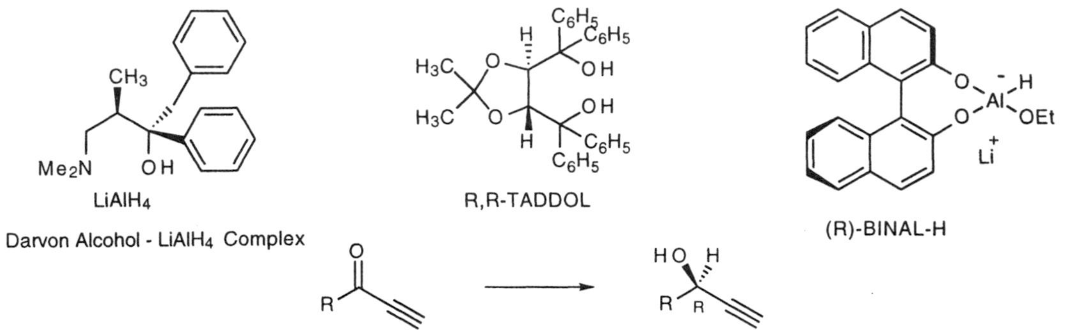

Darvon Alcohol - LiAlH₄ Complex

R,R-TADDOL

(R)-BINAL-H

## b. Chiral Auxiliaries

Some of the initial efforts at enantioselective synthesis involved early chiral auxiliaries, which were attached to achiral substrates. In 1904, McKenzie studied reductions of (-)-menthyl phenylglyoxylate (27). With aluminum amalgam, the α-keto group was cleanly reduced to the corresponding α-hydroxy ester. Hydrolysis gave (-)-menthol and enantiomerically enriched R-(-)-mandelic acid. Using (-)-menthyl phenylglyoxylate, Prelog added methylmagnesium bromide in a diastereoselective manner (28). Hydrolysis of the ester produced enantiomerically enriched R-(-)-atrolactic acid. Conversely, (-)-menthyl pyruvate upon exposure to phenylmagnesium bromide and hydrolysis of the product, yielded enantioenriched S-(+)-atrolactic acid. This body of work later became codified as Prelog's rule (Scheme 6).

Scheme 6. Classical Chiral Auxiliary and Prelog's Rule

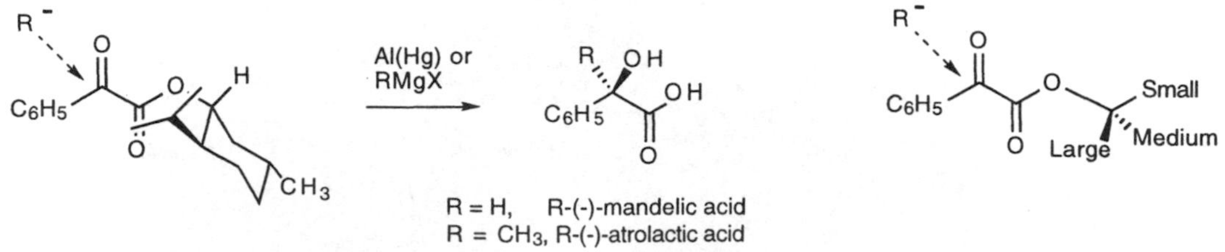

R = H,    R-(-)-mandelic acid
R = CH₃, R-(-)-atrolactic acid

One of the most scalable and versatile chiral auxiliary classes has been the oxazolidinones of Evans and coworkers (29). S-Valinol or S-phenylalaninol were converted into the oxazolidinones with diethyl carbonate. The corresponding boron enolates were exposed to aldehydes to give high yields of the chiral *erythro*-aldols. In the same manner (+)-norephedrine was transformed into its oxazolidinone. Formation of the boron enolate and low temperature reaction produced high yields of the *threo*-aldols. Saponification with lithium hydroxide in aqueous hydrogen peroxide then afforded the chiral β-hydroxy carboxylic acids and the auxiliaries were recovered intact and suitable for recycle (30) (Scheme 7).

Similarly, the lithium enolate of the (+)-norephedrine oxazolidinone was alkylated with high diastereoselectivity at -35°C. The ester amide was then converted into the anticonvulsant candidate CI-1008 (31) by a Parke-Davis group. Ultimately, CI-1008 was prepared in four stages from isovalderaldehyde and diethyl malonate, using an S-(+)-mandelic acid resolution. This proved to be eight-fold less expensive than the chiral auxiliary route.

Scheme 7. Evans Chiral Auxiliaries

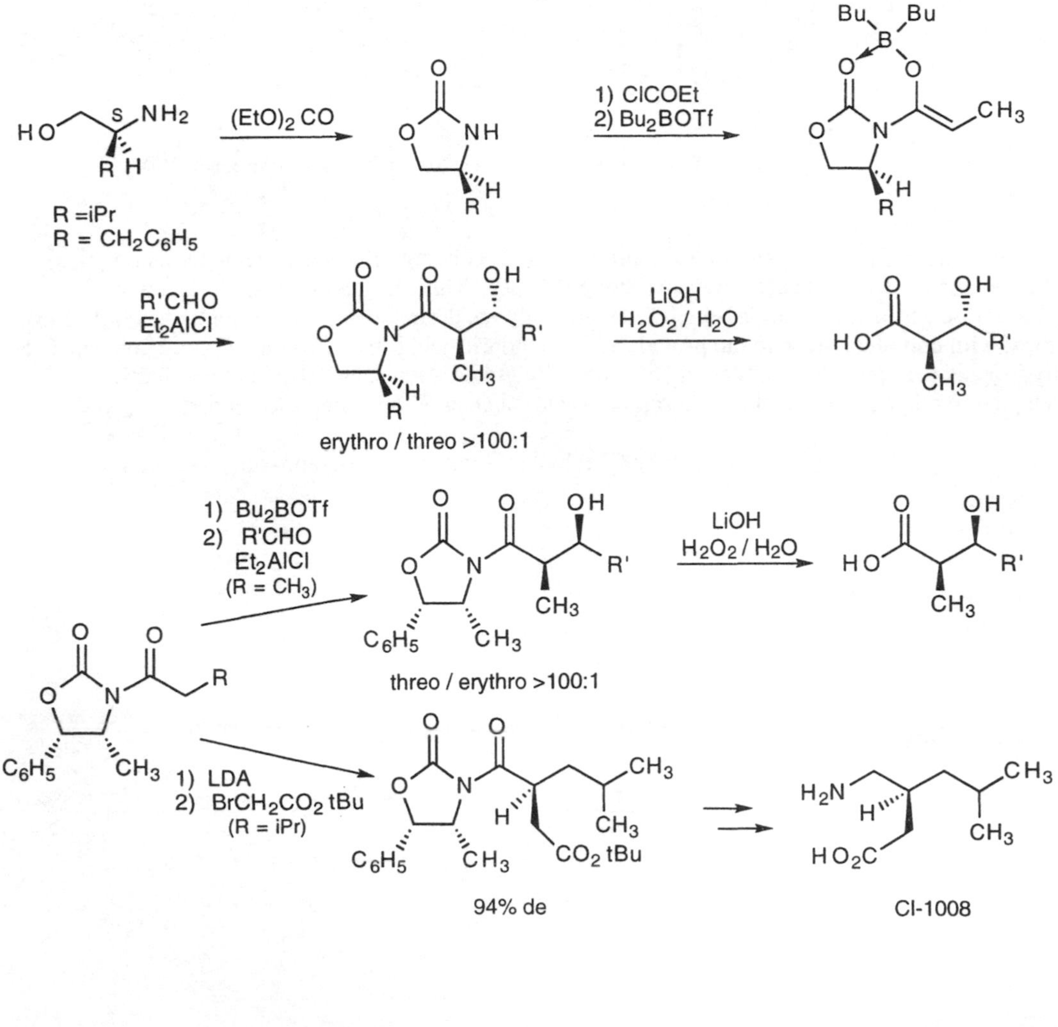

Another variation of a highly diastereoselective aldol reaction employed R-(+)-2-hydroxy-1, 2, 2-triphenylethyl acetate as its lithium enolate with aldehydes (32). The corresponding β-hydroxy esters were hydrolyzed to give β-hydroxy carboxylic acids of 92-94% ee. Both R- and S-HYDRA diol reagents were readily made and recycled. Using this methodology, chiral drugs such as the hypolipidemic Mevacor® (Merck) (33) and the anti-obesity agent Xenical® (Roche) (34) were synthesized in high yields and % ee purities (Scheme 8).

Scheme 8.  HYDRA Chiral Auxiliary for Aldol Reactions

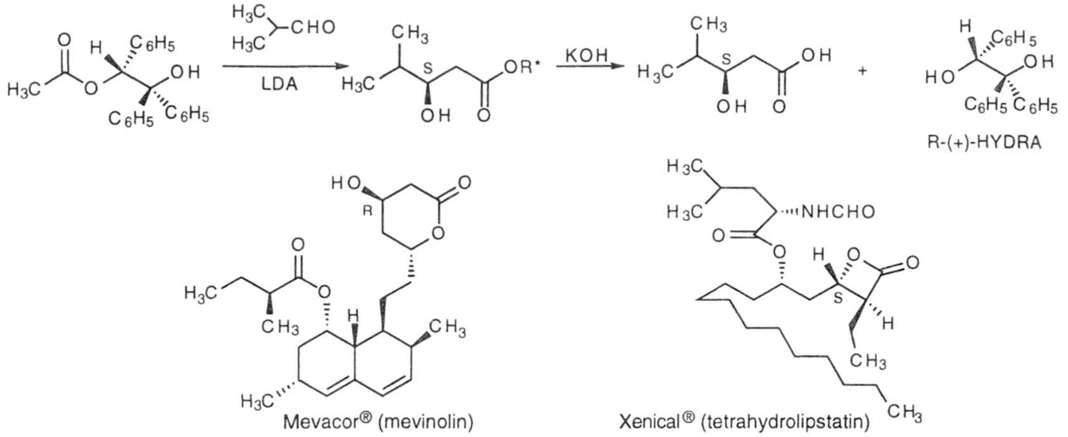

R-(+)-HYDRA

Mevacor® (mevinolin)

Xenical® (tetrahydrolipstatin)

A technically feasible use of R, R-tartaric acid as chiral auxiliary was demonstrated in the Zambon process for S-naproxen manufacture (35, 36). This elegant process employed a diastereoselective bromination (88-90% de) and only the desired diastereomer underwent a 1, 2-aryl shift, with complete inversion to provide 99% ee tartaric acid ester. Bromine hydrogenolysis and hydrolysis then gave S-naproxen in 75% overall yield on a ton scale. This process has been adapted to other commercial α-arylpropionic acids such as S-ibuprofen (Scheme 9).

Scheme 9.  Zambon Technical Synthesis of S-Naproxen

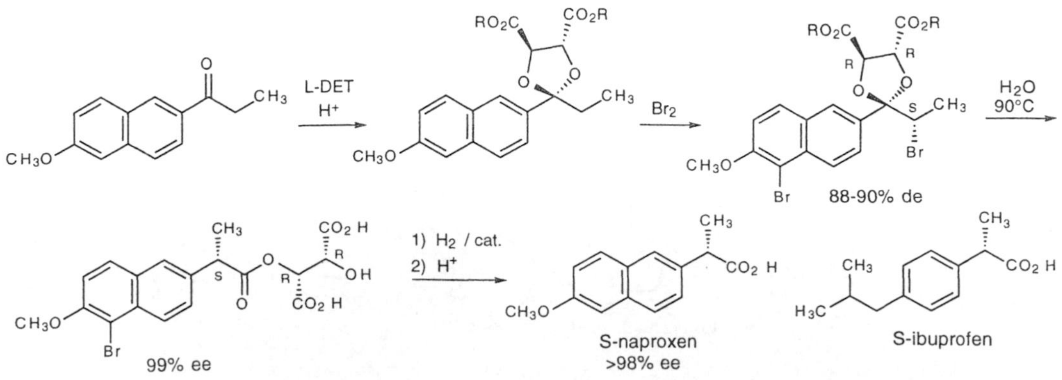

88-90% de

99% ee

S-naproxen
>98% ee

S-ibuprofen

Merck chemists have developed an interesting process for the preparation of some chiral α-arylpropionic acids using inexpensive alcohols as recyclable chiral auxiliaries (37). (±)-Ibuprofen was desymmetrized by conversion to the prochiral ketene. Several lactic acid esters were each added to the ketene carbonyl and reprotonation proceeded with very high diastereoselectivity (Scheme 10).

Scheme 10. Prochiral Ketene Reactions with Lactic Acid Derivatives

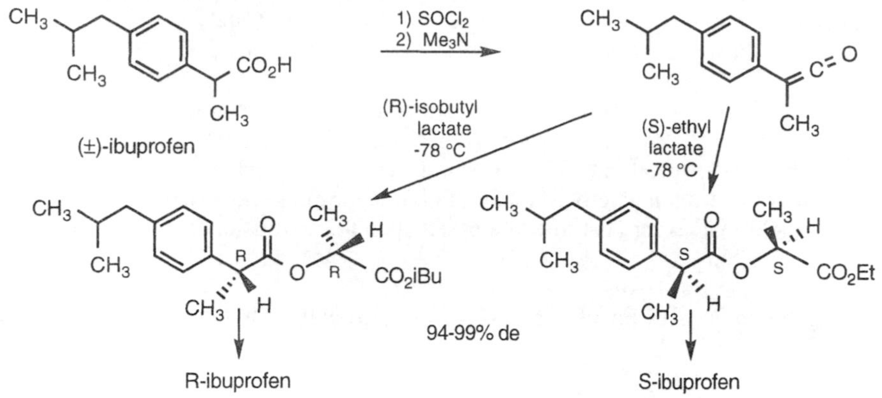

## c. Chiral Catalysts

The creation of chiral catalysts for highly enantioselective synthesis has blossomed both intellectually and technically, leading to cost effective processes. Following the efforts of Kagan, Noyori and other pioneers (38), the Knowles group at Monsanto carried out highly enantioselective hydrogenations of N-acylaminoacrylic acids using the chiral ligand R, R-DIPAMP as part of the chiral Rh(I)-catalyst (39a). In this manner, a variety of natural and unnatural D- and L-amino acids were efficiently formed. As a result, the synthesis was commercialized for L-DOPA (39b), an anti-Parkinson drug sold as Larodopa® (Roche). A large number of related chiral hydrogenations have been devised since this time (40). Recent efforts from the Burk group (41), produced a variety of unnatural amino acids with very high % ee's utilizing the DuPHOS ligands (Scheme 11).

Scheme 11. Enantioselective Catalytic Hydrogenations Using Chiral Phosphine Ligands

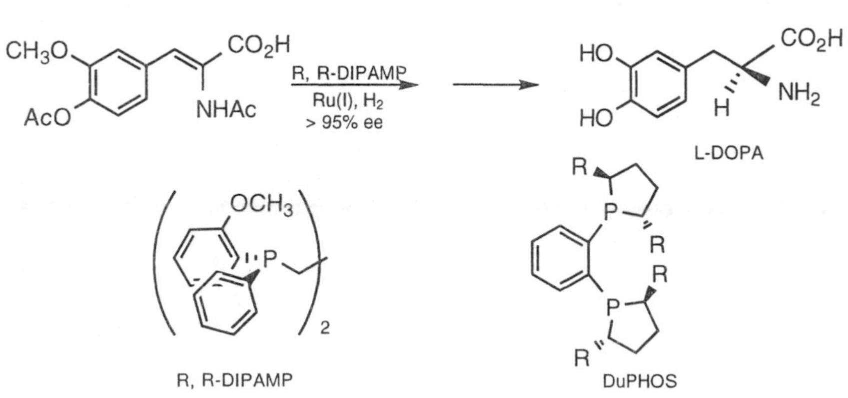

Possibly the most impressive commercial enantioselective synthesis to date has been exemplified in the manufacture of (+)-citronellal, (-)-l-menthol and a variety of other high value-added chiral terpenes from isoprene. This remarkable accomplishment succeeded through the efforts of top R&D staff at the Institute of Materials Science (Okazaki), Nagoya University, Osaka University and the Takasago Perfumery Company (42). Isoprene was converted to N, N-diethylgeranylamine or N, N-diethylnerylamine. Depending upon the catalyst, either substituted amine was transformed into R-(+)- citronellal. A zinc bromide catalyzed ene reaction then gave (-)-isopulegol and catalytic hydrogenation afforded pure (-)-l-menthol. Catalytic turnover numbers exceeded 50,000 for the asymmetric olefin isomerization processes (Scheme 12).

In the case of natural (-)-l-menthol, this commercial product was extracted from Brazilian, Chinese and Indian mint oils and sold for $18-20 per pound in 1992 (43). At that time, Takasago offered USP synthetic (-)-l-menthol at $11 per pound. Thus, a formerly small flavor and fragrance company in Japan, operating in an environment of high labor rates and a strong currency, was able to dominate the world market for (-)-l-menthol based upon this exquisite enantioselective chemistry.

Scheme 12. Technical Synthesis of (+)-Citronellal and (-)-l-Menthol

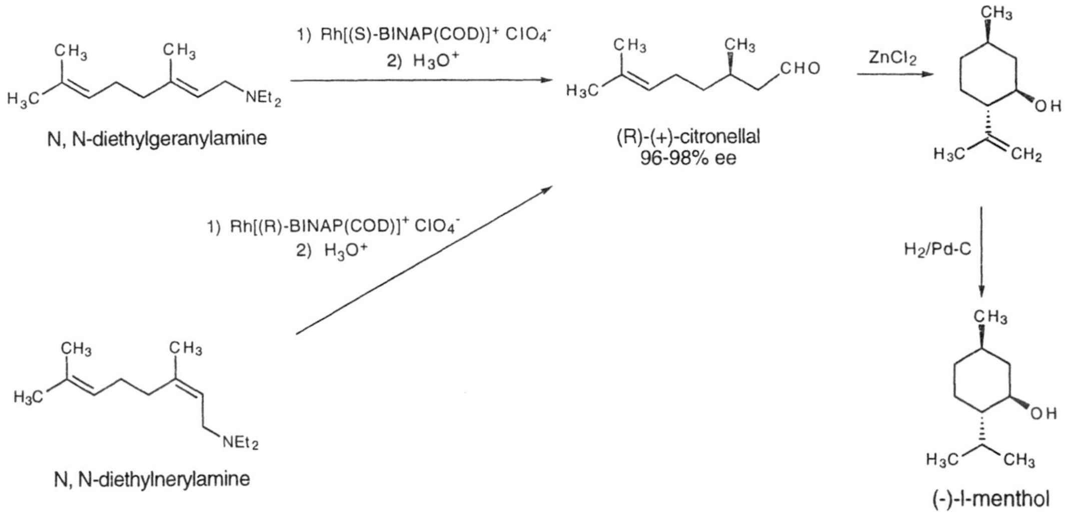

The Sharpless epoxidation of allylic alcohols with tBuOOH oxidant and chiral Ti (IV)-tartrate ester catalysts constituted a similar and major breakthrough in the field of practical enantioselective synthesis (44). These chiral catalysts discriminate between the *re-* and *si-* faces of mono-, di- or tri-substituted allylic alcohols, providing chiral epoxides with high enantiomeric purities. Additionally, the E-substituted allylic alcohols generally reacted more efficiently than their Z-counterparts. This chemistry has been so widely applied that the reaction has attained name recognition status (45). As a practical example, Schering chemists produced a short synthesis of the broad spectrum antibiotics thiamphenicol and florfenicol, exploiting the Sharpless protocol with L-diisopropyl tartrate (46) (Scheme 13).

### Scheme 13.  Sharpless Epoxidation to Thiamphenicol  and Florfenicol

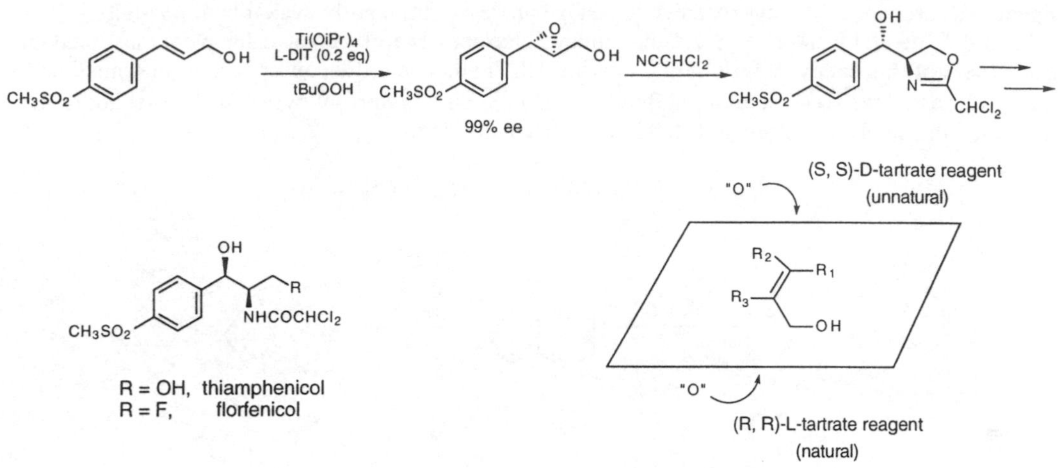

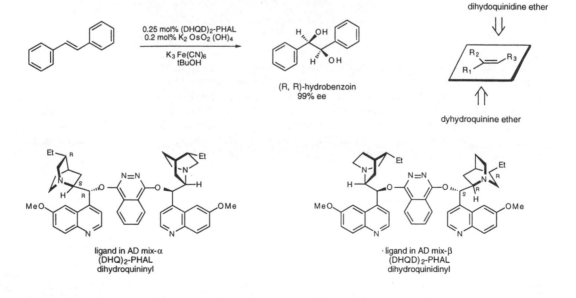

The Sharpless team has also developed an asymmetric version of the OsO$_4$-catalyzed *cis*-dihydroxylation of alkenes which provides absolute stereochemical control at two adjacent carbon centers (47).  The now commercially available AD mix-α and AD mix-β cocktails, contain 0.2% K$_2$OsO$_2$(OH)$_4$, 1% chiral ligand and the stoichiometric oxidant K$_3$Fe(CN)$_6$.  These recipes are used to effect highly enantioselective *cis*-dihydroxylations of many *trans*-disubstituted, trisubstituted and terminal olefins.  They are true ligand-catalyzed reactions in that the chiral ligands increase the reaction rates over the uncatalyzed processes.  The asymmetric dihydroxylation technology invented by Sharpless is proprietary.  The cost of licensing this technology may impact its incorporation into commercial manufacturing operations (48) (Scheme 14).

### Scheme 14.  Sharpless Dihydroxylation Chemistry

Jacobsen has discovered an exciting group of chiral salen Mn(III) reagents that promote the asymmetric epoxidation of unfunctionalized *cis*-olefins (49). The epoxidation is catalytic in the chiral salen Mn(III) reagent and uses commercial bleach as the stoichiometric oxidant. The chiral ligand was prepared by an inexpensive two-stage process from readily available di-*tert*-butyl-salicylaldehyde and R, R- or S, S-1,2-diaminocyclohexane. Merck chemists have demonstrated the robustness of this catalytic system in preparing 1S, 2R-indene oxide, an important intermediate in the synthesis of the HIV protease inhibitor Crixivan®. 3-Phenylpropylpyridine N-oxide addition increased the catalyst activity and stability (50) (Scheme 15).

Scheme 15. Jacobsen Epoxidation Chemistry

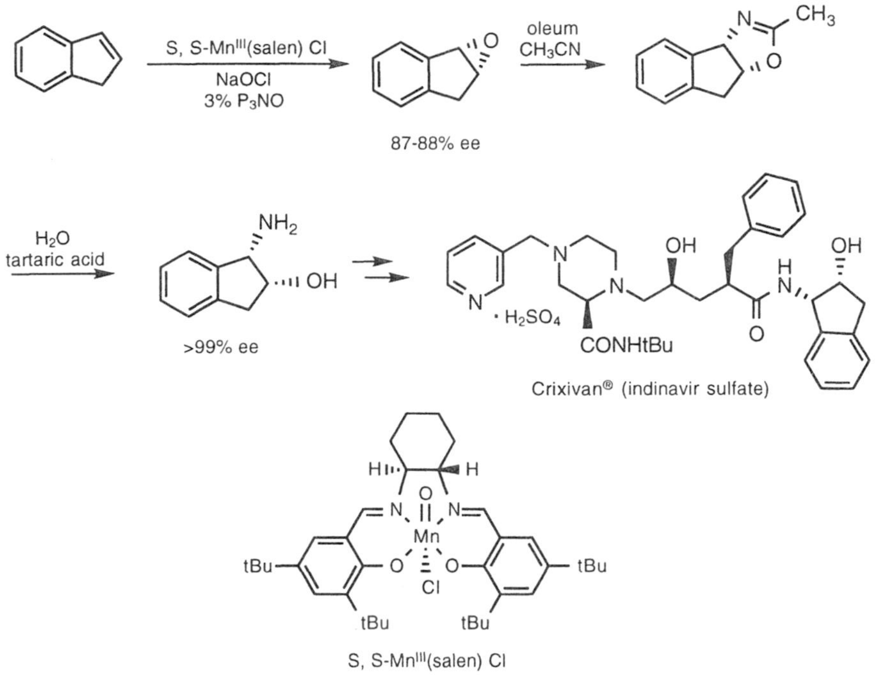

Trost demonstrated desymmetrizations of prochiral substrates or racemic chiral substrates with chiral palladium catalysts (51). In the first example, the desymmetrization of the *meso* -2-ene-1,4-diol diester was catalytic in both Pd(0) and chiral ligand and proceeded in >98% ee. In the second example, treatment of a racemic allyl ester with a catalytic amount of Pd (0) and the chiral ligand generated a common π-allyl palladium intermediate that underwent a highly asymmetric acylation, providing the allylic propionate ester in 98% ee and 95% yield. This ester is an intermediate in the synthesis of the antitumor agent phyllanthocin (52) (Scheme 16).

Scheme 16. Trost Desymmetrization Methodology

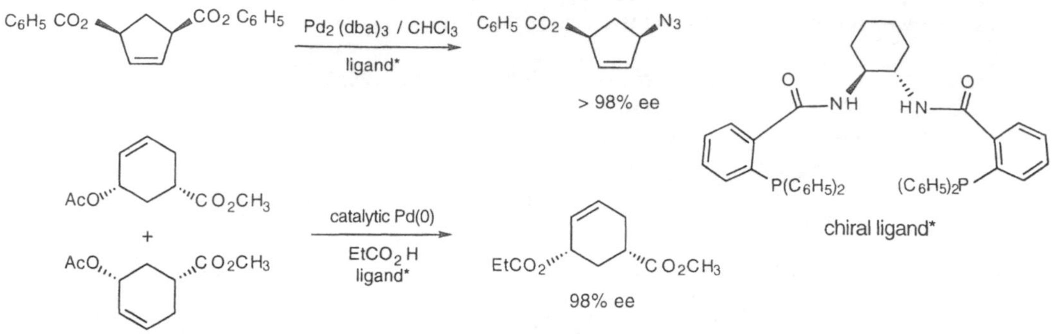

An L-proline catalyzed Michael-aldol reaction was carried out by Hajos and Parrish at Roche (53, 54). This novel chemistry produced a general and highly enantioselective route to a variety of steroid hormones and anti-fertility drugs (Scheme 17).

Scheme 17. Enantioselective Annulation in a Roche Steroid Synthesis

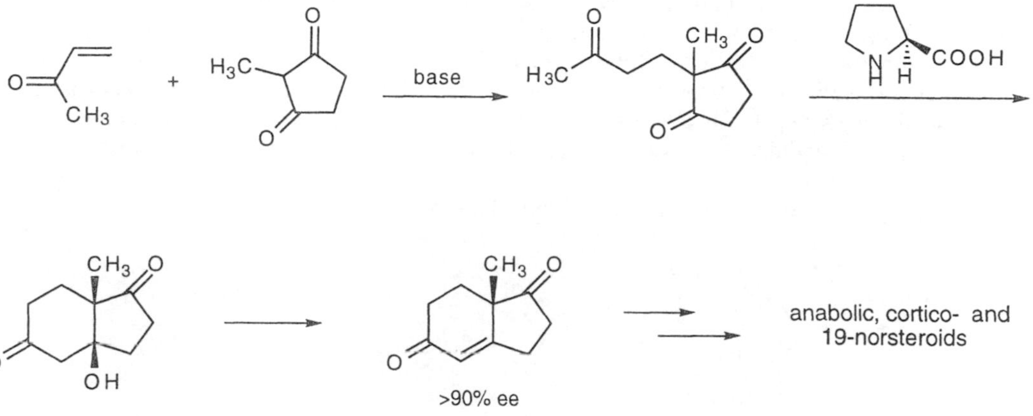

### d. Chiral Adjuvants

The term "chiral adjuvants" is used to refer to chiral agents that are complexed with achiral organometallic reagents and dramatically enhance the nucleophilicity and reactivity of those reagents. Other adjuvants such as the chiral phase transfer agents can provide both a chiral environment and increased basicity and nucleophilicity. Still other adjuvants can be chiral bases, chiral Lewis bases, chiral solvents and even photochemical processes (55).

Metal Chelators: Good examples were shown by the Soai group (56). Diethylzinc compounds react very slowly or not at all with aldehydes in non-polar solvents. However, in the presence of amines the corresponding diethylzinc complexes become much more nucleophilic and reactive. L-Proline can readily be converted to S-(+)-DPMPM. Using this adjuvant or its lithium salt in the presence of aldehydes and dialkylzinc species afforded the corresponding S-secondary alcohols with high % ee values. Similarly, the related L-proline derived reagent or its lithium salt gave

R-secondary alcohols with similar % ee values. A number of other chiral amino alcohols, including the di-n-butyl norephedrine derived agents 1S,2R-(-)-DBNE and 1R, 2S-(+)-DBNE, have acted as highly enantioselective adjuvants for dialkylzinc reagents and aldehydes (Scheme 18).

Scheme 18. Soai Dialkylzinc Adjuvants

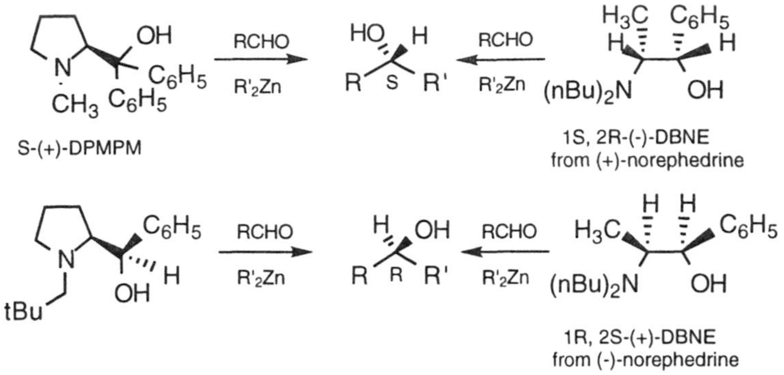

S-(+)-DPMPM

1S, 2R-(-)-DBNE
from (+)-norephedrine

1R, 2S-(+)-DBNE
from (-)-norephedrine

One of the more studied chiral adjuvant classes are the oxazaborolidines and oxazaborines created by Itsuno (57) and Corey (58). Borane solutions in tetrahydrofuran react sluggishly with ketones at ambient temperatures. However, addition of chiral adjuvants such as S-prolinol, B-methyl oxazaborolidine, or the crystalline oxazaborolidine-borane complex, resulted in rapid and highly enantioselective reductions on a wide variety of prochiral ketones. Merck workers have devised an alternative synthesis of these adjuvants (59). Shown below are highly enantioselective reductions of drug candidate intermediates from Merck (17) and from Pfizer (60) (Scheme 19).

Scheme 19. Itsuno - Corey Oxazaborolidine Adjuvants

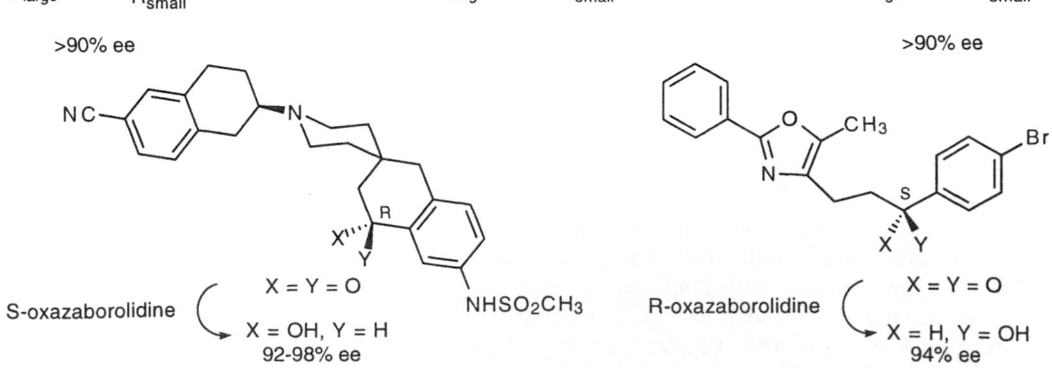

S-oxazaborolidine

>90% ee

R-oxazaborolidine

>90% ee

S-oxazaborolidine

X = Y = O

X = OH, Y = H
92-98% ee

NHSO$_2$CH$_3$

R-oxazaborolidine

X = Y = O

X = H, Y = OH
94% ee

Chiral Phase Transfer Agents: Another group of chiral adjuvants are chiral quaternary ammonium salts and chiral crown ethers used as phase transfer agents (61). By carrying out molecular modeling and a systematic study of this alkylation reaction, both a high chemical yield and a high % ee were achieved for the Merck diuretic candidate indacrinone (62). It should be noted that related cinchona alkaloid phase transfer agents can be degraded under such strongly basic conditions to give unwanted chiral byproducts (63) (Scheme 20).

Scheme 20.  Enantioselective Phase Transfer Alkylation

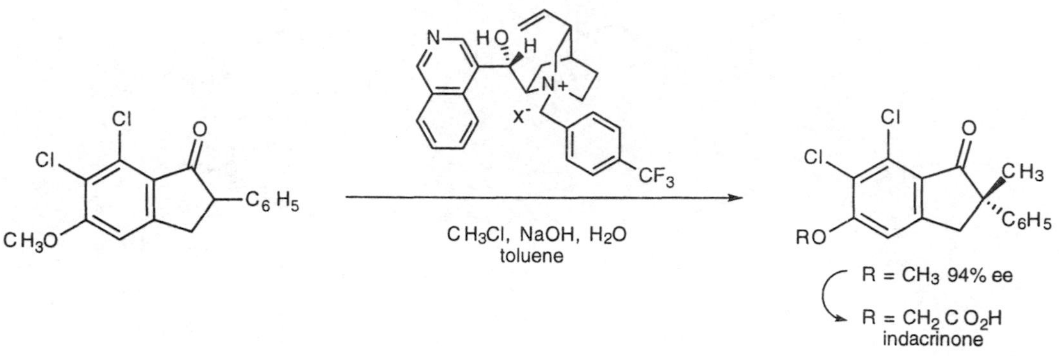

Chiral Bases: Chiral bases such as chiral amines and lithium amides have been utilized to good effect in enantioselective synthesis (64). Two distinct processes seem evident. The first process is one in which the base selects between enantiotopic protons in a kinetically controlled deprotonation of prochiral substrates (65). The second process is one in which the chiral base acts as a strong base, deprotonating a substrate to give a prochiral enolate (66, 67). The complex of the enolate and the chiral base then determine the outcome of the process (Scheme 21).

Scheme 21. Enantioselective Deprotonations

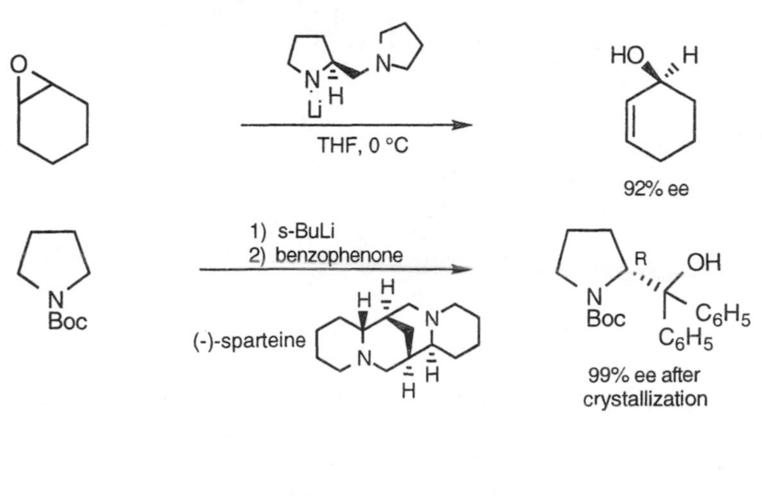

Chiral Lewis Bases: The Denmark group has created several chiral HMPA-like Lewis bases that rapidly accelerated the reactions between trichlorosilyl enolates and aldehydes to afford chiral aldols with 85-95% ee values (68). No chiral auxiliaries were involved (Scheme 22).

Scheme 22. Chiral HMPA-Like Lewis Bases

1:61 *syn / anti*
*anti* 93% ee

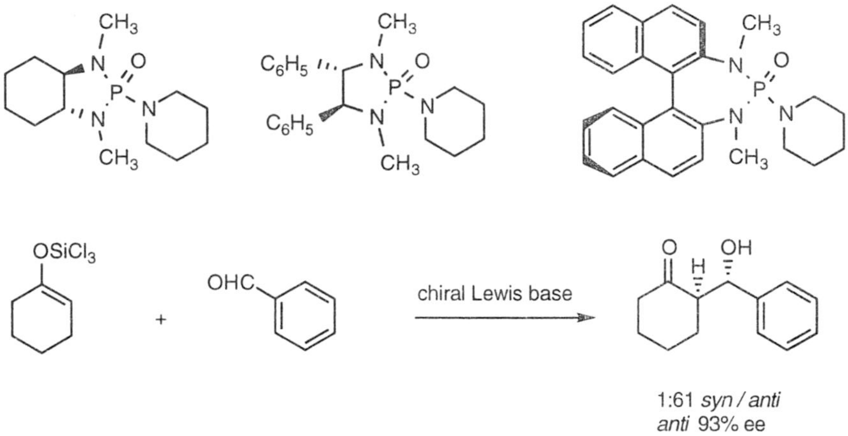

## e. Dynamic Kinetic Resolutions

An enantioselective reduction of the 1,5-benzothiazepine ketone proceeded via a dynamic kinetic resolution to provide the key intermediate for the calcium channel blocker Cardizem® (69). A group at Tanabe Seiyaku prepared the chiral reducing reagent by combining sodium borohydride and (S)-*tert*-leucine in THF. The *cis*-2S, 3S-isomer was isolated in high yield and 95% de, suggesting rapid epimerization at C-2, followed by selective reduction of the 3-ketone (Scheme 23).

Scheme 23. Dynamic Kinetic Resolution Approach to Diltiazem Hydrochloride

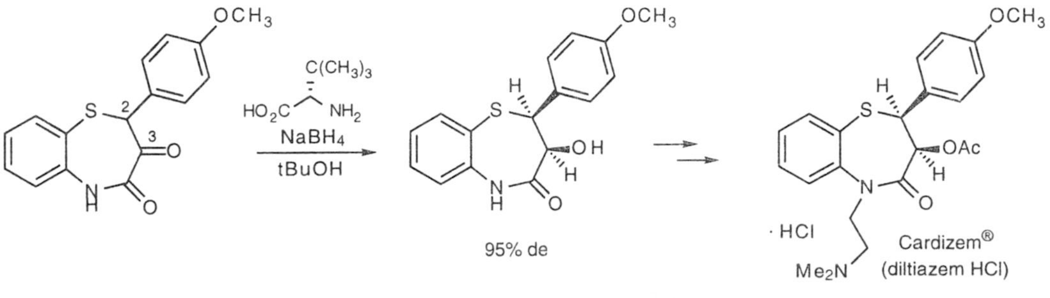

95% de

Cardizem®
(diltiazem HCl)

Merck chemists produced a crystallization induced dynamic resolution during the synthesis of the anti-obesity clinical candidate L-364,718 (70). The racemic 3-amino-1, 4-benzodiazepine was treated with a trace of an aldehyde and S-camphorsulfonic acid. As the S, S-salt crystallized, the R-aminoketone equilibrated to the S-aminoketone via Schiff base formation (Scheme 24).

Scheme 24. Dynamic Resolution via Schiff Base Formation and Equilibration

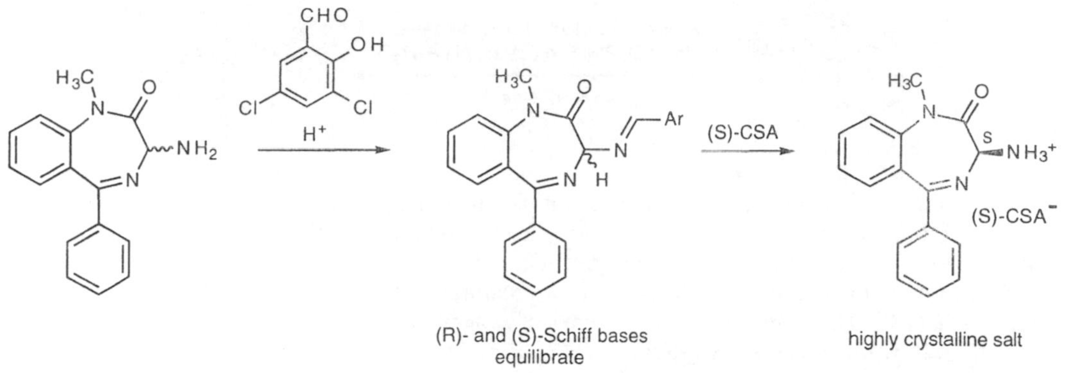

(R)- and (S)-Schiff bases
equilibrate

highly crystalline salt

## f. Enzymatic Catalysis

The use of enzymatic catalysts in organic synthesis is well defined and documented (71). Such biocatalysis has been effected with whole cells of animals, microorganisms and plants or isolated enzymes, purified to various degrees of homogeneity. Enzymes represent a very broad range of efficient chemical catalysts. They are classified into six main groups (Scheme 25).

Scheme 25. Classification of Enzyme Types

1.  hydrolases (hydrolysis of amides, esters, glycosides and lactones)

2.  isomerases (carbon-carbon bond migrations, E / Z isomerizations and racemizations)

3.  lyases (additions to $\pi$ bonds)

4.  oxidoreductases (reversible oxidations and reductions)

5.  synthetases (formation and breaking of C-C, C-N, C-O, C-S and phosphate ester bonds)

6.  transferases (transfer of acyl, glycosyl or phosphoryl from one molecule to another)

Significant issues for enzymatic catalysis include the effective life of the enzyme, as well as recovery, recycle and turnover. Enzymes can be immobilized by adsorption on dialysis membranes or by copolymerization onto solid supports. In such cases the life of the enzymes is extended and made commercially viable. The flexible, yet highly chiral, nature of enzyme active sites often produces excellent chemo-, stereo- and enantioselectivity. Enzymatic catalysis is regularly employed in processes that require discrimination between enantiotopic groups, or discrimination between enantiomers leading to kinetic resolutions.

Some potentially useful enzymatic reactions need cofactors such as NADP-NADPH or ATP which are expensive. Practical enzymatic syntheses require a method of cofactor recycling, usually with a second enzymic system or a second cofactor. In the example shown, the cofactor FMN managed the recycling of a trace amount of the costly cofactor $\beta$-NAD (72) (Scheme 26).

Scheme 26.  Enantioselective Enzymatic Oxidation and Lactonization

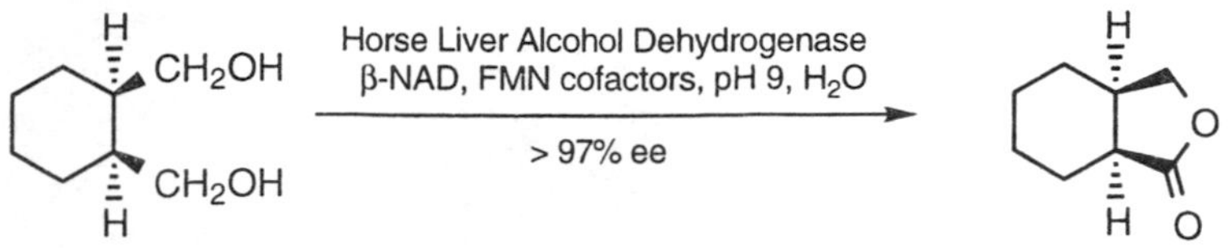

In the case of baker's yeast, the enzyme complex including cofactors is very cheap.  However, the regeneration of cofactors requires a second redox reaction to take place.  Actively fermenting baker's yeast, with sucrose as an energy source, has been used for the enantioselective reduction of a variety of carbonyl compounds.  Reductions of prochiral ketones yielded chiral alcohols with high % ee's (73).  The stereochemical outcomes of enzymatic reductions with baker's yeast generally follow the model shown (Scheme 27).

Scheme 27.  Enzymatic Reduction with Baker's Yeast and the Enantioselection Rule

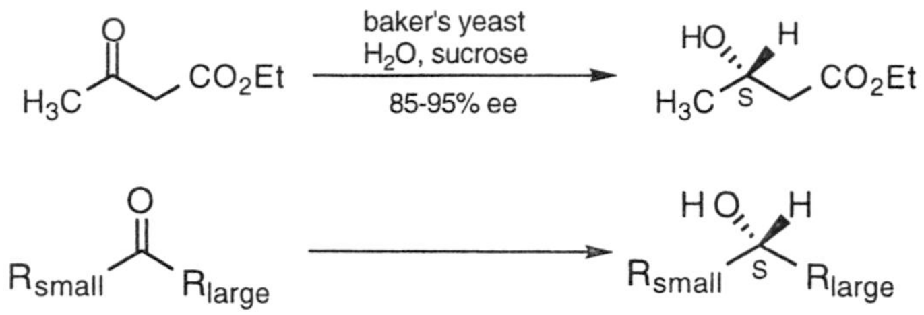

Hydrolytic enzymes in aqueous or organic solutions have been used to effect kinetic resolutions of racemic esters or desymmetrization of prochiral substrates in an industrial setting. Many lipases (74) and esterases (75) are inexpensive and do not require cofactors.  A number of enantioselective hydrolyses and kinetic resolutions have been successfully carried out on scale using P-30 and P-800 Amano lipase preparations (74).  With 0.5 gram of pseudomonas lipase P-800, immobilized between membranes in a bioreactor, multi-kilograms of 98% ee hydroxy acetate were obtained by a Glaxo Wellcome group (76).  No appreciable loss in enzyme activity was observed after 15 runs (Scheme 28).

Scheme 28.  Enzymatic Hydrolytic Desymmetrization

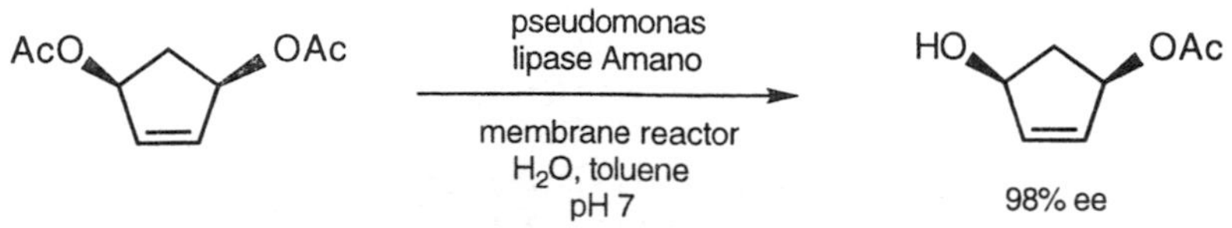

## g.  Catalytic Antibodies

Catalytic antibodies, a new type of designer enzymes, open up the possibility of creating catalysts for a wide variety of chemical reactions on demand.  With the advent of monoclonal antibody technology, such antibody enzymes or abzymes can now be made on a significant scale.  By raising such antibodies using haptens mimicking a number of known transition states, antibody catalysts have been generated that carry out the hydrolysis of amides, carbonates and other esters. Additional catalytic reactions include concerted Claisen and oxy-Cope rearrangements, Diels-Alder cycloadditions, lactonizations, photo-cycloreversions and redox reactions (77). Significantly, peptides can be formed in an aqueous environment with some catalytic antibodies (78).

A typical catalytic antibody is raised against a synthetic antigen or hapten that is bound to a large carrier molecule. The hapten may or may not exhibit chirality. In a given reaction sequence, if the product is chiral, the transition state leading to it will also be chiral. Therefore, the binding interaction between a chiral catalytic antibody and two enantiomeric reactants will be different, leading to useful diastereoselectivity. Catalytic antibodies have been raised against achiral and single enantiomer haptens as well as against racemates and mixtures of diastereomers. Remarkably, a majority of individual catalytic antibodies, selected on the basis of good hapten binding and reaction against a test substrate, nonetheless afforded high 90-98% enantiomer excesses in kinetic resolution reactions (79).

Phosphonate ester haptens model the hydrolytic transition state and have been employed in raising monoclonal antibodies with catalytic activity for carbonate, ester and lactone hydrolysis. Using racemic haptens allowed the formation of antibodies specific for both enantiomeric substrates in a single experiment. In one such study, the Janda group produced eleven antibodies against a single racemic hapten. Of the eleven, nine antibodies were specific for the R-enantiomer and two targeted the S-enantiomer (80a). The Schultz team raised eight antibodies against the hapten shown in Scheme 29. Of the eight, three were specific for R-alanine esters and four reacted with the S-alanine esters (80b).

Scheme 29. Hydrolytic Catalytic Antibody Hapten and Target Ester

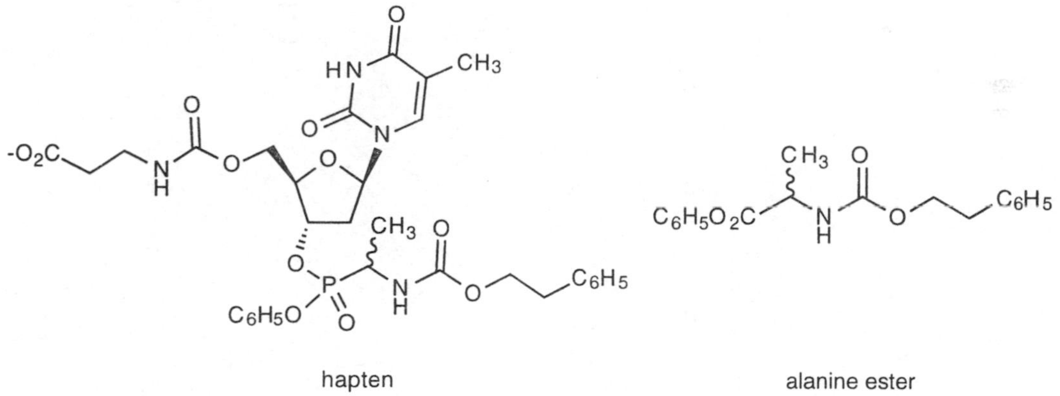

hapten                    alanine ester

In terms of preparative synthetic chemistry, the biggest obstacle to overcome is the requirement of close to equal weight ratios of substrate and antibody protein catalyst (81). Three parameters are paramount for future research---catalyst efficiency must be improved, the scope of reactions that can be catalyzed should be broadened and more efficient methods for generation, identification and scale-up of high turnover catalytic antibodies must be developed.

## III. Design of Enantioselective Synthesis

### a. Retrosynthetic Strategies

As has been seen in this and other recent reviews, a number of highly enantioselective methods are now available which utilize chiral reagents, auxiliaries, catalysts, adjuvants, enzymes or catalytic antibodies. Employing the well known Corey concepts of retrosynthetic analysis (82), "logic tree" reductions of stereochemical complexity lead to simpler achiral or prochiral target precursors. It is desirable to devise a number of retrosynthetic trees that point to a variety of prochiral intermediates. Screening of potential enantioselective processes can then help dictate the most likely and cost effective ways forward, as exemplified by related routes to the oncology candidate GI147211C from Comins (83) and a Glaxo Wellcome group (84) (Scheme 30).

Scheme 30. Retrosynthetic Analysis for the Camptothecin Analog GI147211C

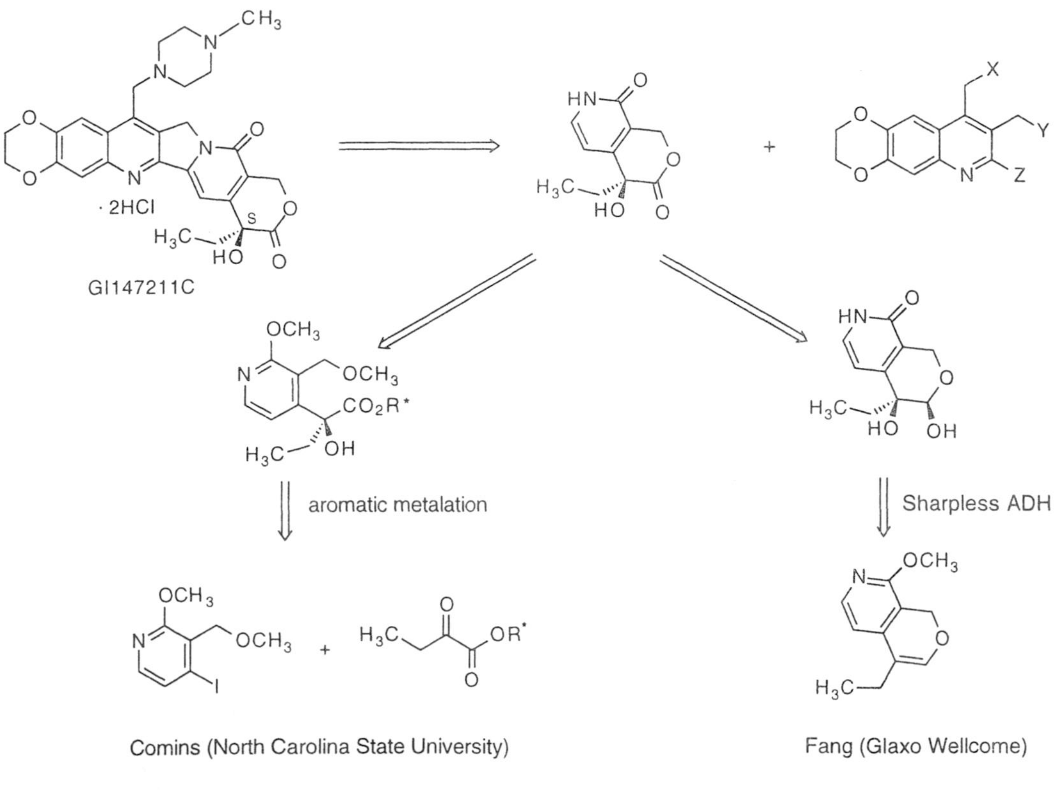

Comins (North Carolina State University)                    Fang (Glaxo Wellcome)

### b. Combinatorial Approaches

Efforts are now underway to apply combinatorial chemistry (85) to the discovery of new enantioselective methods and enantioselective catalysts (86). For related processes, two strategies have been employed. The Still strategy targeted a known ligand structure to which various arrays of binding elements were created, grouped spatially and evaluated (87). The second strategy involves the design of potential ligand libraries without predefined binding sites. This strategy is carried out using diverse functionality and conformational restrictions, resulting in a range of coordination environments and catalytic sites.

The Jacobsen group has already demonstrated the discovery of novel coordination complexes using the second strategy (88). An initial library of ca. 12,000 tagged synthetic ligands was created from four variable components, D- or L- amino acids, chiral "turn" elements and amino acid capping reagents. Tag photolysis and analysis showed that the six strong nickel (II) binders had a strong structural consensus. The same library was then evaluated for binding with Fe (III). A different set of structures was obtained. Related studies are focused on which types of chiral coordination complexes act efficiently as enantioselective catalysts (Scheme 31).

Scheme 31. Combinatorial Chemistry for Chelation and Catalyst Design

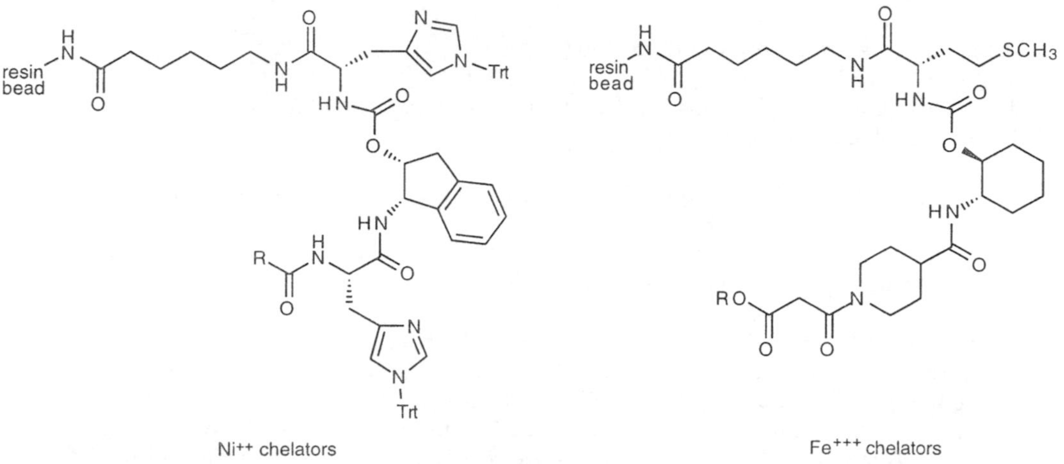

Ni++ chelators

Fe+++ chelators

## IV. Summary

To design a technical route of manufacture, process chemists must consider the chirality of the target molecule, the number of stages, the availability and cost of starting materials and reagents, the reaction temperatures, the solvent volumes, the purifications and waste streams, as well as the proprietary status of reagents, intermediates and processes. Despite the barriers, these practical molecular architects will continue to make important strides in optimizing processes. In the past, a major factor in producing single enantiomer drugs has been the cost of goods. Thanks to the discovery of highly enantioselective scalable reactions, these costs are more manageable. With the advent of such highly enantioselective methodologies as well as paradigmatic combinatorial chemistry, accessible informatics (89), affordable robotics and concepts of statistical design (90), process chemists will undoubtedly move to new levels of creativity and productivity.

## V. References

1    (a) Stinson, S. C. *Chem. & Eng. News*, **1995**, *73* (41), 44; (b) *ibid*, **1994**, *72* (38), 38; (c) *ibid*, **1992**, *70* (39), 46; (d) Borman, S. *ibid*, **1990**, *68* (28), 9.

2   (a) Mori, K. *Biosci. Biotechnol. Biochem.*, **1996**, *60* , 1925; (b) Crosby, J., *Pestic. Sci.*, **1996**, *46*, 11; ( c) Tombo, G. M. R.; Bellus, D. *Angew. Chem. Int. Ed. Engl.*, **1991**, *30*, 1193.

3   Vandak, D.; Sturdik, E. *Chem. Listy*, **1993**, *87*, 709; See also: Stix, G. *Sci. Amer.*, **1997**, *276* (5), 28.

4   *Physician's Desk Reference*, 51st Ed., Medical Economics Co.: Montvale, NJ, **1997**.

5   Stinson, S. C. *Chem. & Eng. News*, **1997**, *75* (22), 28; For updated FDA guidance on stereochemistry, the Internet address is: http://www.fda.gov/cder/guidance/stereo.html.

6   (a) Eliel, E. L.; Wilen, S. H. *Stereochemistry of Organic Compounds*, Wiley: New York, **1994**; (b) Eliel, E. L. *Stereochemistry of Carbon Compounds*, McGraw: New York, **1962**.

7   Eliel, E. L. *Tetrahedron*, **1974**, *30*, 1503.

8   Jacques, J.; Collet, A.; Wilen, S. H. *Enantiomers, Racemates and Resolutions*, 2nd Ed., Krieger: Malabar, FL, **1991**.

9   (a) Sheldon, R. A. *Chimia*, **1996**, *50*, 418; (b) *J. Chem. Technol. Biotechnol.*, **1996**, *67*, 1.

10  (a) Caddick, S.; Jenkins, K. *Chem. Soc. Rev.*, **1996**, *25*, 447; (b) Noyori, R.; Tokunaga, M.; Kitamura, M. *Bull. Chem. Soc. Jpn.*, **1995**, *68*, 36.

11  (a) *Advanced Asymmetric Synthesis*; Stephenson, G. R., Ed., Chapman & Hall: London, **1996**; (b) *Chemical Synthesis: Gnosis to Prognosis*; NATO ASI Series E, Volume 320, Chatgilialoglu, C.; Snieckus, V., Ed., Kluwer: Dordrecht, **1996**; (c) *Reductions in Organic Synthesis*; ACS Symposium Series 641, Abdel-Magid, A. F., Ed., ACS: Washington DC, **1996**; (d) Ager, D. J.; East, M. B. *Asymmetric Synthetic Methodology*; CRC Press: Boca Raton, FL, **1995**; (e) Atkinson, R. S. *Stereoselective Synthesis*; Wiley: New York, **1995**; (f) Seyden-Penne, J. *Chiral Auxiliaries and Ligands in Asymmetric Synthesis*; Wiley: New York, **1995**; (g) Lo, T.- L. *Symmetry: A Basis for Synthetic Design*; Wiley: New York, **1995**; (h) Noyori, R. *Asymmetric Catalysis in Organic Synthesis*; Wiley: New York, **1994**; (i) *Stereocontrolled Organic Synthesis*; Trost, B. M., Ed., Blackwell: Oxford, **1994**; (j) Nogradi, M. *Stereoselective Synthesis: A Practical Approach*; VCH Publishers: New York, **1994**; (k) *Stereoselective Synthesis*; Ottow, E.; Schoellkopf, K.; Schultz, B. G., Ed., Springer-Verlag: New York, **1993**; (l) Rahman, A.; Shah, Z. *Stereoselective Synthesis in Organic Chemistry*; Springer-Verlag: New York, **1993**; (m) *Catalytic Asymmetric Synthesis*; Ojima, I., Ed., VCH Publishers: New York, **1993**; (n) Koskinen, A. M. P. *Asymmetric Synthesis of Natural Products*; Wiley: New York, **1993**; (o) *Asymmetric Synthesis*; Aitken, R. A.; Kilenyi, S. N., Ed., Blackie Academic & Professional: Glasgow, UK, **1992**.

12  (a) *Chirality in Industry II*; Collins, A. N.; Sheldrake, G. N.; Crosby, J., Ed., Wiley: Chichester, UK, **1997**; (b) Sheldon, R. A. *Chirotechnology, Industrial Synthesis of Optically Active Compounds*; Marcel Dekker: New York, **1993**; (c) *Chirality in Industry*; Collins, A. N.; Sheldrake, G. N.; Crosby, J., Ed., Wiley: Chichester, UK, **1992**; (d) *Specialty Chemicals: Innovations in Industrial Synthesis and Applications*; Pearson, B., Ed., Elsevier: London, **1991**; (e) Davies, H. G.; Green, R. H.; Kelly, D. R.; Roberts, S. M. *Biotransformations in Preparative Organic Chemistry: The Use of Isolated Enzymes and Whole Cell Systems in Synthesis*; Academic Press: London, **1989**.

13  (a) Brown, H. C.; Zweifel, G. *J. Amer. Chem. Soc.*, **1961**, *83*, 486; (b) Brown, H. C. *US Patent 3,254,129* (May 31, 1966), *Chem. Abst.*, **1966**, *65*, 7216b.

14  Partridge, J. J.; Chadha, N. K.; Uskokovic, M. R. *Org. Synth. Coll. Vol VII*, **1990**, 339.

15  (a) Chadha, N. K.; Partridge, J. J.; Uskokovic, M. R., (Hoffmann-LaRoche) ) *US Patent 4,140,708* (February 20, 1979), *Chem. Abst.*, **1975**, *82*, 97752a; (b) *US Patent 3,933,892* (January 20, 1976), *Chem. Abst.* **1976**, *85*, 62725d.

16  (a) Jones, M. *Chem. Brit.*, **1988**, 1122; (b) Biggadike, K.; Borthwick, A. D.; Evans, D.; Exall, A. M.; Kirk, B. E.; Roberts, S. M.; Stephenson, L.; Youds, P. *J. Chem. Soc. Perkin Trans. I*, **1988**, 549; (c) Biggadike, K.; Borthwick, A. D.; Exall, A. M.; Kirk, B. E.; Roberts, S. M.; Youds, P.; Slawin, A. M. Z.; Williams, D. J. *J. C. S. Chem. Commun.*, **1987**, 255.

17  (a) Shinkai, I. *Pure Appl. Chem.*, **1997**, *69*, 453;  (b) King, A. O.; Mathre, D. J.; Tschaen, D. M.; Shinkai, I. In *Reductions in Organic Synthesis*; ACS Symposium Series 641, Abdel-Magid, A. F., Ed., ACS: Washington DC, **1996**, pp. 98-111; (c) Shinkai, I., King, A. O.; Larsen, R. D. *Pure Appl. Chem.*, **1994**, *66*, 1551.

18  (a) Brown, H. C. (Aldrich - Boranes, Inc.) *US Patent 4,772,752* (September 20, 1988), *Chem. Abst.*, **1989**, *111*, 39631s; (b) Brown, H. C. (Aldrich - Boranes, Inc.) *US Patent 4,713,380* (December 12, 1987), *Chem. Abst.*, **1987**, *106*, 214124z.

19  (a) Midland, M. M.; Graham, R. S. *Org. Synth. Coll. Vol. VII*, **1990**, 402; (b) Midland, M. M. In *Asymmetric Synthesis*, Vol. 2, Morrison, J. D., Ed., Academic Press: New York, **1983**, pp 45-69.

20  (a) Brown, H. C.; Kamachandran, P. V. In *Reductions in Organic Synthesis*; ACS Symposium Series 641, Abdel-Magid, A. F., Ed., ACS: Washington DC, **1996**, pp. 1-30, 84-97; (b) *J. Organometal. Chem.* **1995**, *500*, 1; (c) *Pure Appl. Chem.*, **1994**, *66*, 201.

21  "Darvon alcohol" and its enantiomer are sold as Chirald® and ent-Chirald® by Aldrich Chemical Co., Milwaukee, WI.

22  (a) Yamaguchi S.; Mosher, H. S. *J. Org. Chem.*, **1973**, *38*, 1870; (b) Cohen, N.; Lopresti, R. J.; Neukom, C.; Saucy, G. *J. Org. Chem.*, **1980**, *45*, 582.

23  Beck, A. K.; Dahinden, R.; Kuhnle, F. N. M. In *Reductions in Organic Synthesis*; ACS Symposium Series 641, Abdel-Magid, A. F., Ed., ACS: Washington DC, **1996**, pp 52-69.

24  Seebach, D.; Dahinden, R.; Marti, R. E.; Beck, A. K.; Plattner, D. A.; Kuhnle, F. N. M. *J. Org. Chem.*, **1995**, *60*, 1788.

25  (a) Noyori, R.; Tomino, I.; Tanimoto, Y.; Nishizawa, M. *J. Am. Chem. Soc.*, **1984**, *106*, 6709; (b) Noyori, R.; Tomino, I.; Yamada, M.; Nishizawa, M. *ibid*, **1984**, *106*, 6717; (c) Noyori, R., (Ono Pharmaceutical Co.) *German Offen. DE 2,940,336* (April 24, 1980), *Chem. Abst.*, **1981**, *94*, 30250q.

26  For a review of chiral stannane reagents see: Marshall, J. A. *Chem. Rev.*, **1996**, *96*, 31.

27  McKenzie, A. *J. Chem. Soc.*, **1904**, *85*, 1249.

28  (a) Prelog, V. *Helv. Chem. Acta*, **1953**, *36*, 308; (b) Prelog, V.; Meier, H. L. *ibid*, **1953**, *36*, 320.

29  (a) Gage, J. R.; Evans, D. A. *Org. Synth.*, **1989**, *68*, 77; (b) *ibid*, **1989**, *68*, 83.

30  (a) Ager, D. J.; Prakash, I.; Schaad, D. R. *Aldrichimica Acta*, **1997**, *30*, 3; (b) *Chem. Rev.*, **1996**, *96*, 835. For a related review of the tin(II) triflate catalyzed aldol reactions see: Mukiayama, T. *Aldrichimica Acta*, **1996**, *29*, 59.

31  Hoekstra, M. S.; Sobieray, D. M.; Schwindt, M. A.; Mulhern, T. A.; Grote, T. M.; Huckabee, B. K.; Hendrickson, V. S.; Franklin, L. C.; Granger, E. J.; Karrick, G. L. *Org. Process R&D*, **1997**, *1*, 26.

32  Braun, M.; Graef, S. *Org. Synth.*, **1993**, *72*, 38.

33  Lynch, J. E.; Shinkai, I.; Volante, R. P. (Merck and Co.) *U.S. Patent 4,611,081* (September 9, 1986), *Chem. Abst.*, **1987**, *106*, 18119n.

34  Barbier, P.; Schneider, F.; Widmer, U. *Helv. Chim. Acta*, **1987**, *70*, 1412.

35  Harrington, P. J.; Lodewijk, E. *Org. Process R&D*, **1997**, *1*, 72.

36  Giordano, C.; Villa, M.; Ponassian, S. In *Chirality in Industry*; Collins, A. N.; Sheldrake, G. N.; Crosby, J., Ed., Wiley: Chichester, UK, **1992**, pp. 303-312.

37  Larsen, R. D.; Corley, E. G.; Davis, P.; Reider, P. J.; Grabowski, E. J. J. *J. Am. Chem. Soc.*, **1989**, *111*, 7650.

38  (a) Noyori, R. *Asymmetric Catalysis in Organic Chemistry*; Wiley: New York, **1994**, pp. 16-94; (b) Takaya, H.; Okta, T.; Noyori, R. In *Catalytic Asymmetric Synthesis*; Ojima, I., Ed., VCH Publishers: New York, **1993**, pp. 1-39.

39  (a) Knowles, W. S. *Acc. Chem. Res.*, **1983**, *16*, 106; (b) Knowles, W. S.; Sabacky, M. J.; Vineyard, B. D. (Monsanto Co.) *US Patent 4,005,125* (January 25, 1977), *Chem. Abst.*, **1977**, *86*, 190463z.

40  Schmid, R.; Broger, E. A.; Cereghetti, M.; Crameri, Y.; Foricher, J.; Lalonde, M.; Mueller, R. K.; Scalone, M.; Schoettel, G.; Zutter, U. *Pure & Appl. Chem.*, **1996**, *68*, 131.

41  Burk, M. J.; Gross, M. F.; Harper, T. G. P.; Kalberg, C. S.; Lee, J. R.; Martinez, J. P. *Pure Appl. Chem.*, **1996**, *68*, 37.

42  (a) Kumobayashi, H. *Recl. Trav. Chim.* Pays-Bas, **1996**, *115*, 201; (b) Tani, K.; Yamagata, T.; Akutakawa, S.; Kumobayashi, H.; Takehomi, T.; Takaya, H.; Miyashita, A.; Noyori, R.; Otsuka, S. *J. Amer. Chem. Soc.*, **1984**, *106*, 5208.

43  For "bid" and "asked" weekly prices on (-)-l-menthol and a wide variety of commercial chemicals see *Chemical Market Reporter*, Schnell Publishing Co., New York.

44  (a) Johnson, R. A.; Sharpless, K. B. In *Catalytic Asymmetric Synthesis*; Ojima, I., Ed., VCH Publishers: New York, **1993**, pp. 103-158; (b) Hanson, R. M.; Koo, S. Y.; Sharpless,

K. B. (Massachusetts Institute of Technology) *US Patent 4,900,847* (February, 13, 1990), *Chem. Abst.*, **1987**, *106*, 66496e and patents cited therein; (c) Rossiter, B. E. In *Asymmetric Synthesis*, Vol. 5, Morrison, J. D., Ed., Academic Press: New York, **1985**, pp. 193-246; (d) Finn, M. G.; Sharpless, K. B. *ibid*, Vol. 5, **1985,** pp. 247-308.

45  *The Merck Index*; 12th Ed., Merck & Co.: Whitehouse Station, NJ, **1996**, Organic Name Reactions - 347: The Sharpless Epoxidation.

46  (a) Wu, G.; Schumacher, D. P.; Tormos, W.; Clark, J. E.; Murphy, B. L. *J. Org. Chem.*, **1997**, *62*, 2996; (b) Nagabhushan, T. L. (Schering Corp.) *US Patent 4,235,892* (November 25, 1980), *Chem. Abst.* **1980**, *94*, 139433c.

47  (a) Johnson, R. A.; Sharpless, K. B. In *Catalytic Asymmetric Synthesis*; Ojima, I., Ed., VCH Publishers: New York, **1993**, pp. 227-272; (b) McKee, B. H.; Gilheany, D. G.; Sharpless, K. B. *Org. Synth.*, **1991**, *70*, 47.

48  Sharpless, K. B.; Beller, M.; Blackburn, B.; Kawanami, Y.; Kwong, H.- L.; Ogino, Y.; Shibata, T.; Ukita, T.; Wang, L. (Massachusetts Institute of Technology) *US Patent 5,516,929* (May 14, 1996), *Chem. Abst.*, **1996**, *112*, 75971g and patents cited therein.

49  Jacobsen, E. N. In *Catalytic Asymmetric Synthesis*; Ojima, I., Ed., VCH Publishers: New York, **1993**, pp. 159-202.

50  Senanayake, C. H.; Smith, G. B.; Ryan, K. M.; Fredenburgh, L. E.; Liu, J.; Roberts, F. E.; Hughs, D. L.; Larsen, R. D.; Verhoeven, T. R.; Reider, P. J. *Tetrahedron Lett.*, **1996**, *37*, 3271.

51  (a) Trost, B.M.; Van Vranken, D. L. *Chem. Rev.*, **1996**, *96,* 395; (b) Trost, B. M. *Acc. Chem. Res.*, **1996**, *29*, 355.

52  Trost, B. M.; Organ, M. G. *J. Am. Chem. Soc.,* **1994**, *116*, 10320.

53  Hajos, Z. G.; Parrish, D. R. *Org. Synth. Coll. Vol. VII*, **1990**, 363.  See also: Buchschacher, P.; Fuerst, A.; Gutzwiller, J. *ibid.*, **1990**, 368.

54  For reviews on the enantioselective formation of C-C bonds and quaternary carbon centers see: (a) Meyers, A. I. In *Stereocontrolled Organic Synthesis*; Trost, B. M., Ed., Blackwell: Oxford, **1994**, pp. 145-175; (b) Fuji, K. *Chem.Rev.*, **1993**, *93*, 2037; (c) Sawamura, M.; Ito, Y. In *Catalytic Asymmetric Synthesis*; Ojima, I., Ed., VCH Publishers: New York, **1993**, pp. 367-388; (d) Revial, G.; Pfau, M., *Org. Synth.*, **1991**, *70*, 35.

55  Inoue, Y. *Chem. Rev.*, **1992**, *92*, 741.

56  (a) Soai, K.; Niwa, S. *Chem. Rev.*, **1992**, *92*, 833; (b) Soai, K.; Ookawa, A.; Kaba, T.; Ogawa, K. *J. Am. Chem. Soc.,* **1987**, *109*, 7111.

57  Itsuno, S.; Sakurai, Y.; Ito, K.; Hirao, A.; Nakahama, S. *Bull. Chem. Soc. Jpn.*, **1987**, *60*, 395 and references therein.

58  (a) Corey, E. J.; Bakshi, R. K.; Shibata, S. *J. Am. Chem. Soc.*, **1987**, *109*, 5551 and following papers; (b) Corey, E. J. (Harvard College) *US Patent 4,943,635* (July 24, 1990), *Chem. Abst.*, **1989**, *111*, 153614p.

59  Xavier, L. C.; Mohan, J. J.; Mathre, D. J.; Thompson, A. S.; Carroll, J. D.; Corley, E. G.; Desmond, R. *Org. Synth.*, **1996**, *74*, 50.

60  Quallich, G. J.; Blake, J. F.; Woodall, T. M. In *Reductions in Organic Synthesis*; ACS Symposium Series 641, Abdel-Magid, A. F., Ed., ACS: Washington DC, **1996**, pp 112-126.

61  O'Donnell, M. J. In *Catalytic Asymmetric Synthesis;* Ojima, I., Ed., VCH Publishers: New York, **1993**, pp. 389-411.

62  Dolling, U.- H.; Hughes, D. L.; Bhattacharya, A.; Ryan, K. M.; Karady, S.; Weinstock, L. M.; Grabowski, E. J. J. In *Phase Transfer Catalysis*; ACS Symposium Series 326, Starks, C. M., Ed., ACS: Washington DC, **1987**, pp. 67-81.

63  Hughes, D. L.; Dolling, U.- H.; Ryan, K. M.; Schoenewaldt, E. F.; Grabowski, E. J. J. *J. Org. Chem.*, **1987**, *52*, 4745.

64  (a) Simpkins, N. S. *Pure Appl. Chem.*, **1996**, *68*, 691; (b) Cox, P. J.; Simpkins, N. S. *Tetrahedron: Asymmetry*, **1991**, 2, 1.

65  (a) Koga, K. *Pure Appl. Chem.,* **1994**, *66*, 1487; (b) Yasukata, T.; Koga, K. *Tetrahedron Asymmetry*, **1993**, *4*, 35; (c) Asami, M. *Bull. Chem. Soc. Jpn.*, **1990**, *63*, 721.

66  Nikolic, N. A.; Beak, P. *Org. Synth.*, **1996**, *74*, 23.

67  Park, Y. S.; Beak, P. *J. Org. Chem.*, **1997**, *62*, 1574 and references therein.

68  Denmark, S. E.; Wong, K.- T.; Stavenger, R. A. *J. Amer. Chem. Soc.*, **1997**, *119*, 2333.

69  Yamada, S.; Mori, Y.; Morimatsu, K.; Ishizu, Y.; Ozaki, Y.; Yoshioka, R.; Nakatani, T.; Seko, H. *J. Org. Chem.*, **1996**, *61*, 8586.

70  Reider, P. J.; Davis, P.; Hughes, D. L.; Grabowski, E. J. J. *J. Org. Chem.*, **1987**, *52*, 955.

71  (a) Azerad, R. *Bull. Soc. Chim. Fr.*, **1995**, *132*, 17; (b) Theil, F. *Chem. Rev.,* **1995**, *95*, 2203; (c) Wong, C.- H.; Whitesides, G. M. *Enzymes in Synthetic Organic Chemistry*, Pergamon: Oxford, UK, **1994;** (d) West, S. In *Specialty Chemicals: Innovations in Industrial Synthesis and Applications*; Pearson, B., Ed., Elsevier: London, **1991**, pp. 499-507; (e) Boaz, N.W. and Laumen, K., *ibid*, **1991**, pp. 509-518.

72  Jones, J. B.; Jakovac, I. J. *Org. Synth. Coll. Vol. VII*, **1990**, 406. For alternative cofactor regeneration see: Matos, J. R.; Wong, C.- H. *J. Org. Chem.*, **1986**, *51*, 2388.

73  (a) Crocq, V.; Masson, C.; Winter, J.; Richard, C.; Lemaitre, G.; Lenay, J.; Vivat, M.; Buendia, J.; Prat, D. *Org. Process R&D*, **1997**, *1*, 2; (b) Seebach, D.; Sutter, M. A.; Weber, R. H.; Zueger, M. F. *Org. Synth. Coll. Vol. VII*, **1990**, 215; (c) Mori, K.; Mori, H. *Org. Synth.*, **1989**, *68*, 56.

74  *Lipases for Resolution and Asymmetric Synthesis*; Amano Enzyme Co., Lombard, IL, **1996**.

75  Deardorff, D. R.; Windham, C. Q.; Craney, C. L. *Org. Synth.*, **1995**, *73*, 25.

76  (a) Bray, B. L.; Goodyear; M. D.; Partridge, J. J.; Tapolczay, D. J. In *Chirality in Industry II*; Collins, A. N.; Sheldrake, G. N.; Crosby, J., Ed., Wiley: Chichester, UK, **1997**, pp. 41-78; (b) Lovelace, T. C.; Bray, B. L.; Mook, R. A.; Partridge, J. J. *211th ACS National Meeting*; New Orleans, LA, **1996**, ORGN 012.

77  (a) Lavey, B. J.; Janda, K. D. In *Antibody Expression and Engineering*; ACS Symposium Series 604, Wang, H. Y.; Imanaka, T., Ed., ACS: Washington DC, **1995**, pp. 123-137; (b) Scanlan, T. S. In *Annual Reports of Medicinal Chemistry*, Vol. 30, Bristol, J. A., Ed., Academic Press: San Diego, CA, **1995**, pp. 255-264.

78  Jacobsen, J. R.; Schultz, P. G. *Proc. Natl. Acad. Sci. USA*, **1994**, *91*, 5888.

79  (a) Keinan, E.; Sinha, S. C.; Shabat, D.; Itzhaky, H.; Reymond, J.- L. *Acta. Chem. Scand.*, **1996**, *50*, 679; (b) Lerner, R. A.; Barbos, C. F. *ibid*, **1996**, *50*, 672; (c) Kirby, A. J. *ibid*, **1996**, *50*, 203.

80  (a) Janda, K. D.; Benkovic, S. J.; Lerner, R. A. *Science*, **1989**, *244*, 437; (b) Jacobsen, J. R.; Prudent, J.R.; Kochersperger, L.; Yonkovich, S.; Schultz, P.G. *ibid*, **1992**, *256*, 365.

81  Jacobsen, E. N. and Finney, N. S. *Chem. & Biol.*, **1994**, *1*, 85.

82  Corey, E. J.; Cheng, X.- M. *The Logic of Chemical Synthesis*; Wiley: New York, **1989**.

83  Comins, D. L.; Baevsky, M. F.; Hong H. *J. Am. Chem. Soc.*, **1992**, *114*, 10971; Comins, D. L.; Baevsky, M. F. (North Carolina State University) *US Patent 5,475,108* (December 12, 1995), *Chem. Abst.*, **1996**, *124*, 146559q.

84  (a) Fang, F. G.; Lowery, M. W.; Xie, S. (Glaxo Wellcome) *US Patent 5,491,237* (February 13, 1996), *Chem. Abst.*, **1996**, *124*, 146558p; (b) Fang, F. G.; Bankston, D. D.; Huie, E. M.; Johnson, M. R.; Kang, M.- C.; LeHoullier, C. S.; Lewis, G. S.; Lovelace, T. C.; Lowery, M. W.; McDougald, D. L.; Meerholz, C. A.; Partridge, J. J.; Sharp, M. J.; Xie, S. *Tetrahedron*, accepted for publication, **1997**.

85  (a) Borman, S. *Chem. & Eng. News*, **1997**, *75* (8), 43; (b) Baum, R. *ibid*, **1996**, *74* (7), 28; (c) Thompson, L. A.; Ellman, J. A. *Chem. Rev.*, **1996**, *96*, 555; (d) Armstrong, R. W.; Combs, A. P.; Tempest, P. A.; Brown, S. D.; Keating, T. A. *Acc. Chem. Res.*, **1996**, *29*, 123.

86  Burgess, K.; Lim, H.- J.; Porte, A. M.; Sulikowski, G. A. *Angew. Chem. Int. Ed. Engl.*, **1996**, *35*, 220.

87  Burger, M. T.; Still, W. C. *J. Org. Chem.*, **1995**, *60*, 7382.

88  Francis, M. B.; Finney, N. S.; Jacobsen, E. N. *J. Amer. Chem. Soc.*, **1996**, *118*, 8983.

89  For concepts of using the World-Wide Web as a chemical information tool see: Murray-Rust, P.; Rzepa, H. S.; Whitaker, B. J. *Chem. Soc. Rev.*, **1997**, *26*, 1.

90  Krieger, J. H. *Chem. & Eng. News*, **1997**, *75* (19), 30.

# Practical Catalysts for Asymmetric Synthesis: Ligands from Natural Sugars for Rh-Catalyzed Asymmetric Synthesis of *D*- and *L*- Amino Acids

**Timothy A. Ayers**
*Hoechst Marion Roussel, Inc., 2110 E. Galbraith Road, Cincinnati, OH, 45215 USA*

**T. V. RajanBabu**
*Department of Chemistry, The Ohio State University, Columbus, OH, 43210 USA*

Key Words: asymmetric hydrogenation, amino acids, diphosphinites, Rh-catalysts

## Introduction

The Rh-catalyzed asymmetric hydrogenation of acetamidoacrylates **1** has become an important process for the preparation of natural and unnatural amino acid derivatives **2** (eq. 1). A plethora of useful chiral ligands including DIOP, DIPAMP, Chiraphos, BINAP, DuPHOS and others has been developed over the past quarter century which provide good to excellent enantioselectivities.[1,2] The preparation of unnatural amino acids with high enantioselectivity continues to be important to the pharmaceutical industry as these species can serve as intermediates in the production of bulk drug substances. In addition, these unnatural amino acids are especially useful in combinatorial chemistry as more diversity is achievable and compounds which may provide better pharmacological stability can be prepared. Thus, the development of new and better methods for the preparation of unnatural amino acids will continue to be of importance from both an academic and industrial perspective.

*This chapter and the work described herein is dedicated to Professor Rudiger Selke, who pioneered the use of carbohydrate-based ligands in asymmetric catalysis.*

Scheme 1

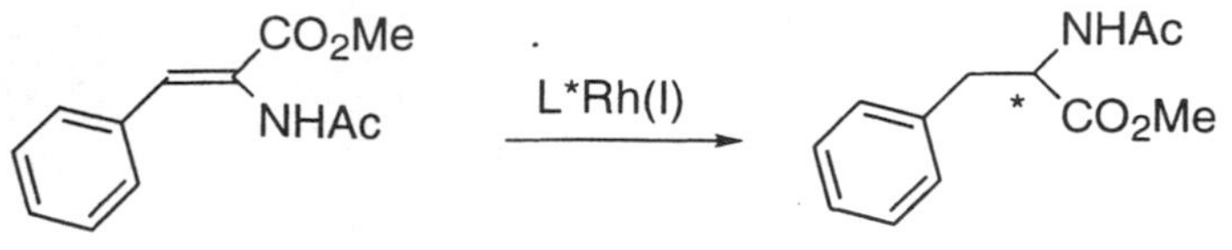

(eq. 1)

Chiral Ligands (L*)

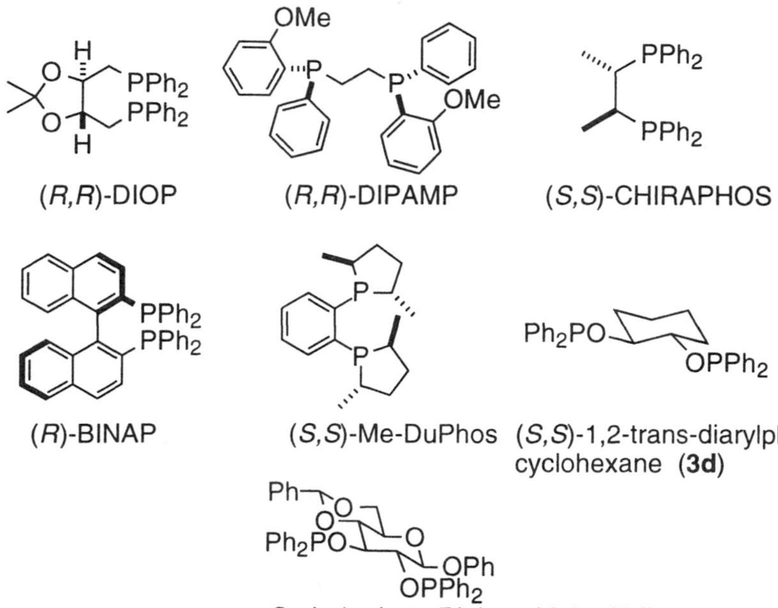

(*R,R*)-DIOP            (*R,R*)-DIPAMP          (*S,S*)-CHIRAPHOS

(*R*)-BINAP         (*S,S*)-Me-DuPhos   (*S,S*)-1,2-trans-diarylphosphinoxy-
                                       cyclohexane (**3d**)

Carbohydrate Diphosphinite (**4d**)

The first report of a diphosphinite in the Rh-catalyzed asymmetric hydrogenation reaction which appeared in 1975 involved the use of a simple cyclohexyl ligand **3d**.[3]  In this case, a moderate ee of 69% for the hydrogenation of α–acetamidocinnamic acid was obtained.  In 1978, several groups began to explore the use of sugar-derived diphosphinites, ligands with a higher degree of conformational rigidity than the simple system **3d**.[4]  These studies showed that good to excellent ee's could be achieved using ligands **4d** for the preparation of simple *L*-phenylalanine derivatives.[4,5]  The major conclusion that can be drawn from these early studies is that ligand **4d**, a glucose derived 2,3-diphosphinite provides the highest ee's for the simple *L*-phenylalanine synthesis; we (vide infra) and others[6] have found severe limitations to the application of this ligand to the synthesis of substituted phenylalanine and heteroarylalanine derivatives as well as phenyl alanine derivatives with nitrogen substituted with groups other than acyl.  This chapter will describe how an understanding of electronic effects and the relative juxtoposition of vicinal diphosphinites on a carbohydrate scaffolding led to the development of an efficient process for the synthesis of both *L*- and *D*- amino acids using one of the least expensive ligand systems, namely those derived from *D*-glucose.[7]

## Asymmetric Catalysis and Carbohydrate Ligands

In the early 1990's, recognizing the importance chirality plays in pharmaceutical and agricultural products, a group was established at DuPont's Central Research and Development to develop a program in asymmetric catalysis. As members of this group, Dr. T. V. RajanBabu and Dr. A. L. Casalnuovo decided to explore the asymmetric hydrocyanation of vinylarenes[8] for which little precedent existed in the literature. With the reported low ee's to date,[9] and limited mechanistic information to rely on, a highly flexible ligand system which was readily available (chiral pool) and readily modifiable (ease of synthesis for changes in the gross structure as well as changes in the ligating element) was desired. This empirical approach was chosen, because the ability to predict *a priori* the steric and electronic influences of a particular chiral ligand on the enantioselectivity is often difficult even in reactions where considerable mechanistic details are available. Typically $c_2$-symmetric ligands are utilized and the enantioselectivity results from predominant product formation through one of the diastereomeric intermediates. In these cases, where the structure of this diastereomeric intermediate can be related to the absolute configuration of the product, models of the spatial requirements of the ligand are extremely useful in building better catalysts.[10-16] However, the catalytic cycle may involve many, often reversible diastereoselective transformations at the metal center (i.e. coordination, oxidative addition, insertion and reduction elemination). An empirical method of ligand design could quickly determine some of the critical features of the ligand and, this in turn could lead to the development of better catalysts. To this end, diphosphinites derived from carbohydrates were chosen as ligands for the Ni-catalyzed asymmetric hydrocyanation of vinylarenes. The chemistry of carbohydrates is well established and allows for ease in manipulation of the gross steric environment at the metal center by using different sugars (e.g. glucose, fructose, mannose), changing protecting groups, changing aglycone, and so forth. In addition, a wide variety of diarylchlorophosphines is available by treatment of the appropriate Grignard reagent (ArMgBr) with $Et_2NPCl_2$ to provide $Ar_2PNEt_2$. Subsequent reaction with HCl converts the aminophosphine to the desired $Ar_2PCl$. Incorporation of different aryl groups at the phosphorus substituent provides a synthesis of sterically similar ligands which are electronically differentiated. Now the fundamental processes that occur at the metal center (coordination, oxidative addition, insertion and reductive elimination) which are known to be influenced by the electronics at the metal can quickly be explored to assess this importance of electronic effects on the enantioselectivity.[17-23]

The ease of the diphosphinite synthesis from readily available sugar derivatives is shown in Scheme 2. The 2,3-*D*-glucose diphosphinites **4** are synthesized from the readily avaiable β–phenyl glucopyranoside by protection of the 4- and 6-hydroxyl groups as a benzylidene acetal followed by reaction with an appropriate diarylchlorophosphine to provide the desired diphosphinite. The 3,4-*D*-fructose diphosphinites **5** are prepared by protection of the 1,6-hydroxyl groups as the trityl ethers, separation of the anomers and subsequent reaction with a diarylchlorophosphine. The 3,4-*D*-glucose diphosphinites are synthesized from α–methyl glucopyranoside by selective protection of the 2- and 6-hydroxyls using Ogawa's conditions followed by reaction with a diarylchlorophosphine.[24,25] The ease of these syntheses is readily apparent. Moreover, from an economical perspective the cost of these ligands derived from D-glucose (<$1.00 per kg) makes these ligands very suitable for large-scale production.

Scheme 2

Synthesis of 2,3-*D*-Glucose Diphosphintes

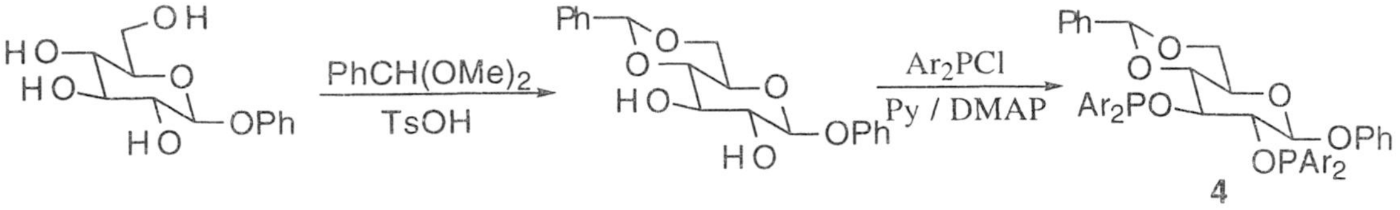

Synthesis of 3,4-*D*-Fructose Diphosphintes

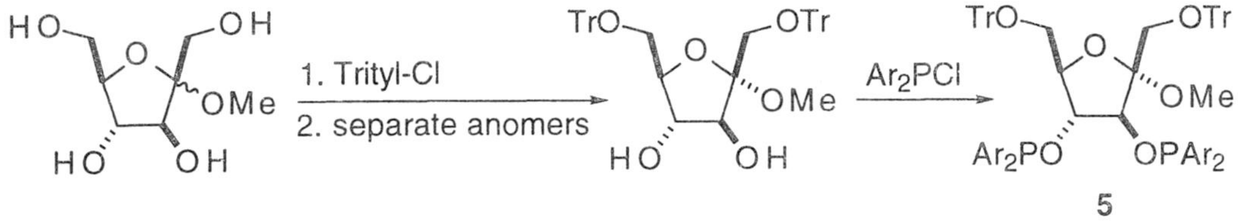

Synthesis of 3,4-*D*-Glucose Diphosphinites

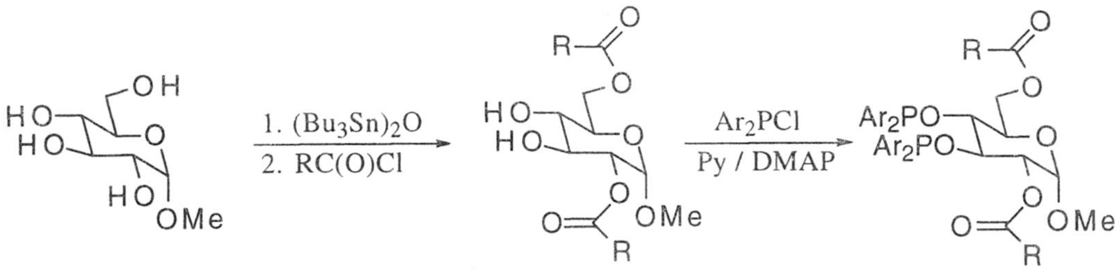

**6.** R = *t*-Bu
**7.** R = Ph

Ar groups

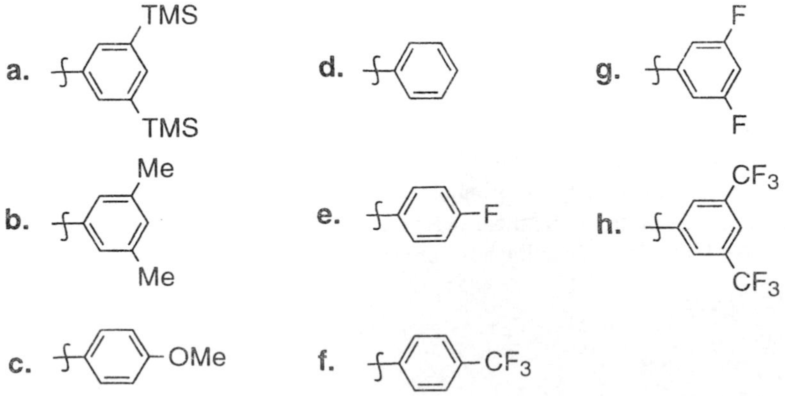

**a.** [ring with TMS, TMS]

**b.** [ring with Me, Me]

**c.** [ring with OMe]

**d.** [phenyl]

**e.** [ring with F]

**f.** [ring with CF₃]

**g.** [ring with F, F]

**h.** [ring with CF₃, CF₃]

## Electronic Effects and Enantioselectivity in the Hydrocyanation Reaction

This section will describe some of our results and conclusions from the asymmetric hydrocyanation reaction which provided the foundation of our understanding of electronic effects in asymmetric catalysis[7,17-20] and later prompted us to explore this influence in the asymmetric hydrogenation reaction. Thus, the objective of our original study was to use the diphosphinite ligands for a highly enantioselective Ni-catalyzed hydrocyanation of olefins such as **8** (eq. 2) in order to develop a process for the preparation of 2-arylpropionitriles like **9**, a precursor to Naproxen. During the course of this study, several important findings were made (Scheme 3).[7,17-20] First, the use of *D*-glucose derived *2,3-diphosphinites* and *phosphites* provided good (up to 60% ee) enantioselectivities for the S-isomer in the Ni-catalyzed hydrocyanation of various vinylarenes, whereas *phosphine* ligands such as BINAP, DIOP, CHIRAPHOS, and DuPHOS gave low enantioselectivity as well as poor turnover rates. Steric tuning (changing of protecting groups and aglycone linkage) of the 2,3-glucose derived ligand showed little effect on the enantioselectivity of the reaction. However, when diphosphinites **4g** and **4h** with highly electron withdrawing aryl groups [Ar = 3,5-$F_2C_6H_3$, 3,5-$(CF_3)_2C_6H_3$] at phosphorus were used, ee's up to 91% were obtained in the hydrocyanation of vinylarenes. Alternatively, when relatively electron-rich phosphinites such as **4b** (Ar = 3,5-$(CH_3)_2C_6H_3$) were used, lower ee's were obtained.

During the course of the screening of carbohydrate-derived diphosphinites in the hydrocyanation reaction, fructose-derived diphosphinites **5** were found to provide the *R*-isomer, albeit with modest ee. In additon, use of the best aryl group, 3,5-$(CF_3)_2C_6H_3$, provided only a marginal increase in the selectivity compared with the Ar = Ph analog (58% vs 43% ee). Not willing to abandon this class of ligands, and recognizing the different reactivitites of the 3- and 4-hydroxyls in the fructose system, we decided to incorporate different phosphorus substituents at these positions. Astonishingly, incorporation of highly electron-withdrawing groups at the 4-position with comparatively electron-rich groups at the 3-position provided much higher enantioselectivities (i.e. **7D**), whereas the reverse substitution (i.e. **7C**) provides little improvement. Thus, a remarkable 95% ee was obtained now for the *R*-isomer using the fructose diphosphinite with the 3-position having Ar = Ph and the 4-position having Ar = 3,5-$(CF_3)_2C_6H_3$.

A detailed study of the mechanism, though not essential in this approach, was undertaken and revealed several fascinating aspects of this reaction which have been published elsewhere.[7] However, the key feature of the mechanism (Scheme 4) of this reaction indicated that the role of the electron withdrawing groups is to increase the rate of reductive elimination ($k_2 >> k_1$) from one of the penultimate intermediates providing an enhancement of enantioselectivity. The rapid development of processes to prepare both the *R*- and *S*-isomers of the Naproxen nitrile with excellent enantioselectivities using these highly tunable carbohydrate-derived diphosphinites demonstrated the power of the empirical approach with the carbohydrate-derived diphosphinites. An unexpected outcome was an understanding of the little-exploited electronic effects as a control element in asymmetric catalysis.

Scheme 3. Ni-Catalyzed Hydrocyanation of **8** using Ligands **4** and **5**[a]

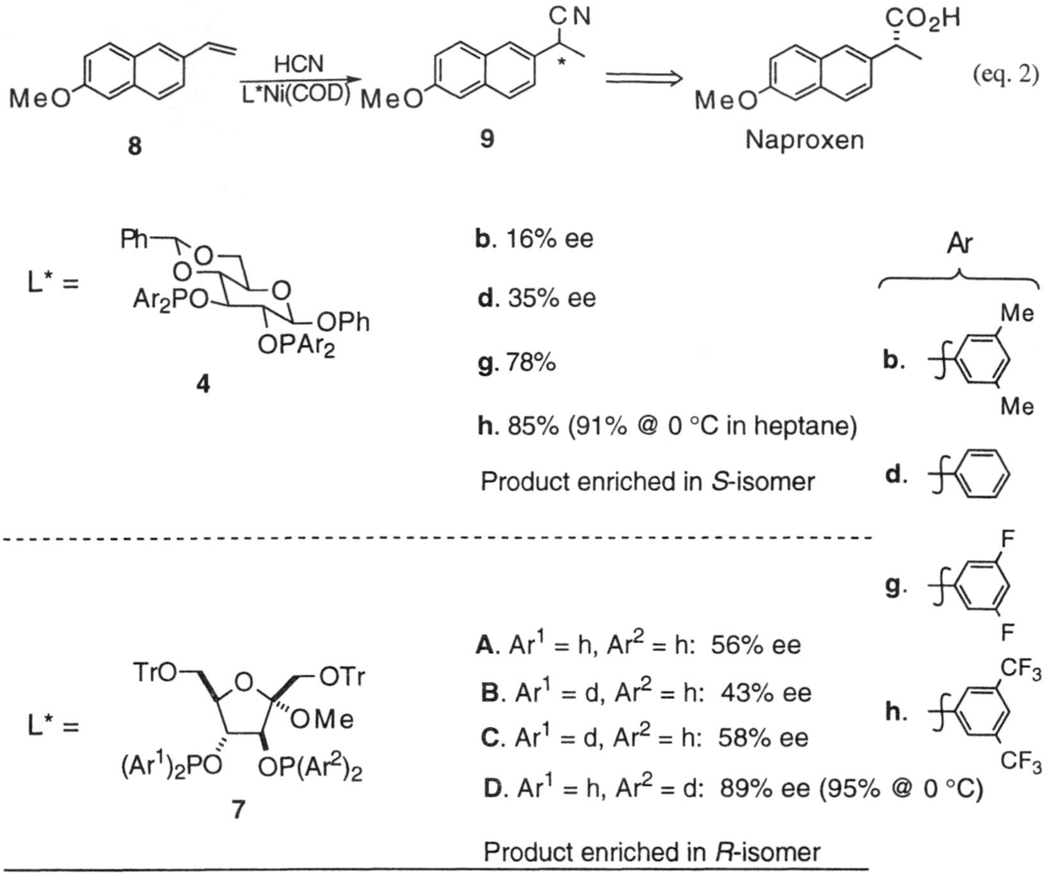
MeO
HCN
L*Ni(COD)
MeO
CN
MeO
$CO_2H$
(eq. 2)
8
9
Naproxen
L* =
Ph
Ar2PO
OPh
OPAr2
4
b. 16% ee
d. 35% ee
g. 78%
h. 85% (91% @ 0 °C in heptane)
Product enriched in S-isomer
Ar
b.
Me
Me
d.
g.
F
F
h.
CF3
CF3
L* =
TrO
OTr
OMe
(Ar1)2PO
OP(Ar2)2
7
A. Ar1 = h, Ar2 = h: 56% ee
B. Ar1 = d, Ar2 = h: 43% ee
C. Ar1 = d, Ar2 = h: 58% ee
D. Ar1 = h, Ar2 = d: 89% ee (95% @ 0 °C)
Product enriched in R-isomer

[a] Reactions run in benzene or toluene using 1 mmol of **8** and 1-5 mol% L / $Ni(COD)_2$.

Scheme 4

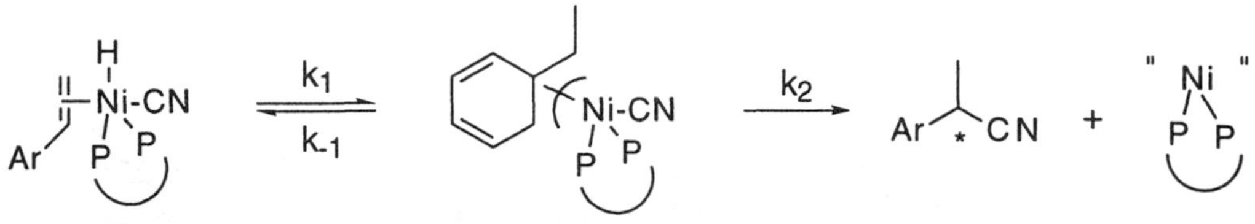
H
Ni-CN
Ar
P
P
k1
k-1
Ni-CN
P
P
k2
Ar
CN
+
Ni
P
P

$k_2 >> k_{-1}$ when diphosphinite has electron-deficient aryl groups

$k_2 \approx k_{-1}$ when diphosphinite has electron-rich aryl groups

### Diphosphinites in Rh-Catalyzed Hydrogenations: Synthesis of *L*-Amino Acids

Following the remarkable electronic effects observed upon using sugar-derived diphosphinites in the Ni-catalyzed hydrocyanation of vinylarenes, other asymmetric reactions using these ligand systems were explored. We were especially interested in the Rh-catalyzed asymmetric hydrogenation,[26] because much mechanistic work has shown that oxidative addition of hydrogen to the minor Rh-acrylate complex is the key step of the process.[1,2] Thus, we hoped to gain further insight on the influence of ligand electronics on enantioselectivity.[23] As stated in the introduction, these simple phenyl-substituted diphosphinite ligands were previously prepared for this purpose. In these studies many different sugars were utilized (e.g. *D*-glucose, *D*-galactose, *D*-mannose) and the diphosphinites were typically placed at the 2,3-position providing the amino acid as the *L*-isomer in the Rh-catalyzed hydrogenation of acetamidoacrylates; however, the 2,3-*D*-glucose derived diphosphinites with all equatorial substituents provided the highest enantioselectivities. Two severe limitations have been found with the simple diphenyl phosphinites (i.e. **4d**, see Table 1) in that substituted aromatic and heteroaromatic amino acids are prepared with lower selectivity and reduction of dehydroamino acids with nitrogen protection groups other than acyl, also provided low enantioselectivities (i.e. Table 1, entry 5.) [6] From a synthetic standpoint, the ability to incorporate easily removable protecting groups such as N-Cbz or N-Boc would be highly desirable. To this end we decided to examine the electronic effects in this reaction using our electronically tuned 2,3-diphosphinites to achieve higher enantioselectivites over a broader range of substrates.[23]

The scouting experiments were performed using 1.0 mmol of substrate[27] and 0.05 mmol of Rh$^+$-catalyst at 30-40 psi of hydrogen pressure in a Fischer-Porter tube. The Rh-catalysts were prepared by reaction of an appropriate Rh$^+$(cyclooctadiene)X$^-$ complex (X = SbF$_6$, BF$_4$, OTf) with 1.05 eq. of the diphosphinite ligand giving the the catalyst L*Rh$^+$(cyclooctadiene)X$^-$. The ee's of the products were determined on the methyl esters (acids were converted to the methyl ester by reaction with diazomethane) by GC using a chirasil S-Val column. The ee of the Cbz-derivatives was determined by HPLC (Chiracel OJ column) on the corresponding amino alcohol after reduction with LiBH$_4$ or by GC on the methyl ester of the acetamido derivative prepared by removal of cbz-group under normal hydrogenation conditions (H$_2$, Pd / C) in the presence of acetic anhydride. The results are shown in Table 1.

In these hydrogenation reactions where the enantioselective step is presumably an oxidative addition of H$_2$ to Rh(I), the use of the relatively electron-rich diphosphinites provide the highest ee's. For example, in the hydrogenation of methyl 2-N-acetylcinnamate using the simple diphenyldiphosphinite **4d**, a 90% ee is obtained. Remarkably, simply by changing the aryl group to 3,5-F$_2$C$_6$H$_3$ decreases the ee to 2%, whereas the comparatively more electron-rich 3,5-(TMS)$_2$C$_6$H$_3$ group increases the ee to 99%.[28] This dramatic effect of going from practically zero to quantitative asymmetric induction simply by changing the electronic nature of the ligand and the metal center is unprecedented and will be discussed later. Moreover, using the electronically-tuned ligand, excellent ee's (over 96%) are observed over a broad range of arylalanine derivatives, whereas the simple diphenyl phosphinites provides marginal to good ee's (85-94%).[29] In general, the use of ligand **4b** with Ar = 3,5-Me$_2$C$_6$H$_3$ is preferred as the synthesis of ligand **4a** is more difficult and provides only slight enhancements (1-2%) of selectivity.[26] Although these increases in selectivity appear small, from a synthetic standpoint, amino acid derivatives with >99% ee can be obtained after one recrystallization in excellent yield when one starts with material with >96% ee. For example, nearly pure samples of the N-acetyl derivatives of phenylalanine 3-thienylalanine, and 3,5-bis-(trifluoromethyl)phenylalanine were prepared in 99.7, 99.5, and 99.3% ee respectively after one recrystallization.

**Table 1.** Enantioselectivity in Hydrogenations Using Catalysts [**4a-h**]Rh⁺(COD)SbF₆⁻

| Entry | Substrate | Ee of Product [Enriched in ($S$)] | | | | | | | |
|---|---|---|---|---|---|---|---|---|---|
| | | 3,5-(TMS)₂ᵃ | 3,5-Me₂ᵃ | 4-(MeO)ᵃ | Hᵃ | 4-Fᵃ | 4-CF₃ᵃ | 3,5-F₂ᵃ | 3,5-(CF₃)₂ᵃ |
| 1. | 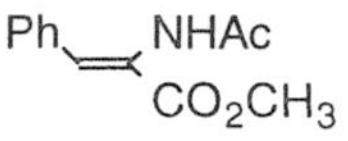 Ph—⧸NHAc / CO₂CH₃ | 99.0 (98.2)ᵇ | 97.4 (94.4)ᵇ | -- | 90.2 (84.7)ᵇ | 81.0 | 2.0 (9.8)ᵇ | 2.0 (6.2)ᵇ | -- (7.2)ᵇ |
| 2. | 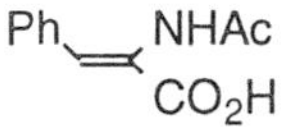 Ph—⧸NHAc / CO₂H | 97.6 | 99.0 (99.7)ᶜ | 93.0 (96.0)ᵈ | 94.0 | 91 | -- | 60.0 | -- |
| 3. | 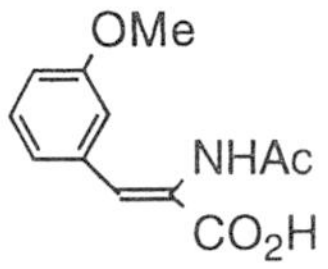 (2-OMe-phenyl)—⧸NHAc / CO₂H | -- | 97.0 | -- | 91.0 | -- | -- | 53.0 | 5.0 |
| 4. | 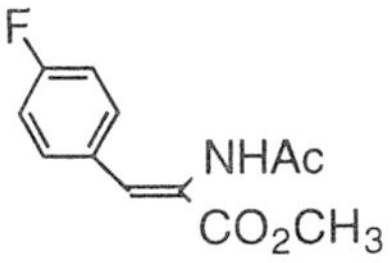 (4-F-phenyl)—⧸NHAc / CO₂CH₃ | 98.7 | 97.2 | 89 | 85.0 (84)ᵇ | 81.0 | -- | 13.0 | 9.0 |
| 5. | 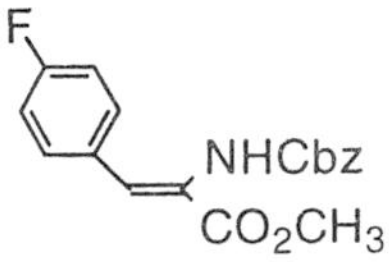 (4-F-phenyl)—⧸NHCbz / CO₂CH₃ | -- | 95.7 | 85 | 62 | -- | -- | <3 | <5 |
| 6. | 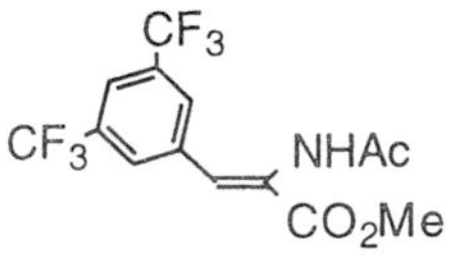 (3,5-(CF₃)₂-phenyl)—⧸NHAc / CO₂Me | 97.1 (96.9)ᵇ | 95.8 (99.3)ᶜ | -- | 85.2 | -- | -- | -- | -- |
| 7. | 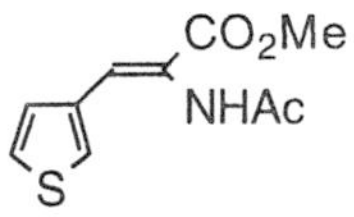 (3-thienyl)—⧸CO₂Me / NHAc | 98.8 | 96.7 (99.5)ᶜ | -- | 86.6 | -- | -- | -- | -- |
| 8. | 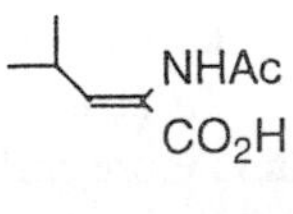 (isopropyl)—⧸NHAc / CO₂H | 83.6 (87.2)ᵉ | 91.0 | -- | 90.0 | -- | -- | 64.4 | 26.0 |
| 9. | 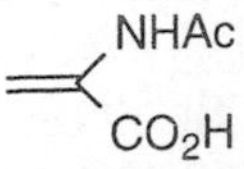 ⧸NHAc / CO₂H | -- | 96.9 | -- | -- | -- | -- | -- | -- |

ᵃ Ar substitution in Ligand **4**. ᵇ [**4**]Rh⁺(COD)BF₄⁻ as catalyst. ᶜ ee upon recrystallization.
ᵈ [**4**]Rh⁺(COD)TfO⁻ as catalyst. ᵉ Ee using methyl ester.

An equally important result is that protecting groups such as Cbz can be tolerated. For example in the reduction of 4-fluorocinnamate using ligand **4b** with the $3,5\text{-}Me_2C_6H_3$ group at phosphorus, an ee of 97% is obtained (entry 5) whereas only a 62% ee is obtained with the diphenyl phosphinite **4d**. With the ability to prepare a variety of substituted phenylalanine derivatives and to incorporate different N-protecting groups, the scope of diphosphinites in the preparation of *L*-amino acids has now been greatly expanded. These results lead to the next question: What about *D*-amino acids?

### Diphosphinites in Rh-Catalyzed Hydrogenations: Synthesis of *D*-Amino Acids

As stated before, we were interested in the Rh-catalyzed asymmetric hydrogenation in an attempt to further understand the influence of electronic effects in asymmetric catalysis. This section will describe the events that led to the development of a practical synthesis of *D*-phenylalanine derivatives. Of course, the use of *L*-glucose-derived 2,3-diphosphintes would provide the D-amino acids; however, the high cost of *L*-sugars makes this approach unattractive. Thus, our initial efforts focused on the 3,4-phosphinites **5** derived from fructose. Even though these ligands provided excellent ee's in the Ni(0)-catalyzed hydrocyanation reaction, only modest ee's (up to 60%) were obtained in the hydrogenation reaction. However, during the course of our work, we began investigating the use of the carbohydrate-derived ligands in other asymmetric reactions. A major breakthrough came when we examined the Ni-catalyzed asymmetric cross-coupling of Grignard reagents and allylic phenyl ethers **10** to provide optically active olefins **11** (Scheme 5). When the glucose derived 2,3-diphosphintes **4** were employed as ligands, the *S*-isomer was obtained with moderate enantioselectivity. During this study, we prepared the glucose derived, 3,4-diphosphinites **6** and **7** as previously described (Scheme 2). A surprise occurred when the 3,4-diphosphinites were used in the cross-coupling reaction in that the *R-isomer* was generated as the

Scheme 5

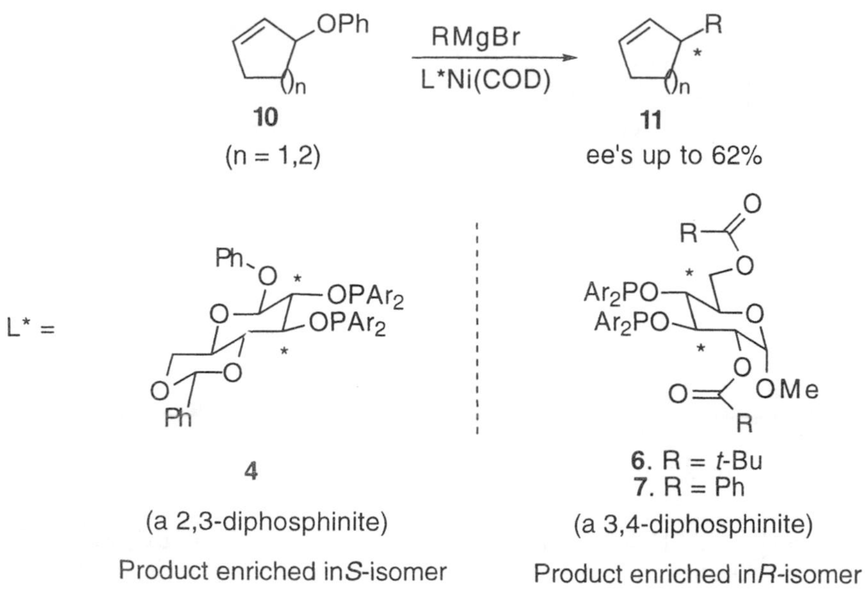

10
(n = 1,2)

11
ee's up to 62%

L* =

4

(a 2,3-diphosphinite)

Product enriched in *S*-isomer

6. R = *t*-Bu
7. R = Ph

(a 3,4-diphosphinite)

Product enriched in *R*-isomer

**Table 2.** Enantioselectivity in Hydrogenations Using Catalysts [**6** or **7**]Rh⁺(COD)SbF₆⁻

| Entry | Substrate | 3,5-Me₂[a] | 4-(MeO)[a] | H[a] | 4-CF₃[a] | 3,5-F₂[a] | 3,5-(CF₃)₂[a] |
|---|---|---|---|---|---|---|---|
| | | | Ee of Product [Enriched in (R)] | | | | |
| 1. | 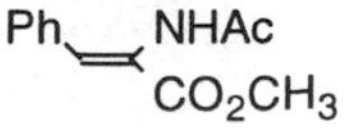 Ph / NHAc, CO₂CH₃ | 96.3 / 93.0[b] / (92.4)[c] | 84.7[b] / (84)[c] | 87.4[b] | 2.0[b] | 1.0[b] / (11.0)[c] | 2.3[b] |
| 2. | 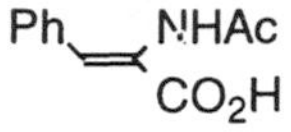 Ph / NHAc, CO₂H | 97.0 | – | – | – | – | – |
| 3. | 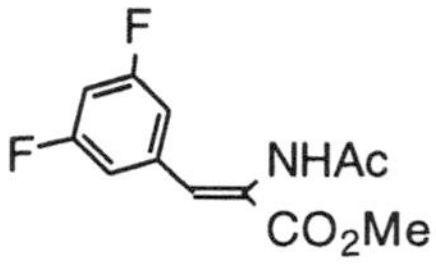 3,5-F₂-C₆H₃ / NHAc, CO₂Me | 96.2 | 85.1 | 73.0 | 2.7 | 3.0 | 5.6 |
| 4. | 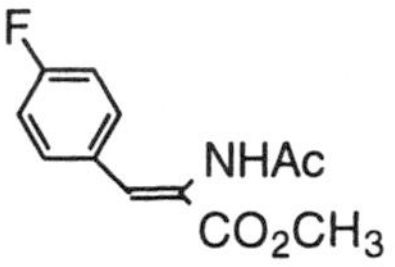 4-F-C₆H₄ / NHAc, CO₂CH₃ | 96.2 / (92.0)[c] | 87.0 | 73.5 | <1 | <1 | 11.0 |
| 5. | 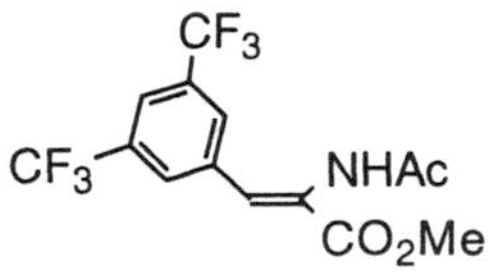 3,5-(CF₃)₂-C₆H₃ / NHAc, CO₂Me | 94.2 / (99.4)[d] | 83.9 | 77.9 | <1 | 3.2 | <1 |
| 6. | 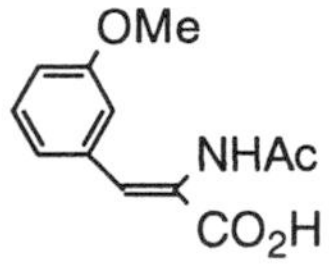 3-OMe-C₆H₄ / NHAc, CO₂H | 95.9 / (93.1)[c] | 85.3 | 73.4 | 2.1 | <1 | 2.3 |
| 7. | 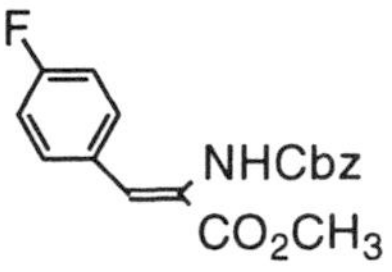 4-F-C₆H₄ / NHCbz, CO₂CH₃ | 90 | – | – | – | – | – |
| 8. | 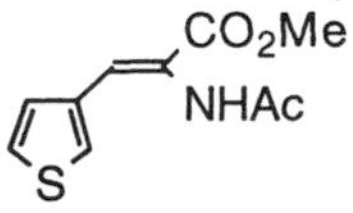 thiophen-3-yl / CO₂Me, NHAc | 97.0 / (99.5)[d] | – | – | – | – | – |

[a] Ar substitution in Ligand **6** / **7**. [b] [**7**]Rh⁺(COD)BF₄⁻ as catalyst. [c] [**6**]Rh⁺(COD)BF₄⁻ as catalyst. [d] Ee upon recrystallization.

major product. This result is readily explained by recognizing the quasi-enantiomeric relationship of the 2,3- and 3,4-positions of glucose-derived diphosphinites and by making the basic assumption that the absolute configuration of the product depends on the local chirality of the two vicinal carbons to which the chelating phosphorus atoms are attached (Scheme 5). Not only does this logic explain this result but application to other sugar diphosphinites should be general.

With this discovery from the cross-coupling reaction, we anticipated a high degree of enantioselectivity in the hydrocyanation reaction (eq. 2) for the R-isomer using now the glucose-derived 3,4-diphosphinite **7**. Unfortunately, hydrocyanation with ligand **7h** with the highly electron-withdrawing 3,5-$(CF_2)_2C_6H_3$ groups at phosphorus provided the R-isomer in only 33% ee. This disappointing result indeed reflects the quasi-enantiomeric relationship of these ligand systems; but one must bear in mind that other conformational features such as the benzylidene acetal may be essential for high ee's with the 2,3-glucose diphosphinites in the hydrocyanation reaction.[30,31] Even though poor enantioselectivity was anticipated in the hydrogenation reaction, the 3,4-diphosphinite ligand **6b** with the phosphorus substituted with Ar = 3,5-$Me_2C_6H_3$ and the corresponding [**6b**]Rh$^+$(COD)BF$_4^-$ complex were prepared. In our first experiment, hydrogenation of α–acetamidocinnamate (eq. 1) with ligand **6b**, a 2,6-dipivaloyl derivative, provided a 92.4% ee for the *D*-isomer (Table 2). We were astonished to find such high enantioselectivity because this same ligand was so poor in the hydrocyanation reaction. Furthermore, this result is a clear reminder that caution should be used when comparing ligands for different asymmetric reactions. With such a good starting point, we knew that slight modification of the ligand system or reaction conditions would increase the enantioselectivity of the hydrogenation reaction. Following our initial report,[26] Selke informed us that he also had prepared *D*-amino acids using ligand **12** with good selectivity.[32] Initially, we looked at simple changes in the reaction parameters such as reaction temperature and solvent. A slight increase in enantioselectivity is observed when the reaction is performed in THF at -10 ∞C (92.4 ee to 94.5% ee) instead of room temperature. A more marked solvent effect is observed in the hydrogenation of methyl α–acetamidocinnamate using catalyst **6b**Rh$^+$(COD)BF$_4^-$ with THF (92.4% ee) being the best solvent. Other solvents provided lower ee's: dimethoxyethane (91.4% ee), toluene (88.1% ee), dibutyl ether (88.4% ee) and MeOH (74.5% ee).

At this stage, it was time to return to the framework and make modifications here. First we changed the ester protecting groups at the 2,6-position and synthesized a series of methyl 2,6-O-dibenzoyl-3,4-glucose diphosphinites **7**. In addition to better ee's, the high crystallinity of the diol precursor as well as of the diphosphinites greatly facilitates the synthesis of these ligands. Once again, we confirmed the significance of the electronic effect. For example, changing from the relatively electron-rich aryl groups (3,5-$Me_2C_6H_3$) to electron-poor aryl groups the ee goes from 96% to practically 0 (Table 2). Of further note, the electronic component is essential for high selectivity (>96%) as the simple diphenyl phosphinites provide only moderate enantioselectivities (73-87%). Use of the counterion SbF$_6^-$ instead of BF$_4^-$ also provided a small improvement in the enantioselectivities. Once again the flexibility of the carbohydrate-derived diphosphinites allowed for an expedient development of a ligand for the synthesis of a wide variety of *D*-phenylalanine derivatives.

| | | | |
|---|---|---|---|
| **12** | **13** | **14** | **15** |

**Table 3.** Hydrogenations with Catalysts $[15]Rh^+(COD)BF_4^-$

| | | ee of Product [Enriched in $(R)$] | |
|---|---|---|---|
| Entry | Substrate | $3,5\text{-Me}_2{}^b$ | $H^a$ |
| 1. | 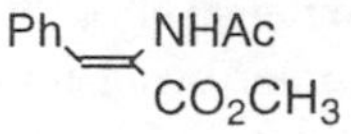 Ph NHAc CO$_2$CH$_3$ | 98.3 (98.4)$^b$ | (94.9)$^b$ |
| 2. | 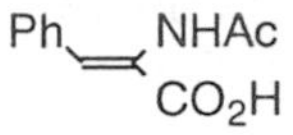 Ph NHAc CO$_2$H | 94.5 | -- |
| 3. | 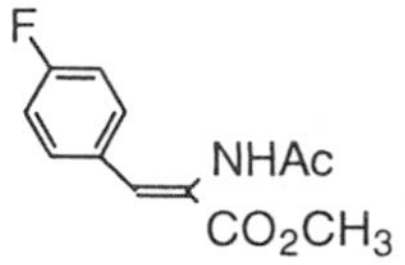 F NHAc CO$_2$CH$_3$ | 97.8 | |
| 4. | 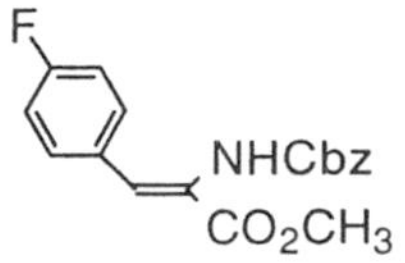 F NHCbz CO$_2$CH$_3$ | 96.0 | |
| 5. | 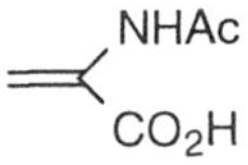 NHAc CO$_2$H | 95.0 | |

$^a$ Ar substitution in Ligand **15**. $^b$ $[15]Rh^+(COD)SbF_6^-$ as catalyst.

To quickly gauge the importance of the sugar back-bone on the enantioselectivity, the 2-deoxy ligand **13b** and mannose derived ligand **14b** (Ar = $3,5\text{-}(CH_3)_2C_6H_3$) were prepared. As expected, the selectivity decreased in both cases (65.1% ee with **13b**, 72.2% ee with **14b**) in the hydogenation of methyl N-acetylcinnamate.

We were also interested in ligands with a β–aglycone substituent. However, lack of direct methods of protecting the 2,6-positions of simple glucose derivatives hindered this effort. The availability of methyl 2-acetamido-2-deoxy β–D-glucose,[33] prompted the preparation of the corresponding diphosphinites **15**. This ligand turned out to be the best for the synthesis of *D*-amino acids, providing ee's over 97%. In addition, this ligand was essential in obtaining high enantioselectivity of 96% (Table 3, entry 4) in the hydrogenation of a N-Cbz-(4-F-Phe-OMe), because the use of ligand **7b** provided only a 90% ee (Table 2, entry 7). Even though the β–methyl N-acetyl-2-amino-2-deoxy glucose diphosphinite provides higher ee's, the facility and the cost of synthesis of the α–methyl 2,6-O-dibenzoyl 3,4-diphosphinite **7b** makes this system the most attractive for the preparation of *D*-amino acids. It should be reminded that upon achieving ee's >96%, a simple recrystallization provides nearly optically pure amino acids derivatives.

Having recognized the quasi-enantiomeric relationship between the 2,3-glucose and 3,4-glucose diphosphinites and the requisite electronic properties of the aryl groups at phosphorus, we have prepared a wide variety of *D*- and *L*-phenylalanine derivatives with ee's > 96% (Scheme 6). Once again, the flexibility of the empirical approach using diphosphinites derived from carbohydrates provided rapid access to both enantiomers in the hydrogenation reaction.

Scheme 6. Other Amino Acids Derivatives Prepared in >96%ee (*R* and *S*)

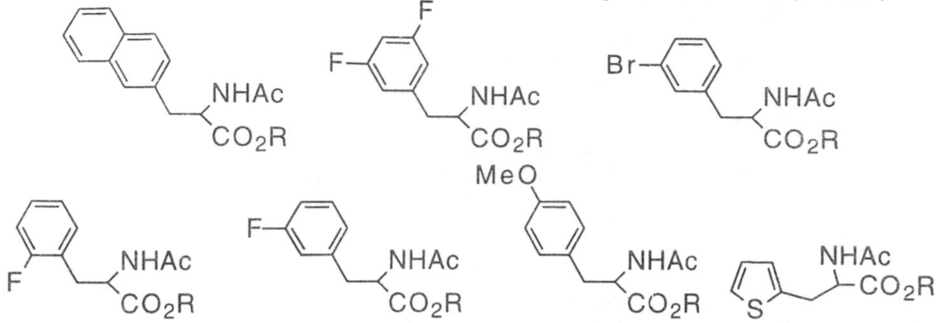

## Aliphatic Amino Acids

Although we have achieved excellent enantioselectivities for the preparation of a wide variety of *D*- and *L*-phenylalanine derivatives, selectivity for simple aliphatic systems remains good at best. Only alanine is prepared in respectable enantioselectivity using these catalysts (Table 1, entries 8 and 9; Table 3, entry 5).[34]

## Origin of Enantioselectivity

The origin of electronic effects as a function of the electronic nature of the P-aryl groups remains speculative at this point. Examination of the Halpern mechanism (Scheme 6) which has been demonstrated to be general for a variety of bidentate phosphine ligands, provides insight into the possible electronic influence of the ligand.[2] The key feature of this well-studied[2,35,36] mechanism involves the reversible formation of the diasteromeric olefin complexes **16** and **17** under ambient conditions whereby oxidative addition of hydrogen to these species is rate determining. Furthermore, reaction of the minor diastereomer **16** is faster (i.e. $k_2^{min} >> k_2^{maj}$) and provides the observed asymmetric induction. Thus, we believe that the *P*- aryl substituents on our ligand system change the reactivity patterns of the major and minor complexes, presumably by increasing the electron-density on the central Rh atom, thereby enhancing the rate of oxidative addition. Indeed, a pronounced role of electronic effects on the course of hydrogen addition to a number of Rh and Ir complexes including Vaska's complex (trans-IrCl(CO)(PPh$_3$)$_2$ has been documented.[37] If such enhancement, for reasons as yet unknown, is more pronounced for the minor, kinetically relevant diastereomeric intermediate, higher enantioselectivity should result. Since the enantioselectivity is also affected, albeit to a lesser extent, by the ratio of major to minor diastereomeric acetamidocinnamate complexes, yet another scenario is that this equilibrium is affected in favor of the kinetically important minor isomer.

Additional information on the electronic influence of the ligands is provided by examination of the ground state differences of the Rh-complexes with the different diphosphinite ligands. The classical approach for determining ligand basicity involves examination of the CO stretching frequencies of the corresponding metal carbonyl complexes which shows the amount of back-bonding of the CO and in turn provides a qualitative assessment of the relative electronics of the system.[38] Although we do not have data for the corresponding Rh-complexes, a series of L*Ni(CO)$_2$ were prepared and the IR spectra were taken. From Table 4, it is seen that the A$_1$ stretching frequency decreases by 34 cm$^{-1}$ from the electron-withdrawing system (Ar = 3,5-(CF$_3$)$_2$C$_6$H$_3$) to the electron-rich system (Ar = 4-MeOC$_6$H$_4$). This is a clear indication that the ligand basicity is changing significantly for the diphosphinites. It is coneivable that in the Rh-

**Table 4.** [4]Ni(CO)$_2$ Carbonyl Stretching Frequencies[a,b]

| Ar on **4** | A$_1$ (cm$^{-1}$) | B$_1$ (cm$^{-1}$) |
|---|---|---|
| 3,5-(CF)$_3$C$_6$H$_3$ | 2038 | 1987 |
| 3,5-F$_2$C$_6$H$_3$ | 2034 | 1983 |
| 4-CF$_3$C$_6$H$_4$ | 2024 | 1968 |
| 4-FC$_6$H$_4$ | 2015 | 1965 |
| H | 2012 | 1947 |
| 3,5-(TMS)$_2$C$_6$H$_3$ | 2006 | 1950 |
| 3,5-Me$_2$C$_6$H$_3$ | 2006 | 1946 |
| 4-MeO-C$_6$H$_4$ | 2004 | 1945 |

[a] Symmetry approximated as C$_2$.  [b] IR spectra of benzene solution.

**Scheme 7.** Abbreviated Halpern Mechanism[2]

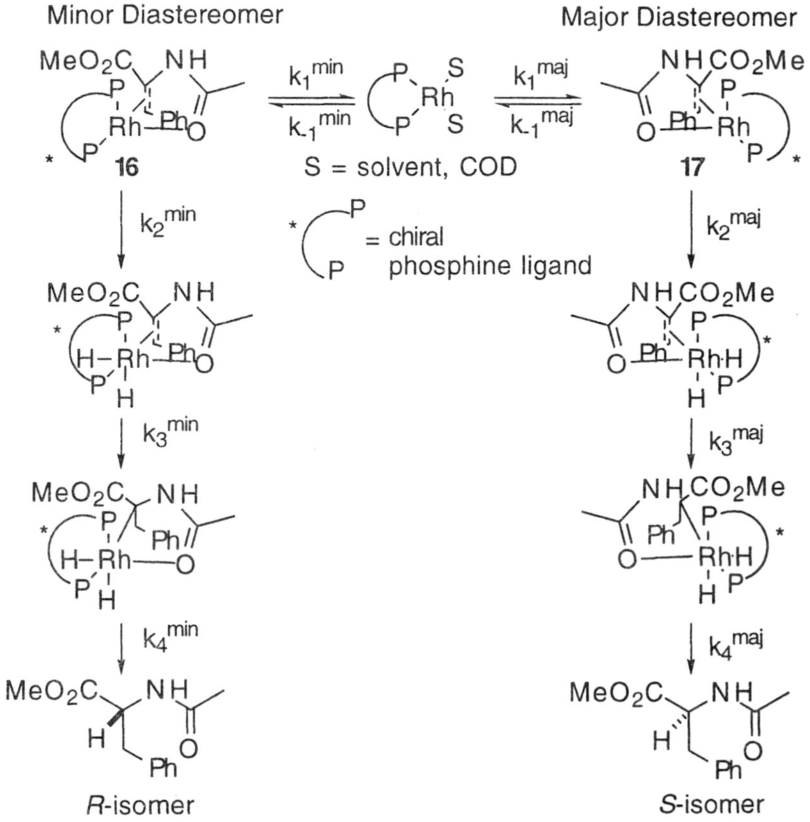

catalyzed asymmetric reductions, the rates of oxidative addition of hydrogen could be affected by electronic changes in the metal.  The margin of change in selectivity (0 to 99% ee) for the diphosphinites compared with the smaller effects (57 to 79% ee) observed with modified DIOP-ligands[23] can also be correlated, because changing the aryl groups in the DIOP series possessing phosphines dominated by the alkyl residue should provide only small changes in the basicity of the ligand. Substrate electronics also has a bearing on the rate of oxidative additon of hydrogen and has been established by examination of the [103]Rh NMR of a variety of aryl-substituted Rh-cinnamate complexes.[39] Thus, the lack of enhancement of selectivity for the aliphatic cases with increasing electron-rich diphosphinites may be the result of an overriding substrate effect with only a small ligand electronic component. Of course, the rates of individual steps may not be affected by the same magnitude for different ligand and substrate structures, so caution should be used in making generalizations of electronic effects. Nevertheless, the electronic flexibility of the carbohydrate diphosphinites allowed for quick assessment of this influence in the hydrogenation reaction and provided the necessary enhancement to achieve excellent ee's for the synthesis of a variety of *D*- and *L*-phenylalanine derivatives.

## Viability of a Commercial Process

The simple ligand system **4d** was investigated for the production of *L*-DOPA using the Rh-catalyzed asymmetric hydrogenation of the requisite acrylate.[40] Even though this ligand has potential, there are several issues which must be addressed. The original Monsanto process for *L*-DOPA has been extensively optimized so that turnovers reaching several thousands have been obtained with excellent enantioselectivity. For a moderately priced pharmaceutical intermediate, to be practical, we determined that at least 10,000 turnovers are required with no recycle of Rh to cover the high cost of the metal. However when lower turnovers (2000) are used or when the ee's are lower, say 90%, a recycle process needs to be developed so the cost of the Rh-metal is not an issue. This is the case with ligand **4d**. The electronically modified ligands should overcome this obstacle and provide higher ee's. In addition, the ability to prepare *D*-amino acids with high enantioselectivity using inexpensive and easily accessible *D*-glucose-derived ligands makes these catalysts very attractive.

The nature of the 7-membered chelate of these diphosphinites compared to other catalysts such as the DuPhos series is noteworthy. For example, it is widely accepted that a 7-membered chelate intermediate can undergo reorganizations within its coordination sphere faster than in a 5-membered one. For this reason catalytic processes that involve 7-membered chelates are relatively faster.[41] Such flexibility has to be balanced against the conformational rigidity at critical stages in the catalytic cycle if high enantioselectivity is desired. In the case of the best sugar-derived vicinal diarylphosphinites both these conditions are met, and the corresponding Rh chelates are extremely active catalysts. Qualitative measurements of the half-lives of the reactions done at room temperature with 0.01 mmol of catalyst per mmol of substrate in 5 mL of THF at 30-40 psi of hydrogen gives a value of less than 10 minutes for the best catalyst **4b**. For a direct comparison, we chose one of the best 5-membered chelates viz., [Me-DuPHOS][COD]Rh[+] complex[1f] and the results are shown in Scheme 8. Using THF as the solvent, at 0.001 equivalents of the respective pre-catalysts, the hydrogenation of *(Z)*-N-acetyl 3,5-bis-trifluoromethylphenyl alanine methyl ester was carried out for 15 minutes and the reaction was terminated by releasing the hydrogen and refilling the reaction vessel with nitrogen. Conversion and ee's were measured by gas chromatography. While the phosphinite catalyzed reaction proceeded to completion giving 96.1% ee, the Me-DuPHOS-mediated reaction was only 10.7% complete (ee 96.2%). However, in making this comparison, it should be noted that in the original studies with the DuPHOS catalysts, the reactions were carried out in alcoholic solvents even though it has been reported that there is little difference between various solvents.[1f] The enantioselectivity for DuPHOS ligands are in general 1-3 % higher for *aromatic* amino acids. This marginal difference in selectivity is more than offset by the ease of preparation of the sugar-derived ligands. In addition, as mentioned earlier, a single recrystallization usually gave nearly optically pure (~ 99.5 % ee) product in cases where we attempted such purifications. There is no reason to doubt that this is the case with other solid amino acids. For *aliphatic* amino acids the sugar-derived ligands are still not competitive with the DuPHOS series, except for alanine. Another important advantage of the DuPHOS ligands is that they are able to reduce both *(Z)-* and *(E)-* isomers of dehydroamino acids with equal facility. Therefore, a mixture of the stereoisomers of these substrates can be used for the production of aliphatic amino acids. For the sugar-derived phosphinites further optimizations are needed before this can be accomplished. For arylalanine derivatives, balance of cost and selectivity advantages of the sugar-derived phosphinite ligands should make them an attractive option for further development.

Scheme 8. Relative Rates of Hydrogenation of a Dehydroamino Acid Ester Using [**4b**]$Rh^+$(NBD) $SbF_6^-$ and [$(R,R)$-Me-DuPHOS]$Rh^+$(COD) $TfO^-$ in THF

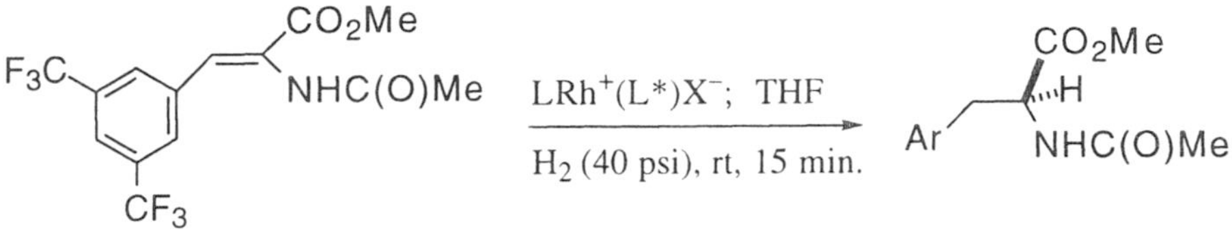

| L* | L | X | Conversion | ee |
|---|---|---|---|---|
| **4b** | NBD | $SbF_6$ | 100 | 96.1 |
| $(R,R)$-Me-DuPHOS | COD | OTf | 10.7 | 96.2 |

## References

1.  (a) Takaya, H.; Ohta, T.; Noyori, R. "Asymmetric Hydrogenation," in *Catalytic Asymmetric Synthesis*; I. Ojima, Ed.; VCH Publishers: New York, 1993; p 1.  (b) Knowles, W. S. *Acc. Chem. Res.* **1983**, *16*, 106. Koenig, K. E. "The Applicability of Asymmetric Homogeneous Catalytic Hydrogenation" in *Aymmetric Synthesis*; Morisson, J. D., Ed.; Academic Press, Orlando, 1985; Vol.5, p 71.  (c) Koenig, K. E. "The Applicability of Asymmetric Homogeneous Catalytic Hydrogenation," in *Aymmetric Synthesis*; J. D. Morisson, Ed.; Academic Press: Orlando, 1985; Vol. 5; p 71.  (d) Chan, A. S. C.; Pluth, J.; Halpern, J. *J. Am. Chem. Soc.* **1980**, *102*, 5952.  (e) Brown, J. M. *Chem. Soc. Rev.* **1993**, 25.  (f) Burk, M. J.; Feaster, J. E.; Nugent, W. A.; Harlow, R. L. *J. Am. Chem. Soc.* **1993**, *115*, 10125.
2.  Landis, C. R. *J. Am. Chem. Soc.* **1993**, *115*, 4040.
3.  Tanaka, M.; Ogata, I. *J. C. S., Chem. Commun.* **1975**, 735.
4.  Cullen, W. R.; Sugi, Y. *Tetrahedron Lett.* **1978**, 1635-1636.  Selke, R. *React. Kinetic. Catal. Lett.* **1979**, *10*, 135-138. Jackson, R.; Thompson, D. J. *J. Organomet. Chem.* **1978**, *159*, C29-C31. For the most recent contributions from the Selke group see: Berens, U.; Selke, R. *Tetrahedron: Asymmetry* **1996**, *7*, 2055.
5.  For leading references to other reports of carbohydrate-derived diphosphinites, see: Yamashita, M.; Naoi, M.; Imoto, H.; Oshikawa, T. *Bull. Chem. Soc. Jpn.* **1989**, 62, 942. Iida, A.;

Yamashita, M. *Bull. Chem. Soc. Jpn.* **1988**, *61*, 2365. Sunjic, V.; Habus, I.; Snatzke, G. *J. Organomet. Chem.* **1989**, *3702*, 295.

6. Kreuzfeld, H.-J.; Döbler, C.; Krause, H. W.; Facklam, C. *Tetrahedron: Asymmetry* **1993**, *4*, 2047. See also: Döbler, C.; Kreuzfeld, H.-J.; Krause, H.; Michalik, M. *Tetrahedron: Asymmetry* **1993**, *4*, 1833. Ochima, J.; Yoda, N.; Yatabe, M.; Tanaka, T.; Kogure, T. *Tetrahedron* **1984**, *40*, 1255. Achiwa, K. *Chem. Lett.* **1977**, 777.

7. Casalnuovo, A. L.; RajanBabu, T. V.; Ayers, T. A.; Warren, T. H. *J. Am. Chem. Soc.* **1994**, *116*, 9869.

8. For an important observation that anti-Marovnikov products are produced in the Ni-catalyzed addition of HCN onto vinylarenes using phosphite ligands, see: Nugent, W. A.; McKinney, R. J. *J. Org. Chem.*, **1985**, *50*, 5370.

9. (a) Elmes, P. S.; Jackson, W. R. *Aust. J. Chem.* **1982**, *35*, 2041. (b) Hodgson, M.; Parker, D.; Taylor, R. J.; Ferguson, G. *Organometallics* **1988**, *7*, 1761. (c) Baker, M. J.; Pringle, P. G. *J. Chem. Soc., Chem. Commun.* **1991**, 1292.

10. For some recent examples see: (a) Trost, B. M.; Van Vranken, D. L. *Chem. Rev.* **1996**, *96*, 395 and references cited therein. (b) Bao, J.; Wulff, W. D.; Rheingold, A. L. *J. Am. Chem. Soc.* **1993**, *115*, 3814. (c) Morken, J. P.; Didiuk, M. T.; Hoveyda, A. H. *J. Am. Chem. Soc.* **1993**, *115*, 6997. (d) Seebach, D.; Devaquet, E.; Ernst, A.; Hayakawa, M.; Kühnle, F. N. M.; Schweizer, W. B.; Weber, B. *Helv. Chim. Acta* **1995**, *78*, 1636 and references cited therein.

11. Mackenzie, P. B.; Whelan, J.; Bosnich, B. *J. Am. Chem. Soc.* **1985**, *107*, 2046 .

12. Consiglio, G.; Waymouth, R. M. *Chem. Rev.* **1989**, *89*, 257.

13. Frost, C. G.; Howarth, J.; Williams, J. M. J. *Tetrahedron: Asymmetry* **1992**, *3*, 1089.

14. Trost, B. M.; Van Vranken, D. L.; Bingel, C. *J. Am. Chem. Soc.* **1992**, *114*, 9327.

15. Johnson, R. A.; Sharpless, K. B. "Catalytic Asymmetric Epoxidation of Allylic Alcohols," in *Catalytic Asymmetric Synthesis*; I. Ojima, Ed.; VCH Publishers: New York, 1993; p 101.

16. Togni, A.; Burckhardt, U.; Gramlich, V.; Pregosin, P.; Salzmann, R. *J. Am. Chem. Soc.* **1996**, *118*, 1031. and references cited therein.

17. RajanBabu, T. V.; Casalnuovo, A. L. *J. Am. Chem. Soc.* **1992**, *114*, 6265.

18. RajanBabu, T. V.; Casalnuovo, A. L. *Pure Appl. Chem.* **1994**, *66*, 1535.

19. RajanBabu, T. V.; Casalnuovo, A. L. *J. Am. Chem. Soc.* **1996**, *118*, 6325.

20. Casalnuovo, A. L.; RajanBabu, T. V. The Asymmetric Hydrocyanation of Vinylarenes. In *Chirality in Industry II*. Collins, A. N.; Sheldrake, G. N.; Crosby, J., Ed.; Wiley: New York, 1997, pp 309-333.

21. RajanBabu, T. V.; Ayers, T. A. *Tetrahedron Lett.* **1994**, *35*, 4295.

22. See for example: (a) Jacobsen, E. N.; Zhang, W.; Güler, M. L. *J. Am. Chem. Soc.* **1991**, *113*, 6703. (b) Nishiyama, H.; Yamaguchi, S.; Kondo, M.; Itoh, K. *J. Org. Chem.* **1992**, *57*, 4306. (c) Schnyder, A.; Hintermann, L.; Togni, A. *Angew. Chem. Int. Ed. Engl.* **1995**, *34*, 931. Electronic effects due to different chelating atoms: (d) Frost, C. G.; Howarth, J.; Williams, J. M. J. *Tetrahedron: Asymmetry* **1992**, *3*, 1089. (e) Togni, A.; Breutel, C.; Schnyder, A.; Spidler, F.; Landert, H.; Tijani, A. *J. Am. Chem. Soc..* **1994**, *116*, 4062. (f) Sakai, N.; Mano, S.; Nozaki, K.; Takaya, H. *J. Am. Chem. Soc.* **1994**, *115*, 7033. (g) von Matt, P.; Lloyd-Jones, G. C.; Minidis, A. B. E.; Pfaltz, A.; Macko, L.; Neuburger, M.; Zehnder, M.; Rüegger, H.; Pregosin, P. S. *Helv. Chim. Acta* **1995**, *78*, 265. (h) Rieck, H.; Helmchen, G. *Angew. Chem. Int. Ed. Eng.* **1995**, *34*, 2687.

23. Anecdotal evidence of electronic effects in Rh-catalyzed hydrogenation dates back to some of the original studies done by Kagan, see: Dang, T. P.; Poulin, J. C.; Kagan, H. B. *J. Organomet. Chem.* **1975**, *91*, 105. Hengartner, U.; Valentine, D., Jr.; Johnson, K. K.;

Larschied, M. E.; Pigott, F.; Scheidl, F.; Scott, J. W.; Sun, R. C.; Townsend, J. M.; Williams, T. H. *J. Org. Chem.* **1979**, *44*, 3741.; Yamagishi, T.; Yatagai, M.; Hatakeyama, H.; Hida, M. *Bull. Chem. Soc. Jpn.* **1984**, *57*, 1897. Inoguchi, K.; Morimoto, T.; Achiwa, K. *J. Organomet. Chem.* **1989**, *370*, C9. Werz, U.; Brune, H. A. *J. Organomet. Chem.* **1989**, *365*, 367. Morimoto, T.; Chiba, M.; Achiwa, K. *Chem. Pharm. Bull.* **1992**, *40*, 2894. Inoguchi, K.; Sakuraba, S.; Achiwa, K. *Synlett* **1992**, 169.

24. Ogawa, T.; Matsui, M. *Tetrahedron* **1981**, *37*, 2363.

25. For large scale production of a-methyl 2,6-dibenzoylglucopyranoside, direct acylation of a-methyl glucopyranoside with picoline as base avoids the use of the toxic tin reagent (work by R. Shapiro).

26. A preliminary account of this work has been published: RajanBabu, T. V.; Ayers, T. A.; Casalnuovo, A. L. *J. Am. Chem. Soc.* **1994**, *116*, 4101.

27. The dehydroamino acid derivatives are readily available in large quantity, see: Schmodt, U.; Griesser, H.; Leitenberger, V.; Lieberknecht, A.; Mangold, R.; Meyer, R.; Riedl, B. *Synthesis* **1992**, 487. Herbst, R. M.; Shemin, D. *Org. Synth.,* Coll. Vol II, **1943**, 1.

28. The required [3,5-(TMS)$_2$C$_6$H$_3$]$_2$PNEt$_2$ for this ligand was prepared as reported: Trost, B. M.; Murphy, D. J. *Organometallics* **1985**, *4*, 1143.

29. Selke has reported ee's approaching 99% using ligand **4d**.

30. Selke has obtained high ee's using 2,3-glucose diphosphinites with no protecting groups at the 4- and 6-positions, see: Selke, R. *J. Organomet. Chem.* **1989**, *370*, 249.

31. Even though this trend should be general for a variety of sugars, the importance of conformational changes (i.e. fructose-derived diphosphinites) can also greatly affect the enantioselectivity.

32. Selke R. Presented at the International Conference on Circular Dichroism, Bochum, Germany, 1991, Abstract p. 305. We thank Dr. Selke for this private communication.

33. Sample provided by Dr. S. Sabesan, DuPont.

34. DuPHOS provides the best ee's for these substrates, see ref. 1f.

35. McCulloch, B.; Halpern, J.; Thompson, M. R.; Landis, C. R. *Organometallics* **1990**, *9*, 1392. Brown, J. M.; Chaloner, P. A. *J. Chem. Soc., Chem. Commun.* **1978**, 321. Brown, J. M.; Chaloner, P. A. *J. Chem. Soc., Chem. Commun.* **1980**, 344. Bircher, H.; Bender, B. R.; von Philipsborn, W. *Magn. Res. Chem.* **1993**, *31*, 293.

36. Kagan, H. B. "Asymmetric synthesis using organometallic catalysts," In *Comprehensive Organomeallic Chemistry*; G. Wilkinson, F. G. A. Stone and E. W. Abel, Ed.; Pergamon Press: Oxford, 1982; Vol. 8; p 463. Knowles, W. S. *Acc. Chem. Res.* **1983**, *16*, 106. Oliver, J. D.; Riley, D. P. *Organometallics* **1983**, *2*, 1032. Bogdan, P. L.; Irwin, J. J.; Bosnich, B. *Organometallics* **1989**, *8*, 1450. Brown, J. M.; Evans, P. L. *Tetrahedron* **1988**, *44*, 4905. Pavlov, V. A.; Klabunovskii, E. I.; Struchkov, Y. T.; Voloboev, A. A.; Yanovsky, A. I. *J. Mol. Catal.* **1988**, *44*, 217. Seebach, D.; Plattner, D. A.; Beck, A. K.; Wang, Y. M.; Hunziker, D.; Petter, W. *Helv. Chim. Acta.* **1992**, *75*, 2171. Sakuraba, S.; Morimoto, T.; Achiwa, K. *Tetrahedron: Asymmetry* **1991**, *2*, 597. Michalik, M.; Freier, T.; Schwarze, M.; Selke, R. *Magn. Reson. Chem.* **1995**, *33*, 835.

37. Kunin, A. J.; Johnson, C. E.; Maguire, J. A.; Jones, W. D.; Eisenberg, R. **1987**, *109*, 2963. Johnson, C. E.; Eisenberg, R. *J. Am. Chem. Soc.* **1985**, *107*, 3148. Burk, M. J.; McGrath, M. P.; Wheeler, R.; Crabtree, R. H. *J. Am. Chem. Soc.* **1987**, *110*, 5034. Sargent, A. L.; Hall, M. B.; Guest, M. F. *J. Am. Chem. Soc.* **1992**, *114*, 517.

38. Ugo, R.; Pasini, A.; Fusi, A.; Cenini, S.; *J. Am. Chem. Soc.* **1972** *94*, 7364. Tolman, C. A. *J. Am. Chem. Soc.* **1970**, *92*, 2953. Jolly, P. W. Lewis-base Nickel Carbonyl Complexes. In *Comprehensive Organometallic Chemistry: The Synthesis, Reactions and*

*Structure of Organometallic Compounds*; Wilkenson, G., Ed.; Pergamon: Oxford, 1982: Vol. 6, pp 28-30.

39. Bender, B. R.; Koller, M.; Nanz, D.; von Philipsborn, W. *J. Am. Chem. Soc.* **1993**, *115*, 5889.

40. Vocke, W.; Haenel, R.; Floether, F. U. *Chem. Tech. (Leipzig)* **1987**, *39*, 123.

41. Oliver, J. D.; Riley, D. P.; *Organometallics* **1983**, *2*, 1032. For a study of rates of hydrogenation using 5-, 6- and 7-membered rhodium chelates see also: Landis, C. R.; Halpern, J. *J. Organomet. Chem.* **1983**, *250*, 485; Descotes, G.; Lafont, D.; Sinou, D.; Brown, J. M.; Chaloner, P. A.; Parker, D. *Nouv. J. Chim.* **1981**, *5*, 167.

# Chiral (Salen)Mn(III) Complexes in Asymmetric Epoxidations: Practical Synthesis of *cis*-Aminoindanol and Its Applications to Enantiopure Drug Synthesis

**Chris H. Senanayake**
*Chemical Research & Development, Sepracor Inc., 111 Locke Drive, Marlborough, MA 01752*

**Eric N. Jacobsen**
*Department of Chemistry and Chemical Biology, Harvard University, Cambridge, MA 02138*

Key words: catalyst design, (Salen)Mn(III) complex 7, asymmetric epoxidation, practical processes, Ritter-type technology, optically pure *cis*-aminoindanol, enantiopure drugs, Crixivan®, (S)-oxybutynin, (R,R)-formoterol, (R)-ketoprofen

## Introduction

Scientists have documented, and the world has come to appreciate, that enantiomers of a compound can interact differently in biological systems because enzymes, receptors, and other binding sites recognize with a specific chirality. Many studies have shown that two enantiomers of a chiral drug usually display different biological activity, and sometimes one enantiomer is detrimental. For the last ten years, the chemical community has realized that preparation of enantiopure materials is critically important and many groups have devoted time to the development of new asymmetric methods. However, finding general and practical chiral catalysts and chiral building blocks for asymmetric synthesis are significant challenges in organic synthesis.[1]

In many cases, it has been recognized that chiral aminoalcohols serve as versatile chiral reagents in a variety of asymmetric processes for the generation of enantiopure materials.[2] The rigid benzocycloalk-1-ene derived *cis*-1-amino-2 alcohols represent a chemically and biologically appealing subclass of these aminoalcohols. It has been disclosed in the literature that the constrained aminoindanol platform plays a crucial role in the synthesis of enantiopure drugs[2d,e] (Scheme 1).

---

Crixivan is a registered trademark of Merck & Co., Inc.

**Scheme 1.**

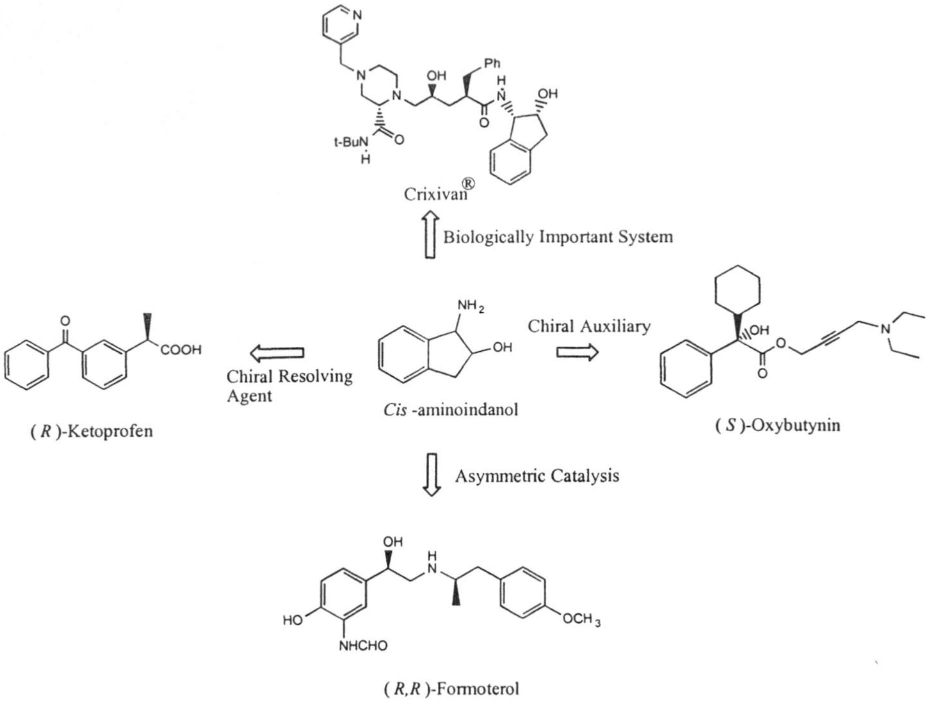

Crixivan®

⇑ Biologically Important System

(*R*)-Ketoprofen     ⇐ Chiral Resolving Agent     *Cis*-aminoindanol     Chiral Auxiliary ⇒     (*S*)-Oxybutynin

⇓ Asymmetric Catalysis

(*R,R*)-Formoterol

   While a great deal of emphasis has been placed upon the synthesis of rigid benzocycloalk-1-ene derived *cis*-1-amino-2-alcohols,[3] practical processes remained elusive until recently.[4] The chiral 1,2 dioxygen benzocycloalkanes have become available by asymmetric epoxidation of the corresponding prochiral olefins **1**.[5] These adducts were utilized as excellent precursors for the synthesis of *cis*-aminoalcohols **2** using appropriate selective amination processes (Scheme 2). Research in this field of unfunctionalized asymmetric epoxidation has been driven both by the recognized utility of optically active epoxides in organic synthesis[6] and by widespread interest in the mechanism of oxygen atom transfer from metals to alkenes.[6,7] As a result, the development of synthetic catalysts that provide practical access to such compounds has constituted a major emphasis in the field of asymmetric catalysis and recently chiral (salen)Mn-based complexes were developed as catalysts for asymmetric epoxidations of prochiral olefins.[6] This chapter focuses on the development of a practical chiral (salen)Mn catalyst for asymmetric epoxidation for unfunctionalized alkenes, and a practical application of (salen)Mn-based epoxidation methodology to the synthesis of enantiopure *cis*-aminoindanol and its application to drug synthesis.

**Scheme 2.**

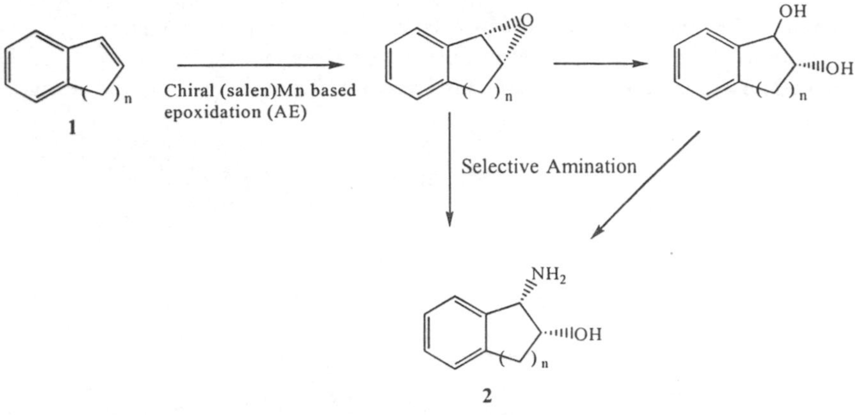

Chiral (salen)Mn based epoxidation (AE)

Selective Amination

## Chiral (Salen)Mn-Based Asymmetric Epoxidations

The first example of asymmetric epoxidation of simple olefins catalyzed by a synthetic porphyrin complex was described in 1983 by Groves.[8] Since then, several types of chiral coordination compounds have been uncovered which mediate enantioselective oxo transfer to simple olefins. The most significant progress has been made with porphyrin derivatives of iron and manganese **3**, and salen complexes of manganese **4** (Figure 1).[6] Both classes of coordination compounds are sterically well defined and kinetically non-labile and, thus, provide an appealing matrix for ligand design.

## (Salen)Mn(III) Complexes

Kochi and coworkers discovered that achiral salen complexes of chromium and manganese catalyze the epoxidation of unfunctionalized alkenes by iodosylbenzene, thus demonstrating that salen complexes can mimic the monooxygenase activity of heme proteins in the same manner as synthetic porphyrins.[9] However, in contrast to porphyrin derivatives, salen complexes bear tetravalent, thus potentially stereogenic, carbon centers in the vicinity of the metal binding site. Stereochemical communication in epoxidation can be enhanced, at least in principle, as a result of the proximity of the reaction site to the ligand dissymmetry.[10] Also, the synthesis of salen complexes from chiral diamines is generally highly efficient and extremely straightforward, and extensive screening of ligand structural types has been possible.

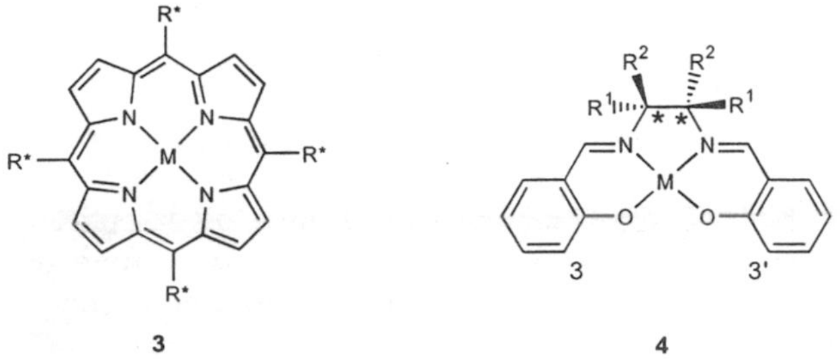

**Figure 1.** Generalized Structures of Chiral Porphyrin (**3**) and Salen (**4**) Complexes

Chiral manganese salen complexes have been developed recently that are effective and practical catalysts for the asymmetric epoxidation of a variety of alkenes (Scheme 3).[11] Systematic variation of the steric and electronic environment of the complexes has led to the discovery of catalysts that are particularly effective for the epoxidation of various important classes of olefins.

**Scheme 3.**

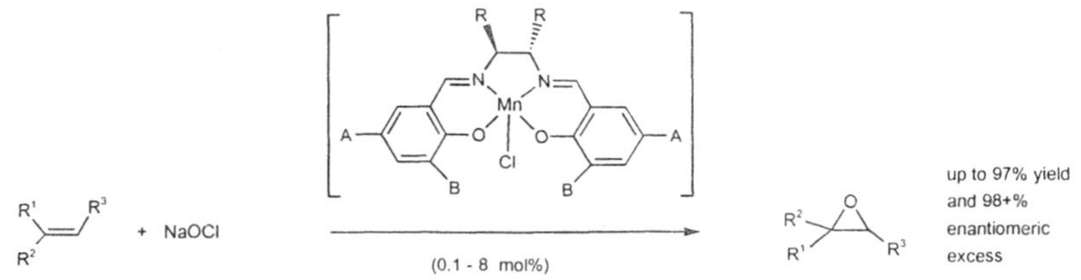

up to 97% yield
and 98+%
enantiomeric
excess

(0.1 - 8 mol%)

## Synthesis of (Salen)Mn Complexes

The preparation of chiral $C_2$ symmetric salen ligands is achieved in a practical manner and in excellent yield by the condensation of appropriately substituted 1,2-diamines with two equivalents of a salicylaldehyde derivative (Scheme 4). Chiral salen complexes of all of the first-row transition metals have been prepared, with Mn(III) derivatives displaying superior selectivity and the highest turnovers in epoxidation of most alkenes.[6] The synthesis of (salen)Mn(III) complexes is readily accomplished by heating an ethanolic solution of a salen ligand to reflux with two equivalents of $Mn(OAc)_2 \cdot 4H_2O$ in air.[11c] The intermediate (salen)Mn(III)OAc complex is converted to the corresponding (salen)Mn(III)Cl complex upon work-up with brine, and precipitated as a dark brown air- and moisture-stable powder by addition of water. This procedure is general for the preparation of Schiff base complexes derived from 1,2- 1,3- and 1,4-diamines, and it has been applied to the synthesis of over 100 optically active complexes.[12,6] Less straightforward procedures have been used for the preparation of analogous cationic complexes bearing non-coordinating counterions such as $PF_6^-$.[9a,11a]

**Scheme 4.**

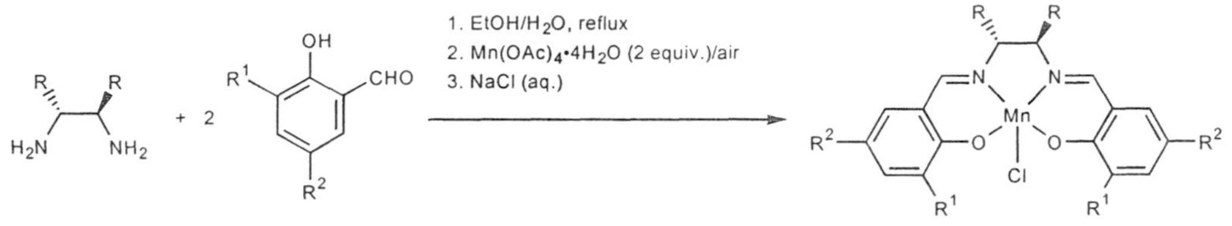

1. EtOH/H$_2$O, reflux
2. Mn(OAc)$_4 \cdot$4H$_2$O (2 equiv.)/air
3. NaCl (aq.)

Because racemic diamines can often be resolved in a straightforward manner with tartaric acid, mandelic acid, or camphorsulfonic acid, both enantiomers of a variety of chiral salen derivatives are readily accessible. These resolving agents are available at low cost in optically pure form as either optical antipode. The synthesis of the commercial catalyst **7** from the commodity chemicals 2,4-di-*t*-butylphenol and 1,2-diaminocyclohexane was effected on multi-hundred kilogram scale according to the method in Scheme 5.

## Scheme 5.

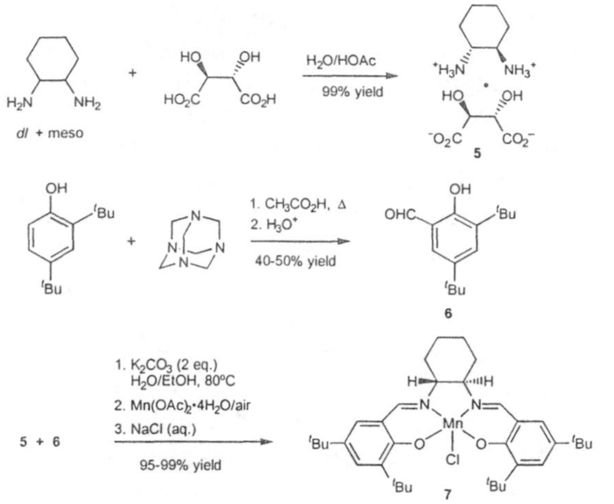

## Ligand Design

In the simplest terms, the incorporation of only two structural properties into the ligand system are required to attain good enantioselectivity in olefin epoxidation by (salen)Mn complexes: 1) a dissymmetric diimine bridge derived from a $C_2$ symmetric 1,2-diamine, and 2) bulky substituents on the 3 and 3' positions of the salicylide ligand (see Figure 1 for the numbering scheme in salen complexes). Scheme 6 illustrates the epoxidation of *cis*-β-methylstyrene by various catalysts. Similar levels of enantioselectivity are attainable with catalysts derived from either 1,2-diphenylethanediamine or 1,2-cyclohexanediamine. The unsubstituted 1,2-diphenylethanediamine derived complex **8** effects epoxidation with negligible levels of selectivity, but incorporation of *t*-butyl substituents at the 3 and 3' positions, as in **9**, improves epoxidation dramatically to 84% ee. The presence of larger groups, as in **10**[13] and **11**,[14] result in only slightly higher epoxidation selectivity.

## Scheme 6.

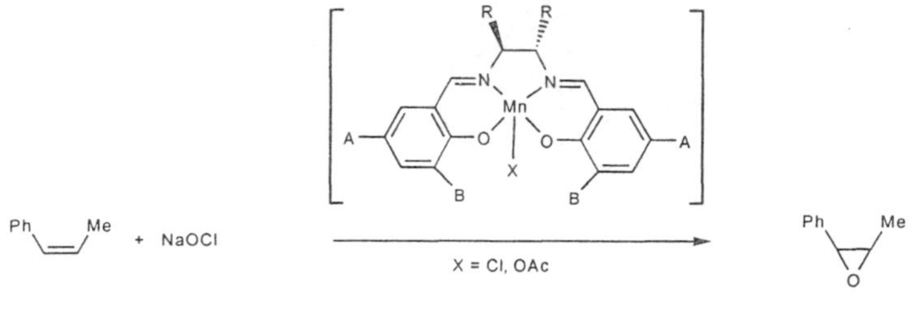

| Catalyst | R,R | A | B | ee (%) |
|---|---|---|---|---|
| 8 | Ph,Ph | H | H | <10 |
| 9 | Ph,Ph | H | $^t$Bu | 84 |
| 10 | Ph,Ph | H | * | 86 |
| 11 | Ph,Ph | H | ** | 89[a] |
| 12 | (CH$_2$)$_4$ | Me | $^t$Bu | 80 |
| 7 | (CH$_2$)$_4$ | $^t$Bu | $^t$Bu | 90 |
| 13 | (CH$_2$)$_4$ | OMe | $^t$Bu | 86 |
| 14 | (CH$_2$)$_4$ | NO$_2$ | $^t$Bu | 46 |
| 15 | (CH$_2$)$_4$ | $^i$Pr$_3$SiO | $^t$Bu | 92 |

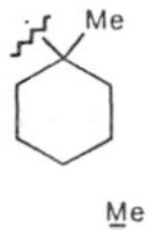

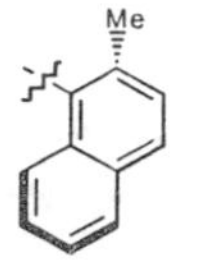

The presence and identity of substituents on the 5 and 5' positions of the salicylide ligand can also have a significant, although generally less important, influence on epoxidation enantioselectivity. In general, electron-donating or sterically demanding substituents improve selectivity. Thus, catalyst **7** affords higher selectivity than the less hindered complex **12**, and OMe-substituted catalyst **13** is more selective than sterically similar complexes such as **14** bearing electron-withdrawing groups. A happy marriage of steric and electronic ligand properties is apparently attained in complex **15**, which is the most selective epoxidation catalyst uncovered to date for most substrates.[15] However, the considerably greater synthetic accessibility of complex **7** has made it the most widely used catalyst in this class.[11c]

The ligand substituent effects outlined above may be rationalized qualitatively within the side-on approach transition state model. As represented in Figure 2, a *cis*-olefin can approach the oxo ligand of a (salen)Mn complex in a side-on manner from various directions. The presence of bulky salicylide substituents at the 3 and 3' and 5 and 5' positions presumably blocks several competing pathways (c-e) and induces olefin approach in the vicinity of the diimine bridge via approaches a or b, such that stereochemical communication between catalyst and substrate is maximized.[16]

Electronic effects on enantioselectivity are subtle, but may be understood in terms of changes imparted by the substituents on the reactivity of the oxo intermediates.[17] Electron withdrawing groups such as ($NO_2$) destabilize the Mn(oxo) intermediate relative to oxidation, whereas electron donating groups such as (OMe) attenuate the reactivity of the oxidant. In accordance with the Hammond postulate, more reactive oxidants effect epoxidation via a more reactant-like transition structure, with greater separation between substrate and catalyst and concomitantly poorer steric differentiation of diastereomeric transition structures. A milder oxidant would be expected, in turn, to effect epoxidation via a more product-like transition state. Secondary kinetic isotope effects in the epoxidation of styrene were measured and found to be consistent with this hypothesis.[17b]

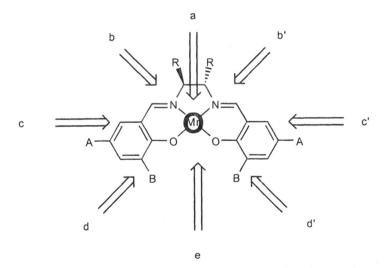

**Figure 2**. Side-on approaches of an alkene to a (salen)Mn(oxo) intermediate. The oxo ligand is oriented out of the page.

Katsuki and co-workers have examined catalytic epoxidations with (salen)Mn catalysts, such as **16**, which bear stereogenic centers on the 3 and 3' positions of the salen ligand.[14,16,18] Although higher enantioselectivities in the epoxidation of *cis* olefins are obtained with more synthetically accessible *t*-butyl-substituted catalysts **7** and **15**, catalysts such as **16** exhibit higher enantioselectivity with *trans* olefins (Scheme 7). By comparison, catalysts **7** and **15** afford very low selectivities (<20% ee) with *trans* disubstituted substrates. Epoxidation of *trans*-stilbene takes place with 61% ee using catalyst **17**, which bears no stereogenic centers on the diimine bridge. It was concluded from this observation that the stereochemical elements at the 3 and 3' positions of catalysts such as **11** and **17** have little influence on the epoxidation of *cis* olefins, but constitute dominant factors in the epoxidation of *trans* alkenes.[19]

**Scheme 7.**

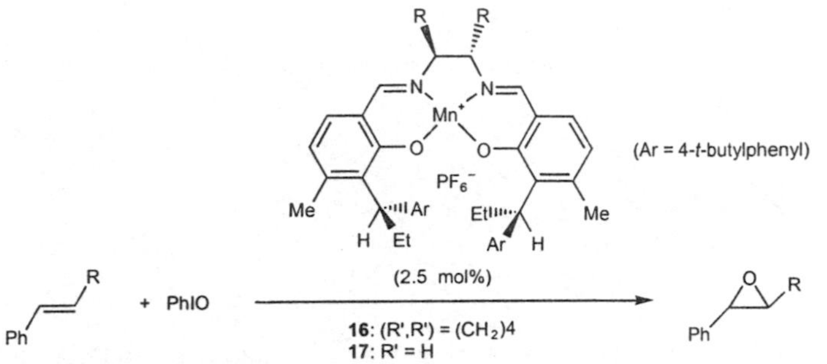

| Catalyst | R | ee (%) |
|----------|-----|--------|
| 16 | Ph | 62 |
| 16 | Me | 55 |
| 17 | Ph | 61 |

## Epoxidation Method

As is the case of metal-porphyrin-catalyzed reactions, a variety of stoichiometric oxidants are effective for (salen)Mn-catalyzed epoxidations.[20] Iodosylarene derivatives were first examined with both achiral and chiral (salen)Mn(III) catalysts,[9a,11a,18] but aqueous sodium hypochlorite was later found to be an equally effective stoichiometric oxidant in epoxidations employing these catalysts.[13] Conditions for NaOCl epoxidations with (salen)Mn(III) catalysts have been developed involving a two-phase system, with an aqueous phase containing commercial bleach and an organic phase composed of a solution of substrate and catalyst in a suitable solvent.[13] Dichloromethane is especially convenient for small-to-moderate scale laboratory procedures, but solvents such as chlorobenzene, 1,2-dichloroethane, *tert*-butyl methyl ether, ethyl acetate, or toluene may be used with generally only minor effects on epoxidation rate, yield, and enantioselectivity. The reactions are typically complete within a few hours at room temperature or at 0 °C. Work-up is accomplished by phase separation and epoxide isolation by recrystallization, distillation, or chromatography.

In general, complete substrate conversion can be achieved using 0.1-5 mol% of catalyst. In certain cases, the addition of pyridine *N*-oxide derivatives serves to improve both catalyst turnovers and enantioselectivities.[20] This is particularly true for substrates that undergo epoxidation sluggishly with the (salen)Mn(III)-based systems, such as conjugated esters and simple alkyl-substituted olefins.[22] In epoxidations employing bleach as oxidant, the catalyst resides entirely in the organic phase of the two-phase reaction medium, so water-insoluble pyridine *N*-oxide derivatives bearing hydrophobic substituents, such as the commercially available lipophilic phenylpyridine *N*-oxide, are most effective.

## Substrate Scope

### *Cis*-Disubstituted Olefins

Cyclic and acyclic *cis*-disubstituted olefins are among the best substrates for the (salen)Mn-catalyzed epoxidation reaction. A predictive stereochemical mnemonic for the epoxidation of *cis*-olefins by catalyst 7 is provided in Figure 3.[20] Especially high enantioselectivity is observed in the epoxidation of 2,2-dimethylchromene derivatives, which seem to combine several steric and electronic elements that lead to enhanced selectivity (Figure 4).

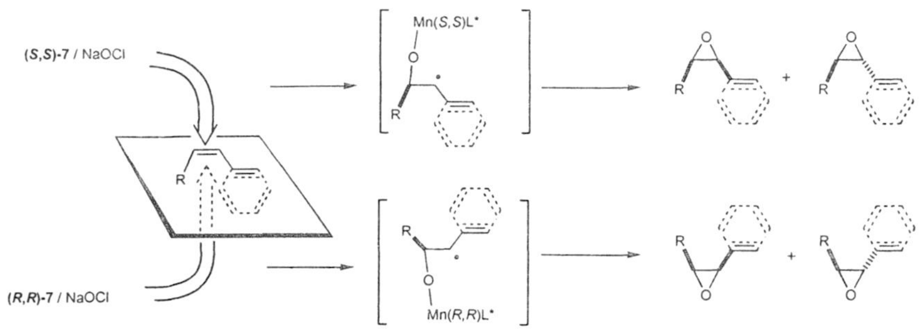

**Figure 3.** Stereochemical mnemonic for epoxidation of conjugated aryl-, alkenyl-, and alkynyl-substituted *cis* olefins by 7 and related catalysts.

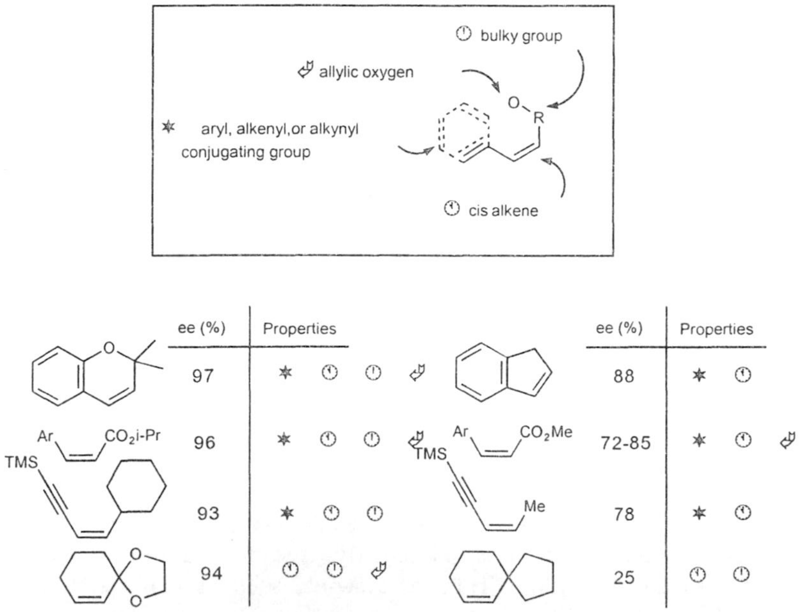

**Figure 4.** Substrate properties favoring high enantioselectivity in epoxidations with 7.

Terminal Olefins

For terminal olefins such as styrene, the *trans* pathway discussed above results in partitioning to enantiomers. Indeed, the epoxidation of *cis*-β-deuteriostyrene with catalyst **7** was shown to generate *cis*- and *trans*-2-deuteriostyrene oxide in an 8:1 ratio.[20] The diminished enantioselectivity observed with styrene (60-70% ee) relative to sterically similar *cis*-disubstituted olefins (e.g. 90% ee for *cis*-β-methylstyrene) can be attributed, at least in part, to "enantiomeric leakage" via a *trans* pathway.

Very recently, anhydrous, low temperature methods have been developed for effecting (salen)Mn-catalyzed asymmetric epoxidation reactions. Improved enantioselectivities are attainable for most substrates under these low temperature conditions, but the effect is especially pronounced in the case of terminal olefins. As illustrated in Scheme 8, epoxidation of styrene at -78 °C is complete in 30 min and affords epoxide with 86% ee when carried out in the presence of catalyst **18** and MCPBA/*N*-methylmorpholine *N*-oxide.[20] This is the highest enantioselectivity attained to date in the epoxidation of styrene.

**Scheme 8.**

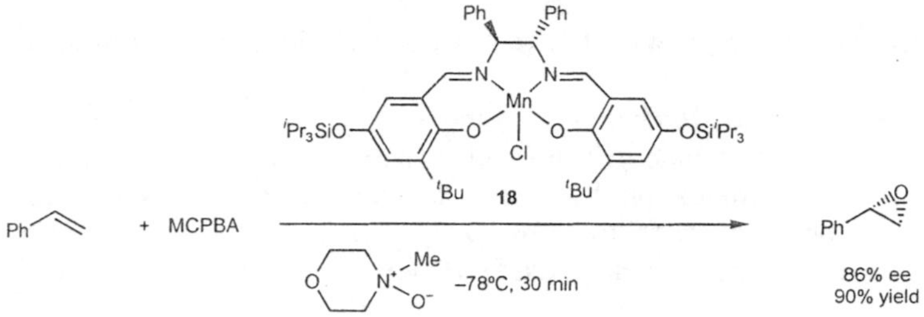

Trisubstituted Olefins

Given the low selectivities generally obtained in the epoxidation of *trans*-disubstituted olefins by (salen)Mn complexes, it was somewhat unexpected that a wide range of trisubstituted alkenes are outstanding substrates for asymmetric epoxidation (Scheme 9). The absolute chemistry of the epoxide products runs contrary to the stereochemical model presented above for *cis*-disubstituted olefins (Figure 5). A qualitative transition state model has been suggested wherein trisubstituted substrate reacts with the metal oxo via a skewed, side-on approach.[23] The distortion of trisubstituted alkenes from planarity due to $A_{1,2}$ and/or $A_{1,3}$ interactions may be critical in this context and would account for the low enantioselectivity of planar, *trans*-disubstituted alkenes.

**Scheme 9.**

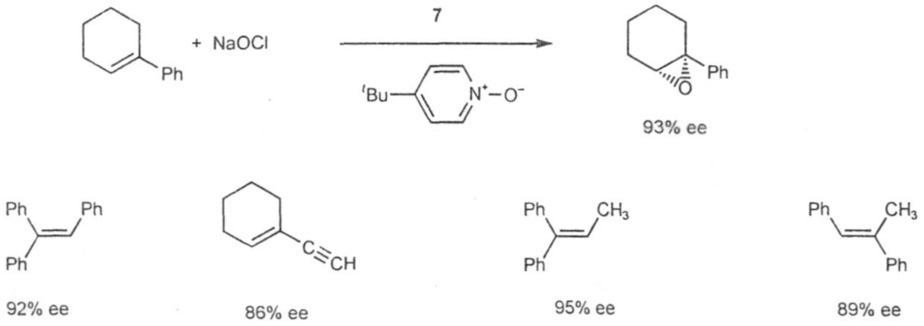

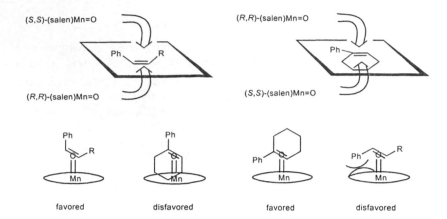

**Figure 5.** Stereochemical mnemonic for epoxidation of phenylcyclohexene and proposed skewed side-on approach transition structure geometries for olefin epoxidation.

## (Salen)Mn-Based Epoxidation Approach to the Synthesis of Chiral *Cis*-1-Amino-2-indanol

Recent literature revealed that the Merck group[4a,b,c] and Sepracor group[4d] independently developed two practical processes for (1S)-amino-(2R)-indanol. These two groups have demonstrated the practicality of asymmetric epoxidation by using the complementary (salen)Mn catalyst **7** (MnLCl)[5a,6,24] for indene epoxidation, followed by taking advantage of either the C-1 or C-2 chiral transfer process of the C-O bond of indene oxide **20** for the preparation of enantiopure (1S)-amino-(2R)-indanol.

### Sepracor Process for *Cis*-Aminoindanol

As illustrated in Scheme 10, the Sepracor group demonstrated that (1R, 2S)-indene oxide can be prepared from inexpensive indene in the presence of 1.5 mol% of (R,R)-MnLCl/13% NaOCl in dichloromethane to provide an 83% yield with an 84% enantiomeric excess. The chiral indene oxide was then subjected to the nucleophilic opening with ammonia to provide *trans*-aminoindanol, which was without isolation transformed to its benzamide under Schotten Baumann conditions to give 83% ee, which was crystallized to afford >99.5% ee. The optically pure *trans*-benzamidoindane was then converted to the optically pure benzaoxazoline **21** by simply exposing it to 80% $H_2SO_4$, followed by addition of water to give *cis*-1-amino-2-indanol.[4d] The overall yield for the preparation of optically pure (1R,2S)-**22** was 40% from indene, which has been demonstrated on a multi-kilogram scale.

### Scheme 10.

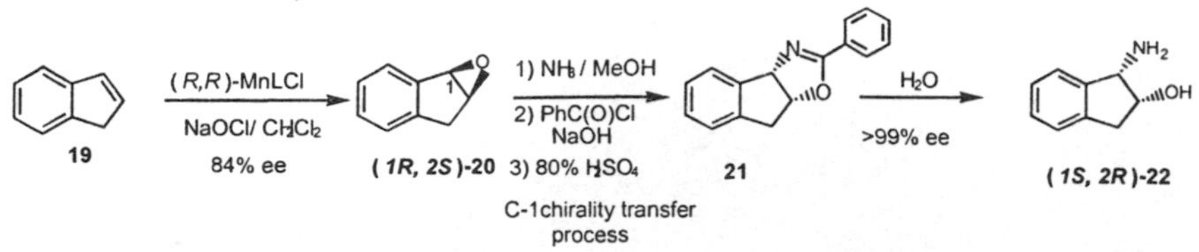

19  (R,R)-MnLCl / NaOCl/ CH₂Cl₂  84% ee  →  (1R, 2S)-20  1) NH₃ / MeOH  2) PhC(O)Cl NaOH  3) 80% H₂SO₄  C-1 chirality transfer process  →  21  H₂O  >99% ee  →  (1S, 2R)-22

## Merck Process for *Cis*-Aminoindanol

A complimentary approach to the synthesis (1*R*, 2*S*)-**22** has been developed by the Merck group by utilizing (*S,S*)-**7** catalyst in hypochlorite media to provide (1*S*,2*R*)-indene oxide. This intermediate is converted without isolation to *cis*-aminoindanol in a stereo- and regioselective manner using Ritter-type technology (Scheme 11). Several key issues have been addressed and resolved in both (salen)Mn-based asymmetric epoxidation[4c] and Ritter-type technology[4a] in order to develop a large-scale process for the synthesis of optically pure *cis*-aminoindanol.

### Scheme 11.

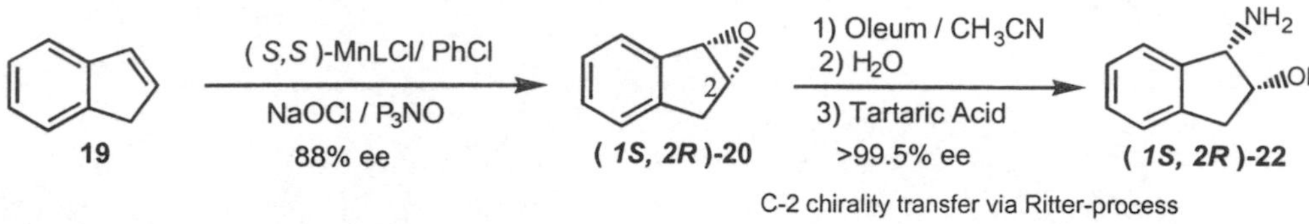

C-2 chirality transfer via Ritter-process

## Practical Asymmetric Epoxidation of Indene

As discussed earlier in this chapter, chiral manganese-salen complexes are effective catalysts for the asymmetric epoxidation of unfunctionalized olefins. In these systems, the addition of appropriate *N*-oxide derivatives serves to both activate and stabilize the catalyst systems.[9a,25] Recently, the Merck group illustrated that with the addition of an axial ligand, such as commercially available phenyl propyl pyridine *N*-oxide (P$_3$NO)[24a] to the (*S,S*)-**7**-NaOCl-PhCl system, a highly activated and stabilized catalyst for indene epoxidation resulted.[4c] Furthermore, catalyst loading could be reduced to <0.4 mol%. Kinetic studies indicated that the active catalyst was Mn$^V$-oxo **23** species[4c] and the true oxidant was hypochlorous acid (HOCl).[26] The slow step of the epoxidation process was identified to be oxidation of Mn$^{III}$ species **24** to Mn$^V$-oxo **23**[26] (Scheme 12). During the Merck development of the epoxidation of indene, it was observed that NaOCl decomposed throughout the course of the reaction, thereby presenting problems with reagent stability and stoichiometry. A secondary oxidation process was also observed which provided isonicotinic acid and benzoic acid via benzylic oxidation of P$_3$NO[4c] (Scheme 13). The decomposition pathway could be minimized by increasing the hydroxide concentration in the hypochlorite. With proper adjustment of the hydroxyl ion concentration of commercial 2M NaOCl from 0.03-0.18 to 0.3M, the hypochlorite can be stabilized and the secondary oxidation minimized (eq. 2). This reagent mixture has been demonstrated on a multi-kilogram scale to afford indene oxide in 89% yield with an optical purity of 88% ee.[4c]

## Scheme 12.

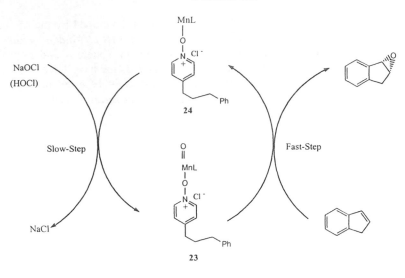

$$[HOCl] = \frac{K_{eq}}{K_w} \frac{[H_2O][OCl^-]}{[OH^-]} \qquad (1)$$

## Scheme 13.

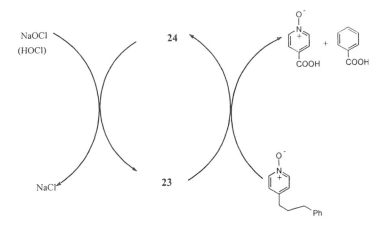

**Commercial Method of Preparation of NaOCl**

$$Cl_2 + NaOH \text{ (excess)} \longrightarrow NaOCl + NaOH \text{ (residual)} + NaCl \qquad (2)$$

**Lowering the pH of NaOCl**

### Application of Ritter Reaction for *Cis*-1-Amino-2-Indanol Synthesis

Styrene oxide has been shown to afford poor yields of regioisomeric oxazolines when exposed to the conditions of the sulfuric acid mediated-Ritter reaction.[27] The Merck group demonstrated that when indene oxide is subjected to the Ritter conditions (acetonitrile/ 97% $H_2SO_4$), methyl oxazoline **28** was formed as the major product in moderate yield.[4a] Several factors have been pointed out by a low temperature NMR study of this Ritter chemistry. As depicted in Scheme 14, at -40 °C the epoxide formed a 1:1 mixture of the methyl oxazoline **28** and sulfate **25**. While warming the reaction to 22 °C, the sulfate was simply converted to oxazoline. The proposed mechanism for *syn*-selective oxazoline formation was an acid-induced ring opening of indene oxide to produce nitrilium species **27** via C-1 carbenium ion **26**, which undergoes a thermodynamically driven equilibration process for the formation of the *cis*-5,5-ring derived oxazoline. In this Ritter process, two roadblocks for product formation were noted as: (a) polymerization via carbenium ion; and (b) the hydrogen shift process from the recipient carbenium ion to form 2-indanone (>12%). The Merck group has demonstrated that the by-product formation processes can be suppressed by stabilizing the carbenium ion by simply introducing a catalytic amount of sulfur trioxide to the Ritter mixture. As depicted in Scheme 15, sulfur trioxide captured the epoxide to form sulfate intermediate **29**, which then undergoes product formation. In addition, the chirality of the epoxide was effectively transferred from the C-2 position to the C-1 of the amino alcohol. By utilizing the Ritter acid as an oleum (21% $SO_3$-$H_2SO_4$), a highly practical and cost-effective process was developed for the conversion of chiral indene oxide to chiral *cis*-1-amino-2-indanol (>80%, yield).[4a,28]

### Scheme 14.

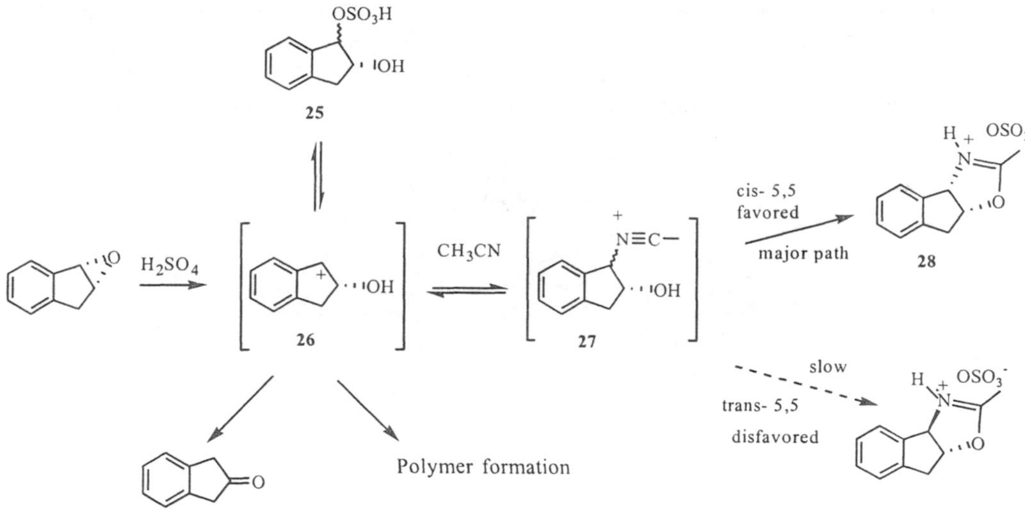

**Scheme 15.**

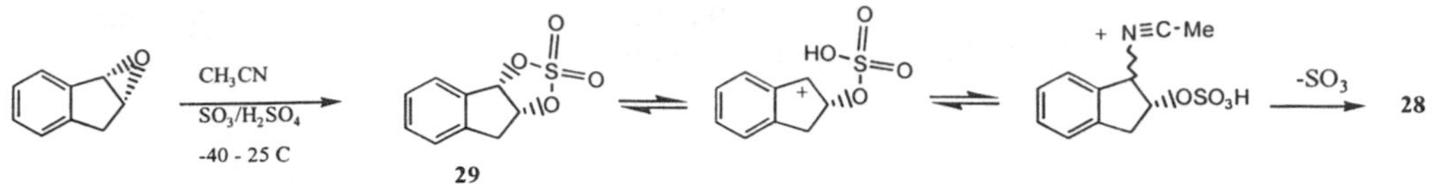

**Applications of *Cis*-1-Amino-2-Indanol in Drug Synthesis**

*Cis*-1-Amino-2-Indanol Contains HIV Protease Inhibitors

The significance of HIV protease inhibitors in the treatment of acquired immunodeficiency syndrome (AIDS) is now well established.[29] In the early 1990s, the Merck group developed a series of novel HIV-PR transition state isosteres which were comprised of constrained *cis*-1-amino-2-indanol moeity.[30] After several structural modifications, Merck's orally active HIV Protease inhibitor, Crixivan® was developed and is one of the leading drugs for the treatment of AIDS to date.[31] The single enantiomeric Crixivan® has five stereogenic centers, and interestingly, four stereocenters derive either directly or indirectly from the indane backbone.[28]

Askin and co-workers have demonstrated that rigid tricyclic aminoindanol acetonide can be utilized as a chiral platform for diastereoselective alkylation of the (Z)-lithium enolate of amide **30** with amino epoxide **31** to give >98 de in the synthesis of HIV-1 protease inhibitor[32] **33** (Scheme 16). This alkylation strategy has been utilized in a highly diastereoselective fashion in the synthesis of the orally active HIV protease inhibitor, Crixivan®.[28,33] As depicted in Scheme 17, amide **30** was allylated with a high degree of induction (97:3 of diastereoselectivity) by reacting with lithium hexamethyldisilazide/allyl bromide at -30 °C to give excellent yields. Pro-(2R) diastereomer **34** was then exposed to the buffered-iodohydrine process which resulted in formation of the 2,4-*syn* adduct **35** with outstanding selectivity (97% de). Upon subjection of **35** to the basic media, epoxy **36** was afforded, followed by conversion to enantiopure Crixivan® by a series of manipulations with high yields.[33]

**Scheme 16.**

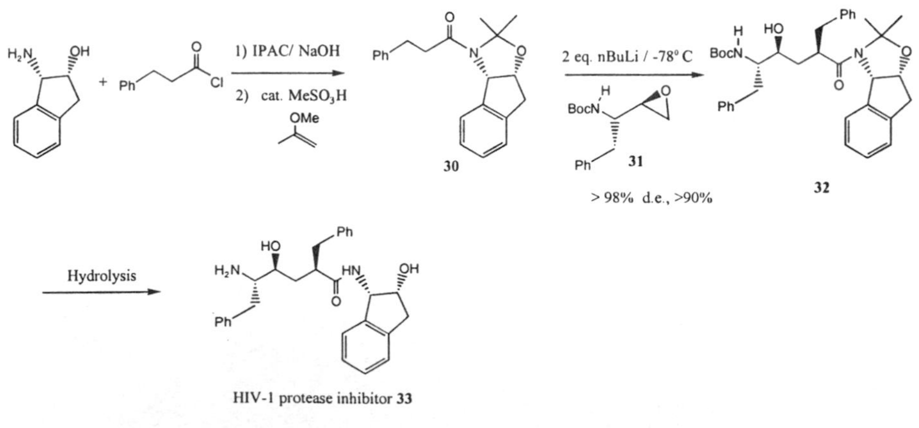

HIV-1 protease inhibitor **33**

## Scheme 17.

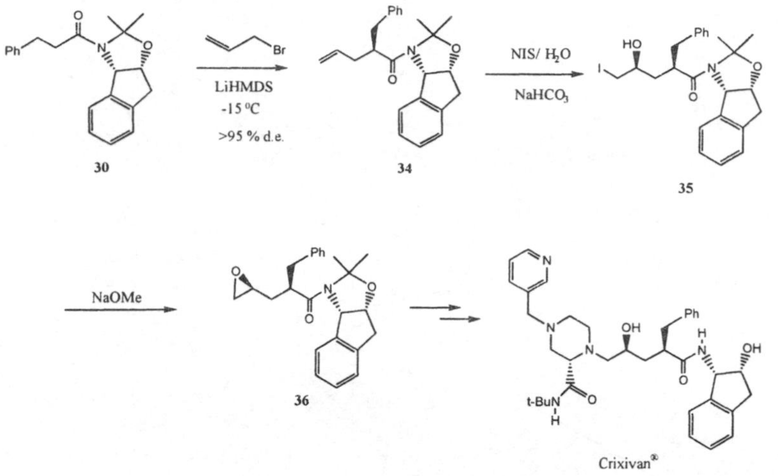

## Applications of *Cis*-1-Amino-2-Indanol as Chiral Auxiliary in the Asymmetric Synthesis of (*S*)-Oxybutynin

Rigid *cis*-aminoindanol and its derivatives have become useful and effective chiral auxiliaries for several asymmetric synthetic processes because it fulfills most of the requirements for an excellent chiral auxiliary category (high induction, ease of recovery and preparation). Recently, the Sepracor team has demonstrated that constrained *cis*-aminoalcohols is a highly defined chiral auxiliary for the production of key subunit of optically pure tertiary α-hydroxy acid of (*S*)-oxybutynin. Racemic oxybutynin (Ditropan®), a widely prescribed muscarinic receptor antagonist for the treatment of urinary incontinence, contains a tertiary hydroxyl moiety that exhibits classical antimuscarinic side effects, such as dry mouth.[34] However, preliminary biological results suggest that (*S*)-oxybutynin displays an improved therapeutic profile compared to its racemic counterpart, and is currently in clinial trials. Interestingly, most of the muscarinic receptor antagonists are comprised of tertiary α-hydroxy acid as a key component.[35] Sepracor's strategy for the synthesis of tertiary α-hydroxy acid was to take advantage of the C-1 amine or C-2 alcohol of *cis*-aminoindanol as a chiral handle and examine the diastereoselective cyclohexyl or phenyl Grignard addition process to the appropriate keto-moiety for generation of (*S*)-acid **37**. As shown in Scheme 18, Sepracor has generated the optically pure (*S*)-acid **37** via either diastereoselective phenyl or cyclohexyl Grignard addition to appropriate ketoester or ketoamide, followed by removal of aminoindanol unit.[36]

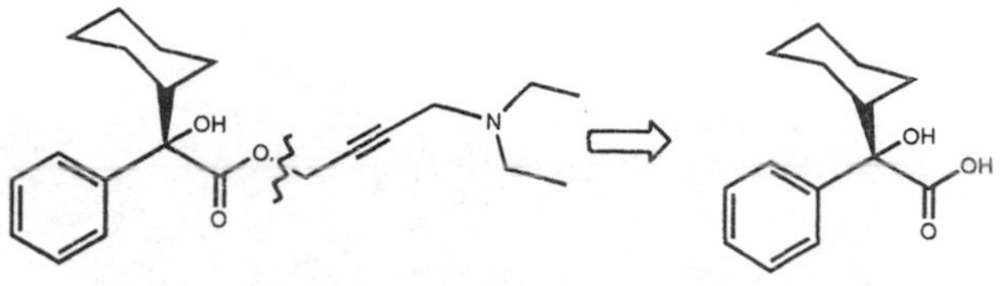

(*S*)-Oxybutynin          (*S*)-Acid **37**

**Figure 6.**

---

Ditropan is a registered trademark of Hoechst Marion Roussel.

## Scheme 18.

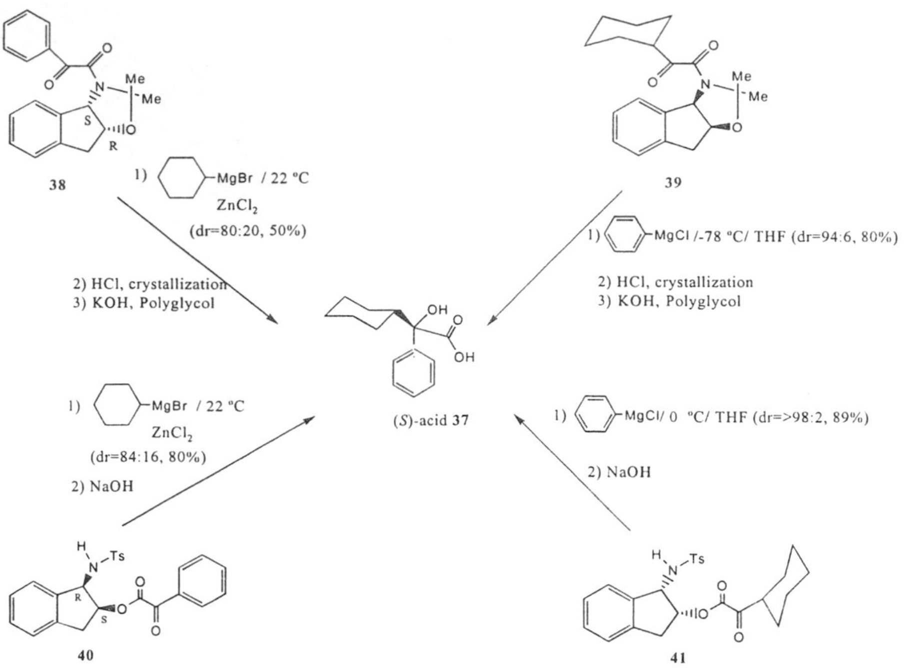

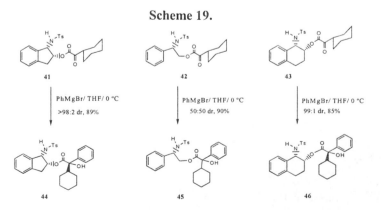

The following requirements were important to obtain high diastereoselectivities: (a) either C-1 amine derived ketoamide **39** or C-2 ketoester **41** with the phenyl Grignard addition process that required magnesium coordination; and (b) either phenyl derivative **38** or **40** were essential for a zinc coordinated transition state in the cyclohexyl Grignard addition process. Interestingly, *N*-tosyl-derived ketoesters provide the most expedient avenue to the preparation of optically pure (*S*)-acid **37**. Encouraged by the outstanding stereocontrolled outcome of the indane derived tosylate ketoester **41**, the following conformationally related ketoesters **42** and **43** were examined. As shown in Scheme 19, phenyl glycinol derivative **42** with PhMgBr/THF at 0 °C provided no selectivity. On the other hand, homologous six-membered analogue **43** gave comparable results (99:1, dr) to the indane system. However, due to the ready availability of *cis*-aminoindanol over tetralin aminoalcohol, Sepracor's current process utilizes the *N*-tosyl *cis*-aminoindanol protocol in the production of enantiopure (*S*)-acid **37** for (*S*)-oxybutynin.[36]

## Scheme 19.

## Application of *Cis*-Aminoindanol Oxazaborolidines for (*R,R*)-Formoterol Synthesis

One of the most economical ways of generating single enantiomers is by utilizing asymmetric catalysis. Recently, more emphasis has been devoted toward the design and development of cost-effective chiral catalysts that display a high degree of enantioselectivity and reactivity in the production of enantiopure materials.[37] Oxazaborolidines are an extremely important class of catalytic enantioselective reducing agents, which have been utilized in the reduction of several functional groups (for example, keto and imine), to produce high levels of enantiomeric excesses. Since the extraordinary discoveries by Itsuno[38] and Corey,[39] an enormous amount of chiral β-aminoalcohols have served as backbones of oxazaborolidines.[40] Most of the enantiopure β-aminoalcohols have been synthesized from natural sources (example, amino acids); however, extensive synthetic manipulations were required for the synthesis of unnatural antipodes. Therefore, development of highly selective oxazaborolide catalysts from readily accessible β-aminoalcohols from practical technologies, such as asymmetric epoxidations (AE)[5] and simple Ritter-type chemistry,[4a,b,41] will provide practical catalytic reducing agents for the reduction of carbonyl groups.[42] The Sepracor group has shown that optically pure *cis*-1-amino-2-indanol is an extremely effective enantioselective catalyst for the borane reduction of several important α-haloketones.[43] They reported that the B-Me catalyst **48** provided higher induction over B-H catalyst **47**. For example, enantioselective reduction of α-chloroacetophenone at -20 °C, catalyst **48** gave 96% ee, while catalyst **47** gave 91.7 % ee at 23 °C.

Currently, the Sepracor group is utilizing the indane-catalyst in the synthesis of optically pure formoterol.[44] Formoterol is a long acting and extremely potent β₂-adrenergic receptor agonist, with a fast onset of action. It is used as a bronchodilator in the therapy of asthma and chronic bronchitis.[45] The (*R,R*)-enantiomer of formoterol has been shown to be 1000 times more potent than the (*S,S*)-enantiomer.[46] During the development of an asymmetric synthesis of (*R,R*)-formoterol, a reproducible and practical reducing agent amenable for large-scale preparation of chiral bromohydrin **50** from bromo-ketone **49** was required (Scheme 20).

**Scheme 20.**

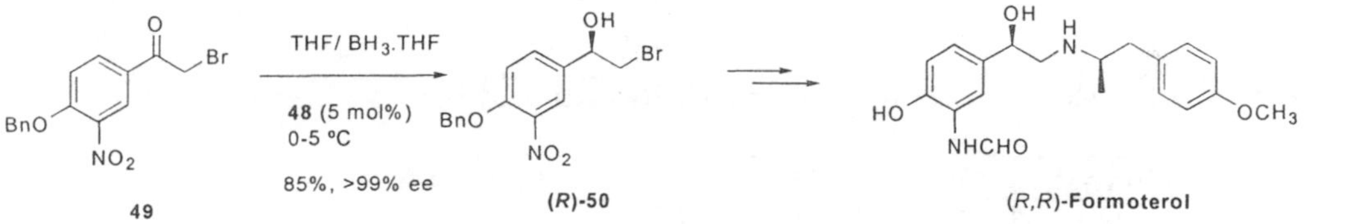

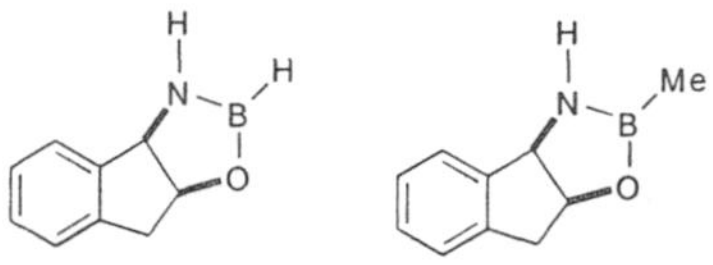

The earlier stage of development for chiral reduction of **49**, 20 mol% B-methyloxazaborolidine **48** was used as the catalyst and BH₃•THF (0.7 eq ) as the reducing agent.[46] This particular catalyst had been prepared by reacting (1*R*)-amino-(2*S*)-indanol with trimethylboroxine, followed by an azeotropic distillation with toluene. The high cost of the trimethylboroxine and the additional handling prompted the study of the reduction with *in situ* generated B-H oxazaborolidine **47**. Selected results of this study are summarized in Table 1.[47]

While the highest selectivities are achieved by using catalyst **48** at -10 °C, the ee's were lower (i.e. 93% vs. 96%) when using *in situ* prepared **47**. As illustrated in Table 1, the boron source did not have a profound effect on the enantioselectivity of catalyst **47** (Table 1, entries 1,2). On the other hand, the B-H catalyst (**47**) gave higher selectivities with $BH_3 \bullet$ THF than with $BH_3 \bullet Me_2S$ (Table 1, entries 7,9). The optimum temperature for B-Me catalyst **48** was -10 °C in the presence of either boron source. However, the optimal temperature for the B-H catalyst was dependent on the boron source. When $BH_3 \bullet Me_2S$ was used, the optimal temperature was 25 °C, for $BH_3 \bullet$ THF it was 0 °C. In addition, the rate of the ketone addition of the B-H catalyst system did not have a severe effect on the outcome of the enantioselectivity. In this study, it was clear that each catalyst has its own optimal condition with respect to temperature, boron source, and additives.

After optimization of the asymmetric reduction process, catalyst **47** was chosen because it was easier to handle, the preparation was less time consuming and no expensive reagents were involved. More importantly, bromohydrin **50** was isolated directly from the reaction by crystallization, and thus, enriched in its enantiopurity. This reduction process has been demonstrated on a multi-kilogram scale to afford (*R*)-bromohydrin **50** in 85% yield with an optical purity of >99% ee. The compound (*R*)-**50** has been utilized as the key building block for enantiopure synthesis of (*R*,*R*)-formoterol.[47, 44b]

## *Cis*-1-Amino-2-Indanol as a Chiral Resolving Agent for Production of (*R*)-Ketoprofen

During the past few decades, chemists have spent an enormous amount of time in the development of efficient and practical processes for chiral α-arylpropanoic acids, which are used as anti-inflammatory medicaments. One of the key approaches to preparing optically pure forms of the α-arylpropanoic acid isomers is by resolution of a racemate. A number of chiral amines are known for their resolution of chiral acids on a commercial scale. Notable examples include brucine, quinidine, cinchonidine, morphine, ephedrine and (1-naphthyl) ethylamine. Some of these chiral amines are expensive and are often difficult to recover. For example, ketoprofen has been resolved using (-)cinchonidine; however, this method has several practical limitations.[48]

**Table 1: Enantioselectivities of the Asymmetric Reduction of Bromo-Ketone 49**

| Entry | Catalyst | Mol | Boron Source | Temperature | ee (*R*) |
|-------|----------|-----|--------------|-------------|----------|
| 1 | 48 | 10 % | $BH_3 \bullet$ THF | -10 °C | 95 % |
| 2 | 48 | 10 % | BMS | -10 °C | 96 % |
| 3 | 48 | 10 % | BMS | 0 °C | 90 % |
| 4 | 48 | 10 % | BMS | 25 °C | 90 % |
| 5 | 48 | 10 % | BMS | -10 °C | 32 % |
| 6 | 48 | 10 % | BMS | 0 °C | 82 % |
| 7 | 47 | 10 % | BMS | 25 °C | 90 % |
| 8 | 47 | 10 % | $BH_3 \bullet$ THF | -10 °C | 87 % |
| 9 | 47 | 10 % | $BH_3 \bullet$ THF | 0 °C | 93 % |
| 10 | 47 | 10 % | $BH_3 \bullet$ THF | 25 °C | 89 % |
| 11 | 47 | 5 % | $BH_3 \bullet$ THF | 0 °C | 93 % |
| 12 | 47 | 1 % | $BH_3 \bullet$ THF | 0 °C | 91 % |

All reactions were run at a concentration of 0.3 M in THF. The ketone was added to the mixture of the catalyst and borane over a 2 h period. The reaction yields are >98%.

Recently, the Sepracor group has demonstrated that enantiopure *cis*-1-amino-2-indanol is highly effective in the diastereomeric resolution of racemic ketoprofen.[49] It was discovered that water plays a crucial role in the resolution process and catalytic amounts of water (<3.8 wt%) are required in acetonitrile to obtain high yields of the preferred diastereomer. In addition, it was demonstrated that the unwanted isomer can be recycled and enantiopure *cis*-1-amino-2-indanol can be easily recovered. The *cis*-1-amino-2-indanol mediated resolution process is extremely productive and cost-effective, and is Sepracor's current manufacturing process for either enantiomer of ketoprofen (Scheme 21). [49]

**Scheme 21.**

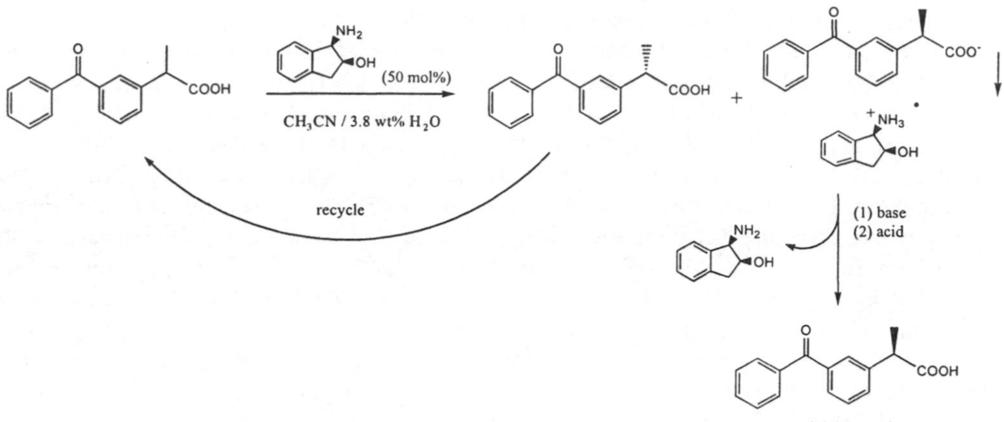

## References

This chapter is based in part on a comprehensive review of "Applications of *cis*-1-amino-2-indanol in Asymmetric Synthesis" by Senanayake, C. H. *Aldrichimica Acta*, **1998**, *31*, 3 (reference 2e) and in part on a comprehensive review by Jacobsen, E. N. in *Comprehensive Organometallic Chemistry II*, Vol. 12, Wilkinson, G.; Stone, F. G. A.; Abel, E. W.; Hegedus, L. S., Eds.; Pergamon: New York; 1995, pp. 1097-1135.

1.  (a) Noyori, R. *Asymmetric Catalysis in Organic Synthesis*; John Wiley and Sons: New York, 1994. (b) Ojima, I. *Catalytic Asymmetric Synthesis*; VCR Press: Berlin, 1993.
2.  (a) Ager, D. J.; Prakash, I.; Schaad, D. R. *Chem. Rev.* **1996**, *96*, 835. (b) Ager, D. J.; East, M. B. *Asymmetric Synthetic Methodology*; CRC Press: Boca Raton, 1995. (c) Deloux, L.; Srebnik, M. *Chem Rev.* **1993**, *93*, 763. (d) Soai, K.; Niwa, S. *Chem. Rev.* **1992**, *92*, 833. (e) Senanayake, C. H. *Aldrichimica Acta* **1998**, *31*, 3. (f) Ghosh, A. K.; Fidanze, S; Senanayake, C. H. *Synthesis* **1998**, 937.
3.  (a) Lutz, R. E; Wayland, R. L. *J. Amer. Chem. Soc.* **1951**, *73*, 1639. (b) Hassner, A.; Lorber, M. E.; Heathcock, C. *J. Org. Chem.* **1967**, *32*, 540. (c) Ghosh, A. K.; Mckee, P. S.; Sanders, W. M. *Tetrahedron Lett.* **1991**, *32*, 711. (d) Thompson, W. J.; Fitzgerald, M. K.; Holloway, E. A.; Emini, P. L.; Daker, B. M.; McKeever, W. A.; Huff, R. J. *J. Med. Chem.* **1992**, *35*, 1685. (e) Takahashi, M.; Koike, R.; Ogasawara, K. *Chem. Pharm. Bull.* **1995**, *43*, 1585. (f) Takahashi, M.; Ogasawara, K. *Synthesis* **1996**, 954. (g) Boyd, R. D.; Sharma, N. D.; Bowers, N. L.; Goodrich, P. A.; Groocock, M. R.; Blacker, A. J.; Clarke, D. A.; Howard, T.; Dalton.

H. *Tetrahedron Asymmetry* **1996**, *7*, 1559. (h) Lakshman, M. K.; Zajc, B. *Tetrahedron Lett.* **1996**, *37*, 2529. (i) Ghosh, A. K.; Kincaid, J. F.; Haske, M. G. *Synthesis* **1997**, 541.

4.   (a) Senanayake, C. H.; Roberts, F. E.; DiMichele, L. M.; Ryan, K. M.; Liu, J.; Fredenburgh, L. E.; Foster, B. S.; Douglas, A. W.; Larsen, R. D.; Verhoeven, T. R.; Reider, P. J. *Tetrahedron Lett.* **1995**, *36*, 3993. (b) Senanayake, C. H.; DiMichele, L. M.; Liu, J.; Fredenburgh, L. E.; Ryan, K. M.; Roberts, F. E.; Larsen, R. D.; Verhoeven, T. R.; Reider, P. J. *Tetrahedron Lett.* **1995**, *36*, 7615. (c) Senanayake, C. H.; Smith, G. B.; Fredenburgh, L. E.; Liu, J.; Ryan, K. M.; Roberts, F. E.; Larsen, R. D.; Verhoeven, T. R.; Reider, P. J. *Tetrahedron Lett.* **1996**, *37*, 3271. (d) Gao, Y.; Hong, Y.; Nie, X.; Bakale, R. P.; Feinberg, R. R.; Zepp, C. M. US Patent #5,599,985, **1997**. (e) Gao, Y.; Hong, Y.; Nie, X.; Bakale, R. P.; Feinberg, R. R.; Zepp, C. M. US Patent #5,616,808, **1997**.

5.   (a) Jacobsen, E. N.; Zhang, W.; Muci, A. R.; Ecker, J. R.; Deng, L. *J Am. Chem. Soc.* **1991**, *113*, 7063. (b) Halterman, R. L.; Jan, S. *J. Org. Chem.* **1991**, *56*, 5254. (c) Jeffrey, A. M.; Yeh, H. J. C.; Jerina, D. M.; Patel, R. T.; Davey, J. F.; Gibson, D. T. *Biochem.* **1975**, *14*, 575. (d) Zeffer, H.; Imuta, M. *J. Org. Chem.* **1978**, *43*, 4540. (e) Boyd, R. D.; Sharma, D. N.; Smith, E. A. *J. Chem. Soc. Perkin Trans. I.* **1982**, 2767. (f) Allain, E. J.; Hager, L. P.; Deng, L.; Jacobsen, E. N. *J. Am. Chem. Soc.* **1993**, *115*, 4415. (g) >99% ee of indene oxide prepared via enzymatic resolution method see: Zhang, J.; Reddy, J.; Roberge, C.; Senanayake, C. H.; Greasham, R.; Chartrain, M. *J. Ferment. Bioeng.* **1995**, *80*, 244. (h) Zeffer, H.; Imuta, M. *J. Am. Chem. Soc.* **1979**, *101*, 3990. (i) Hirama, M.; Oishi, T.; Ito, S. *J. Chem. Soc., Chem. Commun.* **1989**, 665. (j) Hanessian, S.; Meffre, P.; Girard, M.; Beaudoin, S.; Sanceau, J.; Bennani, Y. *J. Org. Chem.* **1993**, *58*, 1991 and references therein. (k) Indene has been converted to *cis*-indene diol >99% ee with dioxigenas, Greasham, R. et al. unpublished results. (l) Allen, C. R. C.; Boyd, D. R.; Dalton, H.; Sharma, N. D.; Brannigan, I.; Kerley, A. N.; Sheldrake, G. N.; Taylor, S. C. *J. Chem. Soc., Chem. Commun.* **1995**, 117 references therein. (m) Sasaki, H.; Katsuki, T.; Irie, R.; Hamada, T.; Suzuki, K. *Tetrahedron* **1994**, *50*, 11827.

6.   Jacobsen, E. N. *Catalytic Asymmetric Synthesis*; Ojima, I., Ed., VCH: New York, 1993; Chap 4.2, pp 159-202.

7.   Johnson, R. A.; Sharpless, K. B. *Catalytic Asymmetric Synthesis*; Ojima, I., Ed., VCH: New York, 1993; Chap 4.1, pp 103-158.

8.   Myers R. S.; Groves, J. T. *J. Am. Chem. Soc.* **1983**, *105*, 5791.

9.   (a) Srinivasan, K.; Michaud, P.; Kochi, J. K.; *J. Am. Chem. Soc.* **1986**, *108*, 2309. (b) Samsel, E. G.; Srinivasan, K.; Kochi, J. K. *J. Am. Chem. Soc.* **1985**, *107*, 7606.

10.  Early examples of the preparation and use of chiral salen complexes: (a) Cesarotti, E.; Pasini, A.; Ugo, R. *J. Chem. Soc., Dalton Trans.* **1981**, 2147, and references therein. (b) Nakajima, K.; Kojima, M.; Fujita, J. *Chem. Lett.* **1986**, 1483.

11.  (a) Zhang, W.; Loebach, J. L.; Wilson, S. R.; Jacobsen, E. N. *J. Am. Chem. Soc.*, **1990**, *112*, 2801. (b) Jacobsen, E. N.; Zhang, W.; Muci, A. R.; Ecker, J. R.; Deng, L. *J. Am. Chem. Soc.* **1991**, *113*, 7063. (c) Larrow, J. F.; Jacobsen, E. N.; Gao, Y.; Hong, Y.; Nie, X.; Zepp, C. M. *J. Org. Chem.* **1994**, *59*, 1939.

12.  Jacobsen, E. N. unpublished results.

13.  Zhang, W.; Jacobsen, E. N. *J. Org. Chem.* **1991**, *56*, 2296.

14.  Sasaki, H.; Irie, R.; Katsuki, T. *Synlett* **1993**, 300.

15.  Chang, S. B.; Heid, R. M.; Jacobsen, E. N. *Tetrahedron Lett.* **1994**, *35*, 669.

16.  Analyses of the factors influencing stereocontrol within approaches a and b,b' in Figure 2 are provided in reference 22 and in: Hosoya, N.; Hatayama, A.; Yanai, K.; Fujii, H.; Irie R.; Katsuki, T. *Synlett* **1993**, 641.

17. (a) Jacobsen, E. N.; Zhang W.; Güler, M. L. *J. Am. Chem. Soc.* **1991**, *113*, 6703. (b) Palucki, M.; Finney, N. S.; Pospisil, P. J.; Güler, M. L.; Ishida, T.; Jacobsen, E. N. *J. Am. Chem. Soc.* **1998**, *120*, 948.

18. Irie, R.; Noda, K.; Ito, Y.; Matsumoto, N.; Katsuki, T. *Tetrahedron: Asymmetry* **1991**, *2*, 481.

19. Hosoya, N.; Irie, R.; Ito Y.; Katsuki, T. *Synlett* **1991**, 691.

20. Palucki, M.; Pospisil, P. J.; Zhang, W.; Jacobsen, E. N. *J. Am. Chem. Soc.* **1994**, *116*, 9333.

21. (a) Irie, R.; Ito, Y.; Katsuki, T. *Synlett*, **1991**, 265. (b) Deng, L.; Jacobsen, E. N. *J. Org. Chem.* **1992**, *57*, 4320.

22. Deng, L.; Furukawa, Y.; Martínez, L. E.; Jacobsen, E. N. *Tetrahedron* **1994**, *50*, 4323.

23. Brandes, B. D.; Jacobsen, E. N. *J. Org. Chem.* **1994**, *59*, 4378.

24. (a) Available from Aldrich Chemical Co., Inc. (b) Multikilogram quantities are available from ChiRex, Dudley, United Kingdom.

25. Jacobsen, E. N.; Larrow, F. J.; *J. Amer. Chem. Soc.* **1994**, *116*, 12129, and references cited therein.

26. Hughes, D. L.; Smith, G. B.; Liu, J.; Dezeny, C. G.; Senanayake, C. H.; Larsen, R. D.; Verhoeven, T. R.; Reider, P. J. *J. Org. Chem.* **1997**, *62*, 2222.

27. Ritter, J. J; Minieri, P. P. *J. Amer. Chem. Soc.* **1948**, *70*, 4045. (b) Review of Ritter reaction see: Bishop, R. *Comprehensive Org. Syn.* **1991**, *6*, 261.

28. Industrial Asymmetric Synthesis of single enantiomer Crixivan see: Reider, J. P. *Chimia*, **1997**, *51*, 306. Reference therein.

29. Stein, D. S.; Flish, D. G.; Bilello, J. A.; Prestion, S. L.; Martineau, G. L.; Drusano, G. L. "A 24-week Open-Label Phase I/II Evaluation of the HIV Protease Inhibitor MK-639" *AIDS* **1996**, *10*, 485. (b) MacDougall, D. S. Indinavir: Lightening the Load. *J. Int. Assoc. Physicians AIDS Care* **1996**, *2*, 6.

30. Lye, T. A.; Wiscount, C. M.; Guare, J. P.; Thompson, W. J; Anderson, P. S.; Darke, P. L.; Zugay, J. A.; Emini, E. A.; Schleif, W. A.; Quintero, J. C.; Dixon, R. A.; Sigal, L. S.; Huff, J. R. *J. Med. Chem.* **1991**, *34*, 1228.

31. Vacca, J. P.; Dorsey, B. D.; Schleif, W. A.; Levin, R. B.; McDaniel, S. L.; Darke, P. L.; Zugary, J. A.; Quintero, J. C.; Blaby, O. M.; Roth, E.; Sardana, V. V.; Schlabach, A. J.; Graham, P. I.; Condra, J. H.; Gotlib, L.; Holloway, M. K.; Lin, J. H.; Chen, I.; Vastag, K.; Ostovic, D.; Anderson, P. S.; Emini, E. A.; Huff, J. R. *Proc. Natl. Acad. Sci. USA* **1994**, *91*, 4096.

32. Askin, D.; Wallace, J. P.; Vacca, J. P.; Reamer, R. A.; Volante, R. P.; Shinkai, I. *J. Org. Chem.* **1992**, *57*, 2771.

33. (a) Maligres, P. E.; Rossen, K. U.; Cianciosi, S. J.; Purick, R. M.; Eng, K. K.; Reamer, R. A.; Askin, D.; Volante, R. P.; Reider, P. J. *Tetrahedron Lett.* **1995**, *36*, 2195. (b) Maligres, P. E.; Weissman, S. A.; Upadhyay, S. J.; Cianciosi, S. J.; Reamer, R. A.; Purick, R. M.; Sager, J.; Rossen, K. U.; Eng, K. K.; Askin, D.; Volante, R. P.; Reider, P. J. *Tetrahedron* **1996**, *52*, 3327.

34. Yarker, Y. E.; Goa, K. L.; Fitton, A. *Drug Aging* **1995**, *6*, 243.

35. (a) Carter, P. J.; Blob, L.; Audia, V. A.; Dupont, A. C.; McPherson, D. W.; Natalie Jr, K. J.; Rzeszotarski, W. J.; Spagnuolo, C. J.; Waid, P. P.; Kaiser, C. *J. Med Chem.* **1991**, *34*, 3065. (b) Tambute, A.; Collet, A. *Bull, De La Chimi. Fr.* **1984**, *1-2*, II-77. (c) Kiesewetter, D. O. *Tetrahedron: Asymmetry* **1993**, *4*, 2183. (d) McPherson, D. W.; Knapp, F. F. *J. Org. Chem.* **1996**, *61*, 8335. (e) Bugno, C.; Colombani, S. M.; Dapporto, P.; Garelli, G.; Giorgi, P.; Subissi, A.; Turbanti, L. *Chirality* **1997**, 721. (f) Atkinson, E. R.; McRitchi, D. D.; Schoer, L. F. *J. Med. Chem.* **1997**, *20*, 1612.

36. Senanayake, C. H.; Fang Q. K.; Grover, P.; Bakale, R. P.; Vandenbossche C. P.; Wald, S. A. *Tetrahedron Letters* **1999**, 0000.

37. Industrial applications, for example: (a) the BINAP-Rh complex-catalyzed enantioselective isomerization of diethylgeranylamine in the production of (-)-menthol: Akutagawa, S. Practical asymmetric synthesis of (-)-Menthol and Related Terpenoids. In *Organic Synthesis in Japan: Past Present, and Future*; Noyori, R.; Hiraoka, T.; Mori, K.; Murahashi, S.; Onada, T.; Suzuki, K.; Yonemistsu, O., Eds.; Tokyo Kagaku Dozin: Tokyo, **1992;** p 75. (b) Jacobsen's (salen)Mn catalyst for epoxidation of indene in the synthesis of HIV protease inhibitor Crixivan, see: references 4c and 28.

38. Itsuno, S.; Sakurai, Y.; Ito, A.; Hirao, A; Nakahama, S. *Bull. Chem. Soc. Jpn.* **1987**, *60*, 395.

39. Corey, E. J.; Bakshi, R. K.; Shibata, S. *J. Am. Chem. Soc.* **1988**, *110*, 1968.

40. (a) Wallbaum, S.; Martens, J. *Tetrahedron Asymmetry* **1992**, *3*, 1475; (b) Deloux, L.; Srebnik, M. *Chem Rev.* **1993**, *93*, 763.

41. Senanayake, C. H.; DiMichele, L. M.; Liu, J.; Larsen, R. D.; Verhoeven, T. R.; Reider, P. J. *Tetrahedron Asymmetry* **1996**, *7*, 1501, and references cited therein.

42. Stoichiometric amounts of aminoalcohols see: Didire, E.; Loubinoux, B.; Tombo, G, M.; Rihs, G. *Tetrahedron* **1991**, *35,* 6631.

43. (a) Hong, Y.; Gao, Y.; Nie, X.; Zepp, C. M. *Tetrahedron Lett.* **1994**, *35,* 6631. (b) Gao, Y.; Hong, Y.; Zepp, C. M. US Patent #5,495,054, **1996**.

44. (a) Hett, R.; Fang, Q. K.; Gao, Y.; Hong, Y.; Butler, H. T.; Nie, X.; Wald, S. A. *Tetrahedron Lett.* **1997**, *38*, 1125. (b) Hett, R.; Fang, Q. K.; Gao, Y.; X.; Wald, S. A.; Senanayake, C. H. *Organic Process Research and Development* **1998**, *2*, 96.

45. Nelson, H. S. *N. Engl. J. Med.* **1995**, *333*, 499.

46. Trofast, J.; Österberg, K.; Källström, B. L.; Waldeck, B. *Chirality* **1991**, *3*, 443.

47. Hett, R.; Senanayake, C. H.; Wald, S. A. *Tetrahedron Lett.* **1998**, *39*, 1705.

48. Manimaran, T.; Potter, A. A. US Patent #5,162,576, **1992**.

49. Van Elkeren, P.; McConville, F. X.; Lopez, J. L. US Patent #005677469A, **1997**.

# DRUG SUBSTANCE FINAL FORM

# AND

# PROCESS SAFETY

# Chemical Development of the Drug Substance Solid Form

**Michael D. Thompson**
*Process Chemistry, Rhône-Poulenc Rorer, Collegeville, Pennsylvania 19426*

**Jean-René Authelin**
*Process Chemistry, Rhône-Poulenc Rorer, 94403 Vitry Sur Seine Cedex, France*

Key words: polymorphism, solid state, crystallization, recrystallization, RG 12525

RG 12525 is an active leukotriene antagonist which belongs to a series of (2-quinolinylmethoxy)phenoxy tetrazoles.[1,2] Conveniently prepared using a four step linear synthesis, final material is obtained during a tetrazole forming step.[3] Initial studies showed that the target molecule would crystallize in two different polymorphic forms.[4] After determining which polymorphic form was desired for further development, a process was defined to prepare that form (form II). After successfully manufacturing in one pilot plant, the process was transferred to a second pilot plant. Unfortunately on larger scale, form II was no longer obtained as a pure polymorph. This unexpected result was difficult to understand from the available data. Therefore, a more thorough and somewhat exhaustive study into solid state properties was required.

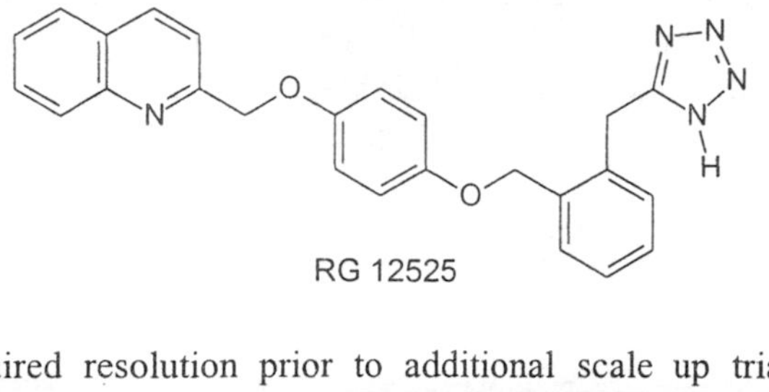

RG 12525

Several issues required resolution prior to additional scale up trials. Was truly enough information available to know which form was better for future development? Perhaps a mistake had been made or additional information was required to make a more informed decision. Were all of the crystallization parameters understood? Were the proper controls in place to ensure that the desired form could be reproducibly prepared? These questions were addressed systematically to find a suitable resolution to the problem. Listed below are issues important to the study and development of processes involving the preparation of solid forms.

## General Principles of Polymorphism

### Terms

Many solid state terms are used loosely by process chemists, so before any meaningful discussion of polymorphism or crystallization can occur, a definition of solid state terms related to process chemistry is required. Critically important is the difference between solids existing in a crystalline state and those in an amorphous state. A *crystalline* compound is a solid that exists in a defined crystal lattice. Evidence of crystallinity can be observed from a defined x-ray diffraction pattern or by microscopic birefringence properties. *Amorphous*[5] or non-crystalline solids do not have regularly shaped crystal lattices and molecules are spaced in undefined or random arrangements. Occasionally amorphous materials result from rapid solvent removal or lyophilization. The solvent strip is sometimes so rapid that the solid cannot reorganize itself into a favorable crystal arrangement. Amorphous solids often exhibit a glass transition as observed by differential scanning calorimetry (DSC). Also, their x-ray diffraction pattern is less defined and typically quite diffuse compared to that of a crystalline form.

For those solids that are crystalline, different solid phases can exist in polymorphic and hydrate (solvate) forms. A *polymorph*[6] is a crystalline solid phase of a compound that can exist in at least two such states yet has the same chemical composition for each phase. These solid phases differ in their crystal packing within the crystal lattice. Carbon can exist in more than one crystal packing, hence more than one polymorphic form. Two easy to recognize polymorphic forms of carbon are diamond and graphite. A *hydrate*[5] exists when water is a necessary part of the crystal lattice. A variation of a hydrate, known as a *solvate,* occurs when a specific solvent is present as a part of the crystal lattice. Hydrates (solvates) can be either stoichiometric or non-stoichiometric. Non-stoichiometric solvates are known as *clathrates.*[5] Levels of hydration can vary significantly and some compounds may have multiple hydrate forms with different levels of water present in the crystal. Often the term *pseudopolymorphism* is used to distinguish hydrates (or solvates) from polymorphs.

Crystals can exhibit themselves in different habits based on a variety of factors associated with crystal growth. A *crystal habit*[5] is the outward shape of a crystal as observed at a microscopic level in the form of plates, rods, needles, blades, irregular shapes, etc. Habits are distinguished from polymorphs in that the x-ray pattern for each is identical because the crystal packing is the same. Different crystal habits for a given compound can result using different crystallization processes.

*Crystallization* is the process by which a chemical is converted from a liquid solution into the solid crystalline state. This principle will be discussed in a later section.

### Solid State Screening

The search for polymorphism has not always been systematic and frequently polymorphic discovery has been serendipitous. Although for some time it has been recognized that polymorphic choice is important to drug testing and development, approaches to solid form screening have not always been consistent.

Early in the life of a project, methodical screening of solid forms should occur. This screening can be elaborate, and one is never certain that all polymorphic forms have evidenced themselves. The best way to ensure that screening reveals the maximum number of forms, especially those stable at room temperature, is to make sure that screening methods are diverse. Knowledge of the different possible solid forms for a given substance will assist the process chemist in focusing on the preparation of a specific and desired solid form.

Examples of polymorphic screening are presented in the literature.[6-9] One method is to screen by optical microscopy.[6-8] Byrn suggests a systematic approach based on a variety of testing.[9] The primary focus of these screening techniques should be to find the stable form at room temperature. The following considerations and methods are useful in any search for new polymorphs.

Purity

Since impurities can influence the formation of polymorphs, it is important to work with pure material. For initial screening experiments, this is not always possible, and repeat screening with pure material may be necessary. When purer compounds are used in the screening process, the researcher will be more confident in the results.

Aged Suspensions

The use of aging suspensions is quite useful in searching for the stable solid form. This can also be helpful in crystallizing an amorphous solid. The technique involves forming a suspension of available solid in a variety of pure solvents or solvents mixed with stoichiometric amounts of water then allowing them to age under storage conditions for up to 30 days or even longer. After sufficient time, the stable solid form should appear assuming that a solvate is not formed. Forcing conditions can be employed including the use of ultrasound, heating/cooling cycles (not allowing the solid to completely dissolve), and the use of seed from a related chemical. If a new form is discovered, the aged suspension technique should be repeated using the new form.

Thermal Methods

The use of thermal methods such as hot stage microscopy or DSC can reveal different solid forms.[8] A typical experiment would involve heating a solid until fusion (verifying that there is no notable decomposition), then cooling at different rates (slow/medium/fast) to determine if a solid recrystallizes. If a new solid is formed, it should be reheated but not to a complete melt. Once cooled, a solid's analysis method should be used such as x-ray diffraction, to determine if the new solid form is indeed different. Fast cooling tends to make metastable forms, and slow cooling tends to form thermodynamically stable forms. Other experimental methods include heating to just below the fusion point and holding to determine if a solid state change occurs, or subliming followed by examination to conclude if the sublimed solid is any different.[6]

Solvent Mediated Crystallization

One of the most common forms of polymorphic screening includes solvent mediated crystallization. A variety of solvents of different polarity should be employed for crystallization or recrystallization experiments. Parameters varied should include temperature, cooling rate, concentration, seed, water content, solvent composition (from pure to mixed solvents), stirring, pH and ionic composition. New solids generated from thermal methods or other techniques can be used as seed crystals for solvent mediated crystallization.

## Variation of Humidity

Hydrates can be observed through solvent mediated crystallization experiments (usually by adding water as a co-solvent) but are also discovered through a variety of humidity experiments. Each solid form that is discovered (anhydrous or hydrated) should be placed in a humidity controlled microbalance (moisture balance). At varying humidity a determination of whether water is sorbed or desorbed is made and observed through the generation of humidity isotherms.[5] If water is sorbed, one must determine if that water is incorporated into the crystal lattice (absorbed) or is surface water (adsorbed). If a hydrate is observed, it is important to find out if intermediate levels of hydration are also observed through these experiments.[5] When more than one hydrated form is observed, the system can become quite complex, and must be studied in detail.

Hydrates (solvates) may exist in one or more hydration (solvation) stoichiometry, or they may be non-stoichiometric. In stoichiometric hydrates, water molecules are a necessary part of the crystal lattice. Drying of these materials is accompanied by a change in crystalline form. In non-stoichiometric hydrates, water molecules are present in the crystal lattice, but when eliminated through drying, the original crystal lattice remains intact. As a result, certain sites in the crystal lattice will be found "empty" and the stoichiometry of the hydrate is incomplete.

## RG 12525

In the case of RG 12525 two forms, form I and form II, were discovered through solvent mediated crystallization screening. Aged suspensions and thermal methods were also performed confirming the presence of two forms. Humidity experiments did not reveal hydrate forms. Later, specific recrystallization processes to generate each solid form will be discussed.

## Solids Analysis

Concurrent with screening to find new solid forms is the need to find an appropriate analytical method to distinguish between solid forms. A variety of analytical methods are used to identify solid forms. Their use in a particular laboratory may be dependent on available instrumentation. Reviews[8,10,11] have been published on analytical methods for solids which include the use of the following common techniques: spectroscopy (infrared; solid-state carbon nuclear magnetic resonance), x-ray crystallography (x-ray powder diffraction pattern; single crystal x-ray), thermal methods (DSC; hot stage microscopy; solution calorimetry), visual techniques (optical microscopy, electron microscopy, visual), solubility (solubility, solution calorimetry[12]), and bulk properties (bulk density, flow properties, solubility, moisture sorption).

Using one or more of the above techniques, different solid forms will exhibit unique differences that can be used for qualitative assessment. However, mixtures of forms often exist, and to understand polymorphic purity, one of these methods can usually be developed into a quantitative measurement.

In addition to knowledge about the solid form, one should also know key issues associated with particle size. Particle size and particle characteristics might also be important for drug delivery. Techniques used to assess particle size or particle characteristics should be available and may include size distribution and surface area determination.

Hydrates (Solvates)

For hydrates, the level of water can conveniently be determined by Karl Fischer titration. For solvates, solvent levels can be measured by NMR or gas chromatography analysis.

RG 12525

For RG 12525, several methods were found useful in distinguishing each polymorphic form, including optical microscopy, DSC, infrared, powder x-ray diffraction, and solubility. The technique that proved useful for quantitating polymorphic mixtures was DSC. [13]

## Physical Chemical Considerations

### Thermodynamics

Once known, each polymorphic form is made available for a comparative characterization study. Typically the characterization of polymorphs involves a study of the relative thermodynamic system involving two or more forms. The result of this study is usually represented by an energy (Gibbs free energy versus temperature), phase (pressure versus temperature), or solubility (equilibrium concentration versus temperature) diagram describing the polymorphic system.[6]

Polymorphs differ from one another in their crystal lattices and hence their lattice energies. If two forms exist at room temperature, one form has a higher Gibbs free energy (G), and would be classified as the least stable form, or metastable, at that temperature. The form with the lowest G at room temperature would be referred to as the more stable form.

Physical stability of each form is governed in part by thermodynamic principles. At a given temperature and pressure, only one polymorphic form is physically stable; any other forms are metastable, and a scale of greater or lesser stability can be established for these forms. If more than one form exists at the storage temperature, there will be a tendency for a less stable form (higher G) to revert to the more stable form (lower G). No matter what the medium (except for hydrates or solvates), metastable polymorphic forms tend to convert to a more stable form. However, this interconversion is only triggered after the appearance of the first nuclei of the stable polymorph.

With an energy, phase or solubility diagram, one can determine whether the different solid forms are monotropic or enantiotropic with respect to each other. In a monotropic system, only one polymorphic form is more stable at all temperatures studied. For a binary enantiotropic system, one form will be more stable below a transition temperature (the temperature at which the stability of the two forms is reversed), and one form will be stable above that temperature. The transition temperature should be available from the energy or phase diagram. Often, but not always, this temperature lies between room temperature and the melting point. The ideal way of finding that temperature is to develop a complete energy or phase diagram of the two forms.

If an energy or phase diagram is not readily available, analysis of the DSC and infrared data according to Burger's rules[14] will help determine whether the system is monotropic or enantiotropic. The first part of the Heat of Transition rule states that if a solid-state endothermic transition between two forms is observed, the system is enantiotropic (this temperature is not the transition temperature). The second part of the Heat of Transition rule states that if an exothermic transition is observed, the system is related monotropically. Should no transition be observed, Burger's Heat of Fusion rule may be more appropriate. It states that if the higher melting form has the lowest heat of fusion, the system is enantiotropic, otherwise it is monotropic. For a monotropic system, the higher melting is the more stable form.

Pure forms of each polymorph must be used in order for Burger's rules to apply. In practice, a compound's purity can vary significantly in early development, resulting in less precise measurements for the heat of fusion. The same is true for the melting temperature. A misinterpretation of the DSC can result if melting temperatures of the two forms are close. Another potential issue is the presence of a third form that may be metastable (monotropic) with respect to the other forms. This can lead to confusing results if not taken into consideration. Should the substance decompose on melting, these rules do not always apply.

Other rules that can assist in determining the more stable form (at room temperature) include the Rule of Density and the IR Rule.[14] The Rule of Density states that the most dense form when cold is generally the most stable. The IR Rule states that the first infrared band in the spectrum of a hydrogen-bonded molecular crystal is higher for one form than for the others. This form possesses the highest entropy and as a result will be the least stable at low temperature. This rule is generally not applicable when an amide group is part of the molecule.

For a binary enantiotropic system where an energy or phase diagram is not readily available, the transition temperature should be determined. This can usually be done through simple experimentation as described by Sato and Boistelle.[15] At a variety of temperatures, in temperature controlled vessels, a saturated mixture of both forms in a good solvent is allowed to evolve freely. Given a sufficient amount of time (with favorable kinetics this could take minutes to hours, or with unfavorable kinetics this conversion may take days to months), the more stable form at a given temperature will predominate, or become the sole form. The change can be monitored by microscopic examination. Should microscopy be done with a video attachment, time accelerated photography would allow one to watch the conversion process. One can estimate the transition temperature by examining the forms observed at the different temperatures.

With a fundamental understanding of the system thermodynamics, one can design a crystallization process with thermodynamics assisting in formation of the desired form. If one tries to "fight" thermodynamics, the process becomes more difficult to define, and may not be as robust. Once one understands a transition temperature for an enantiotropic system, one can use this data to assist in solid form transitions, especially in isothermal processes. By understanding the specifics of the system, thermodynamics can help determine experimental conditions required to prepare each solid form.

RG 12525

In the case of RG 12525, an energy diagram was developed.[13] The polymorphic system was classified as enantiotropic with the higher melting form being the least stable at room temperature. The transition temperature was thought to reside between 35°C and 70°C, being very difficult to measure due to only minor free energy differences between the two forms (less than 0.2 kcal/mol at room temperature). This information helped develop an understanding of the crystallization problem suggesting that the initial process was not rugged enough to prevent formation of an undesired form due to unfavorable thermodynamics.

**Kinetics**

Understanding the thermodynamics of a polymorphic system is important, but to develop a process to prepare the desired solid form, kinetics must also be considered. Polymorph solid state interconversion is also governed by kinetics. However, kinetics of solid state chemistry is complex and only partially understood.[8] Interconversion of solid forms can be slow due to the fact that some metastable polymorphic forms have a long half-life of duration. For instance, consider the extremely slow conversion of diamond, a metastable form of carbon, to its more stable form, graphite. On the other hand, some metastable forms might never be observed at room temperature, due to their rapid conversion into a more stable form. Predicting the rate of interconversion is not always feasible, and this information is usually derived through experimentation.

Two factors that assist in accelerating solid state interconversions are a liquid interface and the use of seed crystal. When solids are in liquid suspension, the chance of undergoing a solid state interconversion to a more stable form is much greater than for only a partially wetted solid on a filter or a solid in the dry state. The liquid helps mediate the transformation of one form into another. One additional aid to accelerating an interconversion is the presence of seed crystal. This is true for the conversion of a less stable form into a more stable form as described in the above section (Sato and Boistelle experimentation).[15]

Most chemical processes to prepare solids involve solutions or suspensions, so the system is more complex than a neat solid. For instance, a crystallization process often involves supersaturating a solution, which is thermodynamically unstable. In this situation kinetic considerations may dominate. Kinetics of crystallization can be divided into two processes, nucleation, formation of new crystals, and growth, the process whereby the crystal grows, and are discussed at length in the literature.[16] Nucleation and crystal growth are important to polymorph crystallization process. Some factors that can influence crystallization kinetics to form a desired polymorph will be discussed in a later section on crystallization processes, including: solvent choice, concentration, cooling rate, the presence of seed, and the presence of impurities. Other factors that are controlled entirely or in part by crystal growth include the generation of a specific crystal habit (shape)[5] and final particle size produced, each being important to pharmaceutical processing.

RG 12525

The system thermodynamics suggested that the two forms were potentially interconvertable at most temperatures, although the energy of activation and hence the kinetics of conversion from one form into the other was not known. At 30°C, aged suspension experiments indicated that form II was converted to almost exclusively form I in about 10 days.[13] Micronization or grinding caused interconversion from the more stable form I into the less stable form II. Additional experimentation was necessary to determine what chemical processing conditions would allow interconversion of the two forms.

## Principles Governing Form Choice

The final molecular composition and solid form of a drug could make a difference in bioavailability and delivery to the desired active site. Ideally, optimal molecular composition, stereoisomers, and possible salt combinations are known and previously screened in Medicinal Chemistry groups prior to selection of the development candidate. However, all possible solid forms are not always available nor studied at the initiation of a project. As a result some solid form issues may not be well understood. If a compound exists as multiple polymorphs, the exact nature of each polymorph should be studied to understand which is optimal for the pharmaceutical and biological systems. Each solid form has a unique thermodynamic profile and equilibrium solubility, therefore form choice could have an important impact on bioavailability and drug product or drug substance stability.[8]

Properties of the final solid will be of keen interest to colleagues in the pharmaceutical sciences (formulation). They will be interested in studying properties such as: flowability, wetability, cohesiveness, moisture sorption, solubility, dissolution rate, particle size, surface area, and porosity. With proper study, all of the desired drug substance solid state properties including any of the above will be known.

Assuming that a proper screening of polymorphs has taken place and characterization of solid forms has been performed, a choice of solid state or polymorphic form for further development needs to be made. The thermodynamically more stable form at room temperature is usually more desirable. However, if a case to develop a less thermodynamically stable form is made, this form should be kinetically favorable to resist solid state changes (just as diamond is slow to change to graphite).

One risk to choosing a less stable form is that a pharmaceutical tablet can crumble if the solid form should change on storage after tableting. This situation occurred in one pharmaceutical company that had prepared a batch of tablets. When checked at a retest date, all of the tablets had softened and crumbled, with active ingredient having undergone conversion to a more thermodynamically stable polymorphic form. Although the company had a reasonable rationale for form choice, kinetics were not well enough understood to prevent such a solid form change. In another situation, a large batch of a drug substance was being prepared on scale up of a crystallization process, but the desired polymorph was not obtained. Apparently the desired form was less stable and all of the processing parameters were not well enough understood prior to scale up; as a result, the more stable form was generated in the reactor. The presence of seed crystal can help guide the crystallization process, but this does not always prevent the undesired form from being generated.

RG 12525

The molecular composition and solid forms were known for RG12525 when the problem of trying to prepare a pure polymorphic form was encountered. In this case, thermodynamics suggested that the more stable form at room temperature was form I. A rugged process for its formation was required.

## Control of the Crystallization Processes

Typically a synthetic chemist is interested in crystallization as a way to eliminate impurities from a target material. Therefore while developing any crystal forming process, a process chemist should carefully observe purity issues (including decomposition). However, the final crystallization of a drug substance is used to define both its chemical and physical quality including polymorphic state. In addition to finding a means to decrease impurity levels, controlling polymorphism and the physical characteristics of the crystals must also be considered. Two ways of generating solid forms will be discussed. One is solid state interconversion and the other is crystallization. Since crystallization processes have been discussed at length elsewhere in the literature,[16,17] only general coverage of key concepts will be included.

**Solid State Interconversion**

Any discussion of polymorphism is not complete without considering solid state conversions.[8] Many of these approaches are not always practical on large scale but can be quite useful for small scale preparations. Solid phase changes often occur based on forming the more thermodynamically stable form but not always. Most of the time, solid state conversions are slow and may not occur in the desired time frame or with the desired control.

One approach for interconversion of enantiotropic polymorphs involves applying direct heat to a solid above its transition temperature. The thermodynamic tendency is for a solid to convert to a more stable form at the higher temperature. This form is usually not the stable form at room temperature. However, conversion by direct heating of a solid can be a slow process. In addition, the solid may have a tendency to cake, requiring crushing action to break up aggregates and lumps that may form. To accelerate the conversion process, solid can be heated in a stirred suspension with a "non-solvent." The "non-solvent" being a liquid in which the solid has little or no solubility. Presence of seed crystal of the stable form above the transition temperature may even further accelerate interconversion. This is an important principle to keep in mind when considering a simple recrystallization process for enantiotropic polymorphs. If crystallization occurs above the transition temperature, a suspension of the less stable form may result.

In an analogous fashion, the less stable form of an enantiotropic polymorph will interconvert below the transition temperature. Kinetics for direct interconversion are usually much too slow to be practical even on small scale. However, a stirred suspension with a "non-solvent" and seed crystal of the more stable form below its transition temperature will usually cause a solid interconversion into the more stable form at room temperature. This type of process may be slow, even days, being dependent on the kinetics of the system. Should one choose a polymorphic form that is less stable at room temperature, when the crystallization medium is cooled, prior to filtration, a thermodynamically unstable mixture will be present. This could result in a solvent mediated transformation or solid state interconversion into an undesired form.

Another type of solid form interconversion involves changing a solvate or hydrate to a non solvated or anhydrous form. One way is direct removal of solvent or water by heating in an oven (with or without vacuum). Unfortunately, this can lead to caking and lumping so that the process is not practical. An alternate approach is through a stirred suspension of solid in a "non-solvent." Heat is usually required to remove crystal bound solvent or water, and sometimes forcing conditions are required such as azeotropic distillation or co-distillation.

Solid forms may be sensitive to physical manipulation such as grinding or micronization. Mechanical energy is enough to cause solid state interconversions. This type of conversion process can be made practical on large scale depending on the ease with which conversion takes place. This type of manipulation can also be important to defining final particle size.

RG 12525

RG 12525 was quite sensitive to physical manipulation in that form I was partially converted to form II by micronization. However, this conversion was consistently incomplete. Other solid state interconversion techniques were too slow to be practical for RG 12525.

## Crystallization

Crystallization is the process by which a chemical is converted from a liquid solution into the solid crystalline state. To describe a crystallization process, it is best to begin by describing a simple two component system. This would involve dissolving a substrate solid into a solvent by heating, after which cooling is initiated to cause recrystallization of the solid. Each item that is important to this simple process will be discussed one at a time including solvent choice, concentration, cooling rate, the presence of seed and purity of the solid.

Solvent Choice

Solvent choice is based on several factors including safety, practical levels of solubility (hot and cold), cost, and effectiveness. Water, simple alcohols, and simple hydrocarbons (toluene, ethyl acetate, heptane, etc.) should be relatively inexpensive and relatively safe to use. Examples of solvents one might consider unsafe for use on large scale include benzene, carbon tetrachloride, carbon disulfide, or diethyl ether. For practical reasons, the actual choice of solvent depends on the ability to generate a fairly concentrated solution when heated to boiling temperatures (10-25%) and low solubility when cooled (0.05-2.5%).

Should the chosen solvent cause solvolysis, decomposition, or enrich impurities, its effectiveness should be questioned, and an alternate should be considered. On large scale because times tend to be longer, impurities formed by the solvent can be magnified from that observed on smaller scale. To verify that impurity levels do not increase, the solution should be incubated for long periods of time at high temperatures, followed by careful purity analysis.

Concentration

The chosen concentration will be dependent on the substrate solubility at different temperatures. At high temperature, high levels of solubility are required; 10-25% is typical. At lower temperatures, in order to recover substrate in reasonable yield, solubility should be about 0.05-2.5%. In screening solvents for a simple recrystallization process, the development of a solubility curve (Figure 1) is quite useful to visualize what happens but can also be time consuming to generate.[16]

Solubility is a measure of concentration at given temperatures. The x-axis of the curve is temperature, and the y-axis is concentration. Saturation, reached when a thermodynamically favorable amount of substance is dissolved into solvent, is indicated by a curve of equilibrium temperature versus equilibrium concentration. To the left of the saturation line, the solution becomes metastable or supersaturated, unstable conditions where solid might spontaneously fall out of solution. In the metastable zone, crystallization is typically driven by thermodynamics and favorable crystal growth can occur. However, some forcing parameters may be required such as adding seed crystal. The width of the metastable zone in the solubility curve can be an important factor in solvent choice.

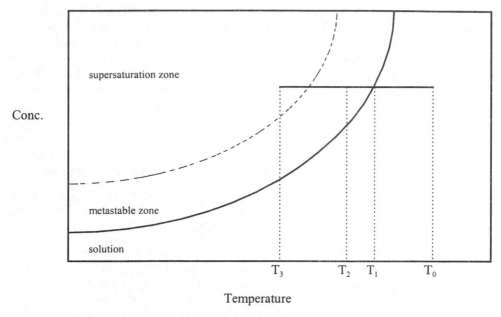

Figure 1. Solubility curve of concentration versus temperature. The solid horizontal line in the middle of the chart represents a solution being cooled - at $T_0$ a solution exists; at $T_1$ the saturation equilibrium is reached; at $T_2$ the metastable zone is reached where more favorable conditions for crystallization exist; at $T_3$ a supersaturation condition exists where spontaneous crystallization occurs to the most accessible solid form.

Cooling Rate

The slower a solution is cooled, the metastable zone will be reached where the mixture has a tendency to crystallize slowly. By bordering the saturation line and remaining in the metastable zone, crystallization will take place with greater control, and the thermodynamically more favored polymorph should crystallize. Once in the metastable zone and crystallization is initiated, holding at that temperature will allow the mixture to reach equilibrium before cooling is reinitiated. With faster cooling, a solution may reach the supersaturation zone and may even become highly supersaturated prior to crystallization. This is a thermodynamically unfavorable situation and spontaneous crystallization will usually result. Since kinetics tend to predominate, crystals are generated in the most convenient form available to the solid. Usually there is little control over the crystallization process.

Typically for slow crystallization, the resulting particle size is larger than if crystallization is fast. Tight control over cooling rate may be required to obtain a desired crystal size. Other factors that can impact particle size distribution include type and speed of agitation and substrate concentration.

Presence of Seed

Seeding is not always required but can be helpful for control of a crystallization process. If no seed is included, initial crystal formation (usually primary nucleation) occurs in a saturated solution in the supersaturation zone. However, one might want to avoid nucleation because it can be irreproducible and sensitive to impurities. Also, nucleation can cause fast crystallization resulting in small, difficult to filter crystals. Seeding is used to control the crystallization process thereby avoiding nucleation. That is, once the desired crystals are present as seed, they will serve as a surface for crystal growth for similar crystals, either on the surface itself or from parts that slough off the original surface. When seed is added to a saturated solution in the metastable zone, controlled crystallization usually results. To develop a crystallization process, one would seed in the metastable zone and cool slowly enough to never reach the supersaturation zone. Seed should not be added to a solution before it reaches the saturation limit, otherwise there is a possibility of having the seed crystal dissolve. The size and amount of seed crystal will also have an impact on the final crystal habit and particle size.[17]

Purity of the Solid

Although not all crystallizations of solids are sensitive to the presence of impurities, impurities even to parts per million can influence a crystallization process. This is quite significant if the solid is crystallized from a reaction mixture where equimolar or even higher amounts of chemical substances and by-products are present. For simple crystallizations, a substance must be pure during the investigation phase in order to develop a good process. Once developed, the sensitivity of that process to impurities can be determined through impurity spiking experiments, or by checking different purity grades of the solid substrate.

For a slower more controlled crystallization, isolated impurity levels tend to decrease and the overall purity of the resulting crystals is usually better than the initial solid. As discussed above, particle size has a tendency to be larger for this type of crystallization, but the addition of growth impurities can help in targeting a specific particle size.[16]

Examples of how the presence of impurities or additives during the crystallization process influences polymorph formation indicate that a metastable form can be generated with reproducibility by deliberately including impurities or additives.[18]

## Crystallization of Polymorphs From Liquid Solutions

With a basic understanding of general crystallization principles and knowing the thermodynamics for a polymorphic system, the process chemist can usually develop a good crystallization process that consistently delivers the desired solid form. The process can be defined considering items discussed above for general crystallization principles. Also to be considered are the addition of co-solvent or antisolvent, the level of hydration (solvation), and the adjustment of pH by acids or bases. Final drying and mechanical treatment (milling, micronization) can be used to achieve the desired chemical composition (desolvation) and any particle size changes; however, there is always the concern of polymorphic conversion to an amorphous solid. Other issues should also be considered during the study of isolated solids: filtration, drying, storage, and safety/industrial hygiene.

The Fusion Process

The simplest type of conversion from one solid form into another is fusion or melting. This would involve heating a solid until it melts then cool to crystallize. This type of approach is useful in screening for polymorphism. Should the melt temperature be high enough, decomposition can occur though. A more practical way to convert forms is to include a solvent for the purpose of recrystallization.

Recrystallization

A simple conversion from one polymorphic form into another involves recrystallization. A typical recrystallization will generate the desired polymorphic form in acceptable purity. This is usually done by heating a mixture of solid and solvent to dissolution (close to boiling temperatures). This is then cooled at a specific rate to the metastable zone (at a temperature that crystals of the acceptable form will fall out of solution). If seed is required, it will be added in the metastable zone. Should the solid be an enantiotropic polymorph, the crystallization temperature should be below the transition temperature to obtain the more stable form at that temperature and at room temperature. The concentration of the solution is adjusted as required to ensure that premature crystallization does not happen. If one desires the metastable form at room temperature, crystallization can be done through fast cooling (kinetics) and possible seeding with the less stable form, or can be done above the transition temperature (thermodynamics). Seed crystals may or may not be required. Quick cooling followed by fast filtration may be required before a liquid mediated solid transition to the more stable form occurs.

Modified Recrystallization Techniques

A variation of recrystallization is to replace one solvent with another through azeotropic distillation or other means of removing solvent or water (extraction, sieves, etc.). The solution can be cooled directly, or diluted with a "non-solvent," resulting in solid formation. This process is usually kinetically controlled and the ability to predict the polymorphic form resides in the individual crystal.

The generation of a desired enantiotropic polymorph can be controlled by a variety of techniques. The recrystallization technique involves changing the temperature with a fixed concentration. Another way is to change the solvent composition while remaining isothermal. This could involve adding a "non-solvent" to force the desired solid out of solution. A typically non-polar solvent like heptane may be useful for this purpose. This technique leads to highly supersaturation conditions resulting in a kinetic product, which can be either polymorphic form, depending on the specific polymorph.

Addition of Ionic Modifiers

Since the solubility of many organic acids or bases is highly dependent on a specific pH, any modification of pH will drastically alter the solubility profile. If one modifies the above recrystallization techniques where an acid or base is used instead of a "non-solvent," the result may be a highly supersaturated solution, and kinetic crystallization can result. Control of crystallization would be highly dependent on the rate of pH change rather than any change in temperature. This technique is most successful under isothermal conditions. One could imagine that the x-axis of the solubility curve represents pH rather than temperature. For the solution, changes in pH would be similar to changes in temperature as it moves from a true solution into the metastable zone and supersaturation zone. However one should understand that for acid or base addition, mixing may be incomplete resulting in "pockets" of high or low pH.

A modification of this approach is to change the ionic composition under isothermal conditions. Usually this requires an aqueous medium. Depending on the type of addition, and the addition rate, this can lead to controlled crystallization conditions or to highly supersaturated conditions. Again experimentation is necessary to determine which polymorphic form is generated.

Hydrates (Solvates)

Many of the above techniques are quite adaptable to the preparation of hydrates and solvates. To form a hydrate, water will need to be present, perhaps in stoichiometric amounts or more likely, higher levels. This would involve the dissolution of a solid in either in an aqueous solvent or an organic solvent followed by addition of water. Seeding or nucleation may also be an important factor to be considered. Solvates will usually result from a normal recrystallization where solvent becomes part of the crystal lattice. Hydrates (solvates) do not have the same relationship as enantiotropic and monotropic polymorphs.

Amorphous Solids

Should one take the approach of completely distilling or removing solvent from a solution, a solid can result. On a rotary evaporator, the result will sometimes be a "foam" that is typically amorphous. A resulting solid may require physical manipulation such as grinding in order to retrieve. In a similar fashion, lyophilization will usually leave an amorphous solid behind. These techniques are always practical for large scale. For a more practical process, the addition of a "non-solvent" can result in production of an amorphous solid. Another alternate is to use spray drying. This scaleable process involves the generation of small droplets of a solution in a heated environment (with vacuum) to rapidly eliminate solvent. The result can be formation of an amorphous solid.

One method that occasionally converts a hydrate (solvate) to an amorphous material is to remove solvent during the drying process. This can modify the original crystal lattice to an unstable solid that may change in one of several ways. It can collapse into an amorphous form, or it can reorganize into a new stable crystalline form, or the desolvated solid can maintain the original crystal characteristic but in a weakened state. A weakened crystal can be quite fragile and may convert to an amorphous form on physical manipulation.

RG 12525

For RG 12525, the design of a crystallization process was quite successful, but only after visiting many of the above techniques. Ultimately, two processes to generate each polymorphic form were developed.[13] Direct recrystallization was done to obtain each form. In addition, a "titration" process was also developed to prepare each form which involved dissolution in base then adjusting the pH at higher temperature to generate the desired form.

Simple recrystallization procedures developed to generate each form were remarkably similar, but had significantly different outcomes. The exact transition temperature was not determined due to experimental error associated with generation of the energy diagram. This temperature was thought to reside between 35°C and 70°C. Unfortunately, this range was too great to be of much assistance in developing a crystallization process. The solubility curve of RG 12525 in methanol indicated that the metastable zone was quite small and in all likelihood, when crystallized, the crystallization would occur in the supersaturation zone. Because kinetics govern crystallization in this zone, the cooling rate was critical. The presence of seed crystal was very important to prevent spontaneous crystallization of the wrong form.

Knowing that the metastable zone was small, coupled with subtle differences of shape and size of plant vessels and stirring type and speed in a new plant, helped explain why mixtures of both forms were obtained when a different plant was used for RG 12525 crystallization. Study of the RG 12525 crystallization process resulted in the following robust procedures.

Recrystallization to generate form I was performed by heating a 3.8% mixture of solid in methanol above 60°C to bring about dissolution. The resulting solution was cooled to 50°C and seeded with form I. After holding for 1 hour, the mixture was cooled to 0°C and filtered. The hold temperature was critical. It helped bring the mixture to an equilibrium close to the saturation line and hence into the small metastable zone. As the mixture was cooled slowly, equilibrium could be maintained in or close to the metastable zone allowing for exclusive formation of form I. Recovery yields were above 90%.

Recrystallization to generate form II was also performed by heating a 3.8% mixture of solid in methanol above 60°C to bring about dissolution. The resulting solution was cooled to 50°C and seeded with form II. Rather than holding at this temperature, the mixture was cooled immediately to 0°C and filtered. Faster cooling allowed the mixture to remain in the supersaturation zone during the entire crystallization helping to force out the kinetic product, form II. The recovery yield was 75%. The lower yield was probably because the solution was not allowed to come to equilibrium prior to filtration. Speed was important in that there was a risk the some solid would convert to form I through a liquid mediated solid state interconversion.

In searching out alternatives to direct recrystallization, titration variations were tried. This involved taking an aqueous sodium or ammonium salt solution of RG 12525 and adjusting the pH to generate solid RG 12525. A variety of acids, bases, co-solvents and conditions were tried, eventually leading to two successful crystallization procedures, one to generate each form.

The procedure to generate form I involved slowly adding an aqueous sodium salt solution of RG 12525 in aqueous isopropanol to a solution of citric acid containing seed crystal of form I. This technique was kinetically controlled in that a supersaturated condition resulted. Recovery yields were above 90%.

A slightly different procedure was used to generate form II. This involved slowly adding a citric acid solution to an aqueous ammonium salt solution of RG 12525 maintaining the pH at 5.3-5.4 throughout crystallization. This procedure was done in water containing some polyethylene glycol. No seed crystal was present, resulting in highly supersaturated conditions. The kinetic product in this case was form II. The recovery yield was above 95%.

Using a variety of techniques as previously discussed, two types of procedures for the generation of each polymorphic form were developed. Only one was thermodynamically controlled, the recrystallization of form I. The others appear to be controlled by kinetic parameters. In three cases, to assist in the formation of the desired form, seed crystals were used.

## Crystallization From Reaction Mixtures

If desired, the process to generate a specific solid form can be coupled with the final chemistry step. Simple refinement of a crystallization procedure can sometimes result in conditions adaptable to reaction mixtures. Issues to be considered in adapting a crystallization are that impurities will influence the form generated, kinetic conditions result with a quick change in solvent or conditions, and that the work up will be critical to setting up the final isolation conditions which need to be compatible with required crystallization conditions.

## Miscellaneous Solids Formation

Using inverse addition is feasible for many of the above techniques that involve addition of one solution into another to effect crystallization. This type of addition results in a highly supersaturated condition which is usually kinetically controlled. For instance, if one slowly adds a substrate solution into a "non-solvent," immediate precipitation of an amorphous solid can result. Alternatively, one could add a substrate solution at one pH into a buffered solution at another pH that allows for better pH control and crystallization of the desired form.

Particle size has not been discussed at length, but to achieve control of particle size, the crystallization process should be considered. Fast and kinetically driven crystallizations tend to favor small particle size, but the purity will usually not improve. Often larger particle size and higher purity result with slower crystallization processes. However, larger particle size is not always a desirable attribute for pharmaceuticals, so a modifier to limit crystal growth can be included as part of the crystallization process.[16]

## Literature Example: Dirithromycin

Although several examples of polymorph crystallization processes have been published, a recent article on dirithromycin[19] is more useful in illustrating some of the above principles and the difficulties associated with development of a polymorph crystallization process (more clearly appreciated by reading the cited reference). Dirithromycin is a semisynthetic macrolide antibiotic which exists in two known anhydrous polymorphic forms (1 and 2) and nine known stoichiometric solvate forms.[20] Differences in the two polymorphic forms are observed through the x-ray powder diffraction patterns and solid state $^{13}$C NMR spectra, the latter technique being useful to distinguishing between solvate forms. The two anhydrous forms are related monotropically with form 2 being more thermodynamically stable (or else the transition is greater than 120°C) as determined by differential thermal analysis.[19] As a result, form 2 became the chosen form.

In studying the dirithromycin crystallization process[19] two factors were important. One was generation of the desired solid form. The other was purity enhancement. Because most solvents used to purify dirithromycin resulted in the formation of solvates, a single solvent system for purification and generation of the desired solid form was not found. A two step approach became necessary: purification followed by conversion of a solvate to form 2. Unpurified dirithromycin contained numerous impurities and the parent tended to isomerize or degrade under certain conditions, making purification difficult. A thorough screening of solvents and solvent systems to enhance purity revealed that several solvents were useful in reducing impurity levels. Since aqueous acetone had preferred features, it became the chosen solvent system. This system was used to purify dirithromycin but delivered an acetone solvate as the solid form. An additional processing step, a heterogeneous slurry of the acetone solvate in water, converted solid into the desired anhydrous form. This conversion process involved first rapid conversion of the acetone solvate into the kinetic product, metastable form 1, followed by its subsequent conversion into the more thermodynamically stable form 2. These solid form changes appear to be solvent mediated transitions. The final dirithromycin purification and solid form crystallization process involved the steps of dissolution at 55°C, filtration to remove impurities, partial solvent removal, slow addition of water to enhance solids recovery yet provide optimal enrichment of purity, cooling to 40°C, solids filtration, slurrying the acetone solvate in water at 50°C, solids filtration, followed by drying to give purified form 2.

## Conclusions

Solid state issues are common place for pharmaceutical chemicals, and process chemists need to pay an increasing awareness to the generation of final isolated solids. Today, regulatory agencies are keenly interested in making sure that correct studies are being performed to generate and monitor the chosen solid form.[9] Regulatory agencies expect a description of solid state properties of new chemical entities to include a rationale for form choice, its method of manufacture, controls associated with its manufacture, and analytical procedures used ensure that correct form is prepared.

RG 12525 presented an issue of impure solid form on preparation. Using this example, many polymorphic considerations were presented and discussed. The result for RG 12525 was the development of four unique crystallization methods to reproducibly prepare by two methods each solid form. An additional literature example on the process development of dirithromycin solid form is also presented to further exemplify the listed principles.

General principles and a general approach to addressing solid state issues for the process chemist are provided herein. Many of these same principles apply to isolated solid intermediates that exhibit polymorphism. With future advances in instrumentation and as new tools are discovered, even more systematic approaches may be developed which can assist the synthetic chemist in the generation of chemical solids.

## References

1. Huang, F.-C. *Drugs Future* **1991**, *16*, 1121-1127.
2. Welch, M. J.; Nelson, H. S.; Paull, B. R.; Smith, J. A.; Feiss, G.; Tobey, R. E. *Ann. Allergy* **1994**, *72*, 348-352.
3. Huang, F.-C.; Galemmo, Jr.; R. A.; Johnson, Jr.; W. H.; Poli, G. B.; Morrissette, M. M.; Mencel, J. J.; Warus, J. D.; Campbell, H. F.; Nuss, G. W.; Carnathan, G. W.; Van Inwegen, R. G. *J. Med. Chem.* **1990**, *33*, 1194-1200.
4. Goetzen, T. Unpublished Results.
5. Haleblian, J. K. *J. Pharm. Sci.* **1975**, *64*, 1269-1288.
6. Haleblian, J.; McCrone, W. *J. Pharm. Sci.* **1969**; *58*, 911-929.
7. Biles, J. A. *J. Pharm. Sci.* **1962**, *51*, 601-617.
8. Byrn, S. R. *Solid State Chemistry*; Academic Press: New York, 1982.
9. Byrn, S.; Pfeiffer, R.; Ganey, M.; Hoiberg, C.; Poochikian, G. *Pharm. Res.* **1995**, *12*, 945-954.
10. Brittain, H. G.; Bogdanowich, S. J.; Bugay, D. E.; DeVincentis, J.; Lewen, G.; Newman, A. W. *Pharm. Res.* **1991**, *8*, 963-973.

11. Threlfall, T. *Analyst* **1995**, *120*, 2435-2460.

12. Lindenbaum, S.; McGraw, S. E. *Pharm. Manuf.* **1985**, *2*, 27-30.

13. Carlton, R. A.; Difeo, T. J.; Powner, T. H.; Santos, I.; Thompson, M. D. *J. Pharm. Sci.* **1996**, *85*, 461-467.

14. Burger, A.; Ramberger, R. *Mikrochimica Acta* **1979**, 259-271.

15. Sato, K.; Boistelle, R. *J. Crystal Growth* **1984**, *66*, 441-450.

16. Rodriguez-Hornedo, N. In *Encyclopedia of Pharmaceutical Technology*; Swarbrick, J., Boylan, J. C., Eds., Marcel Dekker, Inc.: New York, 1990; v 3, pp 399-434.

17. David, R.; Giron, D. In *Powder Technology and Pharmaceutical Processes*; Chulia, D., Deleuil, M., Pourcelot, Y., Eds., Elsevier: New York, 1994; Chap. 7, pp 193-241.

18. Davey, R. J.; Blagden, N.; Potts, G. D.; Docherty, R. *J. Am. Chem. Soc.* **1997**, *119*, 1767-1772.

19. Wirth, D. D.; Stephenson, G. A. *Org. Proc. Res. & Dev.* **1997**, *1*, 55-60.

20. Stephenson, G. A.; Stowell, J. G.; Toma, P. H.; Dorman, D. E.; Greene, J. R.; Byrn, S. R. *J. Am. Chem. Soc.* **1994**, *116*, 5766-5773.

# The Basic Principles of Thermal Process Safety: Towards a Better Communication Between Safety Experts and Process Chemists

**Wilfried Hoffmann**
*Parke-Davis Pharmaceutical Research, A Division of WarnerLambert Company, Chemical Development, Freiburg, D-79090 Freiburg, Germany*

Key words: Exotherms, reactor heat balance, adiabatic temperature risings, reaction calorimetry, MTSR, runaway scenario

## Introduction

In the last 10-15 years process safety has become an integral part of the development of new synthetic routes and more and more safety relevant information is gathered in typical process development labs. This is in part due to the increase of legal requirements, in another part to the widespread availability of powerful tools for acquiring necessary data, and, of course, to the dramatic consequences for a company an accident today may have.

Many development facilities have now established special process safety departments with state of the art equipment and well trained personnel, who are real experts in their fields. However, as generally observable in other disciplines too, there is the danger that these specialists develop their own language, which more and more slides away from readily being understood by people who actually need this information. It must be pointed out that the process ultimately will be run in a manufacturing environment, and that the operators and supervisors in most cases are not process safety specialists with an expertise in the interpretation of the instrumental raw data. Nevertheless, their decisions in critical situations are the most important ones and it is therefore essential that the communication of the safety relevant characteristics of the process is done on a level, which enables them to make the correct decisions.

The purpose of this chapter is to outline the basic principles of thermal process safety on a level that synthetic organic chemists and operation personnel can readily understand, while giving a sound background of the interplay of heat generation of the process and heat withdrawal of the reactor.

## The Hazards of Exothermicity

### Reaction Enthalpies

Table 1. Summary of some selected data for reaction enthalpies

$H^+ + HO^- \longrightarrow H_2O$        $\Delta H = -60$ kJ/mol

(benzene) $+ SO_3 \longrightarrow$ (benzene)$SO_3H$      $\Delta H = -150$ kJ/mol

(benzene) $+ HNO_3 \longrightarrow$ (benzene)$NO_2 + H_2O$     $\Delta H = -100\text{-}170$ kJ/mol

(aniline)$NH_2 + HNO_2 \longrightarrow$ (benzene)$N_2^+ + 2 H_2O$     $\Delta H = -65$ kJ/mol

(nitrobenzene)$NO_2 + 3 H_2 \longrightarrow$ (aniline)$NH_2 + 2 H_2O$     $\Delta H = -560$ kJ/mol

The majority of industrially useful reactions is exothermic, i.e. the conversion of starting materials to products is accompanied by a release of energy, in almost all cases in the form of heat.

This heat release is commonly expressed as an enthalpy change based on a formular conversion as shown in table 1. As most of these reactions are performed at constant pressure (reactors are open to the atmosphere) the thermodynamic state function enthalpy ($\Delta H$) is preferably used to quantitatively express these exothermicities. As long as only reactions in the condensed phase are considered, which is the huge majority in the pharmaceutical industry, these enthalpies are good approximations for the internal energy changes with non-isobaric reactions as well. The terms heat and enthalpy are used synonymously in this chapter.

According to the current IUPAC definition, the negative sign is used for energies which are given off from the system. As we will see later, many instruments provide data with an opposite sign, i. e. signals presenting exothermal events are positive. Care must be applied when experimental data and tabulated enthalpies are mixed together.

As can be seen, the magnitudes of the reaction heats are in the order of 100 to a few hundreds of kilojoules per mol. Many transformations of organic groups have typical associated enthalpy changes in a quite narrow range of only a few kilojoules per mol, some other reactions have much broader ranges, nitrations for example. Here the actual reaction conditions, like the kind of solvent used and the concentration, are important factors which influence the magnitude of the overall heat release, which may be the sum of many different simultaneous processes.

But what is the use of these numbers? What is the benefit to determine such data with considerable effort? Before we are going to answer this question, a small calculation should be presented to show the relationship between chemical energies and the more common to daily life mechanical energies.

## The Mechanical Equivalent of Reaction Energy

In order to get a feeling for the magnitude of these energies, it is impressive to compare these chemical energies with their mechanical equivalents; i.e. with the mechanical work which can be done with this energy.

Let us consider a medium sized production vessel filled with 10 kmols of a reactant (about 1000 - 2000 kg for a molecular weight of 100 - 200 g/mol). An exothermic reaction with a reaction enthalpy of $\Delta H_r$ = -100 kJ/mol thus will release an energy of $\Delta H$ = 10,000 mol x 100 kJ/mol = 1 Megajoule.

Let us further assume that this reactor is covered with a heavy top lid 1000 kg by weight and that this lid is accelerated against the earth's gravity by this reaction energy. To what height can the lid be lifted and what is its starting velocity? The result of such a calculation is astonishing, mainly because of the magnitude of the mechanical-heat equivalent 1 Joule = 1 Newtonmeter = 1 kg / ms$^2$.

For a moving mass in a potential field the sum of kinetic and potential energy is constant, thus

$$E_{kin} + E_{pot} = \frac{1}{2} m v^2 + m g h = E_{chem}$$

at start: $E_{pot}$ = 0 and $v = v_o$

$$v_o = \sqrt{\frac{2\,E_{chem}}{m}} = \sqrt{\frac{2 \cdot 1,000,000\ kJ}{1000\ kg}} = \sqrt{\frac{2 \cdot 1,000,000 \cdot 1,000\ m^2}{1000\ s^2}} = \underline{\underline{1,400\ m/s}}$$

at highest point: $E_{kin}$ = 0 and $h = h_{max}$

$$h_{max} = \frac{E_{chem}}{m\,g} = \frac{1,000,000 \cdot 1,000\ kg\ m^2/s^2}{1,000\ kg \cdot 9.81\ m/s^2} = \underline{\underline{102\ km}}$$

Of course our presumptions to convert the complete chemical energy into top lid movement are not very realistic (much energy is transformed into energy of deformation) and although it is theoretically impossible to convert 100% chemical energy into mechanical energy, it is a fact that heavy reactor parts were found more than several hundreds of meters away from their original location after a thermal reactor explosion. For some case histories, the reader is referred to the books of Yoshida[1] and Grewer[2].

A chemical reactor therefore should always be considered with respect. It is a power station which may have an enormous destruction potential, and the fate of chemical energy must be carefully recognized before a reaction is run on a larger scale.

## The Fate of Energy - The Reactor Heat Balance

When the problem of the fate of reaction energy is discussed in detail, answers must be given to the heat generation and the heat removal rates as a function of time, based on a well defined compartment.

Typically batch reactions are performed in jacketed reactors and the reaction energy is withdrawn by heat transfer through the reactor wall, so that it is reasonable that the compartment is the inner reactor space separated from the surroundings by the reactor wall. Based on this reactor space, conservation equations for mass and energy must be established. Up to this point this approach is quite general and can be easily refined according to the existing reactor; i.e. heat removal capacity can be enlarged by internal coils, power input from a stirrer can be considered, additional heat losses may be included, or a condenser can be added. For our purposes, however, we have to go in more detail. To limit the complexity to the simplest but necessary degree, we will now make a simplification which is never met strictly with real reactions and this is the ideal batch reactor approximation. This means per definition that, within the reactor space, temperature and concentrations are not a function of the location; i.e. temperature and concentration are only functions of time and not of space. For a well stirred, low viscosity reaction system, as is usually seen in the pharmaceutical industry, this approximation is generally good enough to handle the problems of reaction energy and thermal safety. Mixing problems, which may be responsible for changes in the product assay and impurity profiles during scaling up are this way excluded.

Now, with these approximations in mind, a very simple energy balance can be given. As it is of fundamental importance to understand the meaning of the various abbreviations in the mathematical description, these terms are explained in some detail.

In the verbal balance box the words "per unit time" are given. This means that all these terms are changing with time in the course of a reaction and thus a look to the reactor heat balance at a specified moment in time needs the time resolved data.

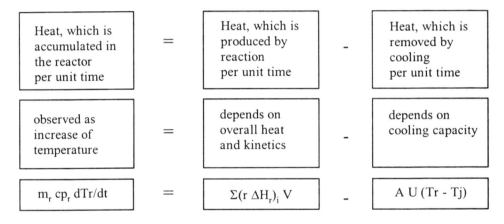

$m_r\, cp_r$ :    The product of the reaction mass ($m_r$) and the specific heat ($cp_r$) is commonly known as heat capacity ($Cp_r$) and includes all material which may change its temperature. All chemicals and solvents have to be added, but also inner reactor parts like stirrer, buffels, temperature sensors etc. For very large reactors, however, the heat capacities of the inserts are usually negligible, representing a few hundred kilos of steel with a small specific heat compared to several tons of organic reaction mass with a much higher specific heat.

$dTr/dt$:    The reaction temperature gradient, that is the change of temperature per time unit. Remember, for the ideal batch reactor, there is only one value at one time for the whole reactor (homogenous reaction temperature, no gradients as a function of space). This temperature change is an important observable, which directly shows the sign of the balance of the heat production and heat removal. In case of an exothermal reaction a positive value indicates that more heat is generated inside the reactor than is withdrawn through the jacket.

r:      The reaction rate, i.e. the change of moles per unit volume and per unit time by chemical reaction. In classical chemical kinetics these rates are represented as products of concentrations. As an example, for a bimolecular reaction $r = k\,[A][B]$, where k is the (bimolecular) rate constant with the dimension [vol]/[moles][time]. [A] and [B] are the concentrations of compound A resp. B in mol/l. The dimension of r is therefore [mol]/[vol][time]. Very often rates are empirically outlined as a power product series of all participated concentrations:
        $r = k\,[A]^a[B]^b[C]^c$... with a,b,c being small not necessarily integer numbers usually called reaction orders. It has to be pointed out that in this case the dimension of the rate constant k has to be chosen in that way that the dimension of r is unaffected.

$r\,\Delta H_r$:   This product is the heat production rate per unit volume, where $\Delta H_r$ is the reaction heat based on a mol change with the dimension [Energy]/[mol]. For more than one reaction running at the same time, all the different rates and reaction heats must be added to get the overall heat production per unit volume. Multiplying this sum with the total volume (which is a function of time in the case of semibatch reactions) will give the overall heat production in the reactor with the dimension [Energy]/[time] = [Power]

A:      This is the total heat exchange area. If we are restricted to pure jacket cooling, this is the wetted surface inside the reactor; that is, A is a function of time in the case of semibatch reactions.

U:      The heat transfer coefficient has the dimension [Power]/[Area][Temperature] and is the most complex term in our simple heat balance. This value is, roughly speaking, the inverse of the heat resistance and can be separated into three different parts.

$$\frac{1}{U} = \frac{1}{h_i} + \frac{d}{\lambda} + \frac{1}{h_o} \quad \text{with d = wall thickness}, \; \lambda = \text{heat conductivity} \quad (1)$$

The first part is the resistance of the reaction mass and the reactor wall, which is often called the inner film coefficient $h_i$, the second part is the resistance of the wall, which depends on the heat conductivity of the wall material and its thickness, the third part then is the outer film coefficient $h_o$, i.e. the resistance of the wall and the cooling medium in the jacket. Both film coefficients are functions of the viscosity and the thermal conductivity of the medium and of the film thickness, which is a function of flow properties (stirrer speed, stirrer and reactor shape, kind of inserts). There are empirical relationships available to correlate film coefficients with dimensionless numbers frequently used in fluid mechanics like Reynolds, Nusselt, or Prandl numbers, one of the best known being the Seider-Tate equation.[3] Some practical problems to access film coefficients for larger reactors are discussed in the literature.[4,5] However, all these relationships are approximations and it should always be in mind that the heat transfer coefficient may change during a reaction.

A U (Tr-Tj):  This correlation can be traced back to Newton[6] and is a good enough approximation for heat flow across a temperature difference when this temperature difference is not too large (Tj being the reactor jacket temperature). As pointed out in the explanation for U, uncertainties in this term can mainly be attributed to a limited knowledge of U.

Now that we have related our heat balance to basic physical parameters, the next step is to look for the availability of the required data, which can be separated into two groups. One set of data can be attributed to the chemistry and the reaction conditions, namely $m_r$, $cp_r$, $\Delta H_r$, and r, the other set is correlated with the reactor in use, namely the cooling capacity A, U, and Tr-Tj.

As this cooling capacity is very difficult to access it is common practice in process safety to deal with conservative approximations. The most conservative approach in handling heat transfer is to reduce U to zero, that is we consider the reactor under adiabatic conditions.

## The Adiabatic Temperature Rise

Starting with the fundamental heat balance equation,

$$m_r cp_r \cdot dTr/dt = \sum (r \Delta H_r)_i \cdot V - A \cdot U(Tr - Tj) \tag{2}$$

we make the assumption that the heat transfer coefficient is zero, which means that there is no heat withdrawn by jacket cooling at all. Thermodynamically speaking, the walls, which separate our system from the surroundings become impervious to heat, which is called an adiabatic system. As was assumed in the derivation of the heat balance that this jacket cooling was the only way to remove heat, the reactor therefore will behave like a perfect Dewar vessel. This simplification has the consequence that the heat balance equation becomes independent from the reactor properties. In practice, in a perfect Dewar vessel, all the heat released by the chemical reaction system will be accumulated; i.e. the temperature gradient is positive as long as the heat production is positive. Thus the reaction temperature will steadily increase throughout an exothermal reaction and the question of overall temperature rise is therefore obvious. The answer can be given by integration of the heat balance equation for the adiabatic case.

As this should be a basic introduction, this integration is presented here in detail, because there are some further simplifications made on the way to the very simple result mostly used as starting equation in discussing adiabatic temperature risings.

Remember the definition of the reaction rate r as the change in moles per time and volume, the adiabatic heat balance equation for a single reaction is then

$$m_r \cdot cp_r \cdot dTr/dt = r \cdot \Delta H_r \cdot V = \left[\frac{dn}{dt}/V\right]\Delta H_r \cdot V = \frac{dn}{dt}\cdot \Delta H_r$$

which after rearrangement gives     $m_r cp_r dTr = dn \Delta H_r$
By integration we obtain

$$\int_{Tr=Tr_0}^{Tr=Tr_0+\Delta T_{ad}} mr\, cp\, dTr = \int_{n=n_0}^{n=0}\Delta H_r dn$$

With the further, not very accurate, assumption that the specific heat as well as the reaction heat $\Delta H_r$ is constant throughout the interval of integration, solving of the integrals is simple and results in

$$m_r c p_r \Delta T_{ad} = -n_o \Delta H_r \quad \text{or} \quad \Delta T_{ad} = \frac{-n_o \Delta H_r}{m_r c p_r} = \frac{\Delta Q}{m_r c p_r} \tag{3}$$

The negative sign for the converted moles compensates the negative sign for the reaction enthalpies in the case of exothermic reactions, and often is omitted in this equation and absolute (positive) values are used for mole and heat changes ($\Delta Q$).

What can be done with this simple relationship? First of all this adiabatic temperature rise $\Delta T_{ad}$ is of much more use to express exothermicities as the reaction enthalpy alone, as here the heat capacity of the reaction mixture and the concentration of the reactands are also considered. Secondly as these adiabatic temperature risings are independent of any reaction rates, only basic reaction data are required which are mostly known (n and mr) or can be estimated from literature data ($cp_r$, $\Delta H_r$).

The specific heat for organic reaction masses usually is between 1.7 and 2.5 J/g K, so that 2 J/g K is a common estimate when no other data are available. In many cases the specific heat of the pure solvent is a good estimation alternative.

For the reaction heats $\Delta H_r$ many reactions have typical values as shown in table 1. An alternative approach is the use of heat of formation data $\Delta H_f^o$. (The superscript o refers to the standard conditions, which are one atmosphere of pressure, 298.15 K, and the standard state at this temperature.)

Thermodynamically, the enthalpy change for a given reaction can be expressed as the sum of the heats of formation of the products minus the sum of the heats of formation of starting materials, each multiplied with their stochiometric coefficients $v$.

$$\Delta H_r = \sum_i v_i \Delta H_f^o \quad \text{with } v_i > 0 \text{ for products and } v_i < 0 \text{ for starting materials} \tag{4}$$

Compilations of heats of formation data are available in the literature.[7,8]

Other methods to obtain heats of formation data include applying group contributions for calculating $\Delta H_f^o$ like the Benson method[9] or directly calculating them by "semi empirical" and even "ab initio" quantum chemical software packages, which have become available to the desk top due to the increasing calculation power of modern personal computers. The drawback of these computational methods is the general lack of consideration of solvent interactions as mostly gas phase values of isolated molecules are obtained. In addition, a reaction path has to be predefined and this can become problematic as nature from time to time does not adhere to our definitions.

Nevertheless, the knowledge of the adiabatic temperature rise is of so much practical use that this should be the first thing one should ask for when dealing with an exothermal reaction.

## Data from Reaction Calorimetry

### Batch reactions

Although some estimations can be made without exact knowledge of all the parameters in the heat balance equation with respect to the shown limitations, as shown in the previous chapter, the best way to get the required data is by measuring them.

An instrument which allows the measurement of heat production rates as a function of time under production-like conditions together with other important reaction parameters like specific heats or heat transfer coefficients is called a reaction calorimeter. The most popular of these instruments in chemical industry being the RC1 from Mettler-Toledo, which is a further development of the bench-scale calorimeter of Ciba-Geigy, designed by Regenass.[10,11]

A typical heat flow curve for a batch reaction obtained from such an instrument is presented in figure 1.

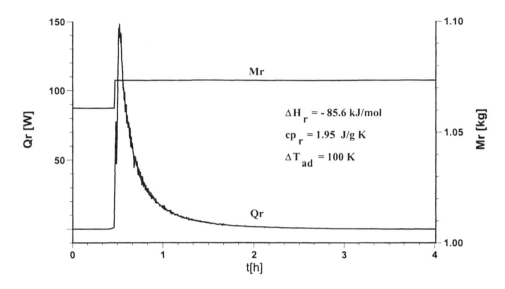

**Figure 1:** Typical heat production shape for a batch reaction (all at once addition of lithium metal to a solution of naphthalene in THF[12] ).

As expected, heat production rate is largest at start and then declines with time to zero. The area under this curve is the total reaction heat, i.e. $\Delta Q = -\Delta H = - n \cdot \Delta H_r$. As heat capacity data are also readily obtained, a single calorimetric determination allows an experimentally based statement on the total adiabatic temperature rise $\Delta Tad$.

This heat flow curve can also be used, at least in simple cases, to obtain kinetic parameters in a similar way than from concentration-time curves. This is not surprising as the heat production term is proportional to the mol change per time and volume. With a constant volume reaction, this is proportional to concentration changes.

For a simple first order batch reaction this analogy is shown below.

$$Q(t) = r \cdot \Delta H_r \cdot V = \frac{(dn_A / dt)}{V} \cdot \Delta H_r \cdot V \qquad \text{where } n_A = \text{moles of substance A}$$

$$\text{with } V \neq f(t): \quad \frac{(dn_A / dt)}{V} = d(n_A / V) / dt = \frac{d[A]}{dt} \quad \text{where } [A] = \text{concentration of A}$$

$$\text{for a first order reaction} \frac{d[A]}{dt} = - k \cdot [A], \text{ there is the solution } [A](t) = [A]_o \cdot e^{-kt}$$

hence

$$Q(t) = - k \cdot [A]_0 \cdot e^{-kt} \cdot \Delta H_r \cdot V = - k \cdot n_{A_0} \cdot \Delta H_r \ e^{-kt} = Q_0 \ e^{-kt}$$

with a complete analogy and $Q_0 = - k \, [A]_0 \, V \cdot \Delta H_r = - k \cdot n_{A_0} \cdot \Delta H_r$

There are several ways to obtain rate constants from measured heat production curves and for simple kinetic cases the agreement with the evaluation based on concentration changes is satisfactory. In reality all these methods have the disadvantage that first a reasonable kinetic model has to be suggested and then the data are evaluated according to this model. This, however, may become increasingly complex with more than one reaction running and may be impossible from heat flow data alone at all. In the last few years, online FT-IR spectroscopy has been developed to simultaneously gather analytical information together with heat measurements.[13]

## Semi batch reactions

Now we will turn to reaction calorimetric data for the more common semi-batch processes, where one reagent is added over a certain time.

A typical heat production curve for a semi-batch process is shown in figure 2.

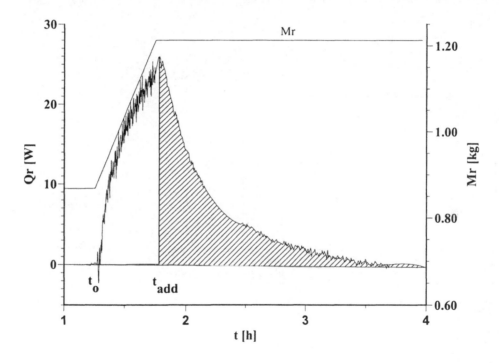

**Figure 2:** Typical heat production curve for a semi-batch reaction.

Compared with figure 1 some characteristics of this semi-batch heat production should be pointed out. Principally the addition time divides the resulting heat production curve into two parts. The part after the addition has been finished (hatched in figure 2) and can be considered a typical batch reaction starting with the reaction mass and the reagent concentrations present at addition time $t_{add}$.

In the dosing part, the maximum heat production rate is generally not at the start of the addition, but increases during the addition. This is easy to understand, as in the case of a reaction of two reagents A and B the feed rate of reagent B (when reagent A is already present in the reactor) is responsible for the increase of the concentration of B, whereas the kinetic parameters are responsible for the decrease of A and B. If the increase of B is larger due to a higher feed rate than the decrease of B by reaction, then certainly the net concentration of B will increase during the addition until its end; and the rate of heat production, which is a function of the concentrations of A and B, will also increase.

This implies that the actual shape of the heat production curve will strongly depend on the ratios of the feed rate and the kinetics. If the feed rate is very large compared to the reaction rate, than the system will behave more and more like a batch reaction.

The other extreme for the ratio of feed rate and reaction rate is a very fast reaction rate, much faster than the feed rate. In this case, every reagent B which enters the reaction mixture is at once converted to product by spontaneous reaction with the reagent A already present in the reactor. There is no increase of B during addition and there is a constant heat production rate, which is directly proportional to the feed rate. A typical reaction of this type is a classical titration. The resulting heat production curve of such a reaction is shown in figure 3.

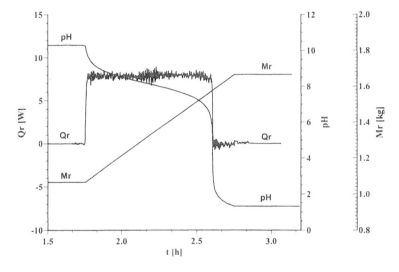

**Figure 3:**   Heat production for titration of 0.5 mol TRIS (tris-hydroxymethylaminomethane) in 1000 ml water with 600 ml 1 n HCl.


**Adiabatic temperature rise of a semi batch reaction**

We have discussed adiabatic temperature risings for batch reactions and derived a simple formula for its calculation with the knowledge of the heat capacity, the reaction heat and the mol changes (eq. 3). This was based on the worst case assumption of a loss of cooling capacity just at the start of the reaction, which would give us the maximum value for $\Delta Tad$. This makes sense with batch mode reactions, where there is no other reaction control once the reagents are mixed. In the case of a semi batch reaction, however, the feed rate is an additional control and we have to consider the dosing phase and its changing heat capacity as well. A loss of cooling capacity near the start of the dosing will allow us to stop the feed and thus reduce the number of reacting moles and therefore reduce the heat being released significantly.

So the answer to the question of the adiabatic temperature rise of a semibatch process is very much dependent on the point in time where the switch to adiabatic conditions may happen.

It is no surprise that this kind of data is also readily obtained by reaction calorimetry. A useful function in this context, which is directly obtained from the calorimetric evaluation, is the so-called conversion function conv(t), defined as the partial integral from time=$t_0$ to time=t, divided by the total heat:

$$\text{conv}(t) = \frac{\int_0^t Q_r(t)\,dt}{\int_{t=0}^{t=t_\infty} Q_r\,dt} = \frac{\int_0^t Q_r(t)\,dt}{\Delta H_{tot}} \quad \text{with} \quad 0 \le \text{conv}(t) \le 1 \tag{5}$$

Obviously this function is 0 at start (no conversion at all) and reaches 1 when the reaction is completed. At any time t, it gives back the fraction of the already released heat, the difference to 1 being the heat, which is to be released that is the heat, which can cause an adiabatic temperature rise. This part of reaction heat is called the accumulated heat $\Delta H_{accu}$.

As an example the hatched part of figure 2, that is the accumulated heat at $t_{add}$, can be evaluated by equation 6, or alternatively with our defined conversion function (equation 7).

$$\Delta H_{accu}(t_{add}) = \Delta H_{tot} - \int_{t=0}^{t=t_{add}} Q_r\,dt \tag{6}$$

$$\Delta H_{accu}(t_{add}) = [1 - \text{conv}(t_{add})] \cdot \Delta H_{tot} \tag{7}$$

Once we know $\Delta H_{accu}$ we can easily adopt the simple formula for the batch case at $t=t_{add}$ by replacing the total reaction heat $\Delta H_{tot}$ by the reduced value of the accumulated heat at the end of addition $\Delta H_{accu}(t_{add})$.

$$\Delta Tad_{accu}(t_{add}) = \frac{\Delta H_{accu}(t_{add})}{mr_0 \cdot cp_0 + m_{dos} \cdot cp_{dos}} = \frac{\Delta H_{accu}(t_{add})}{mr \cdot cp_r} \tag{8}$$

It is evident that at the end of dosing, $\Delta H_{accu}$ is smaller than $\Delta H_{tot}$ and hence also $\Delta Tad_{accu}$ is smaller than $\Delta Tad_{tot}$. This is of course the reason for doing very exothermal reactions preferably in a semi batch mode.

Before we now go one step further and look at the value of $\Delta H_{accu}$ during the dosing, a few remarks to the so-called dosing controlled reactions presented by a titration in figure 3 are necessary. It was pointed out that the kinetics of such reactions are so fast compared to the feed rate that the hatched part of figure 2 has reduced to zero in figure 3. Hence, the reaction is completed at the end of addition and the conversion function is 1. Applying equation 7, there is no accumulation at all, thus $\Delta H_{accu} = 0$ and $\Delta Tad_{accu} = 0$, too. In general once the stochiometric equivalent for a full conversion is added, which can be sooner than $t_{add}$ in the case an excess is being dosed, the conversion function in this case is 1. More strictly these kind of dosing controlled reactions represent the ideal conversion, which is a straight line between 0 and 1 between t = 0 and t = $t_{sto}$ , the stochiometric point.

This definition of the ideal conversion can now be used to define the accumulation function $F_{accu}$ (t) as the difference between the ideal conversion and the real measured conversion.

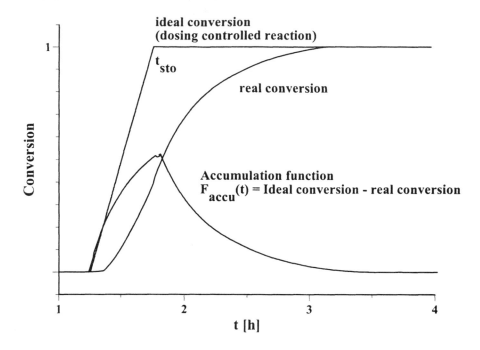

**Figure 4:** The accumulation function for a semi-batch reaction.

As shown in figure 4, the accumulation function has the expected behavior with a value of zero at the start and a maximum somewhere between start and reaction completion. The resulting heat accumulation is then given by equation 9.

$$\Delta H_{accu}(t) = [conv_{ideal}(t) - conv(t)] \cdot \Delta H_{tot} \ = F_{accu}(t) \cdot \Delta H_{tot} \qquad (9)$$

As the adiabatic temperature rise is defined by $\Delta H_{accu}$ divided by the heat capacity, we can now calculate $\Delta Tad_{accu}$ as a function of time throughout the reaction by a slight modification of equation 8.

$$\Delta Tad_{accu}(t) = \frac{\Delta H_{accu}(t)}{mr_o \cdot cp_o + m_{dos}(t) \cdot cp_{dos}} = \frac{\Delta H_{accu}(t)}{mr \cdot cp_r(t)} \qquad (10)$$

As the mass is a function of the dosing, we have to fix, whether in the case of a switch to adiabatic mode, the feed is stopped or not. Generally it is better to stop the feed, although in some cases when the stochiometric point has passed and the maximum total heat can be generated, further addition may increase the heat capacity and thus reduce the adiabatic temperature rise at the cost of a faster reaction rate approaching the higher temperature.

In summary, by applying the provisions of reaction calorimetry, we have derived a general expression for the adiabatic temperature rise for a batch and a semi batch reaction as a function of process time. In the next paragraph we will see the practical use of this approach.

### The Maximum Temperature of the Synthesis Reaction (MTSR)

The concept of the Maximum Temperature of the Synthesis Reaction (MTSR) and the use of the runaway scenario, which will be presented later, was originally outlined by Gigax[14] in an outstanding article. According to Gigax, the MTSR function is defined as the sum of the starting temperature and the maximum adiabatic temperature rise.

$$MTSR = Tr_0 + \Delta Tad_{accu} \text{ (max)} \tag{11}$$

Starting from equation 10 we are now considering the reaction time t, where our resulting adiabatic temperature rise $\Delta Tad_{accu}$ will become a maximum, and add the starting reaction temperature, so that finally we will get a temperature decline rather than a temperature rise.

For the different types of heat production curves these MTSR temperatures are summarized in figure 5.

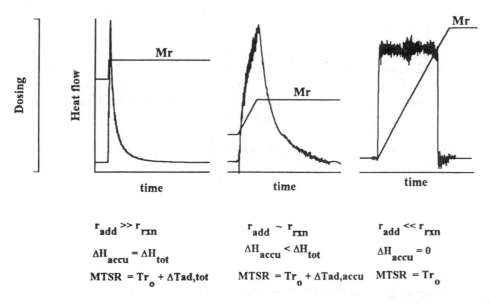

**Figure 5:** Summary of adiabatic temperature risings and MTSR for different types of reactions.

For a batch reaction or a semi batch reaction with a very fast dosing compared to the reaction rate, MTSR is just the sum of the starting temperature and the full adiabatic temperature rise $\Delta Tad_{tot}$, which is the highest MTSR possible. The other extreme with the lowest MTSR is the ideal dosing controlled reaction where MTSR is the starting temperature. The most interesting is the case in between. Here the MTSR is a function of all reaction parameters including the starting temperature and it should be common practice in the process design that MTSR may be as low as possible.

This does not imply necessarily a low starting temperature, just an approach towards a higher starting temperature may be worthwhile. The reason for this is that a higher temperature increases the reaction rate and decreases the accumulated heat, so that MTSR as a sum may decrease in spite of an increase of Tr.[15]

## Secondary Reactions

Up to this point all our data have been obtained by analysis of the heat generation curve produced by a reaction calorimeter. Therefore the obtained adiabatic temperature risings are calculated data only. A critical question therefore is, how reliable these values are. What will really happen with our reaction system on its way to higher temperatures? Is the reaction mixture thermally stable at MTSR or are there other reactions running at these temperatures which have not been detected at the desired synthesis temperatures?

One obvious deviation from reality is an MTSR which is higher than the boiling point of the system. In an open reactor the system will increase its vapor pressure and when the atmospheric pressure is reached the temperature will no longer rise but the reaction heat is used up to convert liquid to vapors. This can be positive in the case of so-called all vapor venting, as the vapors can be re-condensed by a condenser thus allowing additional cooling capacity to be brought to the reactor. However, the formation of gas bubbles can also have a negative effect in the case of generation of a so-called two phase flow, i.e. a gas/liquid mixture is formed, which can fill the whole reactor with foaming, like an over-boiling milk pot. This situation should not be discussed in more detail here and the reader is referred to the available literature.[16]

Another important deviation from the calculated MTSR may happen in the case of secondary reactions. The higher the calculated adiabatic temperature rise, the higher is the chance that our reaction system becomes thermally unstable at these higher temperatures, so that an experimental test for thermal stability is mandatory.

A very fast method is the Differential Scanning Calorimetry (DSC), which is often used as a screening tool to decide if more careful and tedious measurements are required. The instruments of choice for these detailed measurements are adiabatic calorimeters like the ARC[®17] or the PhiTec[®18]. What is typically obtained from these instruments are adiabatic self heating rates as a function of temperature and words often used in the description of these data are phi-factors, onset temperatures, time to maximum rate and also adiabatic temperature risings.

One fundamental parameter, the so-called phi-factor or $\Phi$, is defined as a thermal dilution factor (equation 12).

The reaction heat, which is responsible for heating up the reaction mass under adiabatic conditions does heat up the containment as well. So what is actually measured is the self heat rate of the reaction mixture together with the container and there must be some correction applied to get the temperature excursion with a negligible container heat capacity, as it is the case with large reactors.

With a knowledge of the heat capacities of reaction mixture and containment, the phi-factor $\Phi$ is defined by:

$$\Phi = \frac{mr \cdot cp_r + m_{cont} \cdot cp_{cont}}{mr \cdot cp_r} = 1 + \frac{m_{cont} \cdot cp_{cont}}{mr \cdot cp_r} \tag{12}$$

Obviously, $\Phi$ goes to unity for an infinitely large reactor with a negligible containment heat capacity. Very large real reactors are close to 1 and in the experimental setup of the measurement systems, $\Phi$ usually is between 1.05 and 2. For very highly energetic systems like explosives an even higher $\Phi$ is used. A $\Phi$ of 2 means that an experimentally observed adiabatic temperature rise of 100°C would correspond to a temperature rise of 200°C in a real large reactor, leaving some uncertainties about the not experimentally accessed range between 100 and 200°C temperature rise.

$\Phi$ is often misinterpreted as a measure of the adiabaticity of the measurement system or its goodness. This is not true. The adiabaticity is a measure of how good adiabatic conditions are met, that is heat losses from the measurement system are avoided. As most instruments reduce the heat losses by adjusting the containment temperature by controlled heating, the adiabaticity mainly depends on the dynamics and accuracy of this heating. This has nothing to do with $\Phi$.

Another parameter which causes much confusion in the interpretation of such adiabatic self-heat data is the onset temperature. Strictly speaking the onset temperature is the temperature where the instrument switches from non-adiabatic heat-wait-search mode to adiabatic mode. So the onset temperature somehow reflects the thermal stability of the chemical system and the sensitivity of the measuring system as well. That means the same chemical system is expected to have different onset temperatures when measured with different instruments or even when measured with the same instrument but with a different $\Phi$, slope sensitivity, or temperature step during the heat-wait-search cycles. It is evident that for a given chemical system the lowest $\Phi$-factor and the most sensitive detection limit will give the lowest onset temperature. However, bomb size and bomb material will limit the available $\Phi$ and an increase in the detection sensitivity will also increase the time required to perform such an adiabatic experiment.

By no means, however, should the onset temperature be interpreted as a limit temperature dividing the reaction temperature into a region below the onset temperature, where no dangerous reaction is running and a dangerous region above the onset temperature. We will soon show that another parameter, namely the Time to Maximum Rate (TMR) as a function of temperature, which can be derived from these self heat data, is of much better use when these kind of data are discussed.

A typical curve produced by an ARC is shown in figure 6. The phi-correction in this plot is simply done by multiplying the temperature rise with $\Phi$ and dividing the time by the phi-factor, which is in most cases accurate enough for reactions with large adiabatic temperature risings of more than 100°C.

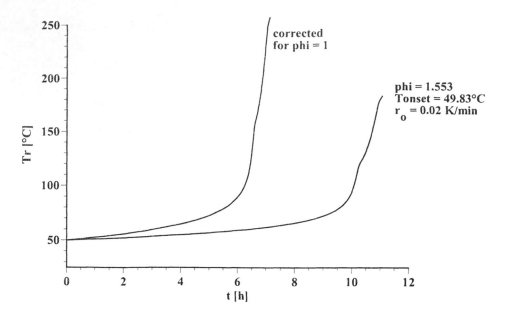

**Figure 6: Temperature - time plot of an adiabatic self heat experiment**

In the case of these dangerous decompositions with a large $\Delta$Tad the acceleration of the rate constant with temperature is the dominating term, so that as a good approximation the change in concentration of starting material can be neglected, which results in a zero order decomposition. This is not astonishing as with a reaction with a $\Delta$Tad of, let's say, 300°C, only 10% of starting material have been consumed at a 30 degrees temperature excursion. At the same time due to the Arrhenius law the rate constant has increased by approximately a factor of 10 and will increase by a factor of 100 for the next 30°C, resulting in the typical shape of a thermal explosion curve with releasing of the majority of the reaction energy in only a few seconds!

The whole theory of the evaluation of self heat rate curves is given in the literature[17] and here only a very basic way is outlined to transform the experimental curves to a statement on the TMR as a function of temperature.

A simple but straightforward approach is to start from the measured data as shown in figure 6 from the time point of maximum rate and looking in the past. This will give a time series as a function of temperature, namely the time to maximum rate (TMR). A plot of the logarithm of TMR data versus 1/T will give a straight line for not too low times. Such a plot allows a very simple correction to $\Phi = 1$ by subtraction of log $\Phi$ from the experimental plot.

When the experimental self-heat rate of figure 6 is plotted this way, figure 7 is obtained.

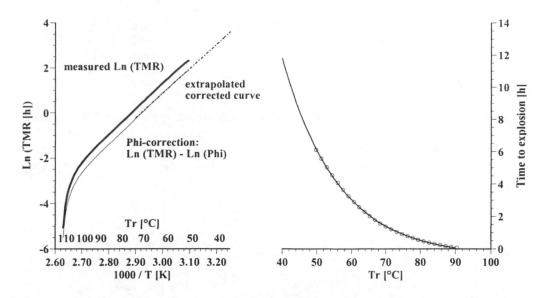

**Figure 7: Plot of Ln (TMR) vs 1/T and Φ-correction for adiabatic self heat experiment.**

As can be seen in the left part of figure 7, there is good straight line behavior to less then a few minutes and only then the plot strongly deviates from linearity. However, these short times are of no interest at all. In the worst case, these times indicate the time to explosion starting from an achievable reaction temperature. It makes a difference between 2 hours and 20 hours, but there is no practical difference between 1 minute and 10 minutes. There may be some discussion on how far away from the experimental range the straight line can be extrapolated to the region of lower temperature. For reaction safety purposes where time scales in the range of a few to several hours are applied some more hours beyond or some 10-20 degrees below the experimental onset temperature may be acceptable. A direct use of such short term data to extrapolate to much longer times for accessing transportation or storage data may not be recommended, although even this may be useful as a first approximation.

The data from the Φ-corrected and properly extrapolated straight line can then be re-plotted to get the desired TMR as a function of temperature, which is the most useful representation of the results of adiabatic self heat rates experiments, as shown in the right part of figure 7.

This representation shows a worst case scenario, i.e. adiabatic conditions with zero order reaction, the time to maximum rate (here synonymously used with time to explosion), that is the available time for counter measures or evacuation. It should be pointed out that this result is within experimental error independent of the Φ-factor used in the experiment and the actual measured onset temperature, so that TMR data should preferably be used when communicating adiabatic self-heat experiments.

However, this is only one aspect. The more practical side of this presentation is to allow a numerical description of the thermal risk based on experiment.

Another more sophisticated evaluation method, which is particular useful, is the kinetic fitting of the measured data to a given model, which can be much more flexible than the conservative zero-order evaluation and which allows a simultaneous fitting of the reaction order.

## The Combination of MTSR and TMR Data

The next straightforward step in an overall experimental approach of the thermal process safety is a combination of the MTSR data mainly obtained by reaction calorimetry and the TMR data obtained by adiabatic self heat rate measurements.

Looking for TMR just at the MTSR will give the so-called runaway scenario, a description of the thermal process safety situation.

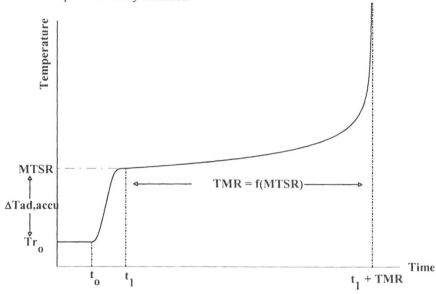

**Figure 8: The Runaway Scenario.**

The use of this temperature vs. time plot in presenting the complex interdependencies of a real chemical reaction system should greatly be encouraged and therefore a detailed guideline to the interpretation should be given here.

First, let us consider a reaction starting at t=0 with a starting temperature of $Tr_0$. From reaction calorimetric investigations we can derive as shown a time $t=t_0$, where the adiabatic temperature rise caused by the accumulated heat is a maximum. Remember, for a batch process this is just at the start, for semi batch processes this is very often when just one equivalent of reagent is added. Changing to adiabatic behavior at $t=t_0$ thus will increase the reaction temperature by the synthesis reaction to MTSR. The time required for this excursion, i.e. $t_1$-$t_0$, can be calculated once the temperature dependence of the rate equations are known, but as the reaction is commonly fast at $Tr_0$ to bring the reaction to completion in less than a few hours, $t_1$-$t_0$ is in most cases less than an hour and significantly shorter than TMR, so that as a first approximation $t_1$-$t_0$ can be ignored.

From the secondary reaction analysis we can now get the TMR just at MTSR to tie together the reaction conditions of the desired synthesis reaction with the resulting numerical measure of the thermal risk in form of the time to maximum rate. What is left now is to point out that the MTSR and the corresponding TMR are coupled. The reaction conditions (reaction temperature, concentrations, feed rate) strongly influence the MTSR. A shift to a higher MTSR will give a shorter TMR and vice versa, TMR can be extended by reduction of MTSR.

In practice, starting with an acceptable TMR will give a fixed MTSR for the process and all other reaction conditions can then be varied and optimized for yield, purity, throughput etc., as is usually done in process development. The value of reaction calorimetry as the tool of choice to experimentally access the MTSR in this approach is indispensable. Many details and comments on this concept are discussed by Stoessel.[19,20]

### What TMR is acceptable?

Reaction calorimetry and adiabatic self heat experiments provide the experimental data to construct the runaway scenario for each of our processes. We have seen that MTSR and TMR data are strongly connected, so that reaction conditions can be chosen in that way to be longer than a certain acceptable limit TMR. But how should this limit be set? This limit TMR is obviously directly related to the acceptable risk and depends on much more factors. In general a TMR larger than the order of the common process time, which should usually be less than a worker's shift, is acceptable. Other factors, which have an influence on the acceptable TMR, are the scale factor, the reactor environment, and the degree of technical equipment available at the reactor. A small pilot plant reactor with a few hundred liters volume may be considered differently than a large production reactor many cubic meters in size. There may be a difference between a reaction, which is performed only a few times in a chemical development environment under careful observation and a reaction which is designed to be run routinely on a large scale for many years. A reactor, which is equipped with temperature alarms and state-of-the-art automatic control of feed, stirrer, and cooling medium can be handled differently than an old fashioned reactor with nothing but a thermometer. Last, but not least, the technical skill and training of the operators may have an influence. So every company must answer the question for its specific acceptable TMR for itself. However, the outlined approach can help to bring these decisions to a scientific and justifiable basis. This approach helps to compare risks and by means of the runaway scenario findings in special thermal hazards labs can be visualized in a way that operators at the reactor can understand.

### Gas Generating Reactions

At the end of this excursion in hazards of exothermal reactions a final chapter should be devoted to gas generating reactions. Ultimately it is a dangerous pressure increase which has the destructive potential and which has been up to now assigned to thermal runaway behavior, which can increase the total pressure by increasing the vapor pressure of the solvent or by producing large amounts of gas through decomposition reactions. On the other hand there are numerous synthetically useful reactions, which produce gases in the desired reaction. These reactions can be exothermal as in the case of substitution reactions of aryldiazonium salts, which produce nitrogen, but there is also a large number of gas producing reactions, where the gas is more or less soluble and escapes from the solution in an overall endothermal reaction step. An example for this kind of reaction is the chlorination of alcohols with thionylchloride most often under reflux conditions, where sulfurdioxyd and hydrogenchloride is generated.

Whereas the first type of reactions can be accessed by the outlined thermal runaway scenario (gas production rate and rate of chemical conversion are synchronous), the latter type of reaction is much more complex. Chemical conversion and rate of gas liberation are not necessarily synchronous but in addition to the chemistry depend on gas solubility, which is a function of temperature, pressure and composition. Parallel to the complexity of these kinds of reactions goes along a much higher experimental effort required to obtain data, which allows an accurate risk assessment. Reaction calorimetry has to be extended to measure under reflux conditions[21] and devices to quantitatively determine gas flow rates have to be added.[22] Latest development in this area is the coupling of a reaction calorimeter with gas-phase FT-IR and a mass spectrometer.[23]

## References

1. Yoshida, T. *Safety of Reactive Chemicals,* Industrial Safety Series Vol 1, Elsevier, Amsterdam, **1987**, Chap. 1, pp 3-22
2. Grewer, T. *Thermal Hazards of Chemical Reactions,* Industrial Safety Series Vol 4, Elsevier, Amsterdam, **1994**, Chap. 7.2, pp 303-341
3. Sieder, E.N.; Tate, G.E. *Ind. Eng. Chem.* **1936**, *28*, 1429
4. Bourne, J.R.; Buerli, M.; Regenass, W. *Chem. Eng. Sci.* **1981**, *36*, 347
5. Choudhury, S.; Utiger,L. *Chem. Ing. Tech.,* **1990**, *62*, 154
6. Newton, I. *Phil. Trans. Roy. Soc.* **1701**, *22*, 82
7. Cox, J.D.; Pilcher, G. *Thermochemistry of Organic and Organometallic Compounds,* Academic Press, London and New York, 1970
8. Pedley, J.B.; Naylor, R.D.; Kirby, S.P. *Thermochemical Data of Organic Compounds* Sec. Edition, Chapman and Hall, London and New York, 1986
9. Benson, S.W. *Thermochemical Kinetics,* Sec. Edition, John Wiley & Sons, New York , 1976
10. Regenass, W. *Thermal and Kinetik Design Data from a Bench Scale Heat Flow Calorimeter,* ACS Symposium Series No. 65, **1978**, 37
11. Regenass, W. *Thermochim. Acta,* **1977**, *20*, 65
12. Hoffmann, W. *Conference Proceedings Scale-up of Chemical Processes,* Brighton, **1994**
13. Landau, R.N.; Penix, S.M.; Donahue, S.M.; Rein, A.J. *Process Development Utilizing Advanced Technologies: the RCI™ Reaction Calorimeter, the Multi-functional SimuSolve™ Software, and the ReactIR™ Reaction Analysis System,* SPIE Vol 1681 Optically Based Methods for Process Analysis, **1992**, 356
14. Gygax, R. *Chem. Eng. Sci.,* **1988**, *43*, 1759
15. Hoffmann, W. *Chimia,* **1989**, *43*, 62
16. Fisher, H.G.; Forest, H.S.; Grossel, S.S.; Huff, J.E.; Muller, A.R.; Noronha, J.A.; Shaw, D.A., and Tilley, B.J.; *The Design Institute for Emergency Relief System (DIERS) Project Manual,* The Design Institute for Emergency Relief Systems of the American Institute of Chemical Engineers, New York, **1992**
17. Townsend, D.I.; Tou, J.C. *Thermochim. A cta,* **1980**, *37*, 1
18. Singh, J. *Thermochim. Acta,* **1993**, 211
19. Stoessel, F. *Chem.Eng.Progress,* **September 1995**, pp. 46-53
20. Stoessel, F. *Chem.Eng.Progress,* **October 1993**, pp. 68-75
21. Nomen, R.; Sempere, J; and Lerna, P.; *Thermochimica Acta ,* **1993**, *225*, 263
22. Hentschel, B.; and Schliephake, V.; *Thermochimica Acta,* **1993**, *225, 239*
23. Am Ende, D. , Presentation on 8th RC1 User Forum, Lugano, **1997**

# DESIGN OF EXPERIMENTS

# AND

# AUTOMATION

# Using Factorial Experiments in the Development of Process Chemistry

Daniel R. Pilipauskas
*Chemical Sciences, Searle, 4901 Searle Parkway, Skokie, Illinois, 60077*

Keywords:  Factorial design, design of experiment, response surface

Identifying a feasible chemical synthesis for the manufacture of a bulk pharmaceutical chemical (BPC) is the first step in the development of process chemistry.  A deep understanding of how process variables effect the chemistry is necessary for the design of a chemical process that can reliably produce BPC with low, predictable levels of impurities as well as with the correct bulk physical properties.  Moreover, due to the business need of bringing products to the market faster, a chemical process must reach this level of performance in as few as three to five years.

To determine the importance of process variables, the chemist must understand the per cent change in yield and impurities for a corresponding change in each process variable.  In the ideal case, this information would be obtained from experimentally determined rate laws for the synthesis and side reactions that relate temperature, concentration, pressure, solvent and temperature effects to yield and impurity levels.  In practice, the process chemist is often faced with the task of quickly scaling up a synthesis without the benefit of a complete mechanistic understanding of the chemistry.  To gain as much knowledge as possible about the chemistry in a limited amount of time requires a good experimental program.

Process chemists learn the importance of process variables through experimentation.  As with most scientists, chemists follow the experimental practice of changing one-factor-at-a-time (OFAT).  While this method has generated most of our chemical knowledge, OFAT experimentation does not easily provide the quantitative information  needed to rank the importance of process variables. Experimental designs that provide an estimate of the effect of process variables are known as two-level factorial designs.[1-4] These designs have been used extensively for studying processes where the underlying principles are not well developed or are extremely complicated, e.g., drug formulation development.  Use of factorial design in synthetic organic chemistry and process chemistry has also been described.[5]

Factorial experiments consist of a systematic variation of two or more variables at once. For a two level experiment, each variable is set to either a high or low value according to a standard pattern. An experimental run is conducted for each possible combination of variable settings. When only a small number of experiments can be run, it is possible to reduce the number of variable setting patterns by a half, a quarter or an eighth. Designs with a reduced number of variable patterns are known as fractional factorial experiments.

Selection of the low and high levels for each variable is important for obtaining meaningful results. If the levels selected are too close together, the calculated effect could be no larger than the experimental noise. Selecting widely separated levels could result in running the chemical synthesis under unrealistic conditions, e.g., above the solvent boiling point, below a reagent's solubility point, or above the decomposition temperature for reagents and reactive intermediates.

Data analysis of factorial designs is essentially a comparison of the experimental response at the high and low settings of each variable. The comparison, or contrast, of the response is an estimate of the effect of the variable. When the experimental response to a variable depends on the setting of a second variable, the two variables are said to interact. The magnitude of this synergistic effect is also obtained as part of the data analysis of factorial experiments. Fractional factorials used to study a number of variables that is large compared to the number of experimental runs do not provide this information.

### Two Level Factorial Designs

To illustrate the construction and analysis of a two level factorial design experiment, a simulated study of the alkaline hydrolysis of 0.05 moles of ethyl 4-nitrobenzoate is used. The experimental response will be $t_{98}$, the time required to consume 98% of the ester. Values for $t_{98}$ are calculated with equation (1). Experimentally determined values for the activation energy $E_a$ and pre-exponential A in acetone/water are used.[6]

$$t_{98} = \frac{1}{A*e^{-(E_a/R*T)}} * \frac{1}{Base_o - Ester_o} * \ln \frac{Ester_o(\ Base_o - 0.98* Ester_o)}{Base_o(\ Ester_o - 0.98* Ester_o)} \tag{1}$$

$Base_o$ is the concentration of the base and $Ester_o$ is the concentration of the ester.

Four factors are studied in this simulated experiment: temperature, the concentration of the ester, the amount of base and stirring (any potential effect of stirring is not actually calculated in this example but is included as a "dummy" variable for the purpose of demonstrating several data analysis concepts) (Table 1). The designation for this design is $2^4$, signifying four factors studied at two levels. A total of sixteen experimental runs are needed (Table 2). The effects are calculated using the Yates analysis (Table 3).[7] If an experimental run were not completed successfully, a least squares fit would be used in place of the Yates analysis.[8]

**Table 1.** Factors And Factor Ranges

| Factor | Low Setting | High Setting |
|---|---|---|
| Temperature | 288°K | 308°K |
| Ester concentration[1] | 0.0025 M | 0.1 M |
| Base | 0.05 moles[2] | 0.06 moles |
| Stirring | 50 rpm | 100 rpm |

[1]Reaction volume of 2 L and 0.5 L for the low and high settings, respectively.
[2]0.050001 was used to avoid dividing by zero in equation (1).

**Table 2.** Design Table and Experimental Results For The $2^4$ Design

| A Temperature (°K) | B Concentration (M) | C Base (moles) | D Stirring (rpm) | $t_{98}$ (minutes) |
|---|---|---|---|---|
| 288 | 0.025 | 0.05 | 50 | 226 |
| 308 | 0.025 | 0.05 | 50 | 55.5 |
| 288 | 0.100 | 0.05 | 50 | 56.5 |
| 308 | 0.100 | 0.05 | 50 | 13.9 |
| 288 | 0.025 | 0.06 | 50 | 51.1 |
| 308 | 0.025 | 0.06 | 50 | 12.5 |
| 288 | 0.100 | 0.06 | 50 | 12.8 |
| 308 | 0.100 | 0.06 | 50 | 0.92 |
| 288 | 0.025 | 0.05 | 100 | 226 |
| 308 | 0.025 | 0.05 | 100 | 55.5 |
| 288 | 0.100 | 0.05 | 100 | 56.5 |
| 308 | 0.100 | 0.05 | 100 | 13.9 |
| 288 | 0.025 | 0.06 | 100 | 51.1 |
| 308 | 0.025 | 0.06 | 100 | 12.5 |
| 288 | 0.100 | 0.06 | 100 | 12.8 |
| 308 | 0.100 | 0.06 | 100 | 0.92 |

**Table 3.** Effects Table For The $2^4$ Design

| Term | Effect | Sum Of Squares |
|---|---|---|
| A | -65.9 | 17369 |
| B | -65.2 | 17027 |
| C | -68.6 | 18849 |
| D | 0 | 0 |
| AB | 38.7 | 5977 |
| AC | 40.7 | 6611 |
| AD | 0 | 0 |
| BC | 40.3 | 3249 |
| BD | 0 | 0 |
| CD | 0 | 0 |
| ABC | -25.3 | 2559 |
| ABD | 0 | 0 |
| ACD | 0 | 0 |
| BCD | 0 | 0 |
| ABCD | 0 | 0 |

The next step is to determine which variables have significant or active effects, i.e., represent an experimental response and not experimental noise. The method typically used is the normal plot of the effects.[9] Inactive effects lie on a line near zero effect, while active effects do not (Figure 1). In this experiment, seven effects are judged to be active. Typically, higher order interactions, such as ABC, are small or inactive, but in the case of simulations, there is no experimental noise in the data to mask the effect.

Success of the normal plot method depends on the experiment having several inactive effects, a condition known as effects sparsity.[10] If this condition is not met, as can often happen in organic chemistry experiments,[11] all effects will lie on the line and no decision about active effects can be made by this method. The lack of effects sparsity can be demonstrated with the same data using a fractional factorial design (Table 4). In this case, the same number of factors are studied but with half the number of experiments (designated as $2^{4-1}$). The result of the smaller number of experimental runs is less information and a smaller number of estimated effects (Table 5). Without these inactive effects to indicate the no effect level or experimental noise, the normal plot method of drawing a line through the effects near zero does not assist the chemist in identifying active effects (Figure 2).

When the normal plot method fails, replicating the experiment to obtain an estimate of the noise can resolve this issue. Other methods, such as using the larger values of the sum of squares (a measure of the amount of variation in the response that is explained by the effect) to provide insight into which effects are active have been reported.[12]

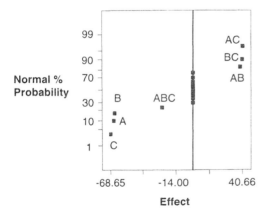

**Figure 1.** Normal Plot For The $2^4$ Design.

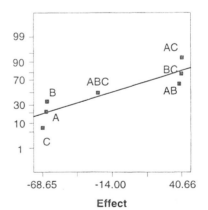

**Figure 2.** Normal Plot For The $2^{4-1}$ Design.

**Table 4.** Design Table and Experimental Results For The $2^{4-1}$ Design

| A | B | C | D | |
|---|---|---|---|---|
| Temperature | Concentration | Base | Stirring | $t_{98}$ |
| (°K) | (M) | (moles) | (rpm) | (minutes) |
| 288 | 0.025 | 0.050001 | 50 | 226 |
| 308 | 0.025 | 0.050001 | 100 | 55.5 |
| 288 | 0.100 | 0.050001 | 100 | 56.5 |
| 308 | 0.100 | 0.050001 | 50 | 13.9 |
| 288 | 0.025 | 0.06 | 100 | 51.1 |
| 308 | 0.025 | 0.06 | 50 | 12.5 |
| 288 | 0.100 | 0.06 | 50 | 12.8 |
| 308 | 0.100 | 0.06 | 100 | 0.92 |

**Table 5.** Effects Table For The $2^{4-1}$ Design

| Term (Confounded With) | Effect | Sum Of Squares |
|---|---|---|
| A (BCD) | -65.9 | 8684 |
| B (ACD) | -65.2 | 8514 |
| C (ABD) | -68.6 | 9424 |
| D (ABC) | -25.3 | 1280 |
| AB (CD) | 38.7 | 2988 |
| AC (BD) | 40.7 | 3306 |
| AD (BC) | 40.3 | 3249 |

Another consequence of running a smaller number of experiments in the $2^{4-1}$ design is the loss of information. Without the extra runs of the $2^4$ design, the calculation of effects results in the main effects A, B, C, and D being combined with the effects of the three-factor interactions BCD, ACD, ABD and ABC, respectively. This situation is known as confounding or aliasing of effects. Since the higher order interactions are usually small, the judgment of active effects is not compromised by this result. However, in this example the three-factor interaction ABC is relatively large and when confounded with the D effect (0.00) causes the D effect to appear significant (this observation is based solely on the fact that D is a dummy effect).

Confounding is also an issue with the two-factor interactions and is more likely to lead to an ambiguous interpretation than the situation of confounding main and three-factor interactions. In this example design, AB, AC and AD are confounded with CD, BD, and BC, respectively. Without additional experimental runs, the ambiguity cannot be resolved (compare the D and AD terms in Tables 3 and 5).

Once the active effects are identified, plotting the experimental responses as a function of the variable setting is often an informative data analysis method. Since the two-factor interaction effects in this example have similar magnitudes as the main effects, plotting the two-factor interactions is more informative than plotting the main effects individually. The effect plot for the AB interaction illustrates the value of plotting the effects, especially two-factor interactions (Figure 3). Each point represents the average response for the given setting. The slope of the line designated B- represents the response of the chemistry when the concentration is low while

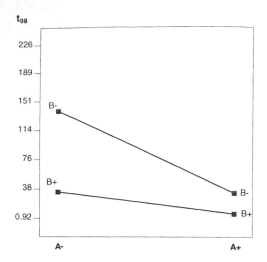

**Figure 3.** Temperature (A)-Concentration (B) Interaction Plot

the B+ line represents the response at the high concentration condition. (The line is used as a guide for the eye. Linearity should not be assumed because simple 2-level designs cannot detect curvature without additional design points). The experimental response $t_{98}$ clearly responds differently to temperature depending on the concentration setting. This type of knowledge is not available from data obtained with OFAT experiments. The effect plots for the AC and BC interactions show similar trends.

Another useful display of this data is a geometric plot (Figure 4). This type of display is helpful when determining the direction in factor space[13] for optimizing process chemistry. Using a cube plot, the process chemist can readily see how reaction time and processing conditions are related. If sufficient knowledge is obtained by the experiment, the chemist can use the cube plot to select the processing conditions that are achievable in the plant while meeting a reaction time criteria.

A more quantitative method for viewing the data is to graph the predicted response against any two factors of interest. Such a plot is known as a response surface and can be rendered in two or three dimensions (Figure 5).[14] These plots are models of the system under study and are only an approximation of the actual response surface.

In the following sections, two process chemistry examples are presented to show how scaleup problems concerning impurity reduction and yield improvement can be addressed using factorial design experiments. This work was conducted shortly after the project entered the chemical development department and after small scale pilot plant runs were conducted to make BPC for clinical supplies and formulation development.

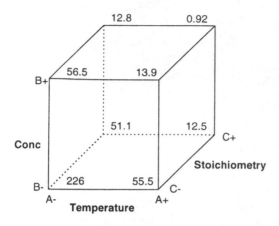
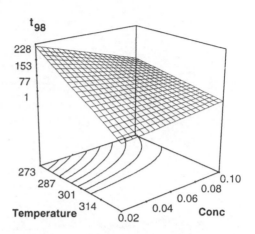

**Figure 4.** Cube Plot Of $t_{98}$ . Factor space was reduced from four dimensions to three by ignoring mixing.

**Figure 5.** Contour and Three Dimensional Plots Of $t_{98}$ . Contour lines represent step changes in yield.

## Critical Variables For Impurity Reduction

Keeping the amount of impurities as low as possible is important for successful down stream purification and isolation operations. When the capacity of these operations to remove impurities is exceeded, the isolated chemical must be either reworked or disposed of as waste. This was the situation that led to the study described in this example.

Two sequential crystallizations were used to purify the BPC (4). When the amount of acid (1) exceeded two weight per cent after the first crystallization, the second crystallization could not reduce the acid in the BPC to acceptable levels. To understand how to reduce the level of impurities in the feedstock to the purification process, a fractional factorial experiment was conducted to identify the important factors for the conversion of acid (1). Since no method was available to accurately measure the amount of the mixed anhydride (2), effects on the first reaction were inferred from the yield of the BPC (4).

**Scheme 1.** Amide Bond Formation

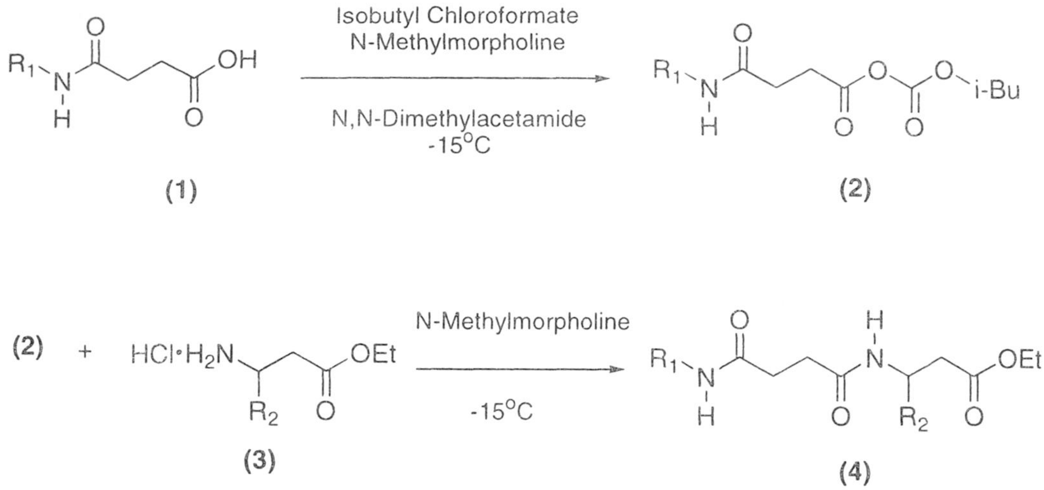

(1)                                                                          (2)

(3)                                                                          (4)

The amount of acid (1) is likely to be controlled by the reagent stoichiometry and side reactions such as formation of the symmetric anhydride of the acid, the reaction between isobutyl chloroformate (IBCF) and the side chain $R_1$, the reaction of the amine (3) at the wrong carbonyl of the mixed anhydride, and combination of unreacted isobutylchloroformate with the amine (3). Factors considered likely to be important were temperature, concentration and reagent stoichiometry. Temperature was held constant at -15°C because lower temperatures would freeze the solvent. Higher temperatures could lead to the decomposition of the mixed anhydride. Five factors were investigated in a $2^{5-1}$ fractional factorial design (Table 6). Main effects and two-factor interactions are confounded only with four and three factor interactions. Process responses were the yield of the amide (4) and the amount of acid (1).

**Table 6.** Factors And Factor Ranges For The $2^{5-1}$ Design

| Factors | Units | Low Setting | High Setting |
| --- | --- | --- | --- |
| Concentration | Molar | 0.18 | 0.50 |
| Isobutyl Chloroformate | Equivalents | 0.95 | 1.05 |
| Amine | Equivalents | 0.95 | 1.05 |
| N-Methylmorpholine (NMM1) | Equivalents | 0.95 | 1.05 |
| N-Methylmorpholine (NMM2) | Equivalents | 0.95 | 1.05 |

**Experimental Protocol**

All reactions were conducted under nitrogen. The mixed anhydride was formed by first adding isobutyl chloroformate to a solution of the acid (1) at -15°C followed by the addition of N-methylmorpholine (NMM1) with a syringe pump. The mixture was held for one hour before adding a solution of the amine hydrochloride (3) and then a second equivalent of N-methyl-morpholine (NMM2). The mixture was then warmed to -5°C and held two hours before warming to 25°C over thirty minutes. The reaction was analyzed for product (4) and the acid (1) with HPLC and the results reported as area % (Table 7).

**Table 7.** Design Table And Experimental Results

| A Conc. (M) | B IBCF (Equiv) | C Amine 3 (Equiv) | D NMM1 (Equiv) | E NMM2 (Equiv) | BPC 4 (Area %) | Acid 1 (Area %) |
|---|---|---|---|---|---|---|
| 0.18 | 0.95 | 0.95 | 0.95 | 1.05 | 82.83 | 15.02 |
| 0.50 | 0.95 | 0.95 | 0.95 | 0.95 | 80.63 | 17.10 |
| 0.18 | 1.05 | 0.95 | 0.95 | 0.95 | 72.10 | 24.10 |
| 0.50 | 1.05 | 0.95 | 0.95 | 1.05 | 81.91 | 15.46 |
| 0.18 | 0.95 | 1.05 | 0.95 | 0.95 | 84.19 | 13.92 |
| 0.50 | 0.95 | 1.05 | 0.95 | 1.05 | 81.28 | 16.74 |
| 0.18 | 1.05 | 1.05 | 0.95 | 1.05 | 80.90 | 17.30 |
| 0.50 | 1.05 | 1.05 | 0.95 | 0.95 | 73.55 | 21.60 |
| 0.18 | 0.95 | 0.95 | 1.05 | 0.95 | 91.23 | 6.16 |
| 0.50 | 0.95 | 0.95 | 1.05 | 1.05 | 76.29 | 19.65 |
| 0.18 | 1.05 | 0.95 | 1.05 | 1.05 | 90.56 | 7.37 |
| 0.50 | 1.05 | 0.95 | 1.05 | 0.95 | 90.19 | 7.46 |
| 0.18 | 0.95 | 1.05 | 1.05 | 1.05 | 88.50 | 9.07 |
| 0.50 | 0.95 | 1.05 | 1.05 | 0.95 | 75.36 | 19.84 |
| 0.18 | 1.05 | 1.05 | 1.05 | 0.95 | 90.38 | 7.30 |
| 0.50 | 1.05 | 1.05 | 1.05 | 1.05 | 88.45 | 9.54 |
| 0.34 | 1.00 | 1.00 | 1.00 | 1.00 | 85.66 | 12.27 |
| 0.34 | 1.00 | 1.00 | 1.00 | 1.00 | 87.74 | 9.98 |
| 0.34 | 1.00 | 1.00 | 1.00 | 1.00 | 85.54 | 11.77 |
| 0.34 | 1.00 | 1.00 | 1.00 | 1.00 | 86.15 | 11.89 |

**Table 8.** Effects Estimate

| Factor | Response BPC 4 | Sum of Squares | Response Acid 1 | Sum of Squares |
|---|---|---|---|---|
| A | -4.11 | 68 | 3.40 | 46 |
| B | 0.95 | 4 | -0.93 | 3 |
| C | -0.41 | 1 | 0.376 | 1 |
| D | 6.71 | 179 | -6.85 | 188 |
| E | 1.62 | 11 | -0.91 | 3 |
| AB | 4.19 | 70 | -3.89 | 61 |
| AC | -2.19 | 19 | 1.64 | 11 |
| AD | -3.48 | 48 | 3.25 | 42 |
| AE | 0.43 | 1 | -0.24 | 0 |
| BC | 0.00 | 0 | -0.03 | 0 |
| BD | 6.10 | 148 | -4.83 | 94 |
| BE | 2.25 | 21 | -1.78 | 13 |
| CD | -0.99 | 4 | 0.90 | 3 |
| CE | 2.26 | 21 | -1.60 | 10 |
| DE | -2.26 | 25 | 2.13 | 18 |

## Discussion

Using the normal plot of effects and sum of squares for the BPC and acid responses, the main effects A and D and the two-factor interactions AB and BD are judged to be active (Table 8). The strong interaction effect BD (isobutyl chloroformate/N-methylmorpholine) for the BPC response is not a surprise (Figure 6b). When the amount of N-methylmorpholine is less than one equivalent (line D-), adding more isobutyl chloroformate (B- → B+) cannot increase the yield because only the salt form of (1) can react. In addition, the excess isobutyl chloroformate actually lowers the yield by reacting with the amine (3). Increasing the amount of N-methylmorpholine (line D+) converts all of the acid to its salt and results in a yield increase with an increase in the amount of isobutyl chloroformate. The acid response follows an opposite trend (Figure 7b).

The AB interaction (concentration/isobutyl chloroformate, Figure 6a) reveals an important concentration effect. A decreased BPC response occurs at low isobutyl chloroformate setting (line B-) as the concentration (A- → A+) is increased while at the the high setting (line B+), little effect is seen. The acid response follows an opposite trend (Figure 7a). A possible explanation for these responses is the formation of symmetric anhydride (5) when isobutyl chloroformate is undercharged. However, anhydride formation does not explain the different BPC and acid responses at the low setting of concentration (see effects at A- in Figures 6a and 7a). Further experimental work is needed to elucidate the underlying reaction mechanism for this effect.

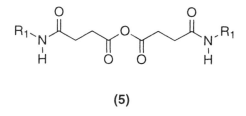

**(5)**

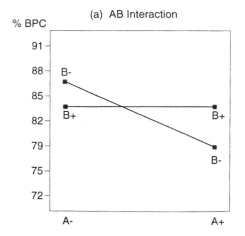
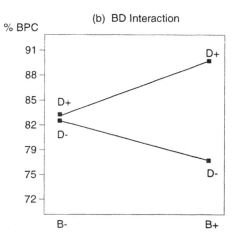

**Figure 6.** Interaction Plots For BPC Response. (a) Concentration (A)-Isobutyl Chloroformate (B); (b) Isobutyl Chloroformate (B)-N-methylmorpholine(D).

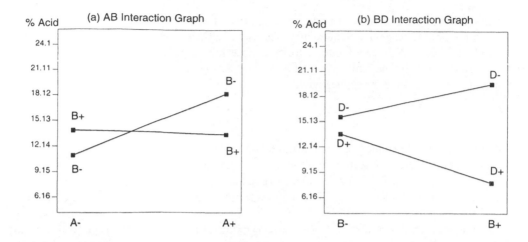

**Figure 7.** Interaction Plots For The Acid Response. (a) Concentration (A)-Isobutyl Chloroformate (B); (b) Isobutyl Chloroformate (B)-N-methylmorpholine(D).

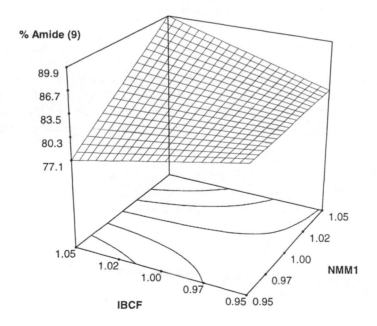

**Figure 8.** Contour Plot Of Amide Yield. Contours represent 2% changes in yield.

Examination of the contour plot of the BPC response (Figure 8) clearly shows the sensitivity of the reaction yield to reagent stoichiometry. The yield is highest when both IBCF and NMM1 are set high (1.05 equivalent each), but the trend towards higher yield will reverse when excess IBCF begins to react with the amine (3). Additional experimental runs are needed to characterize the response surface around the highest yield.

**Conclusion**

The establishment of a quantitative relationship between the amount of the acid impurity (1) and yield of BPC (4) suggests that further development efforts should focus on improving the conversion of the acid to BPC . Further improvements in the conversion of the acid can be gained through increases in the amount of IBCF and NMM1, but the gains are likely to be limited.

The concentration-IBCF interaction (AB effect) indicates that the reaction process is not fully defined by a simple conversion of the acid to the amide through the mixed anhydride (Scheme 1). Side reactions, such as the formation of the symmetric anhydride (5), and physical properties such as the solubility of the acid at reaction temperatures, may be playing a role in reducing the yield.

Finally, the sensitivity of reaction yield and impurity level to stoichiometry (Figure 8) suggests that this process may be too susceptible to variation in stoichiometry. A more robust (less sensitive) production process is needed to avoid the necessity of extremely tight controls.

**Optimizing A Chemical Reaction**

In this example, a synthesis is altered and the yield maximized with the assistance of factorial design experiments. The original β-amino ester synthesis required a reaction temperature of -40°C. The operation involved the generating the enolate (6) and the imine (8) in separate reaction vessels before the enolate was added to the imine. Because the plant did not have large reactors capable of maintaining such low temperatures, many small batches were being run to meet rising clinical demands. To address this issue the preparation was modified to run at higher temperatures.

Conducting stability studies on the intermediates (6) and (8) showed that the low temperature requirement for this reaction was primarily driven by the thermal instability of the enolate (6).[15] If the enolate could be formed and immediately reacted with (8), enolate stability would not be an issue and the reaction could be conducted at a higher temperature.[16]

**Scheme 2.** β-Amino Ester Synthesis

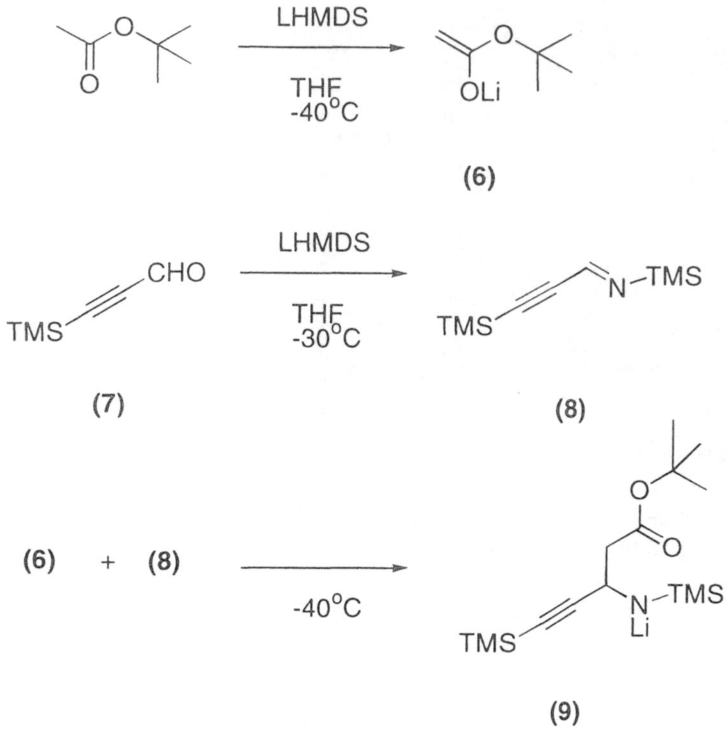

The process was modified by adding the t-butyl acetate (TBA) to the imine (8), followed by lithium bis(trimethylsilyl)amide (LHMDS). Reactions performed at -20 to -10°C gave yields higher than the original process run at these temperatures, but lower than the original process at -40°C. A study was initiated to determine the factors that were important for maximizing the yield. Temperature was considered likely to be important during imine formation and the condensation of the enolate (6) with the imine (8). Preliminary experiments indicated that the amount of t-butyl acetate effected yield as did the amount of LHMDS in imine formation. These four factors were studied in a $2^{4-1}$ fractional factorial design plus replicated center point (Table 9).

**Table 9.** Factors And Factor Ranges

| Factor | Low Setting | High Setting |
|---|---|---|
| Temperature (Imine Formation) | -20°C | -10°C |
| t-Butyl Acetate | 4 Equiv | 8 Equiv |
| LHMDS (Imine Formation) | 0.9 Equiv | 1.1 Equiv |
| Temperature (Condensation) | -20°C | -10°C |

**Table 10.** Design Table And Experimental Results

| A Temp. Imine (°C) | B t-Butyl Acetate (Equiv) | C LHMDS Imine (Equiv) | D Temp. Conden (°C) | Yield (%) |
|---|---|---|---|---|
| -20 | 4 | 0.9 | -10 | 24 |
| -10 | 4 | 0.9 | -20 | 30 |
| -20 | 8 | 0.9 | -20 | 36 |
| -10 | 8 | 0.9 | -10 | 29 |
| -20 | 4 | 1.1 | -20 | 43 |
| -10 | 4 | 1.1 | -10 | 33 |
| -20 | 8 | 1.1 | -10 | 53 |
| -10 | 8 | 1.1 | -20 | 42 |
| -15 | 6 | 1.0 | -15 | 44 |
| -15 | 6 | 1.0 | -15 | 43 |

**Experimental Protocol**

All reactions were conducted under nitrogen. The aldehyde (7) was added to LHMDS in THF at the appropriate temperature. To this mixture was added chlorotrimethylsilane followed by t-butyl acetate. LHMDS in THF was then added at the appropriate temperature. After quenching the reaction in aqueous ammonium chloride, the mixture was analyzed by GC for the amount of β-amino ester. The results are reported as per cent of theory (Table 10).

**Discussion**

Using a normal plot of effects and sum of squares, the effects B (equivalents of t-butyl acetate), C (equivalents of LHMDS), A (temperature during imine formation) and AC are judged to be active (Table 11). The sum of squares values support the selection of these effects as being important. Since the effect of condensation temperature was small, the experimental data is displayed in a cube plot (Figure 9). The data suggests that the starting point for optimizing the yield is at the high level setting for t-butyl acetate and LHMDS and the low setting of temperature for imine formation.

**Table 11.** Effects Table

| Term (Confounded With) | Effect | Sum Of Squares |
|---|---|---|
| A (BCD) | -5.5 | 61 |
| B (ACD) | 7.5 | 113 |
| C (ABD) | 13.0 | 228 |
| D (ABC) | -3.0 | 18 |
| AB (CD) | -3.5 | 25 |
| AC (BD) | -5.0 | 50 |
| AD (BC) | -2.0 | 8 |

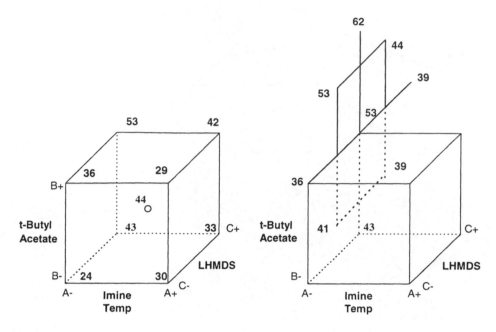

**Figure 9.** Cube Plot Of Per Cent Yield.          **Figure 10.** Central Composite Design Yields.

A series of experiments around the maximum (53%) known as a central composite design (Figure 10) was conducted to map the yield response as a function of the equivalents of t-butyl acetate and the number of equivalents of LHMDS used in imine formation (Table 12). A contour plot of the results (Figure 11) shows the behavior of the process around the maximum yield (center of grid). The reaction process appears to have the highest yield with 1.1 equivalents of LHMDS and twelve equivalents of TBA. Yield improvement diminishes with increasing equivalents of TBA.

**Table 12.** Design Table And Experimental Results

| t-Butyl Acetate (Equivalents) | LHMDS (Equivalents) | Yield (%) |
|---|---|---|
| 6 | 1 | 41 |
| 10 | 1.0 | 53 |
| 6 | 1.0 | 39 |
| 10 | 1.2 | 44 |
| 4 | 1.2 | 43 |
| 12 | 1.1 | 62 |
| 8 | 0.9 | 36 |
| 8 | 1.3 | 39 |
| 8 | 1.1 | 53 |

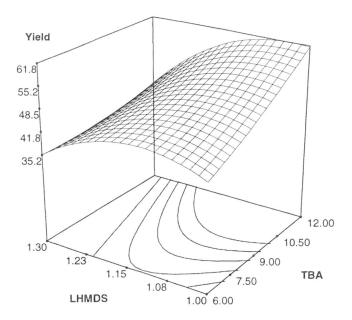

**Figure 11.** Contour Plot Of Yield. Contour lines represent approximately 2% changes.

**Conclusion**

Quantitative estimates for the effects of the experimental variables were achieved with a fractional factorial experiment. With this information, a quick decision was possible to optimize the chemical yield using only two factors, the amounts of LHMDS and t-butyl acetate. Because of the structure of the first factorial design, it was possible to reuse some of the experimental runs in the central composite design experiment. This led to the development of a model of the response surface for yield. Subsequent use of this information allowed for the successful scaleup of the new process.

**Concluding Remarks**

The previous two chemistry problems are typical of the challenges faced by process chemists. Requests to increase the yield, decrease the impurity level, increase the consistency of yield and impurity level, and change the operating conditions to fit the chemistry into plant equipment are typical. With the push to reach the market quickly with cost-effective manufacturing processes, the chemist is faced with responding to these requests faster than ever.[17]

   One approach to achieve an efficient experimental program is through the use of factorial designs. Efficiency gains are achieved by better planning, by a more systematic and thorough approach to studying the problem and by the acquisition of quantitative data about experimental variables and variable interactions. This does not mean that OFAT experiments should be abandoned.

A balanced program of OFAT and factorial experiments is likely to be the most cost-effective approach.[18] OFAT experimentation still plays an important role in any experimental program. OFAT experiments should be used for a quick test of ideas (pass/fail tests) and measuring the effect of one variable. When more than one variable is likely to be critical, discovering two-factor interactions and identifying the important variables is more effectively performed with factorial designs. For this reason, factorial designs, rather than OFAT experiments, are more efficient for optimizing chemical reactions.

## References

1.  Box, G. E. P.; Hunter, W. G.; Hunter, J. S. In *Statistics for Experimenters: An Introduction to Design, Data Analysis and Model Building*; Wiley: New York, 1978.

2.  Morgan, E; In *Chemometrics: Experimental Design*; Wiley: New York, 1995.

3.  Frigon, N. L; Mathews, D. In *Practical Guide to Experimental Design*; Wiley: New York, 1997.

4.  Davies, L. In *Efficiency in Research, Development and Production: The Statistical Design and Analysis of Chemical Experiments*; The Royal Chemistry Society, 1993; Chap 5, pp 72-88.

5.  Carlson, R.; Nordahl, A., *Top. Curr. Chem.*, **1993**, *166*, 1-64; Arroyo, M.; Sinisterra, J. V., *J. Org. Chem.*, **1994**, *59*, 4410-4417; Drouin, J.; Gauthier, S.; Patricola, O.; Lanteri, P.; Longeray, R., *J. Synlett*, **1993**, 791-793; Schwindt, M. A.; Lejon, T.; Hegedus, L. S., *J. Organometallics*, **1990**, *9*, 2814-2819.

6.  Tommila, E.; Hinshelwood, C. N. *J. Chem Soc.*, **1938**, 1801.

7.  Box, G. E. P.; Hunter, W. G.; Hunter, J. S. In *Statistics for Experimenters: An Introduction to Design, Data Analysis and Model Building*; Wiley: New York, 1978; Chap. 10, pp 323-324.

8.  Box, G. E. P; Hunter, W. G.; Hunter, J. S. In *Statistics for Experimenters: An Introduction to Design, Data Analysis and Model Building*; Wiley: New York, 1978; Chap. 14, pp 503-504.

9.  Box, G. E. P ; Hunter, W. G.; Hunter, J. S. In *Statistics for Experimenters: An Introduction to Design, Data Analysis and Model Building*; Wiley: New York, 1978; Chap. 10, pp 329-330.

10. Lenth, R. V. *Technometrics*, **1989**, *31*, 469-473.

11. Sadiris, N. C.; Thanasoulias, N. C., Evmiridis, *Chemom. Intell. Lab. Sys.*, **1991**, *12*, 49-55.

12. Stowe, R. A.; Mayer, R. P. *Ind. Eng. Chem.*, **1966**, *58*, 36-40; Benski, H. C. *J. Qual. Techn.*, **1989**, *21*, 174-178; Box, G. E. P.; Meyer, R. D. *Technometrics*, **1986**, *28*, 11-18.

13. Carlson, R. *Chem. Scr*, **1987**, *27*, 545-552.

14. Box, G. E. P.; W. G. Hunter; J. S. Hunter In *Statistics for Experimenters: An Introduction to Design, Data Analysis and Model Building*; Wiley: New York, 1978; Chap. 14, pp 453-509.

15. Sullivan, D.F. *The Reactions of Ester Enolates*; PhD Dissertation, Michigan State University 1974. *Diss. Abstr. Int. B* 1975, 36, 248.

16. Boys, M. L.; Cain-Janicki, K. J.; Doubleday, W. W.; Farid, P. N.; Kar, M.; Nugent, S. T.; Behling, J. R.; Pilipauskas, D. R. *Org. Process Res. Dev.*, **1997**, (submitted).

17. Pisano, G. P. In *The Development Factory*; Harvard Business School Press: Boston, 1997; Chap 3, pp 51-80.

18. Morgan, E; In *Chemometrics: Experimental Design*; Wiley: New York, 1995; Chap. 1, pp 3-4.

# Laboratory Automation in Chemical Development

**Martin R. Owen**
Chemical Development, GlaxoWellcome, Gunnels Wood Road, Stevenage, Hertfordshire,
SG1 2NY England    mro1220@.ggr.co.uk

**Sheila H. Dewitt**
Orchid Biocomputer, 201 Washington Road, Princeton, NJ 08543-2197 sdewitt@orchidbio.com

Key words: Chemical Development; laboratory automation, robotics, experimental design, process development, process optimization, process screening, combinatorial chemistry

## Introduction

*What is impossible to achieve right now, but, if we could do it, would fundamentally change the way we carry out chemical development?*

Automation of laboratory procedures is a glib answer to this question. To achieve this goal, the most obvious challenge appears to be to overcome technical hurdles, such as developing automated filtration, solid dispensing, thermal control and systems software. However, simply to emulate manual methods with automation is not enough. Chemists must develop techniques that are suited to automation, for example, more efficient ways of carrying out established statistical experimental strategies that are not so easy to apply using manual methods. To implement automation effectively in chemical development, we have to undergo a paradigm shift and rethink the existing frameworks of models and strategy.

A similar paradigm shift that is already widely accepted in the pharmaceutical industry is the use of combinatorial chemistry[1] to aid the process of lead generation and lead optimization in drug discovery. Combinatorial methods offer a way of generating large numbers of novel and structurally diverse compounds. In order to maximize the effectiveness of the approach, new practical techniques and strategies have evolved, often involving solid phase chemistry. Strategic techniques such as "split and mix"[2] and deconvolution[3] have evolved to supplement the traditional "one vessel-one compound" approach. Chemical tag-encoding[4] technology has helped with structure determinations. In the absence of a well-designed approach, combinatorial chemistry is nothing more than "a method of increasing the size of the haystack in which to find your needle."[5]

Some perspective articles[6] and reviews[7] of automated synthesis efforts have been published, a few have explored the shift to and incorporation into combinatorial or high-throughput chemistry[8] but none have focused on the implementation of combinatorial chemistry and automation into process development programs.

The necessity for automation becomes apparent, when chemistry development is seen within the broader context of the pharmaceutical industry. When combinatorial chemistry introduces such a paradigm shift, changes across the board are inevitable. Rapid development of both hardware and software technology open up a whole new world of possibilities. High throughput screening and analytical techniques are emerging to transform the rate of data acquisition. Elsewhere, in pilot plant and production, computer driven technology has had a profound effect on improving process control.

Other external forces provide drivers for change. These include the impact of a global marketplace, aggressive pricing, reallocation of resources and the spiraling costs to bring a drug to market. These costs are escalating, due to ever-increasing tighter regulatory, quality and safety standards. For example, studies show that the average number of clinical trials per New Drug Application has doubled over the past ten years. Development chemists must respond to these increased pressures by improving the effectiveness and efficiency of their research. They cannot afford to continue using the same "one reaction at a time" techniques that were appropriate in the past.

The challenge for chemical development is different than that of combinatorial chemistry, where high throughput is a major goal. For the development chemist, the focus is on the process, rather than the product. It is important to produce high quality data efficiently. This data is used to correlate the effect of factors, such as time and temperature, on yield and purity. With the aid of statistical approaches a lot of useful information can be obtained with relatively few experiments. Automation can potentially facilitate this goal and also allow the boundaries of investigation to be extended. The increased level of interest in this area was evident from the first dedicated conference on laboratory automation in Chemical Process R&D.[9]

Automation will inevitably impact on the way we work. Changing the mind-set is fundamental to the introduction of automation into the laboratory. There are many consequences for both the chemist and the analyst. Traditional departmental barriers are broken down and automation's true potential can only be unleashed through application of a multi-disciplinary approach that includes expertise from chemists, analysts, statisticians, engineers, information technologists and others.

Automation should not be carried out for automation's sake, but as a means of achieving well-defined scientific goals. Automation can be justified if the benefits outweigh the costs. In other words, if it can add value to a business. Over-automation should be avoided. The aims of this chapter is to describe some of the significant automation challenges in chemistry development and to address some fundamental questions such as: what, why and how to automate; when is automation appropriate; how automation impacts on the way we work; and what are the issues of project management of new automation systems.

To answer these questions several case studies have been chosen from the literature. These show how groups with different process development goals require different automated solutions. As these and other groups have had to do, GlaxoWellcome has also had to develop novel systems to fulfil automation needs at the time when the range of commercial equipment is limited. The challenges encountered in championing, developing and implementing such a system are described.

### Assessing the value of automation

To demonstrate value, the benefit must outweigh relative costs. Tangible benefits that can be used to determine value are listed below. However, these cost savings are trivial compared to the benefits of cutting the development time of a drug. If a drug could be delivered to the marketplace in a shorter time frame, this effectively extends its patent life. For best selling drugs, this could translate into millions of pounds of extra sales.

### Benefits

Automation projects should be seen to provide a positive return on investment. Some benefits are clearly quantifiable. For example, suppose the process that is currently in production has been developed using traditional techniques. The process could then be re-optimized, now with the aid of automation. Some of the likely quantifiable benefits derived from the resulting improved process are as follows: enhanced throughput by increased yields, reduced reaction times or by increased concentration. Cost reductions may occur through reduced costs of starting materials and plant time, reduction of waste and associated disposal costs or by a more robust process, avoiding reworks or disposal of batches which fail specification.

Automation provides obvious efficiency benefits, such as the ability to do many experiments at once and the capability of running these experiments for twenty-four hours a day, seven days a week. Automation also provides a way of accelerating process investigation, by building in quality upstream using high quality and timely data, in order to make critical decisions. A better process understanding, obtained through automation would reduce the time spent on laborious manual tasks. It would also ensure that an optimized process is taken into the plant right at the start, rather than using a "make do" process because of limited investigation time. Once a process moves towards a production method, it becomes progressively more difficult to introduce radical change in the method.

The value of robust optimization is also determined by the nature of the process. In the bulk chemical industry, optimization is essential, because the products are usually high volume, with low margins. In the pharmaceutical industry, the products have relatively high margins, but low volume. Here a robust process is usually more pertinent than one fully optimized for yield. Increasing pricing pressures from external forces will increase the importance of optimization in this sector as well.

An additional benefit is that operator exposure to chemicals is reduced. A particularly important example of this is the application of robotics in radiopharmaceutical synthesis[10].

The benefits that can be achieved by liberating both chemists and analysts from mundane tasks are harder to quantify. Automation can provide the luxury of allowing the scientists more time to think about what they are doing, or to carry out more practically demanding tasks such as product isolation, route scouting or investigating side reactions. The scope of work can be broadened, in order to explore areas not previously considered feasible, such as screening additional solvents and reagents. It may provide the opportunity to evaluate several drug candidates at once and be in a better position to "pick the winner."

## Costs

The cost of automation can be both visible and invisible. Initially, the most visible and quantifiable costs of developing and implementing are those associated with set-up, such as capital investment and staff resource. Running costs would include the cost of labor and consumables.

Perhaps less visible, at this stage, are the costs associated with complexity of the system and reliability. These would manifest themselves through the time required for training, or to set-up a new piece of chemistry, the downtime of a system, the cost of repairs and depreciation of the systems.

## Choice of automation system for process investigation

The choice of automation system will have a major impact on the project's ability to deliver the expected benefits. To attempt to develop a system to deal with all the specific tasks in the chemical development life cycle, would be technically challenging and prohibitively expensive. Complexity and costs are likely to increase faster than benefits gained. The goal should be to provide the most benefit at the lowest level of complexity.

The task of identifying what and how to automate is determined by several factors which include the scientific goal (e.g. route scouting, reagent screening, factor optimization, scale-up etc.), ease of automation, and the strategic approach (parallel or sequential experimentation).

## Bottlenecks and work-flow

A good starting point in determining why or whether it is necessary to automate is to look at the entire development life-cycle of a drug. Is chemical development a rate-determining step in this process? Within chemical development, where are the current bottlenecks? Can the bottlenecks be reduced by automation? Once one bottleneck has been relieved by automation, other bottlenecks will emerge in the overall work flow. For example, in the Pharmaceutical industry lead generation and lead optimization may be rate determining. If combinatorial chemistry provides the key to doing this more effectively, then we should expect more potential candidates to enter development. As a consequence of this, more chemical methods will require optimization. If this is relieved by automation, then it is likely that the plant capacity may become the next limiting factor. However, shifting the bottleneck to this position is advantageous, because now the processes are optimized and the plant, a high cost resource, is used to its maximum potential. Other alternatives exist for relieving this new bottleneck, such as outsourcing chemical processes.

The bottlenecks in the chemical development life-cycle will now determine what is important to automate. There are two distinct features of chemical development; obtaining process understanding and production of material. In the early phase of the development life-cycle, these two demands compete for resource. Long-term toxicological studies can take several months to complete and so it is important to start these as early as possible. This requires supply of drug, preferably synthesized by the ultimate production method, so that the trials are carried out on material with typical impurity profiles. If the toxicity testing is carried out and the route is changed significantly, then the impurity profile may also change significantly, often necessitating additional toxicological trials. Therefore it is important to evaluate the route options for production as early as possible. To compare routes, stages have to be screened and optimized and there has to be sufficient starting material generated from previous stages to allow subsequent stages to be evaluated. The dilemma is to decide whether to use resources to generate process understanding or make material. This is dependent on which is rate determining in the overall work flow.

**Task analysis**
To evaluate a synthetic route, each stage in the route has to be examined. It is necessary to be clear about what information is required from this evaluation.

There are essentially three main tasks in performing a chemical reaction:

> reaction preparation and processing
> monitoring
> product isolation

Of these, product isolation is often the most resource demanding step. If the process is segmented into studying the reaction first and then the isolation, this reduces resource costs and simplifies the experimental design. It is often preferable to optimize the reaction conditions first by determining the yield in solution quantitatively. Then the work-up need only be examined for the most promising conditions. In addition, the cost of isolation of intermediates may be avoided completely by performing multiple reactions in single process. Such a 'one pot process' places greater emphasis on achieving high yields, product quality and control in each sub-stage in the overall process. The reaction parameters of each sub-stage can be studied independently, without the purification parameters adding complexity into the experimental design.

The correlation between unisolated and isolated product quality should be appreciated. Different conditions may give rise to different impurity profiles. It is likely that some of these impurities may be harder to remove than others. Therefore it is more important to limit the formation of specified impurities in the unisolated product, rather than to optimize overall purity, irrespective of profile.

**Response determination**
In any type of experimentation, it is essential that the method of determining the response (e.g. yield) is appropriate to the task in hand. The method must also be robust and give meaningful results. For example, in chemical development, high performance liquid chromatography (HPLC), is often used for reaction monitoring. This is because it can give a quantitative measure of product yield and quality. Associated impurities can be detected at low levels (typically <0.1%). Since each set of reaction conditions generates different ranges and levels of impurities, HPLC can be used to compare different reaction profiles. If the products and impurities are volatile, gas chromatography (GC), would be more appropriate.

A robust method is essential, as any variation in response should be attributable to changing experimental conditions, rather than to noise or bias in the response determination.

The results should be meaningful. For example, with HPLC, the relative response factors for product, starting material and impurities are likely to be different at different wavelengths. Some of the impurities may not even be detected. If a standard is used and the relative response factor of the product to the standard is known, then product yield can be quantified, irrespective of impurity profile.

Here too, there is a resource issue. Is it appropriate to develop quality response determination methods such as HPLC for processes which may not come to fruition? Would it be better to use a rough and ready method such as thin layer chromatography for 'quick and dirty' screening? The penalty of adopting the latter approach is that critical decisions may be based on limited and misleading information.

For isolation and future processing, it is important that the product has suitable physical properties so that handling requirements such as filtration rates and flow characteristics can be controlled. If the measured responses can correlate with the handling requirements, these can then be optimized.

## Goals of process investigation

The stage of the process investigation cycle determines what degree of resolution is necessary.
Initially, a screening exercise is necessary to select the most appropriate solvent/reagent combination and what factors critically effect the outcome of the method. The task is then to optimize the factors and finally to ensure that a robust process is in place. At the preliminary stages of exploration, it may simply be a matter of determining qualitatively what works and what doesn't work.

For the initial phase, although it may initially appear that there are infinite combinations of solvents and reagents to be investigated, safety, environmental and economic constraints such a price and availability limit the choice.

Carlson[11] suggests a method of choosing a sub-set of solvents from a larger candidate set by using a technique known as principal component analysis. For example, he characterizes each solvent by nine physical properties. A statistical technique was used to summarize the properties into two numbers T1 and T2. T1 is roughly the measure of the polarity of the solvent and T2 is roughly a measure of the polarizing ability. So T1 is an inherent property and T2 is a measure of how the solvent will react to electronic forces. When T1 is plotted against T2 solvents, a two-dimensional representation of the distribution of solvents in space is obtained. The rationale is that the proximity of two solvents is representative of their similarity. Pragmatically, further improvements to a process may be possible by examining "neighbors" of a good solvent. Alternatively a diverse range of solvents can be selected to assess the scope of the reaction.

In this way each substrate/reagent/solvent combination can be defined as a point in *reaction space*. Each reaction also has its own parameters of *experimental space*, such as reaction time, temperature, stoichiometry and concentration. Each reaction will have its own 'optimum' experimental conditions.

To screen and compare reactions within the reaction space fairly, they should be carried out under their own optimal experimental conditions, rather than under one set of standard conditions. If standard conditions are applied, then the reaction that works best under these conditions will be selected, rather than the best reaction. There is rarely the resource available to screen much more than standard conditions using manual methods. Automation of small-scale experiments make rigorous screening much more feasible. As a result, the scientific validity of screening will be enhanced.

Optimum conditions are often referred to as if there is a single set of experimental parameters that produces the optimum result such as the best yield. However this definition is just too simplistic in the industrial world. There are other responses that are just as important and are often at conflict with optimizing yield.

One example is product quality. It is of no benefit to push the yield up by 5%, if the resulting batch of drug substance fails specification on purity as a result. The batch would have to be reworked to pass specification, which is likely to reduce or eliminate such benefits. The real target should be to optimize the yield, whilst remaining within the quality specification limits.

Economic and environmental benefits are also important to the chemical development chemist. Minimizing the input of starting materials, reagents or solvents can reduce the material costs and waste disposal. Throughput can be improved by reducing reaction time or increasing concentration. Working at temperatures closer to ambient (i.e. at -10°C, rather than -70 °C; or at 60 °C rather than 120 °C) can reduce energy costs.

Another key feature of process investigation is control and robustness. The goal of the optimization study is to select a set of experimental parameters, which best match the specified desirability criteria (e.g. Yield > 90%, level of impurity A < 2%, reaction time < 18h). This should be followed by a robustness study, to see what the effect of minor variation of these parameters would be. It is usual to define parameters, with process deviation ranges, which the plant operator should work within. The tighter the process deviation range, the greater the technical control will have to be. In addition, the tighter the process deviation range is, the harder it is for the operator to remain within those limits.

Therefore to define the optimum conditions is rarely straightforward. There are many trade-offs between responses to be considered and it is important to define and re-evaluate the success criteria for each study.

## Context of laboratory automation to scale-up into production

There are several reasons for working on as small a scale as possible. Many individual reactions can be easily and quickly carried out in a small space. In the preliminary stages of exploration, amounts of starting material are limited.

Small-scale laboratory reactions should provide a basis for scale-up to large-scale laboratory and ultimately industrial-scale production levels. However, some processes such as crystallization, exothermic and phase transfer reactions are very scale dependent, due to bulk transfer and heat transfer effects. In these cases, it is more appropriate to use small-scale experimentation as a guide to identifying the important factors, rather than a predictive tool. It may be necessary to use reaction calorimeters or small plant reactors to obtain the more predictive model.

## Strategic approaches

Traditionally, chemists have used a "one factor at a time" approach to optimization—in other words, assessing the effect of one particular factor by keeping all the other conditions constant. Progress is made by designing the next experiment on the basis of the results of the previous one. Once the value for a particular factor is found to give a maximum response, that factor is held constant and a new factor is investigated.

But while this can be a quick method of improving a single response such as yield, it does not give an overall picture of what is going on. A particular flaw is that it assumes the factors are independent of each other. In reality, of course, factors interact with each other. For example, as every chemist knows, reaction rate increases as the temperature increases. If there are many conflicting goals (such as simultaneously maximizing yield and purity, while minimizing material and waste), then it becomes very difficult to apply this approach efficiently.

In order not to "build a bigger haystack in which to find the needle" the development chemist needs to match automated techniques with strategic approaches. Without doing so, the chemist will simply gain more data, not necessarily more information and knowledge. Corkan and Lindsey suggest that the chemist focuses on synthetic strategies and relegates the implementation of synthetic plans to automation.[12] They distinguish between open and closed loop systems. In open loop systems, data is not evaluated by the robotic system. Instead, the chemist evaluates the data off-line and is responsible for planning of further experiments. Conversely, for closed loop mode, a robotic workstation functions autonomously. Automated reaction planning follows automated data evaluation.

## Experimental design - open loop
A mathematical model of the chemical process is generated on the statistical analysis of a set of many experiments. The design of the set is based on altering several factors simultaneously, in a highly systematic way. The number of experiments that are carried out determines the resolution of the design. A key feature of experimental design is replicated center points, which give an indication of the experimental fit of the model and background noise. With fractional factorial or Plackett Burman designs, the relative importance of factors and their interaction effects can be determined. With high-resolution response surface designs, such as central composite or Box Behnken, predictive models can be generated to determine the best fit for the robust optimization of conflicting goals. Since the reactions can be carried out in parallel, this approach, despite appearing exhaustive, is a very efficient way of generating process understanding using minimum resources. The system operates in open loop mode allowing critical intervention by the chemist to interpret and plan future work.

## Iterative experimentation closed loop
Plouvier and Lindsey have employed a decision tree programming approach.[13] The protocols provide the capability to make decisions about experiments as the data is collected. The yield determines whether additional reagent is added, reactions are truncated where the yield exceeds the threshold and terminated when no activity is found. Thus 'strategic exploration' is possible, enabling rapid pursuit of scientific goals.

The Simplex approach[14] is an alternative iterative approach to optimizing the response of several parameters. The search algorithm is based on an evolutionary optimization procedure. A typical simplex design would be as follows:

Three different experimental conditions are carried out. From this first set, the worst conditions are discarded and a new experiment is performed, which is a reflection of the discarded experiment. From this second set, the worst conditions are discarded and another new experiment is performed. Hence, the experimental conditions evolve towards the optimal region of the response space. If there is a lot of background noise, in other words, if there are other uncontrolled factors contributing to the response variation, it may take some time before it the whereabouts of the true summit becomes apparent. There are also possibilities that two hills exist on the response surface and the lower peak is climbed, rather than the higher.

As with the traditional approach, the Simplex technique is less well suited if there are many conflicting responses to be optimized. The Simplex technique explores less experimental space than experimental design and as a consequence, a less clearer model of the reaction emerges.
The sequential nature of this process mandates the use of a linear sequence of experiments, in closed loop mode. If there is a long reaction time (e.g. greater than 24 hours), the actual time to "climb the hill" can be many days and it is not clear at the outset how much resource will be consumed before the outcome is resolved.

Fuzzy logic and artificial intelligence techniques allow learning via the accumulation of knowledge. Sophistication software systems not only monitor and control processes, but also interpret information and makes decisions based on neural networks. A major characteristic of expert systems is their ability to accept uncertain facts, imprecise knowledge and to work with concepts. Fuzzy controllers are employed in operating and monitoring complex chemical manufacturing processes. These operate on the basis of a set of rules that need only to be roughly specific. An example of this is the work of Moriguchi[15] in the operation of polymerization reactors, where the controller reduces the variability of the target polymer by up to 40%.

## Case studies on automation projects

Several specific case studies from the literature have been chosen to illustrate the issues discussed previously in a chemical development context. Each example presents different strategic approaches and as a consequence, different technological challenges.

### Reaction Optimization

Boettger[16], in common with a number of research groups[17, 7a, 18] has the goal of parallel synthesis in order to optimize chemical reactions. She describes a custom-built Zymark robotic system for achieving this. The assumption is that it is feasible to optimize the reaction first, then to focus on the purification. To prove this, each reaction was monitored by HPLC to obtain quantitative in-situ yields and then the in-process results were related to the purity and yield of isolated product. The robotic in-situ yields correlated very well with the manually isolated yields. Results from the small scale robotic experiments could therefore be used to help identify promising candidates for larger scale manual experiments. An example of a palladium catalyzed coupling of alkyl halides with alkyltributylstannanes is given. The optimized experiments are subsequently evaluated on successively larger scales.

Lindsey[19], using a robotic system which monitors reactions by UV-vis spectroscopy, compares process optimization by both the simplex and a response surface approach. He also exemplifies a decision tree approach to screening.

Table 1: Representative Examples of Automated Process Optimization

| Chemistry | Variables | No. Expts. | Exp. Design | Automation | Ref |
|-----------|-----------|-----------|-------------|------------|-----|
| Base Promoted Synthesis of Styrene | Time Temperature | 4 | | Zymate II Camile 2000 | Weglarz[20] |
| Hydrolysis of Benzaldehyde Dioxolanne | Process Time | 5 | | OMRON P5R | Porte[21] |
| Vinyl Sulfone Synthesis | Solvent Base | 16 | | Zymate II EASYLAB | Fuchs[22] |
| Porphyrin Synthesis | [Pyrrole] [TFA] | 16 | x | Custom | Lindsey[23] |
| Glycine Synthesis | Temperature pH | 18 | x | Logilap® | Porte[24] |
| Hydroxamate Synthesis | [HAP] [DCC], Time | 28 | x | Zymate II EASYLAB | Matsuda[25] |

TFA = trifluoroacetic acid, HAP = hydroxylamine perchlorate, DCC = N,N'-dicyclohexylcarbodiimide

## Crystallization and the Simplex approach

Delacroix et al[26] describe the conception and set-up of an automated laboratory crystallizer. They first define what they consider are the goals for the study and control of crystallization. For example, programmable temperature profiling is considered vital to obtain a controlled shallow rate of cooling in the metastable crystallization zone. The 2L vessel selected is a reactor of height equal to its width ensuring better heat exchange. A stirring rod with a paddle ensures that stirring minimizes sheer forces and will not break the crystals. Four steps are identified in the process sequence; solution preparation, filtration, crystallization and drying. Modules were designed for each of these steps, which enable automatic crystallization of products by either cooling, precipitation or by evaporation of solvent.

A subsequent paper[27] describes how the granularity of a pharmaceutical product can be improved using this technology with the modified Simplex method.[28] The target was to minimize the percentage of particles with a diameter smaller than 630um. By increasing the concentration of the solution and varying the ratio of the two solvents, the initial percentage of 40.1% was reduced to 16% in nine sequential reactions.

## Total synthesis

The goal of Sugawara et al.,[29] is to prepare and isolate derivatives on a several hundred milligram scale. Reactions are carried out in three 100 mL reaction vessels. Additional features of their systems include the following:

1. Extraction/separation funnels, interface detection and passage through drying tubes to remove water from organic fractions.
2. The system is also capable of concentrating solutions and detecting completion of evaporation by monitoring vapor temperature.

3. Preparative HPLC and fraction collector which allows desired fractions to be transferred back into the reaction flasks for further reaction
4. Automated wash and drain of the apparatus after each run.

These additional features are technologically demanding and increase the complexity of the system, but are justifiable in the context of the total solution-phase synthesis goal. In his review of this system, Lindsey[30] records that "a chemist single-handedly operated two machines over the course of one year, performing 2000 reactions and synthesizing 500 new compounds (two to four reactions per synthesis)." Even so, not all aspects of chemistry can be addressed with this system. Sugawara claims they "can perform most of the common reactions used in the pharmaceutical industry that do not require the use of high pressures or the automated handling of solids."

## Miniaturization
The principle of a miniaturized total analysis system is that all reaction preparation, processing and monitoring are incorporated into a fully integrated device. The result, often referred to as a "lab on a chip" can be produced cheaply, using microfabrication techniques by etching micron-sized channels onto silicon or glass wafers. Liquid chromatography, flow injection analysis and electrophoresis have all been performed on samples of nanolitre size.

Chemical reactions, carried out in the manner of flow injection analysis, can take place in vessels the size of a half-inch length of human hair. Multilayer chip arrangements promise the potential of more than one chemical operation, either in parallel or in series. Claims have been made that large-scale production of material could be achieved by replicating a reaction millions of times using chip technology. This so-called "plant on a chip" would provide an interesting alternative to using conventional vessels, as time spent on issues associated with bulk properties and scaling up are avoided.

At the time of writing, miniaturization of total analysis and chip-based chemical synthesis for use in industrial situations have yet to be realized. Many of the technological components, from sample pre-treatment, introduction, detection, data transmission and information management have been developed in isolation. There are efforts to bring together these technologies, identify the additional technologies required to evolve fully working systems and package them into marketable products. In Haswell's review [31] of this subject, he suggests "the political and more importantly, the economic will to support the development of the technology may yet be seen to be the most serious limitation to the growth of the science." Currently, Orchid Biocomputer is developing microfluidic systems for high throughput synthesis.

## Defining automation requirements

There is a balance between designing a general or a specialist system. Vogt[32] argues strongly for the case that only truly necessary items be automated. His experience suggests that the idea of a totally automated laboratory system is a flawed one. His group "only automates only what is truly practical; instead of automating an entire process, only the bottleneck of that a process will be automated."

A similar viewpoint is that of Brown[33] who suggests that intuitively it is "appealing to automate the most complex and difficult operations in the laboratory, these are operations people least like to perform. However, attempting to automate procedures that are difficult to achieve manually is high risk." The recommendation is to use people as part of the work-cell, i.e. leave more complex tasks to be performed by highly skilled people or using semi-automated devices. The consensus is that simpler systems are not only less expensive, but they have quicker development times and have an increased likelihood of success. The law of diminishing returns features strongly in automation development.

Conversely, Lindsey[34] suggests in his review of automation of laboratory synthetic chemistry that "future automated systems will likely include enhanced general-purpose capabilities." Clearly the scientific goal is central here. If the goal is process understanding, then it might be more effective to have one system to investigate small reactions and a different system to carry out larger-scale crystallization experiments. If the goal is total synthesis, then the system must be capable of both reaction preparation and purification.

Both hardware features and software features determine the domain of chemistry that can be explored automatically. The ideal is to be able to take a typical manual procedure and transfer to an automated system with a minimum of modification to the chemical method. To provide cost effective flexible automation the system should be easily programmable and easily configured for a new piece of chemistry that has never been attempted before.

The range of possible chemistries may be constrained by specification limitations such as a restricted temperature range or a lack of inert gas blanketing. Early robotic combinatorial chemistry systems could only carry out reactions at ambient and therefore were limited to reactions such as peptide formation. Improved technical specifications have led to a large diversity of chemistries that can now be automated.

## Hardware features
Precise factor control is paramount in process screening and optimization, compared to combinatorial chemistry. A higher premium is placed on accuracy and precision in all three stages of reaction preparation, control and monitoring.

To reduce the consumption of scarce chemical resources it is preferable to work at reduced scale. The penalty for working at microscale is that a greater technical demand is placed on the system.

Liquid handling techniques are well developed. The choice of automated system determines whether manipulation of materials is achieved by moving tubes and bottles around the working "envelope" to dispensers and balances, or whether fluids are moved and presented to stationary tubes or vials, via syringes or dedicated solvent lines. Fluid delivery systems can be based on syringe pumps with microprocessor controlled stepper motors. Multiple syringes of different volumes create a wide dynamic range of volumes. The ability to control the rate of addition is a desirable feature. Some systems[35] are capable of monitoring the performance of the fluid delivery system by weighing containers before and after a addition of a liquid. The actual delivery is compared to expected weight based on the density of fluid. The addition of solids or gases presents a greater challenge, which some vendors are currently addressing.

Reaction parameter control can be achieved in two main ways. Jacketed vessels for large-scale work typically feature the control expected of a pilot plant vessel; overhead stirring, thermal control and an inert atmosphere. Other features include pH control. For small-scale reactions, a reaction station approach is common, with multiple vessels in a heating or cooling block to enable parallel processing at the same temperature. Some systems offer limited temperature zoning, to give increased flexibility.

Purification is easier to achieve in a system dedicated to a particular piece of chemistry or a series of closely related analogues. For a diverse range of chemistry, the techniques of purification are very varied and hence technically very demanding. Liquid-liquid extractions, filtration, evaporation and crystallization have all been exemplified.

A critical decision in analysis is choosing whether to analyze on or off-line. If samples are stable then off-line analysis would be preferable. However, if there is the possibility of sample degradation, then on-line analysis would be the method of choice. With on-line analysis, both invasive (such as HPLC) and non-invasive methods (such as FTIR) are possible. If multiple sampling is planned, invasive techniques are more suited to homogenous, rather than heterogeneous reactions.

Sample preparation is a key feature of HPLC. Many tasks are required, such as single and serial dilution (a sample is withdrawn from one vial, in another and diluted with diluent, prior to sample injection). Mixing capabilities combined with a time delay to allow for phase separation offer the basics of a liquid-liquid extraction. The needle can be programmed to different heights, when aspirating samples, allowing sampling from either organic or aqueous phases. Vortexing, sonication or multiple aspirating and dispensing can achieve mixing. However emulsions or precipitations can cause insurmountable problems.

Filtration has been accomplished by using a disposable membrane filter. For example, one system aspirates a sample into a holding loop, attaches a filter and then forces sample through the filter into a clean container.[36]

Safety should be considered in automation design. The nature of pharmaceutical synthesis is inextricably linked with the preparation of biologically active compounds. Flammable and toxic solvents are also used. The size of the system will determine whether existing ventilated cupboards can be employed, or whether installation needs to include additional extraction facilities to minimize exposure.

### Software features

As with hardware, the software should be specified appropriate to the scientific goals and the degree of sophistication required. Just as the hardware can limit the scope of the experiments that can be performed, so too can the software, for example, limiting the way the reactions can be monitored. Ideally, the software should be sufficiently versatile to make best possible use of the hardware features.

A versatile menu-driven user interface will allow the operator to load methods and input reaction parameters. A chemistry experiment consists of a series of modular events and generally it is possible to describe each of these events by a macro. This allows set routines such as sampling or washing to be incorporated into many different experiments without necessitating their rewriting. Parameter input, such as the specification of reagents and solvents, stoichiometries, order of addition, rate of addition, may be stepwise or using a spreadsheet.

The function of process control software is to control and monitor the applications. Typically a central multi-tasking computer issues series of instructions leaving the dedicated embedded processors to implement the instructions.

With high throughput screening, optimal scheduling is paramount. With experimental design techniques the emphasis is on obtaining quality rather than quantity of information. There can be three rate determining steps; reaction preparation, reaction processing and reaction monitoring. The relative time taken to perform each step can determine whether the system processes reactions in series, parallel or a by a combination of a serial batches. For systems which involve parallel reactions and serial processing, response measurement is often rate determining. To make best use of the monitoring system, the reactions can be scheduled using a fixed time offset which is equivalent to the analytical time.

For a monitoring system that has a significant run time such as HPLC, the number of data points that can be collected is determined by the chromatography time and the number of experiments. For example, if the HPLC time is 30 minutes, one reaction could be monitored every 30 minutes. If four reactions were placed in parallel each reaction would be analyzed every 2h and for eight reactions the minimum time between sampling would increase to 4h. Conversely, if the chromatography time could be reduced to 10 minutes using "fast HPLC," then for these eight reactions, sampling intervals decrease to 1h 20 min.

For reactions that have long preparation times, this step can become rate-determining. An example would be where a slow addition of reagent over an hour is required. Some systems have scheduling algorithms to maintain system timing. This is reliant on knowing what time is required for the robotic operation of the unit task steps. It may be possible to perform dissimilar applications simultaneously, by interweaving the steps of two or more procedures.

The important goal for data collection, handling and reduction is that the increased amount of data that can be collected should be converted into information. Where possible, this should be automated, otherwise the bottleneck will simply shift to data-processing. The next step is to convert information into useful knowledge that can be used by the chemist to move the project goals forward. This decision analysis can be carried out by the operator or autonomously with more sophisticated software.

The moment automation provides a more rapid method of performing experiments or increased data acquisition, the operator needs to alter the way all this is recorded. What was appropriate for a "one reaction and two analyses a day" type study, may not be feasible once this information is multiplied. Documentation serves two purposes; first, a record and second, a conclusion of the work.

Typical documentation features include recording the following:

> Purpose of study
> Instruction files (method files)
> Parameter input, e.g. specification of reagents and solvents, stoichiometries, order of addition, rate of addition.
> Log file of executed events or errors
> Data-files such as HPLC traces or thermal profiles

Ideally the system should have sufficient integrity to operate unattended for long periods.

Error trapping is a technique to determine what to do when things go wrong. At the very least, the systems should 'fail safe' and simply await operator intervention to recover the error. Hardware error trapping (tactile, switch, beams, weight, vision) and software error trapping are covered in detail by Crook[36]. If an error is detected, the system can respond by providing alarms, diagnosis or computer generated audit trails. Recovery mode is the combination of a hardware error detection method followed by a software check of the hardware and subsequent recovery. Error trapping has also been achieved retrospectively using video recordings.

## Project management of new automation systems

Reaction calorimeters and jacketed vessels are now commercially available. At the time of writing, the author believes that there are few, if any, off-the shelf systems that address the requirements of small-scale process screening and optimization. As the chemical development industry makes its demands known to the vendors more dedicated applications will become apparent. In the meantime many companies have developed their own bespoke systems, often in conjunction with automation vendors, in particular, Zymark.

The planning of a new system demands rigorous analysis of the aims, assessment of commercially available alternatives, careful planning and a clear focus. The client's expectations must be balanced against the realities of resource constraints such as money, time and people.

The development of automation projects can be high risk, due to cost and time required to progress from conception to delivery and implementation. There has to be strong management support to provide the financial backing and resource at all stages of the project. The criteria for success should be established at the onset. Risk can be mitigated or eliminated by prior knowledge of some of the common problems that are associated with development and implementation of automation projects. McDowell[37] outlines a scheme for undertaking a risk analysis and evaluation to assess the degree of risk associated with each of these factors. Risk avoidance and minimization strategies are also included. This risk assessment should be carried out at the start of a project and at intervals throughout the project to re-evaluate and see if any of the factors have changed or new ones have emerged.

The greater the complexity of the system, the higher the risk that something will go wrong. Management of risk approaches should be adopted to chose the simplest approach consistent with supporting the application effectively.

The importance of identifying the scientific goal, degree of complexity and capabilities of existing technology has already been stressed. Consider all options (size, throughput, location), including long-term requirements, to allow for expansion. The correct fundamental design in the automation platform avoids the need for external fixes. The platform chosen should offer adequate flexibility for most applications. There is a wide range of options for automation platforms. Typical examples are shown in Table 2 (no endorsement is implied. For specific claims for brand or model go to manufacturer).

Table 2

| Type | Example model |
|---|---|
| Cylindrical | Zymate (I, XP) |
| Revolute (or articulated robot) | CRS, ORCA |
| Cartesian XYZ | Tecan Genesis |
| Workstation | Bohdan |
| High performance autosamplers | SK233 (DART) |
| Closed system | Argonaut Nautilus |
| Reaction calorimeter | Mettler-Toledo |
| Jacketed vessel | Labmax |

The distinction between robots, workstations and autosamplers is becoming increasingly blurred. The simpler systems have finite boundaries that make them easier to design and market as individual products than robotic systems. Robotic systems are more versatile and adaptable, but involve a higher degree of complexity. Generality and efficiency are inversely related.

The storage for reagents, solvents and analytical vials determines number of samples that can be processed unattended. Holding disposables in magazines and dispensing as required can increase capacity. Combining a robot on a linear track with a second robot can make dramatic improvements to its capacity and productivity.

Whether the automation goal can be achieved through off-the-shelf or purpose built proprietary systems, the partnership of client and vendor is important. For the project to succeed, commitment is required from both parties. It is important to facilitate good communication and obtain data, such as realistic assessment of costs and the time required before system is fully operable. The relationship between performance, time-scales and cost has to be traded off by the project manager. The development and implementation (installation, testing, training, and modification of the system) to obtain a complete solution may take many months.

**Changing needs**

Ideally, the automation system must not only be able to meet immediate needs, but also provide the flexibility to adapt to future needs. This is essential, given the rapid development and life cycle of hardware and communications components. Both application flexibility and configuration flexibility are demanded. An example of application flexibility would be to improve the scheduling. An example of configurational flexibility may be exemplified by new peripherals such as reaction stations with extended temperature ranges or enhanced nitrogen blanketing. A distributed robotic environment permits system evolution.

The current and future availability of integration services such as peripheral interfaces and the impact of software upgrades should be considered. To evolve, continual involvement of technical, scientific and managerial staff is required and supplementary training should be given when required.

In developing a corporate automation policy, consideration should be given to the benefits of standardization of equipment. As with the purchase of any new technology, there is a balance between implementing proven versus cutting edge technology. The advantages of standardization include training, transferability of methodology, ease of maintenance and cost reductions on bulk purchase. A downside of standardization is being tied to a particular vendor who has inferior product development relative to other vendors. Competitive pricing benefits may be restricted. Absolute standardization may be impossible to achieve due to model obsolescence.

## Impact of the new system

The human aspect of implementing automation should not be underestimated. Automation has the potential for radically altering the way the scientist works. Consideration should be given to the morale of the ultimate laboratory users.

Product champions are required to bring in or develop new technology. Without dedicated resource, this will not happen, as it is too easy to focus in on process research project goals and not allow time to look beyond the immediate issues. Championing the introduction and development involves obtaining buy-in from both management and scientist. One of the most effective ways to obtain this is to demonstrate proof of principle. It is important to record usage so that the benefits can be identified. It is also worthwhile examining less successful work in order to improve and further develop the systems, or to recognize that it is time to cut one's losses and move on.

The user-skills demanded in developing new technology include a multi-disciplinary approach and teamwork. It requires a combination of talents from diverse disciplines, which include chemistry, analytical instrumentation, computers and robotics, automated data processing and LIMS (laboratory information management systems- data collection, analysis, reporting, archival). Experience can be drawn from internal or external resources.

Introducing automation into the laboratory involves the scientists acquiring new user skills to fully exploit the system. For example a chemist may need to develop increased analytical skills, knowledge of data-handling using spreadsheets and familiarity with statistics.

## Case study

### Development and introduction of the DART system at GlaxoWellcome

The aim of this case study is to show how an automation project can evolve from proof of principle to the point of becoming a valuable tool in the screening and optimization process. The use of phases and milestones indicates how such a project can be broken down into manageable focused goals and thereby minimizing risk. It is an example of a successful partnership between vendors (Stem and Anachem) and a customer (GlaxoWellcome) to meet a need for a versatile small-scale hands-on parallel synthesis and monitoring system for the general chemist.

The DART system has been specifically developed to address a technological void in the 'tool-kit' of a Process R&D chemist. Although there are several known examples of process optimization robotic systems, these are costly, take up a significant amount of space and have significant associated training and reprogramming costs. At this time (1996), no "off the shelf" package was available and each customer was expected to customize and develop systems to meet their specific needs. Given these drawbacks, it was perceived there was a requirement for an off the shelf package, that was lower cost, took up  less space and was designed for general, rather than specialist use.

The driving force for GlaxoWellcome's entry into the automation field was experimental design. This strategic methodology requires implementation of a set of similar reactions, where all factors are altered simultaneously in a systematic way. Despite the overwhelming theoretical advantages of this approach compared to a traditional "one factor at a time method," this approach had not been exploited to any great extent within Chemical Development in-house. Knowing what the next set of experiments are going to be can be very daunting if the reactions are to be done routinely, repetitively and sequentially by hand. It is these exact features that make the approach unappealing for the manual approach that become beneficial with automation.

Furthermore, experimental design demands good control of both the factors that are studied and those that are held constant. This is not always possible with commonly used laboratory equipment. For example electrical heaters, oil-baths and cooling baths do not give the precision or the range of temperatures that is required.

## Phase I : Proof of principle - Desilylation of trinem
The particular reaction chosen for proof of principle was the desilylation of a silylether to give the corresponding alcohol, a key step in the synthesis of a trinem antibiotic. Although many methods are available, conditions for deprotecting silylethers are often very harsh and when these methods were applied to our stage, they gave complete degradation. A traditional laboratory screening exercise uncovered triethylamine trihydrogen fluoride in N-methylpyrrolidone as the most promising reagent to achieve the desired conversion.

However, under these reaction conditions, a β-lactone impurity gradually forms, and it is difficult to remove during isolation. The amount of impurity that forms is critically dependent on when we stop the reaction  - too early and conversion is incomplete, or too late and the level of impurity becomes unacceptable. Therefore one of the factors that we wished to study was reaction conversion. We were able to manually monitor the reaction progressing during the day, but by the next day the reaction was degrading. Hence, with the reaction under study, the critical time points occurred during the night. Without on-line monitoring by HPLC, we had no knowledge of when the optimum time to halt the reaction should be. In addition there was the safety consideration of handling hydrogen fluoride solutions outside normal working hours.

We had a process we needed to get into pilot plant as quickly as possible, a strategy we wished to evaluate, no robotics and no method of on-line monitoring! Yet, the strategy and methodology of what we were trying to achieve had already been exemplified in an automated way by Kramer (HPLC)[18], Metevier (HPLC)[17], Lindsey (response surface and on-line UV-Vis)[7a], Boettger (Zymark and on-line HPLC)[16] and others. Realistically, the  chances of obtaining funding, purchasing and customizing equipment, modifying the laboratory and undergoing training, prior to the scheduled plant campaign were zero. Necessity demanded a rethink of available options. In the laboratory we already had a commercially available autosampler, the Gilson 231XL[38], that was capable of analytical sample preparation and injection.

### Milestone 1
A much simpler automation goal was addressed; to use this existing autosampler for on-line reaction monitoring, using the trinem desilylation reaction to exemplify the method.

Fortuitously, the features of the desilylation of trinem reaction made this initial proof of principle unexpectedly straightforward.

1.  The reaction was homogenous and hence need not be stirred.
2.  The reaction was at 'room temperature' and wasn't exothermic, so it did not need to be heated or cooled nor require refluxing capabilities.
3.  It was not air or moisture sensitive.

A central composite experimental design[39] was constructed, to study the effect of three factors; equivalents of reagent, time and concentration on product yield, consumption of starting material and the yield of the critical by-product. The reactions specified by the experimental design model were prepared by hand. Aliquots of reactions were placed in the sample tray of the autosampler, the position in the tray determined the time-point at which the reaction was monitored. Therefore, using this semi-automated approach, the required data-points were obtained.

### Milestone 2: Enhancement of the temperature range

The obvious limitation of this methodology was that only ambient reactions could be studied. Since the room temperature varied from the day and night, process control was poor. However, by purchasing a commercially available Peltier block from Gilson, the temperature range was extended from 0 to 40°C. Although this range is restricted, it was ideal for the desilylation reaction under study. The Peltier block proved to offer convenient and accurate temperature control. A second experimental design, incorporating the additional factor, temperature, was completed. In this study, temperature turned out to be the most critical factor. A highly predictive model was obtained and this allowed us to determine a set of optimum conditions (see summary of Case study 1).

A set of these conditions was validated in traditional glassware in the laboratory and subsequently on pilot plant. During the initial pilot plant run, an aliquot was removed from the reaction vessel and placed in the Peltier block. The block containing this unstirred aliquot was set at the same temperature as the stirred, nitrogen-blanketed reaction in the 400L vessel. After 26 hours a second aliquot was removed from the reaction vessel and both aliquots indicated the reaction had proceeded to the same degree of completion. Hence, the validity of the assumption that, for this reaction, stirring and nitrogen blanketing was not required was proven.

## Summary of Experimental Design Case study 1 [40]

| (1) Silylester | (2) Hydroxyester | (3) Lactone impurity |

**Aims**: To use a central composite experimental design, in conjunction with automated on-line HPLC to obtain a predictive model of the reaction process.

**Factors studied**

|  | Range |  |  |
|---|---|---|---|
| Temperature | 10 | 30 | °C |
| Time | 19 | 31 | hours |
| Concentration (amount of solvent) | 3 | 7 | volumes |
| Equiv. of Et$_3$N.3HF reagent | 1 | 1.68 | equivalents |

**Goals**

To maximize the yield of product with the following constraints:

- Control the level of lactone impurity within prescribed specification
- Place additional constraints on the factor ranges to optimize the operating conditions; e.g.

| Constraint |  | Effect on process |
|---|---|---|
| Limit the concentration | < 3.5 volumes | increase throughput reduce waste |
| Limit the Et$_3$N.3HF  reagent | < 1.18 equivalents | reduce material costs, reduce waste |
| Limit the time | < 23 hours | increase throughput |

**Outcome**

A highly predictive model of the reaction was obtained. This mathematical model could be interrogated using the software package to examine the effect on yield and quality under a variety of conditions or constraints. In total, six different scenarios were evaluated (see table 3, column 1). In each case, the model suggests the factor settings required to obtain such an outcome (column 2). When these six different processes were performed experimentally, the actual yield and predicted yields showed a high degree of correlation (on average accurate within 0.2% area/area by HPLC) (column 3).

Table 3

| Target/constraints | Conditions suggested by the model | Product yield* | | Impunity yield* | |
|---|---|---|---|---|---|
|  |  | Predicted | Actual | Predicted | Actual |
| Maximize product yield* No limit on lactone | Temperature   19 °C<br>Time             31h<br>Solvent        3.6 vols<br>Et$_3$N.3HF            1.43 equivs. | 95.3 | 95.8 | 3.3 | 3.3 |
| Maximize product yield Limit on lactone < 2% | Temperature   17 °C<br>Time             31h<br>Solvent        4.8 vols<br>Et$_3$N.3HF      1.51equivs | 94.2 | 94.0 | 1.9 | 1.7 |
| Maximize product yield Limit on lactone < 1.1% | Temperature   17 °C<br>Time             31h<br>Solvent        4.8 vols<br>Et$_3$N.3HF      1.68equivs. | 92.4 | 93.1 | 1.1 | 1.1 |

Table 3 continued

| Target/constraints | Conditions suggested by the model | Product yield* | | Impunity yield* | |
|---|---|---|---|---|---|
| Maximize product yield<br>Limit on lactone < 2%<br>Solvent <3,5 vols | Temperature    14 °C<br>Time                31h<br>Solvent            3.45 vols<br>$Et_3N.3HF$       1.60equivs | 94.2 | 94.2 | 2.0 | 2.0 |
| Maximize product yield<br>Limit on lactone< 2%<br>$Et_3N.3HF$ < 1.18 equiv | Temperature    28 °C<br>Time                19h<br>Solvent            3.5 vols<br>$Et_3N.3HF$       1.17equivs | 93.7 | 93.4 | 1.9 | 2.0 |
| Maximize product yield<br>Limit on lactone< 2%<br>Time < 23h | Temperature    24 °C<br>Time                23h<br>Solvent            3.65 vols<br>$Et_3N.3HF$       1.41equivs | 94.2 | 94.2 | 1.9 | 1.9 |

* 'yield' in this table refers to unisolated yield %area/area by HPLC

### Milestone 3 : Extending the versatility; stirring, inert gas blanketing, wide temperature range

Despite this success, the system was still essentially limited in its ability to deal with the typical conditions of reactions within the department. In order to extend the application, the equipment was modified so that the autosampler was able to aspirate aliquots of the reaction mixture from remote reactions, via tubing and ports located within its working envelope. The reactions could now be carried out in any size and type of vessel. These reactions could be stirred, fitted with condensers and carried out with nitrogen blanketing. The temperature range could now be widened to that typically carried out in the laboratory, -78 to +150°C. Hence reaction profiles of reactions under quite diverse conditions could be generated. New software had to be written[41] by the vendor to accommodate the new routines.

The weak link in this design is that of the connecting tubing, which links the reaction to the autosampler. This blocked if the reaction contained any particulates or showed any tendency to crystallize on cooling. Cross-contamination between sampling was also an issue, as it was impossible to decontaminate the tubing without introducing wash solvent into the reactions. This had not been an issue when the reactions were directly aspirated using the needle, as the needle could be rinsed out to a waste port, between sampling.

### Milestone 4 : Reaction stations
We felt that we required the flexibility of the tubing modifications with the benefits of the Peltier block. Ideally we needed to perform our reactions within the envelope of the autosampler, using a block to a much wider specification. We commissioned two purpose built blocks, from Stem Corporation[42]. We refer to these blocks as the StemHi and the StemLo. The StemHi covers the temperature range from ambient to +150°C. The StemLo covers the temperature range from -30 to +50°C. The two blocks cover the typical range used in the Pharmaceutical industry. Both blocks are capable of carrying out ten reactions in 25x150mm tubes. Each tube is agitated using PTFE stirrer bars which are magnetically coupled to rotating magnets positioned underneath the aluminum block. Inert gas blanketing is achieved by passing nitrogen between two septa.

## Phase II : The DART system (Development Automated Reaction Toolkit)[1]

Phase I afforded us the knowledge, confidence and experience to develop the DART system. We now required an autosampler that had a larger working envelope. To achieve this we switched to the larger autosampler, the Gilson 233XL. A fundamental design criterion was that data precision was more important than throughput.

Figure 1: The DART system

### Milestone 1 : Hardware

Once the reaction stations were built, subsequent modification of the hardware was relatively straightforward. Custom built tables were designed to integrate the autosamplers with the reaction stations. The standard autosampler needle was replaced with a longer septum-piercing needle. The MkII versions of the reaction stations incorporated several enhancements, most notably a communication port to enable control of the reaction station by the autosampler software.

### Milestone 2 : Software

The software proved a more daunting task. The original software for the Gilson 233XL was designed for analytical applications. The predetermined menu of sub-routines made programming very user-friendly. However, these routines were very limiting when it came to performing chemistry on the system. In order to achieve the flexibility that we required, new routines had to be written. David Emiabata-Smith was able to achieve this using the operating language, TurboPascal, at the DOS level but at the expense of ease of programming. With the aid of internal Information Technology support, a Windows based user-interface enhanced the ease of use. Finally, a third party software company was brought in to add even more flexibility, and robustness to the software. Enhanced ease of use is obtained using easily programmable versatile menu driven user interfaces.

### Milestones 3: Testing

Although Stem was responsible for the design and testing of the reaction stations, the specifications do not relate to how well the DART system can perform chemistry. The responsibility of implementation and validation of performance characteristics, with real chemistry lay, with GlaxoWellcome. A typical evaluation is as follows:

## Summary of Experimental design Case study 2 [44]

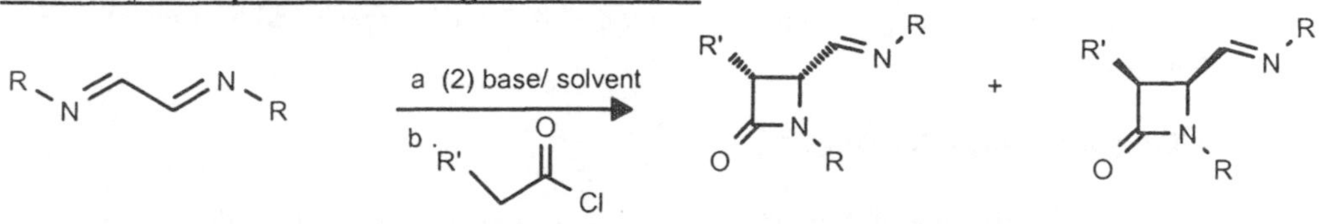

(1) bis-imine           (3) acid chloride     Diastereomer 1

**Aims**: To use the DART system in conjunction with experimental design strategies, to obtain process understanding of this reaction

### Study 1 Order of addition (6 reactions).

**Goal:**
Keeping all other factors constant, look at six permutations of order of addition of bis-imine, base, and acid chloride (as shown in the reaction sequence
**Outcome:**
To obtain maximum yield of diastereoisomer 1, the best sequence was found to be
(1) imine, (2) base, (3) acid chloride

**Study 2 Screening of solvents (16 reactions at two time-points).**

> **Goal:**
> A range of eight solvents (dichloromethane, tetrahydrofuran, isopropylether, ethyl acetate, chloroform, toluene, acetonitrile N-methylpyrrolidone) was screened, under four sets of conditions varying time and temperature.
> **Outcome:**
> This screen indicated that dichloromethane, chloroform and acetonitrile gave the highest yields. For the other solvents, yields were temperature dependent. In general, time was relatively unimportant.

**Study 3 Screening of bases (12 reactions at 2 time-points)**
> **Goal:**
> A range of six bases (triethylamine, trioctylamine, triisopropylamine, 4-methylmorpholine, pyridine, N-methylpiperidine) was screened, under four sets of conditions varying time and temperature.
> **Outcome:**
> This screen indicated that triethylamine, trioctylamine, triisopropylamine gave the highest yields.

**Study 4 Screening of sovent/base array (18 reactions at 2 time-points)**
> **Goal:**
> A array of three solvents (dichloromethane, chloroform and acetonitrile) and three bases (triethylamine, trioctylamine, triisopropylamine) was screened, under four sets of conditions varying time and temperature.
> **Outcome**
> This screen indicated that gave the combination of dichloromethane and triethylamine gave the best result.

**Study 5 Screening of factors (20 reactions)**
> **Goal:**
> Keeping the combination of dichloromethane and triethylamine constant, the following factors were screened for relative importance on their effect of influencing yield in solution.
> - Equivalents of $Et_3N$
> - Equivalents of $R'CH_2COCl$
> - Rate of addition of $R'CH_2COCl$
> - Time
> - Temperature
>
> **Outcome**
> This screen indicated that the following factors were the most critical in effecting yield:
> - Equivalents of $Et_3N$
> - Rate of addition of $R'CH_2COCl$
> - Temperature

**Study 6 and 7 Optimization of the three critical factors (2 x 20 reactions)**
   **Goal:**
   Study 6: To apply a central composite design on the three critical factors (identified by study 5) to maximize the yield.
   Study 7: To repeat study 6 using the same factors, but narrowing the range for each factor.
   **Outcome:**
   These studies revealed the optimal settings for the factors, indicated the degree of robustness and the effective ceiling for the reaction yield using these conditions.

**Overall outcome for studies 1-7**
   These studies showed the versatility and synergy of using experimental design and the DART system to solve typical process investigation targets.

   **Milestone 3: Implementation**
Having satisfied ourselves that the DART system met our expectations at Stevenage, we began to roll-out the system to other sites, at Dartford and Ware in the UK, the US and Italy. Feedback from the chemists and analysts ensures that the system continues to evolve. Anachem plan to market the product under the trade-name SK233[39].

**The future**

The rate of technological change is currently very rapid. This chapter has described some of the challenges that currently exist within Chemical Development. The aim has been to look at the commonalties of risks, costs and benefits of developing and applying automated techniques. The questions that remain are: what do we still need and what are we likely to achieve?

There will always be need to improve systems so that they are easy to use, cost-effective, flexible and reliable. The greater sophistication in software design and communication should improve sample scheduling, data management, system operation and facilitate operator training.

With a serial analytical process such as HPLC, the analysis may take longer than reaction processing, which is carried out in parallel. Advances in analytical chemistry will open up new possibilities; faster throughput, alternative response measurements, submission of samples for a complete battery of analyses.

The execution of combinatorial chemistry strategies has often been limited by the availability of equipment and instrumentation. Automated synthesis in combination with emerging technologies will revolutionize the field of process optimization for more efficient and cost-effective product development efforts.

Automation technology has largely attempted to emulate conventional manual procedures. Today's target may simply be to obtain the versatility of chemistry possible with quick-fit apparatus. Emerging technology such as microfluidics offers the promise of surpassing emulation and open up new possibilities in their own right. The key to scientific growth will be to demonstrate return on investment.

## Acknowledgements

Martin Owen wishes to thank his colleagues at Glaxo Wellcome, in particular David Emiabata-Smith, who was responsible for Phase II of the DART system, Bruce Porteous for development of the software, Derek Crookes for sponsorship of the project, Lai-Wah Lai and Linden Smith for the practical evaluation of the evolving systems.

[1] For a selection of overviews, see *Chem. Rev.* **1997** *97, 2.*

[2] Furka, A. et al., *Int. J. Peptide. Protein. Res.*, **1991**, *37*, 487.

[3] Terret, N.K. et al., *Tetrahedron*, **1995**, *51*, 8135.

[4] Gallop, M. A. et al., *J. Med. Chem.*, **1994**, *37*, 1233.

[5] Floyd, D.F.; Lewis, N.; Whittaker, M. *Chemistry in Britain*, **1996**, *32*, 31.

[6] [a] Bauer, B. E. In *Mol. Diversity Comb. Chem.: Libr. Drug Discovery, Conf* Chaiken, I. M., Janda, K. D., Eds.; American Chemical Society, Washington, D. C.: San Diego, CA, **1996**, p 233-243.

[b] Domanico, P. L. *Chemom. Intell. Lab. Syst.* **1994**, 26, 115-121.

[c] Beugelsdijk, T. J. *GATA* 1991, 8, 217-220

[d] O'Connor, S. J. *Autom. Chem.* **1993**, 15, 9-12.

[7] [a] Lindsey, J. S. *Chemom. Intell. Lab. Syst.* **1992**, 17, 15-45.

[b] Guette, J.; Crenne, N.; Bulliot, H.; Desmurs, J.; Igersheim, F. *Pure & Appl. Chem.* **1988**, 60, 1669-1678.

[c] Auffrett, A. D.; Hirst, W.; Meade, L. G.; Thacker, M. W. *Protides Biol. Fluids* 1986, 34, 15-18

[d] Andrews, R. P.; Summers, C. *Am. Biotechnol. Lab.* **1986**, 4, 30-31.

[e] Newton, R.; Fox, J. E. *Adv. Biotechnol. Processes* **1988**, 10, 1-24.

[f] Newton, R. *Am. Biotechnol. Lab.* **1989**, 7, 41-45.

[g] Kaplan, B. E.; Itakura, K. In Synthesis and Applications of DNA and RNA; Narang, S. A., Ed.; Academic Press, Inc., Orlando FL: Orlando FL, 1987, pp 9-45.

[8] [a] DeWitt, S. H. In *Annual reports in combinatorial chemistry and molecular diversity*; Moos, W. H., Pavia, M. R., Ellington, A. D., Kay, B. K., Eds.; **1997**,1, 69-77.

[b] Hardin, J. H.; Smietana, F. R. *Mol. Div.* **1996**, 1, 270-274.

[c] Fruchtel, J. S.; Jung, G. *Angew. Chem., Int. Ed. Engl.* **1996**, 35, 17-42.

[d] DeWitt, S. H.; Czarnik, A. W. *Curr. Opin. Biotech.* **1995**, 6, 640-645.

[9] *'The Evolution of a Revolution; Laboratory Automation in Chemical Process R&D'*, Symposium organised by Scientific Update, Leeds, UK, Sept. **1997**.

[10] Brodack, J.W. *New Trends in Radiopharnmaceutical Synthesis, Quality Assurance and Regulatory Control*, Plenum Press, New York, **1991**, 317-322.

[11] Carlson, R. *Design and optimization in organic synthesis*. Elselvier, Amsterdam, **1992**

[12] Corkan, L.A.; Lindsey J.S. *Chemom. Intell. Lab. Syst.*, **1992**, *17, 47-74.*

[13] Plouvier, J.C.; Corkan, L.A.; Lindsey J.S. *Chemom. Intell. Lab. Syst.*, **1992**, *17*, 75-94.

[14] Bayne C.K.; Rubin I.B. *Practical Experimental Designs and Optimisation Methods for Chemists*, VCH Publishers, Deerfield Beach, FL, **1986**.

[15] Moriguchi I. et al., *Chem. Pharm. Bull.* **1992**, *40, 930.*

[16] Boetteger, S.D. *Lab. Rob. Autom.* **1992**, *4,* 169.

[17] Metevier, P.; Josses, P., Bulliot, H. Corbet, J.P. Joux, B. *Chemometrics and Intelligent Laboratory Systems: Laboratory Information Management*, **1992**, *17*, 137-143

[18] Kramer, G.W.; Fuchs P.L. *Chemtech*, **1989**, 688.

[19] Corkan, L.A.; Plouvier, J.C.; Lindsey, J.S. *Chemom. Intell. Lab. Syst.*, **1992**, *17*, 95.

[20] Weglarz, T. E.; Atkin, S. C. *Adv. Lab. Autom. Rob.* **1990**, 6, 435-461.

[21] Porte, C.; Canatas, A.; Delacroix, A. *Lab. Rob. Autom.* **1995**, 7, 197-204

[22] Frisbee, A. R.; Nantz, M. H.; Kramer, G. W.; Fuchs, P. L. *J. Am. Chem. Soc.* **1984**, 106, 7143-7145.

[23] Corkan, L. A.; Plouvier, J. C.; Lindsey, J. S. *Chemom. Intell. Lab. Systems* **1992**, 17, 95 105.

[24] Fauduet, H.; Nikravech, M.; Porte, C. *Process Control Qual.* **1996**, 8, 41-53.

[25] Matsuda, R.; Ishibashi, M.; Takeda, Y. *Chem. Pharm. Bull.* **1988**, 36, 3512-3518

[26] Caron, M.; Martin-Moreno C.; Bondiou, J.C.; Borgogne J.P.; Porte C. Delacroix A. *Bulletin de la Societe Chimique de France* **1991**, *1*, 684

[27] Caron, M.. Porte C.; Bourgogne J-P.; Delacroix A. *Process Control and Quality* **1993**, *5*, 17.

[28] Nelder J.A.; Mead; R.J. *Computers* **1965**, *7*, 308-313.

[29] Sugawara, T.; Kato, S.; Okamoto, S. *Journal of Automatic Chemistry*, **1994**, 33-42,

[30] Lindsey, J.S.; *American Laboratory*, **1993**, *25*,17-20.

[31] Haswell S.J. *Analyst* **1997** 122 (1R-10R)

[32] Vogt, D.G. *Journal of Automatic Chemistry*, **1994**, *16*, 231-233.

[33] Brown, R.K. and Proulx, A. Zymark Corporation **1995**

[34] Lindsey J.S. *Chemometrics and Intelligent Laboratory Systems: Laboratory Information Management*, **1992**, *17*, 15.

[35] e.g. Rudge, D. A. *Lab. Autom. Inf. Manage.* **(1997)**, 33(2), 81-86.

[36] Crook, M. *Chemom. Intell. Lab. Syst t*, **1992**, *17*, 3-14.

[37] McDowell, R.D. *Chemom. Intell. Lab. Syst.*, **1993**, *21*, 1-19.

[38] Anachem, Charles Street, Luton, Beds. England LU2 0EB Tel. 01582 456666

[39] Box, G.; Hunter W. G.; Hunter J.S. *Statistics for Experimenters* Chapter 15, John Wiley and Sons 1978

[40] Owen, M.R. *et al. Journal of Process Research* (under compilation)

[41] Jim Stirling of Anachem

[42] Stem Corporation, Woodrolfe Road, Tollesbury, Essex, England, CM9 8SJ, Tel. 01621 868685

[43] Emiabata-Smith, D. *et al. Journal of Process Research* (under compilation)

[44] Owen, M.R. *et al. Journal of Process Research* (under compilation)

# AUTHOR INDEX

# AFFILIATION INDEX

# SUBJECT INDEX

Lightning Source UK Ltd.
Milton Keynes UK
15 April 2010
152847UK00009B/3/A